Handbook of

Control of Fungal Plant Pathogens in Agricultural Biotechnology

Handbook of

Control of Fungal Plant Pathogens in Agricultural Biotechnology

Contributors :
Hye Lin Park,
Youngchul Yoo, *et al.*

Volume 1

KOROS PRESS

KOROS PRESS LIMITED
London, UK

Handbook of Control of Fungal Plant Pathogens in Agricultural Biotechnology
Contributors : Hye Lin Park *and* Youngchul Yoo, *et al.*

Published by Koros Press Limited

www.korospress.com

United Kingdom

Printed in 2017 for Sale in the Indian Subcontinent

Handbook of Control of Fungal Plant Pathogens in Agricultural Biotechnology

ISBN: 978-1-78569-231-4

British Library Cataloguing in Publication Data
A CIP record for this book is available from the British Library

Exclusively distributed by CBS Publishers & Distributors Pvt. Ltd.
Sales & Distribution Rights only for India, Pakistan, Bangladesh, Sri Lanka, Nepal and Bhutan.This book is not to be sold outside these territories.

PREFACE

Fungi are a kingdom of Eukaryotic Organisms. The fungi are heterotrophic organisms characterised by a chintinous cell wall, and in majority of species, filamentous growth as multicellular hyphae forming a mycelium; while some fungal species grow as single cell. Sexual and asexual reproduction is via pores, often produced on specialised structures or in fruiting bodies. For example, yeasts, moulds, and mushrooms are fungi. Fungi are found all over the world, and grow in a wide range of habitats, including desert. Most fungi grow in terrestrial environments, but several species occur only in aquatic habitats. Fungi along with bacteria are the primary decomposers of organic matter in almost all terrestrial ecosystems, worldwide.

Aquatic fungi contribute to the energy flow and productivity of an aquatic and semi aquatic ecosystems by their active role in the utilization and biodeterioration of organic materials. These fungi also possess the ability to parasitize aquatic plants and animals including fishes under certain conditions. Physiology, environmental needs and activity of these fungi vary with their morphological characters. The taxonomy of these fungi can not remain static. The morphological variability of these species ultimately be linked to the genetic control of their characters. A considerable amount of taxonomic studies has been carried out by a number of mycologists all over the world.

This book blends information on classical fundamental aspects with recent development in fungal systematics. The textbook of fungi presents information on the morphology, life cycle and their economic uses in human life. Special attempt has been made on the biological activities of the microbial products. They produce several types of drugs including antibiotics, drugs that reduce high blood pressure. They also produce many of the enzymes used in the food, chemicals and pharmaceutical industries.

This page left intentionally blank.

CONTENTS

This page left intentionally blank.

LIST OF CONTRIBUTORS

Hye Lin Park

Graduate School of Biotechnology, Kyung Hee University, Yongin 446-70, Korea;

E-Mails: hlpark@khu.ac.kr (H.L.P.)

Youngchul Yoo

Department of Plant Molecular Systems Biotechnology & Crop Biotech Institute, Kyung Hee University, Yongin 446-70, Korea; E-Mail: yooyc@khu.ac.kr

Tae-Ryong Hahn

Graduate School of Biotechnology, Kyung Hee University, Yongin 446-70, Korea;

E-Mails: trhahn@khu.ac.kr (T.-R.H.)

Seong Hee Bhoo

Graduate School of Biotechnology, Kyung Hee University, Yongin 446-70, Korea;

E-Mails: shbhoo@khu.ac.kr (S.H.B.)

Sang-Won Lee

Department of Plant Molecular Systems Biotechnology & Crop Biotech Institute, Kyung Hee University, Yongin 446-70, Korea

Man-Ho Cho

Graduate School of Biotechnology, Kyung Hee University, Yongin 446-70, Korea

This page left intentionally blank.

Chapter 1

ANTIMICROBIAL ACTIVITY OF UV-INDUCED PHENYLAMIDES FROM RICE LEAVES

Hye Lin Park[1], Youngchul Yoo[2], Tae-Ryong Hahn[1], Seong Hee Bhoo[1], Sang-Won Lee[2,*] and Man-Ho Cho[1,*]

1 Graduate School of Biotechnology, Kyung Hee University, Yongin 446-70, Korea; E-Mails: hlpark@khu.ac.kr (H.L.P.); trhahn@khu.ac.kr (T.-R.H.); shbhoo@khu.ac.kr (S.H.B.)

2 Department of Plant Molecular Systems Biotechnology & Crop Biotech Institute, Kyung Hee University, Yongin 446-70, Korea; E-Mail: yooyc@khu.ac.kr

* Authors to whom correspondence should be addressed; E-Mails: manhocho@khu.ac.kr (M.-H.C.); swlee6803@khu.ac.kr (S.-W.L.); Tel.: +82-31-201-2127 (M.-H.C.); +82-31-201-3473 (S.-W.L.); Fax: +82-31-201-2156 (M.-H.C.).

ABSTRACT

Rice produces a wide array of phytoalexins in response to pathogen attacks and UV-irradiation. Except for the flavonoid sakuranetin, most phytoalexins identified in rice are diterpenoid compounds. Analysis of phenolic-enriched fractions from UV-treated rice leaves showed that several phenolic compounds in addition to sakuranetin accumulated remarkably in rice leaves. We isolated two compounds from UV-treated rice leaves using silica gel column chromatography and preparative HPLC. The isolated phenolic compounds were identified as phenylamide compounds: *N-trans*-cinnamoyltryptamine and *N-p*-coumaroylserotonin. Expression analysis of biosynthetic genes demonstrated that genes for arylamine biosynthesis were upregulated by UV irradiation. This result suggested that phenylamide biosynthetic pathways are activated in rice leaves by UV treatment. To unravel the role of UV-induced phenylamides as phytoalexins, we examined their antimicrobial activity against rice fungal and bacterial pathogens. *N-trans*-Cinnamoyltryptamine inhibited the growth of rice brown spot fungus (*Bipolaris oryzae*). In addition to the known antifungal activity to the blast fungus, sakuranetin had antimicrobial activity toward *B. oryzae* and *Rhizoctonia solani* (rice sheath blight fungus).

UV-induced phenylamides and sakuranetin also had antimicrobial activity against rice bacterial pathogens for grain rot (*Burkholderia glumae*), blight (*Xanthomonas oryzae* pv. *oryzae*) and leaf streak (*X. oryzae* pv. *oryzicola*) diseases. These findings suggested that the UV-induced phenylamides in rice are phytoalexins against a diverse array of pathogens.

Keywords: rice; UV treatment; phenylamide; phytoalexin; *N-trans*-cinnamoyltryptamine; *N-p*-coumaroylserotonin; *N-trans*-cinnamoyltyramine; *N*-benzoyltryptamine; antimicrobial activity

1. INTRODUCTION

Plants produce antimicrobial secondary metabolites, called phytoalexins, as a defense against pathogens. Diterpenoid and flavonoid phytoalexins are produced in rice in response to pathogens and environmental stress [1–12]. Momilactones A and B isolated from rice seed husks have antifungal activity against *Magnaporthe grisea*, the fungal pathogen that cause rice blast disease [1,13]. Since the finding of momilactones A and B, many phytoalexins have been identified from rice. Oryzalexins A–D, phytocassanes A–D and *ent*-10-oxodepressin are found to be induced by *M. grisea* infection in rice [3,4,9,12]. Phytocassane E was identified in rice infected with *Rhizoctonia solani* [10]. Production of most phytoalexins, including the momilactones and oryzalexins, is induced by UV-irradiation [2,5,7,8]. The production of diterpenoid phytoalexins is also induced by treatment with jasmonic acid, CuCl2 or chitin elicitor [14–16]. Sakuranetin was isolated from UV-treated rice leaves and has antifungal activity against *M. grisea* [6]. Most rice phytoalexins, except sakuranetin, are structurally related and grouped as diterpenoids [17–19]. The flavonoid sakuranetin is a phenolic phytoalexin.

Phenylamides are the conjugated form of phenolic acids and amines and are widely distributed throughout the plant kingdom [20]. They function in growth, development and floral initiation [20,21]. Accumulating evidence shows that phenylamides participate in defense against biotic and abiotic stress in some plant species [22–29]. Production of phenylamide compounds, such as feruloyltryptamine and *p*-coumaroylserotonin, in rice leaves is induced by the infection by the rice brown spot fungus *Bipolaris oryzae* [30]. However, antimicrobial activity of phenylamides against rice pathogens has not been reported yet.

We previously found that in addition to sakuranetin, several phenolic compounds accumulate in UV-treated rice leaves [31,32]. The UV-induced phenolic compounds include *N-trans*-cinnamoyltyramine and *N*-benzoyltryptamine [32]. In this study, we isolated two additional phenolic compounds from UV-treated rice leaves that were also identified as phenylamides. These findings indicated that UV radiation substantially induced the production of phenylamides. Gene expression analysis also showed the induction of arylamine and phenolic acid biosynthetic pathways in UV-treated rice leaves. The UV-induced phenylamides were tested as phytoalexins similar to sakuranetin and other diterpenoid phytoalexins by examining their antimicrobial activity against rice pathogens.

2. RESULTS AND DISCUSSION

2.1. Phenylamides Accumulate in UV-Treated Rice Leaves

Synthesis of most rice phytoalexins including sakuranetin is stimulated by UV irradiation [2,5–8,31]. In a previous study, we analyzed phenolic-enriched extracts from UV-treated rice leaves and found that, in addition to sakuranetin (**1**), accumulation of four potential phenolic compounds **2–5** increased remarkably after UV irradiation (Figure 1A). These phenolic compounds were rapidly accumulated in rice leaves after UV treatment (Figure 1B). Among the UV-induced phenolic compounds, compounds **2** and **3** had been previously identified as *N-trans*-cinnamoyltyramine and *N*-benzoyltryptamine, respectively, suggesting that UV-irradiation induces the synthesis of known phytoalexins and phenylamides in rice leaves [32].

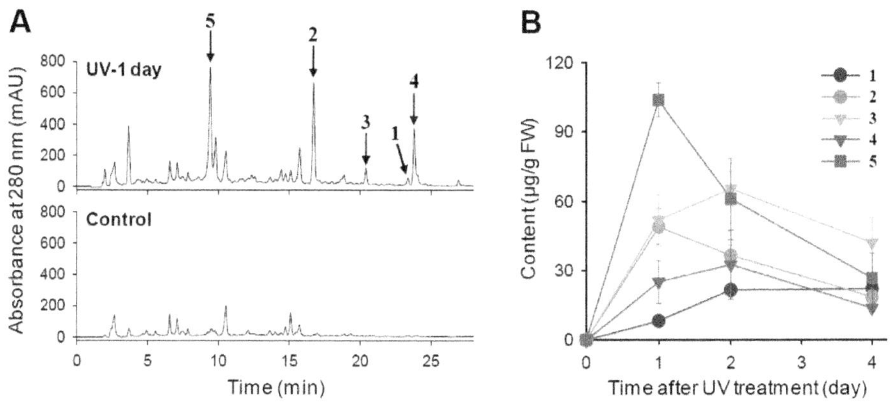

Figure 1. UV Induced accumulation of phenolic compounds in rice leaves. (**A**) HPLC-UV (280 nm) analysis of phenolic-enriched fractions extracted from rice leaves collected at 1 day after UV treatment and untreated control. Phenolic compounds induced in UV-treated rice leaves were designated as **1–5**. (**B**) Contents of phenolic compounds in rice leaves collected at 1, 2 and 4 days after UV treatment. FW; fresh weight.

We isolated compounds **4** and **5** from rice leaves collected 1 day after UV treatment using silica gel column chromatography and preparative high performance liquid chromatography (HPLC). The isolated compounds were identified using NMR spectroscopies and positive electron spray ionization mass spectrometry (ESI MS). Typical proton signals for cinnamoyl and tryptamyl moieties were observed in the ^{1}H-NMR spectrum of compound **4**. In a heteronuclear multiple bond correlation (HMBC) spectrum, a methine proton signal at δ 3.60 (H-11) was correlated with a carbonyl carbon signal in the cinnamoyl moiety at δ 168.8 (C-9›). These findings indicated that a tryptamyl moiety connected to the cinnamoyl moiety via an amino bond. The pseudomolecular ion peak of compound **4** was detected at *m/z* 291.3 $[M+H]^+$ in positive ESI MS. These results identified compound **4** as *N-trans*-cinnamoyltryptamine (Figure 2).

^{1}H and ^{1}H-^{1}H correlation spectra suggested that compound **5** had a similar structure to compound **4**. Compound **5** lacked two proton signals for H-6 and H-4′ of compound **4**. Carbon signals for C-6 and C-4′ in compound **5** were observed at δ 151.3 and δ 160.7, respectively, suggesting hydroxylation of these positions. Thus, compound **5** appeared to be composed of serotonin and *p*-coumaroyl moieties. Correlation between the methine proton signal in the serotonin moiety at δ 3.56 (H-11) and the carbonyl carbon signal in the *p*-coumaroyl moiety at δ 169.4 (C-9›) in the HMBC spectrum indicated that serotonin and *p*-coumaroyl moieties are connected via an amino bond. NMR data of compound **5** were consistent with data of *N*-*p*-coumaroylserotonin, which was identified previously from safflower seeds [33]. The pseudomolecular ion peak of compound **5** observed at *m*/*z* 323.1 $[M+H]^+$ in the positive ESI MS was consistent with *N*-*p*-coumaroylserotonin. These data identified compound **5** as *N*-*p*-coumaroylserotonin (Figure 2). ^{1}H- and ^{13}C-NMR data of compounds **4** and **5** are given in Table S1. Besides **1–5**, a putative phenolic compound at retention time 3.7 min was induced by UV treatment (Figure 1A). This compound is needed to be further isolated and definitively identified.

Our results indicated that UV-irradiation stimulated the synthesis of phenolic compounds in rice leaves and the majority of UV-induced phenolic compounds were phenylamides. Phenylamides are induced in response to *B. oryzae* infection and are suggested to have defense-related function [30]. *N*-*p*-Coumaroylserotonin is found both in UV-treated and *B. oryzae*-infected rice leaves. In addition, feruloyltryptamine and feruloylserotonin have been identified in *B. oryzae* infected rice leaves [30]. These findings suggest that a diverse array of phenylamides is produced in rice leaves in response to biotic and abiotic stress.

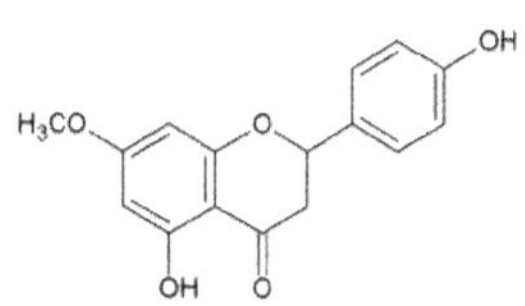

1 Sakuranetin

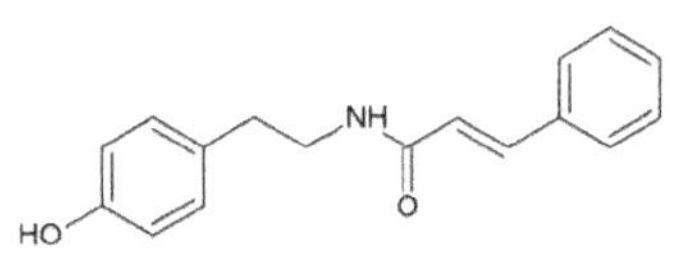

2 *N*-*trans*-Cinnamoyltyramine

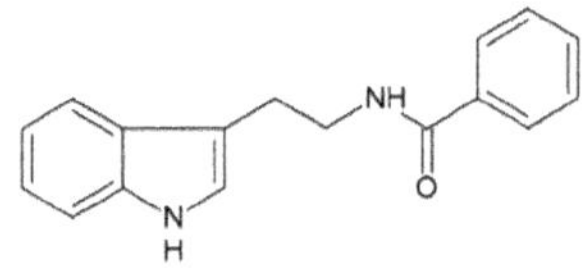

3 *N*-Benzoyltryptamine

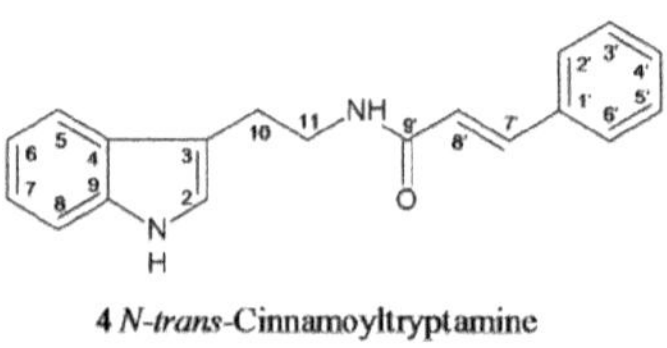

4 *N*-*trans*-Cinnamoyltryptamine

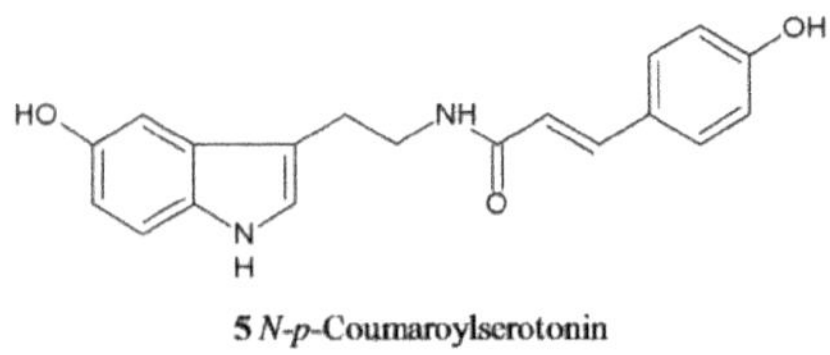

5 *N*-*p*-Coumaroylserotonin

Figure 2. Structure of phenylamides isolated from UV-treated rice leaves. Compounds **4** and **5** were isolated from UV-treated rice leaves and identified as *N*-*trans*-cinnamoyltryptamine and *N*-*p*-coumaroylserotonin using NMR and MS analysis, respectively. Compounds **2** and **3** were previously identified as *N*-*trans*-cinnamoyltyramine and *N*-benzoyltryptamine, respectively [32].

2.2. UV Irradiation Stimulates Phenylamide Biosynthetic Pathways in Rice Leaves

In some plant species, accumulation of phenylamides is induced by pathogens and environmental stress [22–30]. Analysis of salt treated- and *B. oryzae* infected-rice suggests defense-related functions of phenylamides and polyamines [29,30,34]. HPLC analysis of phenolic-enriched fractions obtained from UV-treated rice leaves showed that phenylamides are rapidly synthesized in rice leaves in response to UV irradiation and reach maximum levels about 1 or 2 days after UV treatment (Figure 1B). Similarly, accumulation of phenylamides, such as feruloyltryptamine and feruloylserotonin, in rice leaves is reached the highest levels 1 day after *B. oryzae* infection [30]. The phenolic part of phenylamides might be synthesized from phenylalanine through the phenylpropanoid pathway [31]. Transcriptome analysis of UV-treated rice leaves showed that expression of phenylpropanoid pathway genes encoding phenylalanine ammonia-lyase, cinnamate 4-hydroxylase and 4-coumarate CoA ligase is highly induced by UV irradiation [31]. These genes are most likely involved in the formation of phenolic acid-CoAs that are activated donors of an acyl group in phenylamide production [20,21]. The arylamines in UV-induced phenylamides are tryptamine and tyramine (Figure 2), which are derived from tryptophan and tyrosine, respectively. Tryptophan is synthesized from the shikimate pathway product chorismate, a branch point in aromatic amino acid biosynthesis.

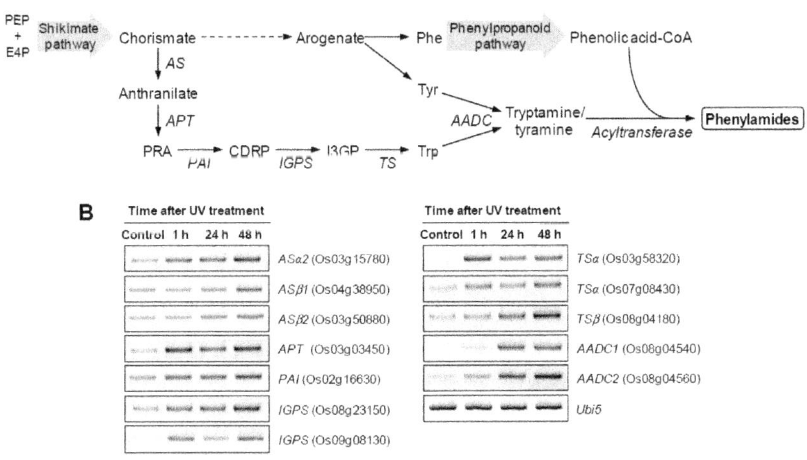

Figure 3. Activation of the phenylamide biosynthetic pathways in response to UV treatment. (**A**) Biosynthetic pathways in the UV-induced phenylamide production. (**B**) UV-induced expression of arylamine biosynthetic genes possibly involved in phenylamide formation. Expression of phenylamide biosynthetic genes in rice leaves 1, 24 and 48 h after UV treatment was analyzed by semi-quantitative RT-PCR. *AS*; anthranilate synthase, *APT*; anthranilate phosphoribosyltransferase, *PAI*; phosphoanthranilate isomerase, *IGPS*; indole-3-glycerol phosphate synthase, *TS*; tryptophan synthase, *AADC*; aromatic amino acid decarboxylase. PEP; phosphoenolpyruvate, E4P; erythrose 4-phosphate, PRA; phosphorubosyl anthranilate, CDRP; 1-(*O*-carboxyphenylamino)-1-deoxyribulose-5-phosphate, I3GP; indole-3-glycerol phosphate.

Five enzymes are involved in tryptophan biosynthesis from chorismate (Figure 3A). Previous transcriptome analysis observed upregulation of a few genes involved in tryptophan biosynthesis in UV-treated rice leaves [31].

We analyzed the expression of genes in all steps in tryptophan biosynthesis using semi-quantitative reverse transcription polymerase change reaction (RT-PCR). The conversion of chorismate to anthranilate is catalyzed by anthranilate synthase (AS) which is composed of two subunits, ASα and ASβ (Figure 3A). Expression patterns in UV-treated rice leaves were determined for two genes for ASα (*ASα1* and *ASα2*) and three genes for ASβ (*ASβ1-3*). RT-PCR analysis showed that *OsASα2* (Os03g15780) and *OsASβ1* (Os04g38950) were upregulated in response to UV treatment and *OsASβ2* (Os03g50880) was slightly upregulated in UV-treated rice leaves (Figure 3B). Expression of *OsASα2* is reported to be induced by elicitation [35]. Upregulation of these genes was also observed in *B. oryzae*-infected rice leaves [30].

The next two steps from anthranilate to 1-(*O*-carboxyphenylamino)-1-deoxyribulose-5-phosphate (CDRP) are catalyzed by anthranilate phosphoribosyltransferase (APT) and phosphoanthranilate isomerase (PAI) (Figure 3A). The expression of *APT* (Os03g03450) was highly induced by UV irradiation and *PAI* (Os02g16630) was slightly increased in UV-treated rice leaves (Figure 3B). Indole-3-glycerol phosphate synthase (IGPS) catalyzes the formation of indole-3-glycerol phosphate from CDRP (Figure 3A) and is important in tryptophan and auxin biosynthesis [36]. RT-PCR analysis showed that expression of two *IGPS* genes (Os08g23150 and Os09g08310) was stimulated by UV treatment (Figure 3B). The last two steps of tryptophan biosynthesis are catalyzed by tryptophan synthase (TS) which is composed of two subunits (TSα and TSβ) (Figure 3A). In the *TSα* and *TSβ* gene family, Os03g58320 and Os07g08430 for TSα and Os08g04180 for TSβ were upregulated in response to UV treatment (Figure 3B). Taken together, RT-PCR results showed that a set of genes involved in tryptophan biosynthesis was upregulated by UV treatment in rice leaves. Tyrosine biosynthesis shares biosynthetic pathways with phenylalanine, which is activated by UV treatment [31].

The arylamines tryptamine and tyramine are formed by decarboxylation of tryptophan and tyrosine, respectively, which is catalyzed by aromatic amino acid decarboxylase (AADC). Putative rice *AADC1* gene (Os08g04540) was expressed in *Escherichia coli* and the resulting enzyme showed tryptophan decarboxylase activity [37]. *AADC1* is upregulated in rice leaves by UV irradiation (Figure 3B). Expression of *AADC1* is increased in rice leaves by *B. oryzae* infection [30]. RT-PCR analysis of UV-treated rice leaves showed that rice *AADC2* (Os08g04560) was induced by UV irradiation (Figure 3B). This result suggested that, *AADC2* is also possibly involved in arylamine formation in UV-treated rice leaves. Phenylamide formation from arylamines and phenolic acid-CoA is likely catalyzed by acyltransferases. Previous transcriptome analysis of UV-treated rice leaves showed upregulation of several acyltransferases possibly involved in phenylamide formation [31]. Our results and previous transcriptome data suggested that arylamine and phenolic acid-CoA biosynthetic pathways are activated by UV irradiation for the phenylamide production in rice leaves.

2.3. UV-Induced Phenylamides in Rice Leaves are Potential Phytoalexins Against Rice Pathogens

2.3.1. Antimicrobial Activity of UV-Induced Phenolic Compounds

As shown in above, UV irradiation stimulated the production of phenylamides in rice leaves in addition to sakuranetin accumulation. To determine whether UV-induced phenylamides were phytoalexins, we examined the antimicrobial activity of phenylamides isolated from UV-treated rice leaves against rice pathogens. The fungal pathogens used were *M. grisea, R. solani,* and *B. oryzae*. To examine the antifungal activity of phenylamides and sakuranetin, the growth of fungal pathogens was measured on potato dextrose agar (PDA) plates containing different concentrations of UV-induced phenolic compounds. Sakuranetin and compound **4** inhibited the growth of the tested fungal pathogens (Table 1). The concentration of sakuranetin to inhibit 50% of mycelial growth (IC50) of *M. grisea* was 6.44 μg/mL. This result was comparable to the previously determined IC50 (5 μg/mL) of sakuranetin to germ tube growth of *M. grisea* [6]. Sakuranetin showed antifungal activity to *R. solani* and *B. oryzae* and had an IC50 of 54.04 μg/mL against *R. solani* and 19.05 μg/mL against *B. oryzae*. Although most phenylamides showed no significant antimicrobial activity against the tested fungal pathogens, compound **4** inhibited the growth of *B. oryzae* with an IC50 value of 26.92 μg/mL (Table 1).

Table 1. IC50 (μg/mL) of the phenolic compounds isolated from UV-treated rice leaves to fungal and bacterial pathogens.

Compound	Fungal Pathogen			Bacterial Pathogen		
	M. grisea	*R. solani*	*B. oryzae*	*Xoo*	*Xoc*	*B. glumae*
1	6.44 ± 1.35	54.04 ± 22.61	19.05 ± 9.83	19.95 ± 6.20	2.36 ± 0.35	8.22 ± 1.49
2	- [a]	-	-	21.96 ± 7.00	3.18 ± 0.44	-
3	-	-	-	34.76 ± 10.98	3.72 ± 0.20	-
4	-	-	26.92 ± 4.74	24.34 ± 5.60	2.45 ± 0.48	41.09 ± 5.96
5	-	-	-	-	54.54 ± 20.84	-

a No significant antimicrobial activity of the phenolic compounds to rice fungal and bacterial pathogens.

Antimicrobial activity of the UV-induced phenolic compounds against rice bacterial pathogens was also examined using *Burkholderia glumae,* which causes bacterial grain rot; *Xanthomonas oryzae* pv. *oryzae (Xoo),* which causes blight; and *X. oryzae* pv. *oryzicola (Xoc),* which causes leaf streak. Sakuranetin and compound **4** inhibited the growth of all tested pathogens to different extents (Figure 4).

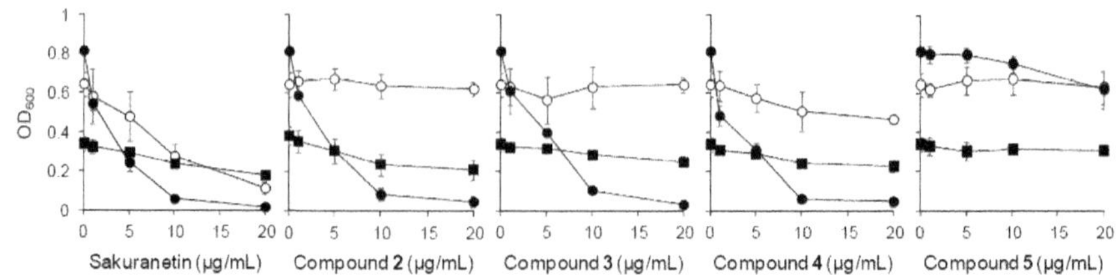

Figure 4. Antibacterial activity of the UV-induced phenolic compounds to bacterial pathogens, *Xoo* (■), *Xoc* (•), *B. glumae* (o).

Compounds **2** and **3** showed antimicrobial activity against *Xoo* and *Xoc* (Figure 4). Compound **5** inhibited *Xoc* growth (Figure 4). Sakuranetin and compound **4** showed the strongest inhibitory effects on *Xoc* growth with IC50 of 2.36 μg/mL for sakuranetin and 2.45 μg/mL for compound **4** (Table 1). Sakuranetin had an IC50 of 8.22 μg/mL against *B. glumae* and 19.95 μg/mL against *Xoo* (Table 1). Compound **4** similarly inhibited *Xoo* growth compared with sakuranetin and showed relatively weak antimicrobial activity to *B. glumae* (Figure 4, Table 1). Compounds **2** and **3** had slightly weaker activity than sakuranetin but inhibited *Xoc* growth with IC50 of 3.18 μg/mL for compound **2** and 3.72 μg/mL for compound **3** (Table 1). Compounds **2** and **3** also inhibited *Xoo* growth (Figure 4). Compound **5** showed relatively weak antimicrobial activity to *Xoc* only with an IC50 of 54.54 μg/mL (Figure 4, Table 1).

2.3.2. UV-Induced Phenylamides Are Potential Phytoalexins in Rice

Although phenylamides have been identified in pathogen-infected and UV-irradiated rice leaves [26,31], no studies have reported on their antimicrobial properties against rice pathogens. We examined the antifungal activity of four phenylamides and sakuranetin isolated from UV-treated rice leaves. Sakuranetin is suggested to be a strong rice phytoalexin against *M. grisea* [6]. The IC50 of sakuranetin against *M. grisea* germ tube growth is 5 μg/mL [6]. The IC50 values for rice diterpenoid phytoalexins against *M. grisea* germ tube growth are 1 to 35 μg/mL [1,3,9,12,17]. Oryzalexin D was reported to inhibit mycelial growth of *M. grisea* with an IC50 of 230 μg/mL [38]. Our result showed that sakuranetin strongly inhibited mycelial growth of *M. grisea* with the IC50 of 6.44 μg/mL (Table 1), which is comparable to its IC50 for the germ tube growth [6]. In addition to activity against *M. grisea*, sakuranetin inhibited *R. solani* and *B. oryzae* mycelial growth (Table 1). *R. solani* mycelial growth was reported to be inhibited by phytocassanes B and C [9]. Sakuranetin also showed antibacterial activity against all tested bacterial pathogens. In particular, sakuranetin almost completely inhibited *Xoc* growth at 10 μg/mL (Figure 4). These findings suggested that sakuranetin is a wide spectrum antimicrobial agent in rice.

Although compound **4** showed no significant inhibition of *M. grisea*, it had antifungal activity against *B. oryzae* (Table 1). The inhibition of *B. oryzae* mycelial growth by compound **4** was comparable to inhibition by sakuranetin.

Compound **4** also showed antimicrobial activity against all tested bacterial pathogens with an IC50 against *Xoo* and *Xoc* similar to sakuranetin (Figure 4, Table 1). These results suggested that compound **4** was a phytoalexin against a wide array of rice pathogens. Compounds **2** and **3** showed no significant inhibitory effects against fungal pathogens but strong antibacterial activity against *Xoc* comparable to sakuranetin and compound **4** (Figure 4, Table 1). Compounds **2** and **3** almost completely inhibited *Xoc* growth at 10 μg/mL (Figure 4). Compounds **2** and **3** also inhibited *Xoo* growth (Figure 4, Table 1). Compound **5** had weak antimicrobial activity against *Xoc* compared to other phenylamides and sakuranetin (Figure 4, Table 1). These results suggested that compounds **2**, **3**, and **5** were also potential phenolic phytoalexins to rice bacterial pathogens.

Phenylamides serve as phytoalexins in some plant species [20,21,25–28]. *N*-*p*-Coumaroylserotonin from bamboo has antifungal activity against *Aciculosporium take*, which cause witches' broom disease [27]. Antimicrobial acyltyramines such as *N*-feruloyltyramine and *N*-*p*-coumaroyltyramine, which act against pathogenic bacteria and fungi, are found in gallic root and pepper leaves [25,26]. *p*-Coumaroylnoradrenaline from tomato has antimicrobial activity against the bacterial pathogen *Peudomonas syringae* pv. *tomato* [28]. Many phytoalexins have been identified from rice and their antimicrobial activity against fungal pathogens, mainly the blast fungus *M. grisea*, has been examined. However, little is known about the antimicrobial activity of phytoalexins against other rice pathogens, in particular, pathogenic bacteria. We examined the antimicrobial activity of UV-induced phenylamides and sakuranetin against rice fungal and bacterial pathogens. These phenolic compounds were antimicrobial agents against fungal pathogens and bacterial pathogens. Thus, our findings strongly suggested that, in addition to sakuranetin, UV-induced phenylamides serve as a phytoalexins in rice against a diverse array of rice pathogens.

3. EXPERIMENTAL SECTION

3.1. Plant Growth Conditions and UV Treatment

Rice plants (*Oryza sativa* L. spp *japonica* cv. *Dongjin*) were grown in a greenhouse at 28 °C during the day and 20 °C at night. Eight week-old rice plants were UV-irradiated in a growth chamber with five 20 W germicidal lamps (maximum emission at 254 nm, 7.5 W UV output, Sankyo Denki Co. Kanagawa, Japan) for 2 h. UV-treated plants were transferred to a greenhouse and leaves were collected at 1, 24, 48, and 96 h after UV treatment for phytochemical and gene expression analyses.

3.2. Analysis, Isolation and Identification of Phenylamides from UV-Treated Rice Leaves

Rice leaves were ground in liquid nitrogen and extracted with 70% methanol (MeOH)-water for 1 h. After centrifugation, the aqueous methanol extract was obtained and dried *in vacuo*. Residue was dissolved and fractionated in ethyl acetate (EtOAc)-water (1:1) mixture to enrich phenolic compounds. The ethyl acetate phase was dried *in vacuo* and residue was dissolved in MeOH and analyzed using a reversed phase HPLC with a Sunfire C18 column (4.6 × 250 mm) (Waters, Milford, MA, USA) with detection at 280 nm using elution conditions reported by Park *et al.* [32].

Phenolic-enriched samples from rice leaves collected 1 day after UV-treatment were applied to a silica gel column to isolate UV-induced compounds. Consecutive elution was performed with hexane–benzene (1:1), hexane–benzene (1:2), hexane–benzene (1:3), benzene–chloroform (1:1), chloroform–EtOAc (1:1), and EtOAc. Eluted fractions were analyzed by reversed phase HPLC as above Fractions containing UV-induced compounds were dried *in vacuo*. Fractions enriched in UV-induced compounds were purified by reversed phase HPLC with a preparative Sunfire C18 column (10 × 150 mm) (Waters) on a linear gradient of 25%–60% acetonitrile in 3% acetic acid-water for 25 min (flow rate 3 mL/min) with detection at 280 nm.

Isolated compounds were identified using NMR spectroscopy and positive ESI-MS. NMR spectra of isolated compounds were recorded in CD3OD on an Avance 600 NMR spectrometer (Bruker, Rheinstetten, Germany) using tetramethylsilane as an internal standard. MS spectra were obtained with the ion source of positive mode electronspray ionization using an Agilent 6410 Triple Quadruple LC/MS System (Agilent Technologies, Santa Clara, CA, USA).

3.3. RNA Isolation and Semi-Quantitative RT-PCR Analysis

Total RNA was isolated from UV-treated and untreated rice leaves using TRIzol reagent (Invitrogen, Carlsbad, CA, USA) and 3 μg of RNA was used to synthesize cDNA using the RNA to cDNA EcoDry kit (Clontech, Mountain View, CA, USA) according to the manufacturer's protocol. Semi-quantitative RT-PCR was performed with gene-specific primers. Primers and PCR conditions for each gene were summarized in Table S2. The ubiquitin gene (*Ubi5*) was the control.

3.4. Antimicrobial Activity

Antifungal activity of phenolic compounds from UV-treated rice leaves against three rice pathogenic fungi (*M. grisea* PO6-6, *R. solani* AG-1 and *B. oryzae*) was examined. Compounds were dissolved in dimethylsulfoxide (DMSO) and added to PDA media (Becton, Dickinson and Company, Sparks, MD, USA) at final concentrations of 1, 5, 10 and 20 μg/mL. The final DMSO concentration in media was 1% (v/v). Mycelial discs (1 mm in diameter) were cut from the edge of 2-day-old cultures of pathogens on PDA plates and placed on PDA plates containing different concentrations of compounds. Pathogenic fungi were incubated at 28 °C for three days and mycelial radius was measured to determine growth inhibition by UV-induced phenolic compounds. Experiments were performed in triplicate.

Pathogenic bacteria (*Xoo*, *Xoc* and *B. glumae*) were grown in NB media (Becton, Dickinson and Company) at 28 °C. Cultures were inoculated in NB media containing 1, 5, 10 and 20 μg/mL compounds. The final DMSO concentration in the media was 1% (v/v). To determine inhibition by compounds, bacterial growth was examined by measuring OD600 during exponential growth using a V-550 UV/Vis spectrophotometer (Jasco, Tokyo, Japan). Experiments were performed in triplicate.

4. CONCLUSIONS

UV irradiation stimulates the production of phytoalexins in rice leaves. Besides known phytoalexins, we isolated and identified phenylamides including *N-trans*-cinnamoyltryptamine and *N-p*-coumaroylserotonin from UV-treated rice leaves. To unravel the role of UV-induced phenylamides as phytoalexins, we examined their antimicrobial activity against rice bacterial and fungal pathogens. UV-induced phenylamide had antimicrobial activity against the tested pathogens, in particular bacterial pathogens. These results suggested that the UV-induced phenylamides are a new class of phytoalexins in rice against a diverse array of pathogens.

Supplementary Materials

Supplementary materials can be accessed at: http://www.mdpi.com/1420-3049/19/11/18139/s1.

ACKNOWLEDGMENTS

This research was supported by a grant from the Next-Generation BioGreen 21 Program (grant numbers: PJ00950604 and PJ008098), Rural Development Administration, Republic of Korea, and the Basic Science Research Program through the National Research Foundation of Korea (NRF) funded by the Ministry of Education, Science and Technology (2012R1A1A2044057).

Author Contributions

M.-H.C., S.-W.L. S.H.B. and T.-R.H. designed the experiments; M.-H.C., H.L.P. and Y.Y. performed the experiments; M.-H.C., H.L.P., Y.Y. and S.-W.L. analyzed the data and wrote the paper. All authors discussed, edited and approved the final manuscript.

Conflicts of Interest

The authors declare no conflict of interest.

REFERENCES

1. Cartwright, D.; Langcake, P.; Pryce, R.J.; Leworthy. D.P.; Ride, J.P. Chemical activation of host defense mechanisms as a basis for crop protection. *Nature* **1997**, *267*, 511–513.
2. Cartwright, D.W.; Langcake, P.; Pryce, R.J.; Leworthy, D.P.; Ride, J.P. Isolation and characterization of two phytoalexins from rice as momilactones A and B. *Phytochemistry* **1981**, *20*, 535–537.
3. Akatsuka, T.; Kodama, O.; Sekido, H.; Kono, Y.; Takeuchi, S. Novel phytoalexins (Oryzalexins A, B, and C) isolated from rice blast leaves infected with *Pyricularia oryzae*. Part I: Isolation, characterization and biological activities of oryzalexins. *Agric. Biol. Chem.* **1985**, *49*, 1689–1694.
4. Sekido, H.; Endo, T.; Suga, R.; Kodama, O.; Akatsuka, T.; Kono, Y.; Takeuchi, S. Oryzalexin D (3,7-(+)-sandaracopimaradiene), a new phytoalexin isolated from blast-infected rice leaves. *J. Pesticide Sci.* **1986**, *11*, 369–372.
5. Kodama, O.; Suzuki, T.; Miyakawa, J.; Akatsuka, T. Ultraviolet-induced accumulation of phytoalexins in rice leaves. *Agric. Biol. Chem.* **1988**, *52*, 2469–2473.
6. Kodama, O.; Miyakawa, J.; Akatsuka, T.; Kiyosawa, S. Sakuranetin, a flavanone phytoalexin from ultraviolet-irradiated rice leaves. *Phytochemistry* **1992**, *31*, 3807–3809.
7. Kato, H.; Kodama, O.; Akatsuka, T. Oryzalexin E, a diterpene phytoalexin from UV-irradiated rice leaves. *Phytochemistry* **1993**, *33*, 79–81.
8. Kato, H.; Kodama, O.; Akatsuka, T. Oryzalexin F, a diterpene phytoalexin from UV-irradiated rice leaves. *Phytochemistry* **1994**, *36*, 299–301.
9. Koga, J.; Shimura, M.; Oshima, K.; Ogawa, N.; Yamauchi, T.; Ogasawara, N. Phytocassanes A, B, C and D, novel diterpene phytoalexins from rice, *Oryza sativa* L.. *Tetrahedron* **1995**, *51*, 7907–7918.
10. Koga, J.; Ogawa, N.; Yamauchi, T.; Kikuchi, M.; Ogasawara, N.; Shimura, M. Functional moiety for the antifungal activity of phytocassane E, a diterpene phytoalexin from rice. *Phytochemistry* **1997**, *44*, 249–253.
11. Okada, A.; Shimizu, T.; Okada, K.; Kuzuyama, T.; Koga, J.; Shibuya, N.; Nojiri, H.; Yamane, H. Elicitor induced activation of the methylerythritol phosphate pathway toward phytoalexins biosynthesis in rice. *Plant Mol. Biol.* **2007**, *65*, 177–187.
12. Inoue, Y.; Sakai, M.; Yao, Q.; Tanimoto, Y.; Toshima, H.; Hasegawa, M. Identification of a novel casbane-type diterpene phytoalexin, *ent*-10-oxodepressin, from rice leaves. *Biosci. Biotechnol. Biochem.* **2013**, *77*, 760–765.
13. Kato, T.; Kabuto, C.; Sasaki, N.; Tsunagawa, M.; Aizawa, H.; Fujita, K.; Kato, Y.; Kitahara, Y. Momilactones, growth inhibitors from rice, *Oryza sativa* L. *Tetrahedron Lett.* **1973**, *39*, 3861–3864.
14. Cho, E.M.; Okada, A.; Kenmoku, H.; Otomo, K.; Toyomasu, T.; Mitsuhashi, W.; Sassa, T.; Yajima, A.; Yabuta, G.; Mori, K.; *et al.* Molecular cloning and characterization of cDNA encoding *ent*-cassa-12,15-diene synthase, a putative diterpenoid phytoalexin biosynthetic enzyme, from suspension-cultured rice cells treated with a chitin elicitor. *Plant J.* **2004**, *37*, 1–8.
15. Nemoto, T.; Cho, E.M.; Okada, A.; Okada, K.; Omoto, K.; Kanno, Y.; Toyomasu, T.; Mitsuhashi, W.; Sassa, T.; Minami, E.; *et al.* Stemar-13-ene synthase, a diterpene cyclase involved in the biosynthesis of the phytoalexin oryzalexin S in rice. *FEBS Lett.* **2004**, *571*, 182–186.
16. Okada, K. The biosynthesis of isoprenoids and the mechanisms regulating it in plants. *Biosci. Biotechnol. Biochem.* **2011**, *75*, 1219–1225.
17. Peters, R.J. Uncovering the complex metabolic network underlying diterpenoid phytoalexin biosynthesis in rice and other cereal crop plants. *Phytochemistry* **2006**, *67*, 2307–2317.
18. Toyomasu, T. Recent advances regarding diterpene cyclase genes in higher plants and fungi. *Biosci. Biotechnol. Biochem.* **2008**, *72*, 1168–1175.
19. Schmelz, E.A.; Huffaker, A.; Sims, J.W.; Christensen, S.A.; Lu, X.; Okada, K.; Peters, R.J. Biosynthesis, elicitation and roles of monocot terpenoid phytoalexins. *Plant J.* **2014**, *79*, 659–678.

20. Edreva, A.M.; Velikova, V.B.; Tsonev, T.D. Phenylamides in plants. *Russ. J. Plant Physiol.* **2007**, *54*, 287–301.

21. Bassard, J.E.; Ullmann, P.; Bernier, F.; Werck-Reichhart, D. Phenolamides: Bridging polyamines to the phenolic metabolism. *Phytochemistry* **2010**, *71*, 1808–1824.

22. Klapheck, S. Polyamines and cinnamoyl-putrescines in normal and sulfur-starved suspension cultures of *Nicotiana tabacum*. *Z. Pflanzenphysiol.* **1983**, *112*, 275–279.

23. Edreva, A.; Yordanov, I.; Kardjieva, R.; Hadjiiska, E.; Gesheva, E. Expression of phenylamides in abiotic stress conditions. *Bulg. J. Plant Physiol.* **1995**, *21*, 15–23.

24. Pearce, G.; Marchand, P.A.; Griswold, J.; Lewis, N.G.; Ryan, C.A. Accumulation of feruloyltyramine and *p*-coumaroyltyramine in tomato leaves in response wounding. *Phytochemistry* **1998**, *47*, 659–664.

25. Fattorusso, E.; Lanzotti, V.; Taglialatela-Scafati, O. Antifungal *N*-feruloyl amides from roots of two *Allium* species. *Plant Biosyst.* **1999**, *133*, 199–203.

26. Newman, M.A.; von Roepenack-Lahaye, E.; Parr, A.; Daniels, M.J.; Dow, J.M. Induction of hydroxycinnamoyl-tyramine conjugates in pepper by *Xanthomonas campestris*, a plant defense response activated by *hrp* gene-dependent and *hrp* gene-independent mechanisms. *MPMI* **2001**, *14*, 785–792.

27. Tanaka, E.; Tanaka, C.; Mori, N.; Kuwahara, Y.; Tsuda, M. Phenylpropanoid amides of serotonin accumulate in witches' broom diseased bamboo. *Phytochemistry* **2003**, *64*, 965–969.

28. Von Roepenack-Lahaye, E.; Newman, M.A.; Schornack, S.; Hammond-Kosack, K.E.; Lahaye, T.; Jones, J.D.G.; Daniels, M.J.; Dow, J.M. *p*-Coumaroylnoradrenaline, a novel plant metabolite implicated in tomato defense against pathogens. *J. Biol. Chem.* **2003**, *278*, 43373–73383.

29. Ishihara, A.; Nakao, T.; Mashimo, Y.; Murai, M.; Ichimaru, N.; Tanaka, C.; Nakajima, H.; Wakasa, K.; Miyagawa, H. Probing the role of tryptophan-derived secondary metabolism in defense responses against *Bipolaris oryzae* infection in rice leaves by a suicide substrate of tryptophan decarboxylase. *Phytochemistry* **2011**, *72*, 7–13.

30. Ishihara, A.; Hashmoto, Y.; Tanaka, C.; Dubouzet, J.G.; Nakao, T.; Matsuda, F.; Nishioka, T.; Miyagawa, H.; Wakasa, K. The tryptophan pathway is involved in the defense responses of rice against pathogenic infection via serotonin production. *Plant J.* **2008**, *54*, 481–495.

31. Park, H.L.; Lee, S.W.; Jung, K.H.; Hahn, T.R.; Cho, M.H. Transcriptomic analysis of UV treated rice leaves reveals UV-induced phytoalexin biosynthetic pathways and their regulatory networks in rice. *Phytochemistry* **2013**, *96*, 57–71.

Chapter 2

CHEMICAL IDENTIFICATION OF FUNGI

CHEMICAL TESTS IN MUSHROOM IDENTIFICATION

Chemical tests in mushroom identification are methods that aid in determining the variety of some fungi. The most useful tests are Melzer's reagent and potassium hydroxide.

Ammonia

Household ammonia can be used. A couple of drops are placed on the flesh. For example, *Boletus spadiceus* gives a fleeting blue to blue-green reaction.

Iron Salts

Iron salts are used commonly in *Russula* and Bolete identification. It is best to dissolve the salts in water (typically a 10% solution) and then apply to the flesh, but it is sometimes possible to apply the dry salts directly to see a color change. For example, the white flesh of *Boletus chrysenteron* stains lemon-yellow or olive. Three results are expected with the iron salts tests: no change indicates a negative reaction; a color change to olive, green or blackish green; or a color change to reddish-pink.

Meixner Test for Amatoxins

The Meixner test (also known as the Wieland Test) uses concentrated hydrochloric acid and newspaper to test for the deadly poisonous amatoxins found in some species of *Amanita, Lepiota,* and *Galerina.*

Psilocin gives a positive Meixner test (100% false positive), therefore the Meixner test cannot differentiate between a hallucinogenic species and a deadly poisonous one.

Melzer's Reagent

Melzer's Reagent can be used to test spores for amyloid, nonamyloid, and dextrinoid.

- Spores that stain bluish-gray to bluish-black are amyloid
- Spores that stain brown to reddish-brown are dextrinoid

This test is normally performed on white spored mushrooms. If the spores are not light colored, a change will not be readily apparent. It is easiest to see the color change under a microscope, but it is possible to see it with the naked eye with a good spore print.

Paradimethylaminobenzaldehyde

In the genus *Lyophyllum* the lamellae usually turn blue with the application of para-Dimethylaminobenzaldehyde (PDAB or pDAB).

Phenol

A 2–3% aqueous solution of phenol gives a color change in some species when applied to the cap or stem.

Potassium Hydroxide

A 3–10% solution of potassium hydroxide (KOH) gives a color change in some species:

- In *Agaricus*, some species turn yellow with KOH, some have no reaction, and one species turns green.
- Distinctive change for some species of *Cortinarius* and Boletes

Schaeffer Reaction

Developed by Julius Schäffer to help with the identification of *Agaricus* species. A positive reaction of Schaeffer's test, which uses the reaction of aniline and nitric acid on the surface of the mushroom, is indicated by an orange to red color; it is characteristic of species in the section *Flavescentes*. The compounds responsible for the reaction were named schaefferal A and B to honor Schäffer.

Two intersecting lines are drawn on the surface of the cap, the first with aniline or aniline water, the second with an aqueous solution of 65% nitric acid. The test is considered positive when a bright orange color forms where the lines cross.

GENERAL PROPERTIES OF FUNGI

Structural-functional Relationships

The fungi are more evolutionarily advanced forms of microorganisms, as compared to the prokaryotes (prions, viruses, bacteria). They are classified as *eukaryotes, i.e.,* they have a diploid number of chromosomes and a nuclear membrane and have sterols in their plasma membrane. Genetic complexity allows morphologic complexity and thus these organisms have complex structural features that are used in speciation.

Fungi can be divided into two basic morphological forms, yeasts and hyphae. *Yeasts*are unicellular fungi which reproduce asexually by *blastoconidia* formation (budding) or fission. *Hyphae*are multi-cellular fungi which reproduce asexually and/or sexually. *Dimorphism* is the condition where by a fungus can exhibit either the yeast form or the hyphal form, depending on growth conditions. Very few fungi exhibit dimorphism. Most fungi occur in the hyphae form as branching, threadlike tubular filaments. These filamentous structures either lack cross walls (*coenocytic*) or have cross walls (*septate) depending on the species. In some cases septate hyphae develop clamp connections* at the septa which connect the hyphal elements.

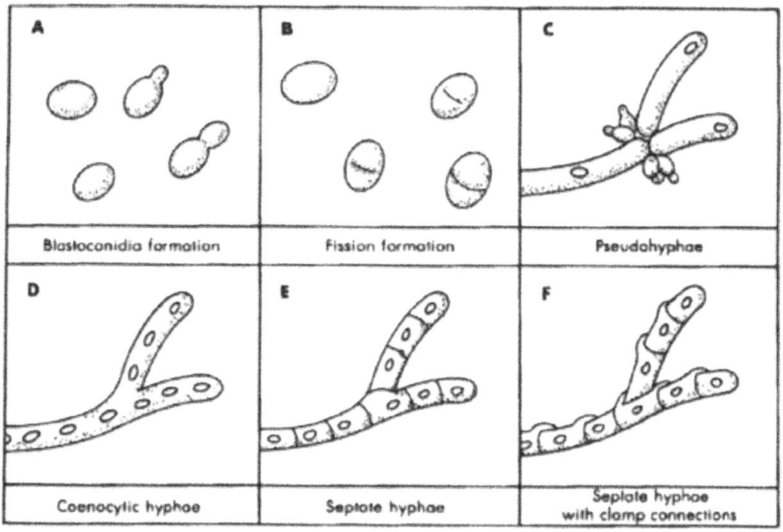

Fig. : A. Yeast cells reproducing by blastoconidia formation; B. Yeast dividing by fission; C. Pseudohyphal development; D. Coenocytic hyphae; E. Septate hyphae; F. Septate hyphae with clamp connections.

A mass of hyphal elements is termed the *mycelium* (synonymous with *mold*). Aerial hyphae often produce asexual reproduction propagules termed *conidia* (synonymous with *spores*). Relatively large and complex conidia are termed *macroconidia* while the smaller and more simple conidia are termed *microconidia.* When the conidia are enclosed in a sac (the sporangium), they are called *endospores.* The presence/absence of conidia and their size, shape and location are major features used in the laboratory to identify the species of fungus in clinical specimens.

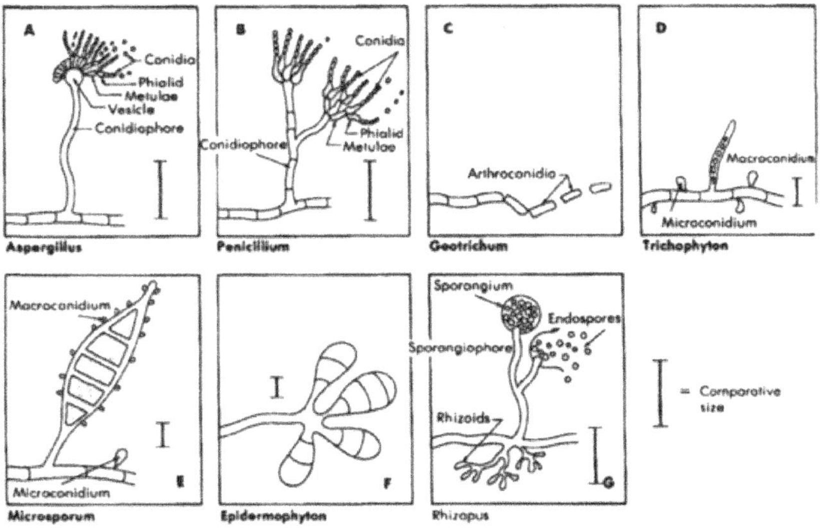

Fig. : A. Aspergillus; B. Penicillium; C. Geotrichum; D. Trichophyton; E. Microsporum; F. Epidermophyton and G. Rhizopus.

Asexual reproduction, via conidia formation, does not involve genetic recombination between two sexual types whereas *sexual reproduction* does involve genetic recombination between two sexual types.

Metabolism

All fungi are free living, *i.e.*, they are *not* obligate intracellular parasites. They do not contain chlorophyll and cannot synthesize macromolecules from carbon dioxidc and energy derived from light rays. Therefore all fungi are heterotrophs, living on preformed organic matter. For medical purposes the important aspects of fungal metabolism are:

1. The synthesis of chitin, a polymer of N-acetyl glucosamine, and other compounds, for use in forming the cell wall. These induce immune hypersensitivity.
2. The synthesis of ergosterol for incorporation into the plasma membrane. This makes the plasma membrane sensitive to those antimicrobial agents which either block the synthesis of ergosterol or prevent its incorporation into the membrane or bind to it, *e.g.* amphotericin B.
3. The synthesis of toxins such as:
   a. *Ergot alkaloids*- these are produced by *Claviceps purpurea* and cause an alpha adrenergic blockade
   b. Psychotropic agents - these include psilocybin, psilocin and lysergic acid diethylamide (LSD)

c. *Aflatoxins* - these are carcinogens produced by *Aspergillus flavus* when growing on grain. When these grains are eaten by humans or when they are fed to dairy cattle and they get into the milk supply, they affect humans.

4. The synthesis of proteins on ribosomes that are different from those found in bacteria. This makes the fungi immune to those antimicrobial agents that are directed against the bacterial ribosome, *e.g.*, chloramphenicol.
5. The ability of certain metabolites to alter morphology of yeast and/or be assimilated by yeast with concomitant clinical identification affects.

POLYPHASIC APPROACH TO MICROBIAL IDENTIFICATION

The selection of method(s) used for microbial identification depends on the type and nature of the microorganism. The method(s) chosen should be well-described in scientific literature and consistent with those currently used in the field of microbial identification and taxonomic classification and they must enable identification of the organisms to the genus and species and, if possible, strain level. The robustness, precision and validity of the methodologies used to identify the microorganism are critical elements in the assessment of the safety of the product.

The choice of methods for microbial identification is at the discretion of the product proponent. However, the Fertilizer Safety Office recommends that applicants adopt an integrated polyphasic approach that includes classical microbiological and phenotypic analysis along with molecular tools, to accurately identify the active microorganism(s).

The strengths and weaknesses of the various identification methods should be taken into consideration, such that the methods chosen complement each other to result in a conclusive and definitive identification of the microorganism, and allow for clear differentiation of the organism from any closely related pathogenic and/or toxigenic species and strains. The methods commonly used in identification and substantiation of taxonomic classification of microorganisms are summarized below.

Phenotypic Analysis

Preliminary analysis in microbial identification often involves one or more phenotypic methods. Phenotypic methods are suitable for microorganisms that are culturable (*i.e.*, can grow as pure culture on artificial media), have well-established growth parameters, and physiological and biochemical profiles.

Analysis of Morphological Traits

These methods utilize colony and cell morphology to obtain an initial identification of a microorganism. This is accomplished through simple isolation and culturing of the microorganism and subsequent visual observation using microscopy. The morphological properties include:

- shape,
- size,
- surface characteristics and pigmentation,
- cell wall characteristics (Gram-staining),
- sporulation characteristics,
- mechanisms of motility, and
- other cellular inclusions and ultrastructural characteristics.

Analysis of Biochemical, Physiological and Metabolic Characteristics

Phenotypic identification methods include the study of the biochemical profile and metabolic properties of a microorganism by testing its growth requirements, enzymatic activities and cellular fatty acid composition.

The biochemical tests use specific growth media, nutrients, chemicals or growth conditions to elicit an observable or measurable biochemical response from the microorganism, thereby enabling its identification and characterization. These tests include: utilization of carbon and nitrogen sources, growth requirements (anaerobic or aerobic; temperature-optimum and range, pHoptimum and range), preferred osmotic conditions, generation of fermentation products, production of enzymes, production of antimicrobial compounds, as well as sensitivity to metabolic inhibitors and antibiotics.

Examples of recognized tests include: phenol red carbohydrate, catalase and oxidase tests, oxidation-fermentation tests, methyl red tests, Voges-Proskauer tests, nitrate reduction, starch hydrolysis, tryptophan hydrolysis, hydrogen sulfide production, citrate utilization, litmus milk reactions, *etc.* Several miniaturized and automated commercial systems are currently available with well-defined quality control procedures that allow for rapid identification of microorganisms.

Analysis of Fatty Acid Methyl Ester Composition (FAME Analysis)

Microorganisms can be identified by analysing the fatty acid profiles of whole cells or cell membranes using gas-liquid chromatography or mass spectrometry. The data on the type, content, proportion and variation in the fatty acid profile are used to identify and characterize the genus and species by comparing it against the fatty acid profiles of known organisms.

The expression of microbial phenotypes is highly dependent on environmental variables (*e.g.*, culture pH, temperature, selective vsnon-selective media, depletion of nutrients, presence of stressors *etc.*), and thus, may introduce inconsistencies in the identification process. The phenotypic methods are only acceptable if the response criteria are sufficient to identify the microorganism with a high level of confidence and distinguish it from phylogenetically close relatives that potentially pose safety concerns. Also, the applicability of the method is based on the robustness of information in reference databases. As such, results from phenotypic methods may require supporting data from other methods to accurately identify a microorganism.

Molecular Methods

Development of molecular methods has greatly improved the ability to rapidly detect, identify and classify microorganisms and also establish the taxonomic relationship among closely related genera and species. Identification, using molecular methods, relies on the comparison of the nucleic acid sequences (DNA, RNA) or protein profiles of a microorganism with documented data on known organisms. The molecular methods are considered sensitive enough to allow detection of low concentrations of viable or non-viable microorganisms in both pure cultures and complex samples (*e.g.*, soil, peat, water *etc.*).

Genotypic Methods

These include methods such as nucleic acid hybridization (Southern blot analysis or Solution-phase hybridization) and amplification-based or polymerase chain reaction (PCR) technologies. The latter consist of sequence comparisons of conserved genomic regions such as 16S or 18S rRNA, or comparisons of restriction fragment length polymorphisms (RFLP), amplified fragment length polymorphisms (AFLP) or G+C % content in the genomic DNA with corresponding data on known organisms.

Reliable genotypic identification requires databases with accurate and complete sequence information from a large number of taxa. The commonly used gene sequence databases include:

1. GenBank;
2. Ribosomal Database Project (RDP);
3. Europe's collection of nucleotide sequence data (EMBL); and
4. Universal Protein Resource (UniProt).

Product proponents are not limited to using the reference materials listed above; these are intended as guidance only.

Some of the limitations associated with genotypic methods also include: difficulties in differentiating between species that share identical and/or similar conserved region sequences, limited information on the quality of sequence data available in public databases and the complexity of taxonomic nomenclature overall. Considering the above, it is important to validate the results of genotypic microbial identification methods with data from other sources (*e.g.*, morphological and/or phenotypic analysis).

Protein Based Methods

Serological methods such as Western blotting, Immuno-precipitation and Enzyme Linked Immuno-sorbant Assay (ELISA) use antibodies to detect specific proteins that are unique and/or characteristic of a microorganism. The applicability of serological methods is dependent on the availability, sensitivity and specificity of the antibodies used. There are commercial kits available for immuno-detection of several microorganisms. Protein based methods also include gel electrophoresis (SDS-PAGE, 2D-gels, *etc.*) that can separate cellular proteins on a defined matrix and identify microbial proteins of interest by comparing with microorganisms with known protein profiles.

Genomics

More recently, complete profiling of the transcriptome, genome, proteome or metabolome have been used to identify and characterise organisms. Several modern technologies such as DNA and protein microarray analyses, mass spectral protein profiling, nuclear magnetic resonance (NMR) spectral analysis, in-silico microbial metabolome platforms are increasingly used in identification and characterization of microorganisms.

The knowledge of the sensitivity and specificity of genomic tools and their application in microbial identification is rapidly evolving. However, challenges related to standardization of genomics methodologies (including optimization of protocols and bioinformatics tools for reliable data annotation, interpretation *etc.*) continue to hinder their applicability in safety (risk) assessment and regulatory decision making. The Fertilizer Safety Office will consider data generated by genomics-methods on a case-by-case basis. However, validation of genomics data using alternate methods is currently required to substantiate the identification and taxonomic classification of an active microorganism(s) in a supplement product.

TAXONOMIC CLASSIFICATION AND NOMENCLATURE:

The taxonomic identification of the microorganism(s) should be based on the currently used and internationally accepted taxonomic classification system. The description of the microorganism(s) in the product and its characteristics must correspond to the characteristics described in standard resources and/or references that are commonly used by the scientific community to validate taxonomic classification. These can include but are not limited to:

- textbooks such as the Bergey's Manual of Systematic Bacteriology;
- The Prokaryotes;
- Applied Microbial Systematics;
- Principles of fungal taxonomy *etc.*;

- online resources such as the Catalogue of Life;
- PubMed Taxonomy and UniProt Taxonomy *etc.*; and
- peer reviewed journals.

The taxonomic name should follow the nomenclature code officially recognized by the International Committee on Systematics of Prokaryotes (ICSP). Applicants should verify the "Approved List of Bacterial Names" to ensure that the nomenclature is in accordance with the latest Validation List developed and updated by the International Journal of Systematic and Evolutionary Microbiology (IJSEM).

Please note that microbial taxonomic classification and nomenclature, particularly for bacteria, is in a constant state of flux as methodologies evolve to generate more reliable information to identify/classify and/or reclassify the current taxonomic scheme. Cross referencing more than one resource/reference will help in validating the current taxonomic designation and classification of a microorganism.

Streamlining the Information Requirements for Taxonomic Identification of Microorganisms

The following is general guidance to streamline the information requirements for identification of microorganisms represented as active ingredients in fertilizers and/or supplements regulated under the *Fertilizers Act*. The Fertilizer Safety Office will consider the information for each application on a case-by-case basis.

Microorganisms that are Exempt from Full Safety Data Requirement

The Fertilizer Safety Office has conducted a comprehensive safety assessment of microorganisms belonging to the Rhizobial group (*Rhizobium* spp., *Bradyrhizobium* spp., *Mesorhizobium* spp., *Sinorhizobium* spp.), *Bacillus subtilis* and Vascular Arbuscular Mycorrhizae (VAM). Based on the currently available scientific literature and well documented history of safe use of these microorganisms in Canada, the Fertilizer Safety Office has exempted these microorganisms from full safety data requirements. However, submission of core safety data is required for this category of microorganisms.

If the Rhizobia, VAM or *Bacillus subtilis* strains used as active ingredient in the inoculant product, are obtained from a recognized culture collection, substantiation of taxonomic identification. If the microorganism was isolated from a natural environment, applicants are expected to adopt a polyphasic approach (*e.g.*, morphological and biochemical methods) to support the identification of microorganisms.

For the Rhizobial group and VAM, identification is required up to the species and, if possible, to strain level. Microorganisms belonging to genus *Bradhy-*

rhizobium also require identification up to the species level, and, if possible, to strain level, in order to differentiate the active ingredient from closely related toxigenic species like *B. elkanii*.

All microorganisms belonging to the genus *Bacillus*, including *Bacillus subtilis* must be identified to the strain level.

Microorganisms Obtained from Culture Collections

If the microorganism is obtained or purchased from a recognized International Culture Collection such as the American Type Culture Collection (ATCC), Germany's Deutsche Sammlung von Mikroorganismen und Zellkulturen GmbH (DSMZ), USDA-ARSNational *Rhizobium* Germplasm Collection (BRCC), or Canadian Collection of Fungal Cultures (CCFC) *etc.*, the following information will be required to substantiate the taxonomic identification of the microorganism.

- a certificate of analysis from the culture collection;
- description of methods used to identify and substantiate the taxonomic designation, and
- results of analysis to demonstrate the purity and true-type of the microorganism and its relatedness or similarities to microorganisms that are detrimental to the health of plants, animals and humans and the environment.

Environmental Isolates

For all environmental isolates, taxonomic designation should be based on a complementary, polyphasic approach that enables identification and substantiation of the microorganism up to species level, and if possible, up to strain level, and allows differentiation from closely related pathogenic and/or toxigenic species and/or strains. This is particularly important for microorganisms belonging to genera where pathogenic and/or toxigenic species and strains are known to exist.

For instance, strain level identification is required for microorganisms belonging to the genera *Bacillus*, in order to differentiate the microorganism from related phylogenetic kins that include pathogenic and/or toxigenic species (*e.g., Bacillus anthracis, Bacillus thuringiensis, Bacillus cereus etc.*), and from potentially pathogenic and/or toxigenic strains within a particular species (*e.g., Bacillus licheniformis etc.*).

FUNGI AS DECOMPOSERS

Certain species of fungi and bacteria are the engines of the process of decomposition. When plants grow and produce new leaves, fruits, and stem wood, they use scarce nutrients like nitrogen to make the new tissues. Over time, without decomposition, so much nitrogen would be locked up in leaves and other tissues that there would not be enough nitrogen available for the plant to make new leaves, stems and wood. The surface of the ground would also be buried by dead leaves and wood lying forever where they fell.

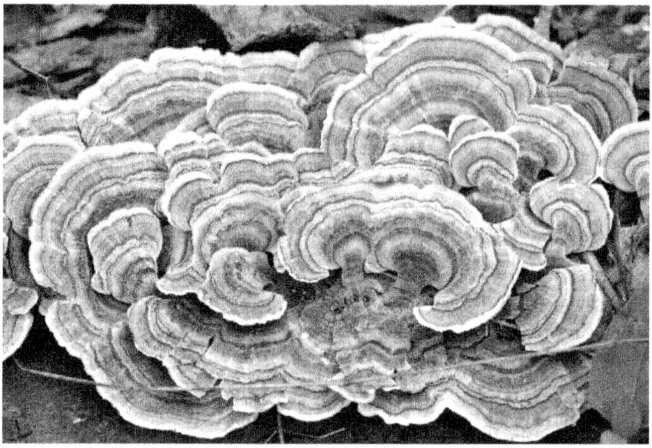

Fig. : Fungi.

Fungi and bacteria are the major organisms decomposing dead leaves and other organic matter. Here, we do not use the word "organic" in the same way it is used by the food industry. In Biology and Chemistry, "organic" describes any material made up of molecules containing carbon and hydrogen atoms. All living things, and a few other surprising substances, are considered organic. Decomposition is a complex process. Organic matter is broken down into carbon dioxide and the mineral forms of nutrients like nitrogen. It is also converted into fungi and bacteria through these organisms feeding on the organic material and reproducing. Scientists call the organisms that decompose organic matter decomposers, saprobes or saprotrophs.

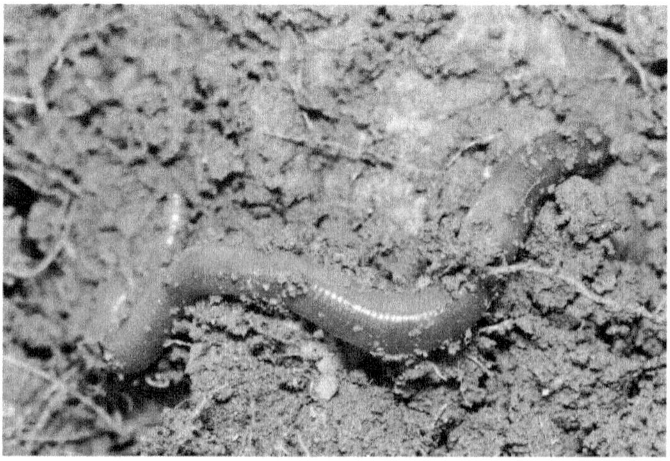

Fig. : Earthworm

Decomposition is made up of a number of subprocesses. Consider the decomposition of leaves. Earthworms and other soil animals break the leaves into smaller pieces in a process called fragmentation. This is an important step, because smaller fragments have more surface area to support the growth of bacteria and fungi. Bacterial growth is especially affected by fragment size, since fungi can penetrate substances more easily than bacteria. Rainwater percolates through the leaves, dissolving and carrying away some of the chemicals in the leaves in a process called "leaching".

The movements of earthworms and other soil animals stir the leaf fragments and mineral soil particles together in a process called "mixing". The result of the interaction of these processes can be seen in the changes in the leaves. The fallen leaves started out whole, and were green, yellow or red. They were reduced to small dark brown shreds as their fragments became heavily colonized by fungi. Finally, at the end of the processes of decomposition, they have become fine black particles of soil organic matter, and their original shape no longer exists.

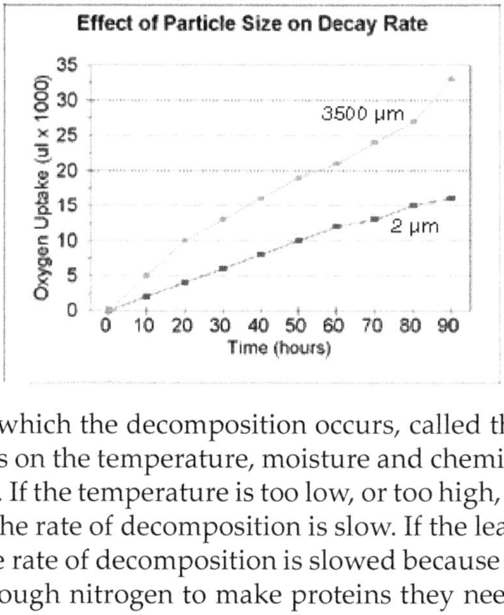

The speed at which the decomposition occurs, called the "rate of decomposition", depends on the temperature, moisture and chemical composition of the organic matter. If the temperature is too low, or too high, fungi and bacteria cannot grow and the rate of decomposition is slow. If the leaves have a low nitrogen content, the rate of decomposition is slowed because fungi and bacteria can not extract enough nitrogen to make proteins they need for growth. The oxygen level is another important factor, since fungi require oxygen for growth. In lakes and other low oxygen environments, fungal growth will be slower and thus decomposition will be slow.

If decomposition could not occur, the nitrogen in dead organic matter would remain locked up. Plant growth would decrease over time as the nitrogen the plants took from the soil was not replaced. This would be a catastrophe, because plant growth supplies all of our food. We need it to feed ourselves directly in the form of grain, fruits and vegetables, and to raise the animals we eat. By understanding the factors controlling decomposition we can protect our food supply, and increase plant growth.

Fungi and bacteria are not restricted to decomposing leaves and other plant materials. They will decompose any dead organic matter, whether it is a cardboard box, paint, glue, pair of jeans, a leather jacket or jet fuel. Remember that jet fuel is made from petroleum, which is made of decomposed microscopic creatures from the oceans of the Mesozoic Era.

Decomposition can be a problem on a large and small scale. The U.S. military wouldn't want fungi to decompose uniforms and tents in a wet environment, or dissolve the glue holding the lenses in a pair of binoculars. Wooden buildings need regular attention to prevent their wood from being invaded by rot. In our own homes, we don't want our food to become moldy. Many times, we want to stop decomposition.

The enormous amount of organic matter produced by decomposition creates problems of storage and disposal. In nature, a forest produces about a pound of dead leaves and wood per square yard every year. This may not sound like much material, but consider the amount that would fall on an area the size of a football field or soccer field. An American high school football field, including end zones, covers 6400 square yards (120 by 53.3 yards wide). A football field-sized area in a forest would be covered by 6400 pounds, or 3.2 tons, of dead organic matter per year. A soccer field covers 8800 square yards (110 by 80 yards wide). An area this size in a forest would be covered by 4.4 tons of dead organic matter. To give an idea of how much material this is, remember that an average-sized car weighs only 1.5 tons.

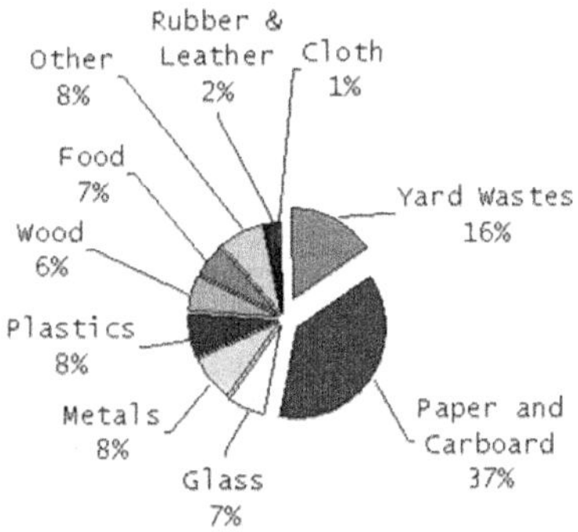

American cities produce even more organic waste (garbage) than forests, and very little of it decomposes and becomes recycled into new plant growth by natural processes. Every resident of Los Angeles produces 7 pounds of garbage per day! This is equal to one ton of garbage per year per person. The disposal and recycling of nutrients in garbage is a large problem because of the huge amount of garbage produced every day. Sanitary landfills are the current solution to the garbage problem.

The first modern sanitary landfill was built in Champaign, Illinois in 1904. The phrase "sanitary landfill" was invented in the 1930s by Jean Vinenz, the commissioner of public works of Fresno, California. By 1945, about 100 American cities had sanitary landfills; by 1960 there were 1400. Before sanitary landfills were invented "dumps" were used to dispose of garbage. Dumps were open, the garbage was not covered, and the garbage was often burned. Dumps smell terrible, are often surrounded by blowing paper, attract flies and provide food for large populations of rodents. Sanitary landfills have less odour, there is no black smoke from burning tires and garbage, and the number of flies and rodents is greatly reduced, hence the term "sanitary".

The largest American sanitary landfill is the Fresh Kills Landfill on Staten Island, New York. It opened in 1948 and covers 3,000 acres, a rectangular area roughly 3 by 4 miles. When it reached full capacity and closed in 2001, its estimated height was 505 feet. If it had been allowed to get any taller, it would have interefered with air traffic.

Landfills are usually lined with several feet of dense clay and then sealed with thick layers of plastic to prevent leaks of hazardous chemicals. The garbage is dumped in rows or piles from 10 to 20 feet high. Bulldozers are used push the garbage into rows and squash large objects. Compactors with 5 foot wide studded rollers are also used to squash the garbage. Squashed garbage takes up less space extending the life of the landfill. Each day, soil, glass, or plastic foam pellets is spread over a landfill to reduce odors and pests.

The soil covering also reduces the amount of rainwater that seeps in. Invading rainwater carries water-soluble chemicals out of the garbage to form liquids called leachates. When leachates pool in the bottom of the landfill, they are pumped out, collected and treated. The treated leachate is handled like sewage. The particles, called sludge, are separated from the liquid and burned, or used as fertilizer, or dumped in the ocean or back into the landfill. If the sludge is considered hazardous, it is shipped to a hazardous waste disposal site.

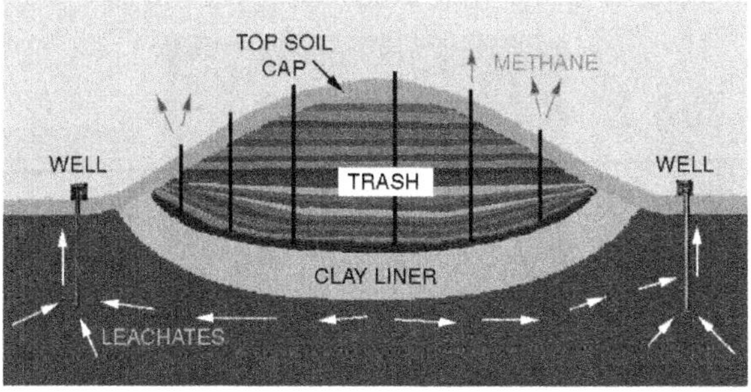

Landfills also produce methane gas. Once landfills are full, they are often capped with a layer of clay. The cap excludes water, reducing the danger of chemicals leaching from the landfill into the surrounding soil and groundwater. The cap also greatly slows the movement of oxygen from the atmosphere to the buried garbage. When the oxygen available below ground is used up by fungi and bacteria, their decomposing activity stops.

Easily decomposed organic substances will continue to be broken down by anaerobic bacteria, which can live where no oxygen exists. Methane gas is one of the products of their decomposing activity. The anaerobic condition (very low to no oxygen) and low moisture level in the garbage stops or greatly slows decomposition. The garbage is essentially "mummified" in this stable environment. Newspapers that have been buried in landfills can still be read 20 years later. Only one-third to one-half of even easily decomposed materials such as lawn, garden and food waste is decomposed after 20 years.

The newest landfills contain a system of perforated pipes to collect methane gas. Methane is one of the main ingredients of natural gas, and is dangerous because it is explosive in high concentration. In April, 2000, a house in Michigan was destroyed when methane seeped from an old nearby landfill into its basement and then exploded. To dispose of methane, it is sometimes released into the air or burned, but it can also be purified and used as fuel. Releasing methane into the atmosphere is not recommended because it is a potent "greenhouse gas". Methane and other greenhouse gases are responsible for destroying part of the ozone layer in our atmosphere, which protects us from harmful solar radiation.

Landfills are expensive to build and run. The cost of landfills can be reduced by making them last longer, which means filling them more slowly. One way to reduce the amount of garbage being deposited in landfills is for homeowners to compost lawn, garden and food waste.

Homeowners usually construct a compost heap from a mixture of lawn and garden waste (grass clippings, leaves, branches) and garbage. Paper and cardboard are sometimes added, but they take longer to decompose. The resulting compost is a soil which contains a high percentage of fine organic matter whose nitrogen and other nutrients can be easily used by plants. Compost can be used to pot plants, or enrich the soil in vegetable and flower beds. By using what we know about decomposition, homeowners can adjust the rate at which the

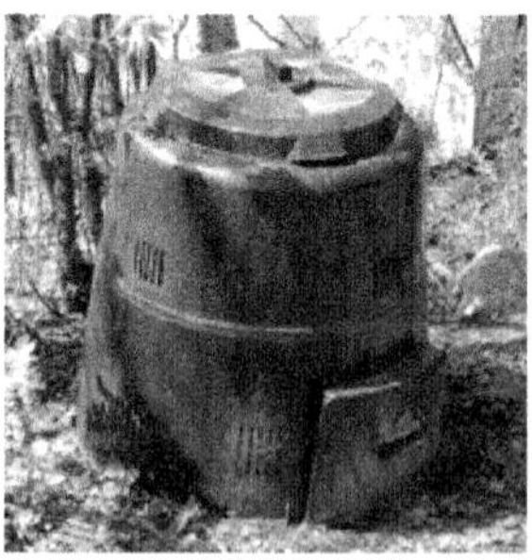

Backyard Compost Bin

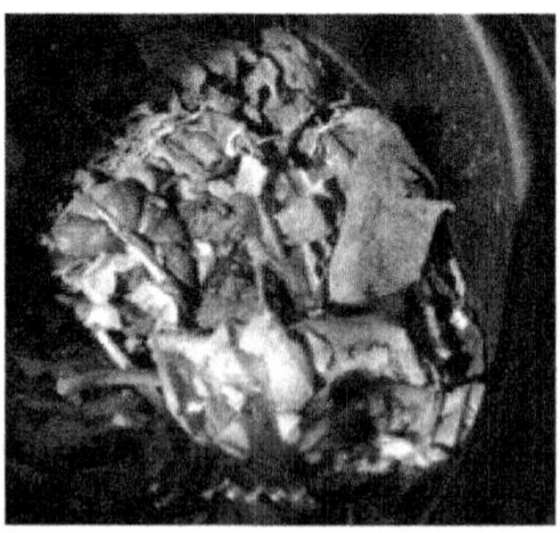

Top of compost pile

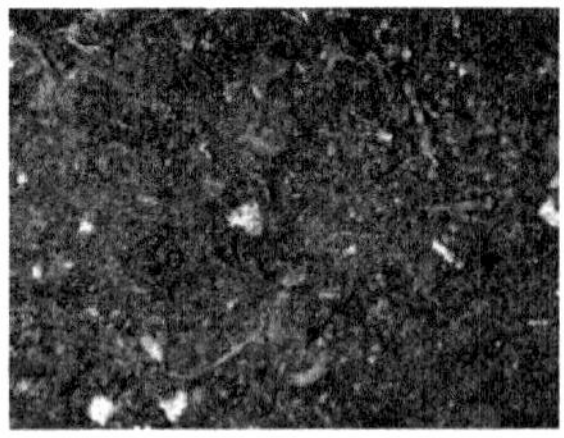

Finished compost

compost heap decays. The rate of decomposition can be increased by adding earthworms, mixing the contents occasionally, adding chemical fertilizer, and the keeping the heap moist. Chemical fertilizer is added when the organic material has low nitrogen levels, because fungi and bacteria also need nitrogen to grow. Mixing the compost is probably the hardest step, because a cubic yard of compost weighs about a ton.

Ants may have discovered composting before us. *Fungus growing ants* collect leaves and inoculate them with a fungus that decomposes the leaves. The ants harvest the fungus for food and discard the decayed leaves.

Chapter 3

ISOZYME ANALYSIS OF FUNGAL TAXONOMY, GENETICS, AND POPULATION BIOLOGY

Applications of Isozyme Analysis

Isozyme analysis can be applied to address a wide variety of biological questions in the fungi. These markers usually are considered to be selectively neutral so are ideal for looking at sources of inoculum for pathogenic fungi or gene flow among populations. The markers can indicate whether specific tissues are haploid or diploid, can aid strain or species identification, and can be used to identify hybridization either in the laboratory or in nature.

Taxonomy and Species Identification

Isozyme analysis was first applied to fungal taxonomy during the 1960s. Since then it has been applied at many taxonomic levels from typing individual strains to delimitation and identification of species. The best enzymes for taxonomic purposes are those that are monomorphic within but different among taxa. A caveat is that isozymes cannot be used to infer phylogenetic relationships. This is because bands with the same migration rate on a gel in fact may not be identical. Furthermore, it is not possible to infer which allele is ancestral or to estimate the number of mutations that cause the isozymes to migrate differently; alleles of similar size may be more different evolutionarily than those with larger migration distances on a gel.

Therefore, clustering algorithms such as neighbor joining that allow for unequal rates of evolution on branches are not appropriate for isozyme data. Instead, isozyme data are analyzed usually by calculating a simple distance coefficient and drawing clusters with the Unweighted Pair Group Method with Arithmetic mean (UPGMA). These analyses can be performed with several computer programs such as NTSYSpc, POPGENE, PHYLIP, or PAUP. Cluster analyses always should be accompanied by bootstrap analysis or some alternative method of indicating the level of statistical support for particular groupings. Unfortunately, bootstrap analysis cannot be performed with NTSYS, but it is available with some of the other computer programs as well as the program WinBoot.

Specific applications of isozyme analysis to taxonomic questions include identifying strains of Trichoderma harzianum , varieties of Verticicladiella wageneri , or anastomosis groups within Rhizoctonia. The technique also can be used to identify fungal cultures to species. Isozyme analyses have confirmed a high level of genetic differentiation among host-associated varieties of Leptographium wageneri and have revealed previously unknown genetic subdivision within various species of Phytophthora. In the rust genus Puccinia, isozyme variation can distinguish among species and also among formae speciales on different hosts.

The most common use of isozyme analysis in fungal taxonomy is to divide isolates into species and to test how well biochemical species identification corresponds to classical taxonomy. Usually, species groups identified by isozyme analysis correspond quite closely with those identified morphologically. However, sometimes two or more taxa are found to be the same genetically and are combined into a single species. The opposite also is a common result of isozyme analysis: single species frequently can be divided into two or more species based on previously hidden genetic differentiation uncovered by isozyme analyses.

Genetics

In addition to simple Mendelian genetics and linkage analysis, isozymes can be used to distinguish hybrid from nonhybrid progeny in both intra- and inter-specific crosses , and to infer the ploidy level of vegetative hyphae. Dimeric enzymes are ideal for this kind of analysis. Isozymes also can be used to analyze parasexual genetics in fungi. In addition to laboratory genetics, isozyme analysis implicated somatic hybridization as the probable origin of a new forma specialis of cereal rust in Australia and identified naturally occurring hybrids among field isolates of Phytophthora species. Estimates of relatedness based on isozyme analysis among strains of Agaricus brunnescens were used to aid the choice of parents in a mushroom breeding program. Most studies have shown normal Mendelian segregation of isozyme alleles (*i.e.*, the isozymes can be considered allozymes), although instances of aberrant segregation have been noted. It is important to remember that no isozyme should be referred to as an allozyme until a proper genetic analysis has been completed.

Population Biology

Gene flow is the movement of individuals among populations and can be estimated indirectly through isozyme analysis as the number of migrants per generation or by analysis of "private" alleles those limited to a single species or population. Different species usually have highly differentiated allele frequencies and a higher number of private alleles compared to individuals from a single panmictic population. The computer program POPGENE, among others, can perform this type of analysis.

Applications of isozyme analysis to epidemiology have included monitoring the spread of specific genotypes of Phytophthora infestans, determining the influence of various evolutionary forces on the composition of inoculated populations of the barley scald pathogen Rhynchosporium secalis , and estimating the size of and boundaries between individuals of the puffball Lycoperdon pyriforme on decaying logs.

The first real attempt to use isozymes to analyze the genetic structure of fungal populations was for Neurospora intermedia by Spieth (1975). That analysis revealed a high level of genetic variation within but low differentiation among populations. Spieth's pioneering work was followed by additional studies within a decade and this has accelerated during the 1990s. Many of these studies have had similar results and demonstrated high gene diversity within but low differentiation among populations of a wide diversity of fungi. Other authors have found varying degrees of subdivision among populations, usually among host-associated forms and, more rarely, among physiological races.

Gene flow analysis of isozyme markers also can indicate when taxa have become sufficiently isolated reproductively to be considered separate species. Existence of separate species was shown conclusively for several closely related groups in the genus Phytophthora which could not have been discovered without the use of molecular markers. A similar approach identified distinct biological species within the mushroom Pleurotus eryngii.

Isozyme analyses of some populations of fungi and oomycetes have revealed extremely low levels of genetic variation reflective of a highly clonal population structure. This occurred for rice-infecting isolates of the rice blast fungus Magnaporthe grisea , asexual populations of cereal rusts , and worldwide populations of the oomycete Phytophthora infestans. Populations of fungi infecting conifer hosts in the western United States also appeared to be highly clonal and specific clones are characterized by fixed heterozygosity.

Differences in levels of genetic variation among locations can help identify the center of origin of a species. The most diverse populations genetically should occur at or near the center of origin, as these populations will have had the longest time in which mutations can accumulate; derived populations usually will contain only a subset of the total species diversity. This idea has been combined with isozyme analysis to indicate that the edible mushroom Agaricus bisporus may be indigenous to North America and that Phytophthora infestans, P. palmivora, and P. megakarya probably originated in central Mexico , southeast Asia , and central Africa , respectively. Very limited genetic diversity indicated that the plant pathogens Puccinia graminis f. sp. tritici, Phytophthora cinnamomea, and Dilophospora alopecuri had only very limited introductions into Australia, although the original source populations for the introductions could not be identified for certain.

ISOZYME

Isozymes (also known as isoenzymes or more generally as Multiple forms of enzymes) are enzymes that differ in amino acid sequence but catalyze the same chemical reaction. These enzymes usually display different kinetic parameters (*e.g.* different K_M values), or different regulatory properties. The existence of isozymes permits the fine-tuning of metabolism to meet the particular needs of a given tissue or developmental stage (for example lactate dehydrogenase (LDH)). In biochemistry, isozymes (or isoenzymes) are isoforms (closely related variants) of enzymes. In many cases, they are coded for by homologous genes that have diverged over time. Although, strictly speaking, allozymes represent enzymes from different alleles of the same gene, and isozymes represent enzymes from different genes that process or catalyse the same reaction, the two words are usually used interchangeably.

Introduction

Isozymes were first described by R. L. Hunter and Clement Markert (1957) who defined them as *different variants of the same enzyme having identical functions and present in the same individual*. This definition encompasses:

(1) enzyme variants that are the product of different genes and thus represent different loci (described as *isozymes*) and

(2) enzymes that are the product of different alleles of the same gene.

Isozymes are usually the result of gene duplication, but can also arise from polyploidisation or nucleic acid hybridization. Over evolutionary time, if the function of the new variant remains *identical* to the original, then it is likely that one or the other will be lost as mutation accumulate, resulting in a pseudogene. However, if the mutations do not immediately prevent the enzyme from functioning, but instead modify either its function, or its pattern of gene expression, then the two variants may both be favoured by natural selection and become specialised to different functions. For example, they may be expressed at different stages of development or in different tissues.

Allozymes may result from point mutations or from insertion-deletion (*indel*) events that affect the DNA coding sequence of the gene. As with any other new mutations, there are three things that may happen to a new allozyme:

- It is most likely that the new allele will be non-functional — in which case it will probably result in low fitness and be removed from the population by natural selection.

- Alternatively, if the amino acid residue that is changed is in a relatively unimportant part of the enzyme (*e.g.*, a long way from the active site), then the mutation may beselectively neutral and subject to genetic drift.
- In rare cases, the mutation may result in an enzyme that is more efficient, or one that can catalyse a slightly different chemical reaction, in which case the mutation may cause an increase in fitness, and be favoured by natural selection.

An Example of an Isozyme

An example of an isozyme is glucokinase, a variant of hexokinase which is not inhibited by glucose 6-phosphate. Its different regulatory features and lower affinity for glucose (compared to other hexokinases), allows it to serve different functions in cells of specific organs, such as control of insulin release by the beta cells of the pancreas, or initiation of glycogen synthesis by liver cells. Both of these processes must only occur when glucose is abundant, or problems occur.

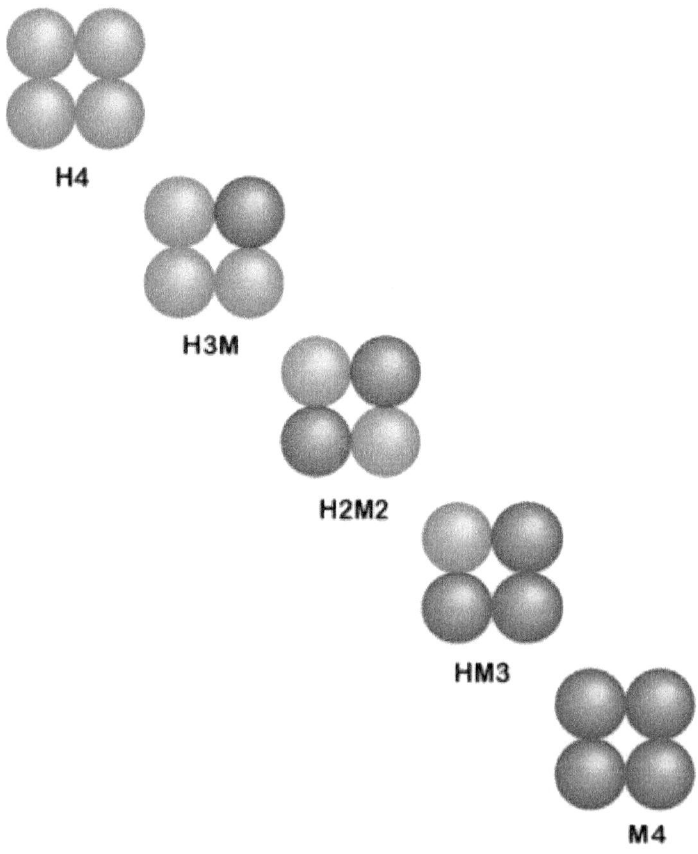

Fig. : The 5 isozyme of LDH.

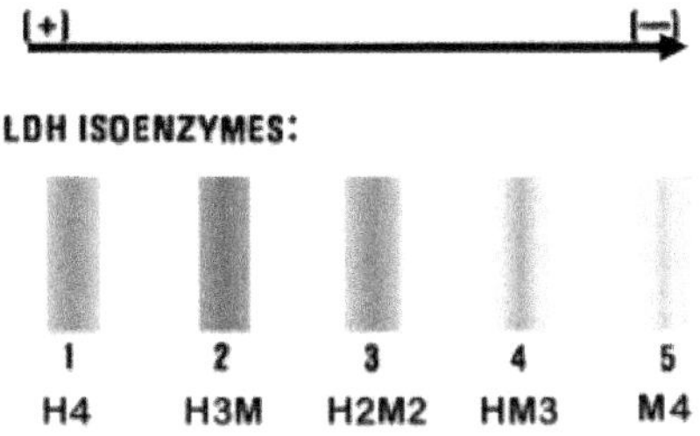

Fig. : The differences between 5 isozyme when use Electrophoresis.

The enzyme Lactate Dehydrogenase is made of two(H-form and M-Form) different sub units, combines in different Permutations and Combinations in depending on the tissue in which it is present as shown in below table,

Type	Composition	Location
LDH_1	HHHH	Heart and Erythrocyte
LDH_2	HHHM	Heart and Erythrocyte
LDH_3	HHMM	Brain and Kidney
LDH_4	HMMM	Skeletal Muscle and Liver
LDH_5	MMMM	Skeletal Muscle and Liver

Distinguishing Isozymes

Isozymes (and allozymes) are variants of the same enzyme. Unless they are identical in terms of their biochemical properties, for example their substrates and enzyme kinetics, they may be distinguished by a biochemical assay. However, such differences are usually subtle (particularly between *allozymes* which are often neutral variants). This subtlety is to be expected, because two enzymes that differ significantly in their function are unlikely to have been identified as *isozymes*.

Whilst isozymes may be almost identical in function, they may differ in other ways. In particular, amino acid substitutions that change theelectric charge of the enzyme (such as replacing aspartic acid with glutamic acid) are simple to identify by gel electrophoresis, and this forms the basis for the use of isozymes as molecular markers. To identify isozymes, a crude protein extract is made by grinding animal or plant tissue with an extraction buffer, and the components of extract are separated according to their charge by gel electrophoresis. Historically, this has usually been done using gels made from potato starch, but acrylamide gels provide better resolution.

All the proteins from the tissue are present in the gel, so that individual enzymes must be identified using an assay that links their function to a staining reaction. For example, detection can be based on the localised precipitation of soluble indicator dyes such as tetrazolium salts which become insoluble when they are reduced by cofactors such as NAD or NADP, which generated in zones of enzyme activity. This assay method requires that the enzymes are still functional after separation (native gel electrophoresis), and provides the greatest challenge to using isozymes as a laboratory technique.

Isoenzymes differ in kinetics (they have different Km and Vmax values).

Isozymes and Allozymes As Molecular Markers

Population genetics is essentially a study of the causes and effects of genetic variation within and between populations, and in the past, isozymes have been amongst the most widely used molecular markers for this purpose. Although they have now been largely superseded by more informative DNA-based approaches (such as direct DNA sequencing,single nucleotide polymorphisms and microsatellites), they are still amongst the quickest and cheapest marker systems to develop, and remain (as of 2005) an excellent choice for projects that only need to identify low levels of genetic variation, *e.g.* quantifying mating systems.

Other Major Examples

1. The cytochrome P450 isozymes play important roles in metabolism and steroidogenesis.
2. The multiple forms of phosphodiesterase also play major roles in various biological processes.

Although more than one form of these enzymes have been found in individual cells, these isoforms of the enzyme are unequally distributed in the various cells of an organism. From the clinical standpoint they have been found to be selectively activated and inhibited, an observation which has led to their use in therapy.

EFFICIENCY AND LIMITATIONS OF ISOZYME STUDIES IN FOREST TREE GENETICS

Studies on the genetics of forest trees have been performed using mainly three categories of simply inherited traits: morphological markers, monoterpene variants and isozymes.

Morphological traits are easy to assess, but their use in genetic studies is limited, since phenotypic morphological characteristics are generally controlled by several genes, and quantitative factors are under strong environmental influence. Thus, contributions of individual genes cannot be identified, and specific genetic differences among individuals are not detectable.

In gymnosperms, single genes are controlling the level of a few monoterpenes. However, genetic interpretation of monoterpene patterns is complicated and analysis of large numbers of samples by gas-liquid chromatography is expensive and time consuming.

Compared to the above, isozymes are relatively easy to handle. Their use was introduced into forest tree genetics some 15 years ago and is now widely applied, based on the hypothesis that the isozymes are the direct products of specific allelic genes. Once identified, statistical methods are available to study their distribution (frequencies) within and among populations.

Allozymes as Gene Markers

Diploid, sexually reproducing organisms receive one complete set of chromosomes from each parent. Accordingly, they have each gene in duplicate; the homologous genes are called alleles. If more than one allele occurs at the locus of a structural gene, the varying enzyme products produced by them can either appear in a single diploid (heterozygotic) organism, or in different (homozygotic) members of a species.

The number of alleles at a given gene locus depends on the species and on the locus itself. Genes are characterized as polymorphic if they comprise two or more alleles. Invariant (monomorphic) genes are not suitable for genetic studies.

To interprete isozyme patterns correctly, their sub-unit composition (number of polypeptides forming the complete enzyme) must be known.

The genes coding for allozymes are identified by an abbreviation of the enzyme name, *e.g.* "Aat" for aspartate amino transferase. If a gene comprises more than one locus (precisely defined site on the DNA molecule), then these loci are indicated by a capital letter (or a number) and linked to the gene name as a suffix, *e.g.* Aat-A or Aat-B (or Aat-1 or Aat-2). The corresponding alleles may be indicated by consecutive numbers such as Aat-Al, Aat-A2 *etc.* (or Aat-1, Aat-1_2).

Allozymes can be used in genetic studies if the following conditions are fulfilled:

- The genotype is phenotypically expressed through given enzyme patterns, which can be observed in the laboratory. A prerequisite for this is that the enzyme forms of interest segregate in a Mendelian pattern ; and
- At least one locus per enzyme is polymorphic.

Specific genotypes can be identified using:

- Examination of progeny resulting from controlled crossings;
- Comparative studies of haploid and diploid tissues of an individual (haploid endosperm and diploid embryo of gymnosperms, or haploid pollen and diploid embryo of angiosperms); or
- By applying Hardy-Weinberg's law in theoretical models.

Tools to Detect Isozymes

Identification of isozymes is based on their high substrate specificity and varying electrophoretic mobilities. If two allelic variants of an enzyme differ in the number of charged amino acids, then their different net charges can be used to separate them by electrophoresis. The primary structure (peptide backbone) of these enzymes is encoded in the two homologous strands of the DNA in the individual and, accordingly, allelic genes can be identified by differently charged allozymes. Unfortunately, however, two thirds of the proteinogeneous amino acids are uncharged, reducing the possibility for identification of existing allelic enzyme variants to less than 30% of the total number present.

However, for those variants which do have a charge, electrophoretic methods are relatively simple to carry out in practice. It is sufficient to place crude cell or tissue extracts on some type of electrophoretic separation medium such as starch gel-, cellogel- or polyacrylamide gel; and subsequently stain the gel using a staining agent which permits only a specific enzyme and its multiple molecular forms to become visible, while leaving all other separated proteins (and enzymes) unstained. Finally the stained enzyme forms are counted and the estimated frequencies are used for further genetic calculations.

APPLICATION OF ALLOZYME RESEARCH

Identification of Seedlots

Differences in allozyme frequencies can be used to determine the origin of a seedlot or of seedlings originating from a seed orchard or from different populations of a given tree species. Particularly in Central Europe, however, identification of provenances of norway spruce (*Picea abies*) is difficult, since a large number of provenances have been introduced and their true origin is unknown.

For identification of spruce provenances, at least 7 seeds from each of 200 trees must be analyzed for determining allozyme frequencies. According to Bergmann, (1975a), equal allozyme frequencies were found in seeds whether they originated from individual trees of a certain population, or from a mixture of seeds of that particular population. Examination of 100 diploid genome probes collected from a seed orchard (or 50 haploid endosperms of a coniferous species) allows detection of alleles occuring at a frequency of 0.03 or higher with a probability of 95 %. However, if serious pollen contamination from non-orchard sources is suspected and isozyme analysis is performed to prove it, it is recommended to double the above sample to 100 haploid endosperms of seeds.

In many countries forests have been divided geographically, topographically and ecologically into seed zones. Each zone is presumed to represent a relatively homogeneous environmental unit, within which the genotypes, in principle, are adapted for use in any other part within the same zone, while seeds from outside the zone are expected to be less well adapted. Restricting seed used for reforestation to those collected from within selected seed zones, is thus expected to ensure that seed will be properly adapted to the conditions at the plantation site. The identity of seedlots can be verified following their collection by using isozyme techniques. However, if the genetic diversity between populations is small ($d_0 < 0.10$), this will constrain the use of isozyme studies in the identification of seedlots from various seed zones.

Identification of Clones

Isozymes can also be utilized to identify closely related individual trees, or clones. Isozyme studies may be performed on winter buds, leaves or current year needles. For clone identification it is sufficient to compare phenotypic isozyme patterns, even without knowledge of their genetic base. In one study, analysis of six to seven variable enzyme systems, comprising 10 to 20 different loci, enabled certification of 50 to 100 clones of a norway spruce (*Picea abies*) multi-clone variety. This application of isozyme studies is potentially of great importance as multi-clone varieties have been developed in some countries as part of a clonal forestry strategy to help avoid reduction of genetic diversity in future generations of artificial populations.

Determination of the Reliability of Controlled Crossing and Levels of Self Fertilization

The reliability of controlled crossing can be assessed by examinating isozyme patterns in seeds resulting from such crosses, comparing the results with those theoretically expected based on the genotypes of the parents. A controlled pollination can be considered a failure, if the seeds in one out of four samples (or, one out of two for coniferous species), contain alleles not present in either of the supposed parents. Experiments carried out with douglas-fir showed that most failures in controlled crosses are due to pollen contamination by foreign pollen.

Furthermore, isozyme gene markers can be used to identify self-fertilizing trees in seed orchards. Since self-fertilization affects the whole genome equally, it can be revealed through investigation of one single polymorphic gene locus. In norway spruce (*Picea abies*), at the leucine amino-peptidase B-locus (Lap-B), self-fertilization occured in two out of four genotypes. The rate of self-fertilization was estimated to be 15 %. Since the genetic diversity of seeds of self-fertilizing trees is reduced, it is generally desirable to identify and eliminate such self-fertilizing individuals from seed orchards.

Identification of Tree Species

Variation in allozyme patterns can be utilized to distinguish different tree species or hybrids. Seeds of two larch species, *Larix decidua* and *Larix kaemp-*

feri were separated by analysis of the mobility of shikimate dehydrogenase enzymes. In *L. kaempferi*, the enzyme migrated faster to the anode than in *L. decidua*. In hybrids of both parental species, both enzyme variants could be indentified.

Oaks comprise a taxonomically complex group. More than 50 tree-like oak-species occur in USA, and the total number of species in this genus is more than 300. Some of these possess unique and easily identifiable morphological and anatomical traits, however several are difficult to identify and differentiate from each other. Taxonomic identification is further complicated by apparent hybridization among a number of species. Over 60 hybrids have been reported in the USA alone..

Investigations of buds and acorn cotyledon tissue originating from various seed sources of northern red oak (*Quercus rubra* L.), black oak (*Q. velutina* Lam.), white oak (*Q. alba* L.), swamp white oak (*Q. bicolor* Willd), bur oak (*Q. macrocarpa* Michx.) and chinkapin oak (*Q. muehlenbergii* Engelm.), showed that most isozyme bands of leucine aminopeptidase, alcohol dehydrogenase, acid phosphatase and aspartate amino transferase were common to two or more of these species. This similarity in isozyme patterns is likely to be a reflection of their close taxonomic (and thus genetic) relationship. However, individuals of some species exhibited distinctive genotypes in the analysis; these could also be readily distinguished through differing phenotypes.

The species *Quercus petraea* and *Q. robur*, could be classified according to their peroxidase patterns. Some populations of *Q. petraea* showed evidence of introgression with *Q. robur*.

Characteristics of Forest Tree Populations

In addition to the above uses, the following genetic characteristics of populations have been studied utilizing allozyme frequencies:

i. the level of genetic variability within a population;
ii. the variation between populations;
iii. possible factors accounting for the observed variation;

The genetic structure of populations, reflected in inter-locus variance in allozyme studies is subject to evolutionary forces such as mutation, selection and random genetic drift. The intra-locus variance identifiable in laboratory studies depends on the sample size and gene frequencies at the locus studied. The degree of heterozygosity (H) in forest tree populations has been found to range from 0.19 to 0.41, with the higher values being typical in conifers.

Allelic variation can be defined as "the expected proportion of heterozygotes per locus per population", as is generally marked h. Calculations of an average value for h in 10 different populations of sitka spruce, an important forest tree native to the west coast of Northern America, yielded values ranging from 0.475 (glucose 6-phosphate dehydrogenase, G6pdh) to 0.001 (phosphoglucomutase, Pgm-2). This wide range in h, with values near zero, is typical for several coniferous species.

The level of genetic diversity in conifers is usually high. This is due to:

(i) divergent selection for macro-geographical adaptation,

(ii) balancing selection for micro-geographical differentiation, and

(iii) a breeding system which facilitates gene flow within and between subpopulations. Heterosis might further promote the maintenance of genetic variability.

Studies of genetic variation in ten IUFRO populations of sitka spruce (Picea sitchensis) showed that:

(i) the amount of genetic polymorphism varied considerably from locus to locus;

(ii) several populations were similar in the amount and pattern of genetic variability, in regard to most of the loci studied; while (iii) some specific loci exhibited large differences between populations.

Despite large geographic distances between seed sources tested, the differences in genetic constitution of sitka spruce did not increase appreciably in proportion with geographic separation.

Genetic differences between populations expose these differently to natural selection at the population level, favouring population selection. This conclusion has been confirmed for several tree species in addition to sitka spruce, like for instance *Picea abies* (L.)Karst. , *Pinus monticola* Dougl. and *Pseudotsuga menziesii* (Mirb.) Franco. In Scandinavia, the frequencies of alleles situated at the locus for acid phosphatase-B, were more frequent in southern norway spruce populations than in northern ones.

Generally, conifers exhibit high levels of heterozygosity. A notable exception to this rule is *Pinus resinosa*. One of the theories explaining the lack of variability in *P. resinosa* is that this lack is the result of a severe population bottleneck, which probably occured during the Pleistocene period, when the distribution of the species was reduced to small refugial populations, as a result of glaciation.

The genetic variability in deciduous trees is generally smaller than in conifers. In 17 *Quercus* populations, the average value of polymorphic loci was 30 %, with 2.32 alleles per polymorphic locus. Low estimates of genetic diversity between populations may indicate that a greater genetic variability resides within populations than between them.

Adaptation of Trees to Different Environments

Trees are long-lived organisms, and must therefore be well-adapted to their native habitats and expected environmental fluctuations within the habitats over long periods of time. This may be accomplished either by a broad adaptation within the population, or by population-specific adaptations to specific site conditions. In norway spruce (*Picea abies*) and douglas-fir (*Pseudotsuga menziesii*), a gene locus coding for acid phosphatase was identified, which showed a significant, environmentally dependent variation. In both species, the Acpalleles coding for single-band enzymes dominate in colder zones of the distribution range of the species, whereas the alleles coding for double-band enzymes occur more frequently in all moderate climatic zones.

The variation between populations can be quantified by a single number, the genetic distance (d_0). Genetically completely different populations are characterized by a value of $d_0 = 1$, while the value is zero for identical populations. In Finnish stands of *Picea abies*, a complete random distribution of genes within stands has been observed using allozymes of leucine aminopeptidase, Lap; peroxidase, Pod, and esterase, Est. However, for a number of other species, a correlation between genetic and geographic distances seem to occur, *e.g. Camellia japonica* L. , *Picea abies* (L) Karst. , *Pinus monticola* Dougl. and *Pseudotsuga menziesii* (Mirb.) Franco. Isolation by distance may be an important factor in population differentiation of these species.

THE CLASSES OF MOLECULAR MARKERS

Linkage maps of many plant species were limited in size until the advent of molecular mapping. The primary difficulty with developing linkage maps was the inability to incorporate many markers into a single stock to be used for genetic analysis. This inability occurred because of the deleterious effects of the expression of all mutant phenotypes in the single stock. Because normal DNA or protein molecules are used to score the genetic material, molecular markers are phenotypically neutral. This is a significant advantage compared to traditional phenotypic markers.

The three most common types of markers used today are RFLP, RAPD and isozymes. Of the three marker types, RFLPs have been used the most extensively. RFLP markers have several advantages in comparison with the RAPD and isozyme markers:

1) they are codominant and unaffected by the environment;
2) any source DNA can be used for the analysis; and
3) many markers can be mapped in a population that is not stressed by the effects of phenotypic mutations.

The primary drawback to RAPD markers is that they are dominant and do not permit the scoring of heterozygous individuals. The weakness of isozyme markers is that each of the proteins that are being scored may not be expressed in the same tissue and at the same time in development. Therefore several samplings of the genetic population need to be made.

- RFLP - Restriction Fragment Length Polymorphism; a molecular marker based on the differential hybridization of cloned DNA to DNA fragments in a sample of restriction enzyme digested DNAs; the marker is specific to a single clone/restriction enzyme combination
- RAPD - Randomly Amplified Polymorphic DNA; a molecular marker based on the differential PCR amplification of a sample of DNAs from short oligonucleotide sequences
- AFLP - Amplified Fragment Length Polymorphism; a molecular marker generated by a combination of restriciton digestion and PCR amplification
- Isozyme - a molecular marker system based on the staining of proteins with identical function, but different electrophoretic mobilities<

RFLP Loci

RFLP analysis is an application of the Southern hybridization procedure. The general principles will be explained here and we will then discuss several papers to obtain a more in depth understanding of the procedure.

Clones, Enzymes and Informative Hybridizations

RFLP markers are defined by a specific enzyme-probe combination. The first step in the analysis is to derive a set of clones that can be used to identify RFLPs. Genomic clones that represent sequences at random are a poor choice as hybridization probes because plant genomes consist of a large percentage of repeated sequences. Thus, many of the clones will contain repeated sequences, and hybridizations with those clones containing repeated sequences generate many hybridization bands that are difficult to analyze genetically.

The two primary sources of these clones for RFLP mapping of plants are cDNA clones and PstI-derived genomic clones. These two clone sources are generally represent expressed genes whice are in low copy number. cDNA clones are DNA copies of expressed genes. PstI clones are based on the suggestion that expressed genes are not methylated. As we saw earlier, GC and GXC methylation is the most prominent form of methylation in plants. The enzyme PstI enzyme is C-methylation sensitive. Therefore, the enzyme will only cut non-methylated sites. If a gene is expressed, then its sequence will not be methylated and will be susceptible to PstI digestion. And because they probably contain expressed sequences, these fragments would have a greater probability of being low copy number.

Once a series of clones are derived, DNA from potential parental genotypes is digested with a series of enzymes and hybridized with the clones. Some of these hybridizations will generate fragments of only one size and are not polymorphic. Other hybridizations will give a distinctive hybridization pattern for each parent. These polymorphisms occur because the sequence of the probe is homologous to restriction fragments of different sizes. Those genotypes that are highly polymorphic are candidates as parents from which a mapping population can be derived.

RAPD Loci

RAPD markers have recently caught the fancy of many individuals in the field of applied plant breeding. This molecular marker is based on the PCR amplification of random locations in the genome of the plant. With this technique, a single oligonucleotide is used to prime the amplification of genomic DNA. Because these primers are 10 nucleotides long, they have the possibility of annealing at a number of locations in the genome. For amplification products to occur, the binding must be to inverted repeats sequences generally 150-4000 base pairs apart. The number of amplification products is directly related to the number and orientation of the sequences that are complementary to the primer in the genome.

Steps of RAPD PCR Amplification of Plant DNA

1. Isolate total DNA from the individuals of interest.
2. Establish the conditions for the amplification of the DNA from your specific species. Several considerations are: $MgCl_2$, primer and dNTP concentrations and the quality and concentration of the target DNA. The goal of RAPD experiments are to compare populations. This requires well to well and run to run consistency. Thus, you need to be sure that each well of the cycler will produce the same result given the same, target DNA, primer and dNTP concentration. Without this control you may not be sure that the amplification polymorphisms are the result of population variability or cycler variability.
3. The conditions used in our lab for the amplification of bean DNA are:
   i. 25 ng DNA
   ii. 200 mM dNTP
   iii. 200 uM primer
   iv. 1.5 mM MgCl2
   v. 1X Promega Taq polymerase buffer
   vi. 1 unit Taq polymerase
   vii. Reaction volume: 10 μl overlaid with 35μ l mineral oil
4. The PCR amplification steps are:
   - 94°C, 1 min for denaturation<
   - 35°C, 30 sec for annealing
   - 72°C, 2 min for extension

This profile is performed 45 times and then the products are completed by a single extension for 7 min at 72°C.

5. The amplification products are then separated on a 2% agarose gel, stained and photographed.
6. Variability is then scored as the presence or absence of a specific amplification product.

For RFLP mapping, you first need to determine the clone/restriction enzyme combinations that are polymorphic between the parents of your mapping population. The analgous experiment with RAPD markers is to identify the primer and reaction conditions that are polymorphic between you parents. These primers and conditions will be used to amplify and score the products of your segregating population.

Isozyme Loci

Isozymes are protein markers. The technique is based on the principal that allelic variation exists from many different proteins. For example, alleles of malic dehydrogenase would both perform the correct enzymatic function, but the electrophoretic mobility of the two may differ. Therefore, two alleles would not migrate to the same location in a starch gel.

The procedures to identify isozyme variation is simple. A crude protein extract is made from some tissue sources, usually leaves. The extracts are next separated by electrophoresis in a starch gel. The gel is then placed in a solution that contains reagents required for the enzymatic activity of the enzyme you are monitoring. In addition, the solution contains a dye that the enzyme can catalyze into a color reagent that stains the protein. In this manner allelic variants of the protein can be visualized in a gel.

Several drawbacks should be noted with regards to isozyme. First, the number of isozyme loci that can be scored is limited. To date, only 40-50 reagent systems have been developed that permit the staining of a particular protein in a starch. Furthermore, not all of these reagent systems work efficiently with all plant species. Therefore, for many species only 15-20 loci can be mapped. A second drawback is tissue variability. Some isozymes are better expressed in certain tissues such as roots, whereas other are best sampled in leaf tissue. Therefore, several samplings of the segregating population are necessary to score all the available isozyme. Because neither of these drawbacks affect RFLP or RAPD loci, isozyme loci are rarely scored today.

AFLP Loci

Amplified Fragment Length Polymorphism (AFLP) is the most recent class of molecular loci. These loci are generated using a procedure that combines restriction digestion and PCR amplification. The power of this procedure is that you can generate a large number of mappable loci with a single amplification. This will help you saturate a region of the genome rather quickly. The drawback is that procedure is a bit time consuming and requires the running of a DNA sequencing gel.

AFLP Procedure

1. Digest sample DNA with restriction enzymes *Eco*RI and *Mse*I.

```
5'----GAATTCN------------------------NTTAA----3'
3'----CTTAAGN------------------------NAATT----5"
                 ||
                 ||
                 \/
      AATTCN------------------------NT
          GN------------------------NAAT
```

2. Anneal *Eco*RI and *Mse*I adaptors to restriction products.

```
??????AATTCN------------------------NTTA??????
??????TTAAGN------------------------NAAT??????
```

(?????? = unknown sequences that are unique for the companies primers)

3. Preselect products by PCR amplfication with "*Eco*RI + A" and "*Mse*I +C" oligonucleotide primers.

```
EcoRI Primer+A
????????AATTCN---------------------NTTA????????
????????TTAAGN---------------------NAAT????????
                                   C+MseI Primer
```

4. Selectively amplify preselect PCR products by using "*Eco*RI + 3" and "*Mse*I +3" oligonucleotide primers.

```
EcoRI Primer+AAC
????????AATTCA----------------------GTTA????????
????????TTAAGN----------------------CAAT????????
                                  AAC+MseI Primer
```

5. Separate fragments by denaturing polyacrylamide gel gelectrophoresis.

MAPPING AND MAPPING POPULATIONS

F_2, backcross, and recombinant inbred are the three primary types of mapping populations used for molecular mapping. An F_2 population is developed by selfing (or intermating for cross pollinated species) among F_1 individuals. These F_1 individuals are developed by crossing two parents that show significant polymorphism for whichever type of loci you are going to score. Backcross populations are developed by crossing the F_1 with one of the two parents used in the initial cross. The major drawback to using F_2 or backcross populations is that the populations are not eternal. Therefore, your source of tissue to isolate DNA or protein will be exhausted at some point in time. You then would have to begin mapping again in another population.

Populations of recombinant inbred lines can be a powerful solution to this problem. Recombinant inbred lines are developed by single-seed selections from individual plants of an F_2 population. (Because of this procedure, these lines are also called F_2-derived lines.) Single-seed descent is repeated for several generations. At this point, all of the seed from an individual plant is bulked.

For example, a $F_{3:4}$ RI population underwent single-seed descent through the F_3 generation, and was bulked to develop the F_4. This population of seed can then be grown to obtain a large quantity of seed of each individual line. Importantly, each of the lines is fixed for many recombination events.

RI population level of inbreeding	% within-line homozygosity at each locus
$F_{3:4}$	75.0
$F_{4:5}$	87.5
$F_{5:6}$	92.25
$F_{6:7}$	96.875
$F_{7:8}$	98.4375
F_{8*9}	99.21875

These lines have several uses. First, they can be used be used to derive a map because it essentially is an eternal F_2 population with unlimited mapping possibilities. Additionally, these lines can be scored for morphological traits (such as disease resistance or flower color) or quantitative traits (such as yield or maturity). This morphological trait data can then be compiled and those traits can be placed on the developing molecular map. These lines are especially powerful for analyzing quantitative traits because replicated trials can be analyzed using identical genetic material. The quantitative trait data can then be used to determine if any molecular markers are closely associated with those traits.

TAXONOMY INTRODUCTION

Fungi are largely microscopic. Their forms diverge in only minor ways right the way across the Kingdom. Their developmental processes are difficult to study meaning that past technologies resulting in a limited understanding of the Kingdom. Even so, a huge diversity of fungi has been described. An even larger diversity is becoming evident through the use of molecular analyses of the fungi in the laboratory from cultures, and various common and uncommon substrates.

Plants and animals have been examined in much greater detail than fungi. However, the use of molecular and developmental tools is enabling greater focus on the fungi. One result of this focus is that the taxonomy of the fungi is being revised at all levels. The Fungal Tree of Life is taking a top down approach, and industrial mycology is forcing a bottom-up approach.

The result to date of FTL is that the over arching arrangement of the Fungi has been revised. The main groups remaining include Basidiomycota, Ascomycota and Glomeromycota. The other divisions are in a state of change. New divisions will be described to accommodate the information arising from the detailed examination of a spread of 200 species.

At the other end of the spectrum, highly detailed examination of economically important fungi is resulting in an extraordinarily important understanding of the stability and diversity of fungal species and sub species. These projects have enormous ramifications for how we develop the new taxonomy of the fungi. Stay tuned for the ride of your life.

Division Ascomycota

Asexual Spores

The fungi form nonmotile *conidia* from the tip or side of a sporogenous (conidiogenous) region or cell. Conidia are not surrounded by a second wall, as in the case of *sporangiospores*. The spores are usually dispersal units and a few may travel great distances. Arthrospores and some conidia may be long-lived. Viability of conidia is reduced rapidly in the presence of UV radiation, and in dry conditions. Other spores have high metabolic rates which limit their viability.

Conidia may arise from conidiogenous regions in several ways. In *blastic* ontogeny, the conidium originates from a narrow portion of the region which swells before being cut off by a septum. In thalloblastic ontogeny, the conidium originates from a broad region of the hypha, and as with blastic ontogeny, the cell swells before delimitation by the septum. In *thallic* ontogeny, the conidia arise from a length of hypha; septa are laid down before each conidium swells. *Arthric* conidia are specialised thallic conidia formed in chains that disarticulate readily.

Conidia mostly germinate by *germ* tube extension. In a few cases, conidia germinate to form *yeast* cells, or to form a further conidium.

Taxonomy

Ascomycota is the largest division of fungi. Classification of the division is in a state of change due to the integration of molecular information. The most recent analyses have proposed a major reconsideration of the group. These analyses benefit from using both molecular, morphological and developmental data. The orders suggested by Schoch et al (2009) include: Sordariomycetes, Laboulbeniomycetes, Leottiomycetes, Lichinomycetes, Lecanoromycetes, Eurotiomycetes, Dothidiomycetes, Arthoniomycetes, Orbiliomycetes, Pezizomycetes, Saccharomycetes, Schizosaccharomycetes, Pneumocystidiomycetes and Noelectomycetes. This hypothesis is based on 434 taxa. As the division contains many fungi, the classification will, inevitably, be expanded and refined as further information is included.

The following only the diversity within the division.

Order 1: Laboulbeniomycetes

The fungi form parasitic attachments to arthropods.

Order 2: Schizosaccharomycetes

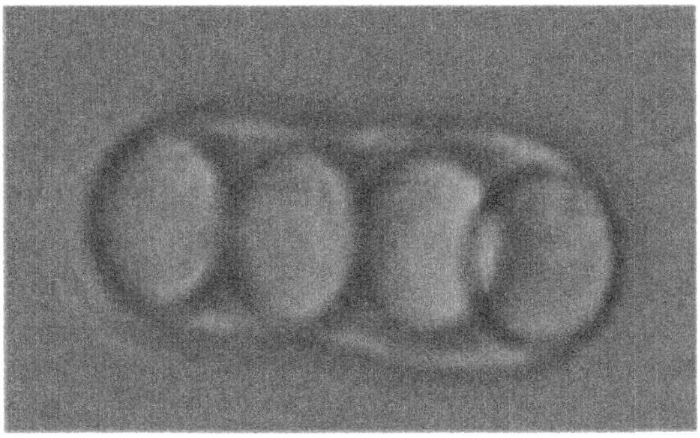

Yeast ascus.

A group lacking ascogenous hyphae and ascomata. The asci formed lack a protective covering if they form from a hymenium. The taxon contains some of the ascomycetous yeasts.

Order 3: Pezizomycetes

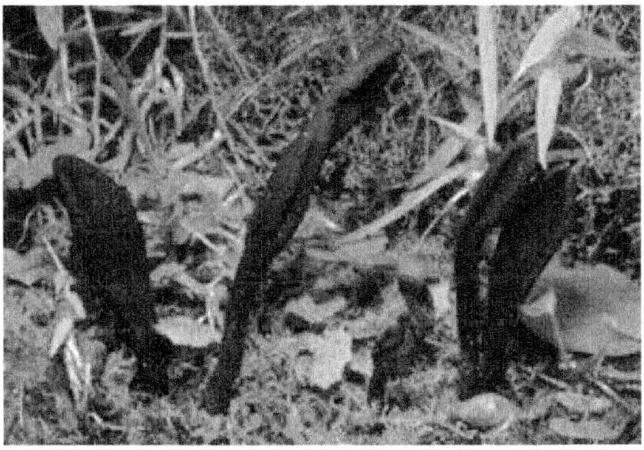

Ascocarp.

A huge group containing fungi which form large and obvious ascomata. The orders are separated on the structure of the *ascus* and the manner of ascus opening. The ascocarp is often seen as a flattened sheet, or one that has variably evolved to a series of cup shaped structures held above or at the surface of the supporting substrate.

Cultural Conditions

Nutritionally, most are either saprobes or weak plant parasites. Others include *lichen* symbionts, *endophytes* of angiosperms, arthropod parasites and a few form *ectomycorrhiza*. Most of these fungi have the capacity to utilise simple sugars as a source of organic energy. Some parasites produce enzymes that enable them to digest structural carbohydrates such as cellulose, chitin and pectin.

Ecology

The vast majority of these fungi are terrestrial, a few are aquatic. Many species are of great importance because they are *pathogens* of plants and animals causing serious disease. Several of the fungi produce metabolites of commercial importance. One yeast is used for production of bread and alcohol. A few filamentous species are the basis of the production of antibiotics, or immune suppressants, or used in food processing and flavouring. A few species are important producers of toxins in animals, especially human foods. Among the taxa are the most xerotolerant organisms known, those that live on dried foods. Many ascomycota form associations with algae called lichens, and others form ectomycorrhizas.

Hyphal Characteristics and Pores

The yeast fungi are found as single cells. The cells divide either by holoblastic budding or fission. Fission fungi may form asci following fusion of cells of opposite mating types. The cells are separated by a septum which is perforate.

The hyphae of Ascomycota have bilayered walls with primary septa at regular intervals along the hypha. The cell is divided by a septum which has a simple pore. The pore may be associated with an electron dense structure known as the Woronin body. Woronin bodies are only found in the hyphal phase of Ascomycota. In true yeasts and thalli transforming to a yeast phase, *Woronin* bodies are not seen.

Sexual Spores and Mating Types

Anastomosis of somatic hyphae of opposite mating type leads to the dikaryotic phase only within the ascoma. The dikaryotic phase is restricted and leads immediately to the formation of the *crozier*, usually within a *hymenium*. The reflexed tip, or crozier, enables mitotic division of the nuclei, delimiting the ascus mother cell in which *karyogamy* is followed immediately by meiosis. In other words, *plasmogamy*, karyogamy and meiosis are closely linked spatially and temporally in Ascomycota.

Haploid ascospores, usually eight, are formed within an ascus immediately following meiosis and one mitotic division. Ascospores are released with the opening or breaking of the wall of the ascus. Asci are sometimes held singlely. Compound structures have hymenia either within an inoperculate spherical structure called a cleistothecium, an ostiolate flask-shaped structure called a *perithecium*, or an open hymenium called an*apothecium*. The compound ascoma may be held within locules of a *stroma*, and the stroma may be *epigeous* or *hypogeous* during spore release. Spore release is usually by active ejection from apothecia and perithecia. Hypogeous ascocarps are usually consumed by animals which disperse the spores during deposition of faeces.

Macroscopic ascocarps are rare. The largest is an apothecium which may reach 15 cm across (*e.g. Peziza*) or 15 cm high (*e.g. Morchella*). Ascocarps rarely exceed 2 to 4 mm in any one dimension.

Division Basidiomycota

Asexual Spores

The usual asexual spores formed by these fungi are *arthrospores*, sometimes called oidia. Few species form blastic conidia. In some cases, such as with some rust fungi, up to 4 different types of asexual spore are formed during the life cycle.

Oidia can have two functions. They may germinate to form mycelium or they may function in the mating process. *Asexual* spores are usually haploid and the hyphae that form following germination are also haploid.

Classes

Taxonomy of the division is in a state of flux. Originally, the fungi were divided on the basis of the morphology of the basidium culminating in the taxonomy of Talbot in 1968, in which three classes were recognised: Holobasidiomycetes, Phragmobasidiomycetes and Teliomycetes. Microscopic and molecular information is now clarifying the relationships within this group, and much work remains. Using available information, Walker proposes the following:

Class 1: Ustomycetes

Known as the smuts and related fungi. Wall adjacent to the pore slightly swollen. Pore usually very small, with banded occlusion and surrounded by lipid bodies. Basidiomata not formed.

Class 2: Unnamed

Known to form stilboid and pycnidioid gasteroid basidiomata, usually less than 5mm high. The septa are simple. Basidiospores statismosporic.

Class 3: Urediniomycetes

Known as the rusts and related fungi. The fungi have well-differentiated probasidia, often persistent. Septa simple, slightly tapered, often with pore occluded with dense material.

Class 4: Unnamed

Exobasidial plant parasites. Have erumpent basidiomata. Simple septa of uniform width with occluded pore. Yeast phase common.

Class 5: Unnamed

Known as the bunts and related fungi. Have septal dolipores lacking parenthisomes.

Class 6: Unnamed

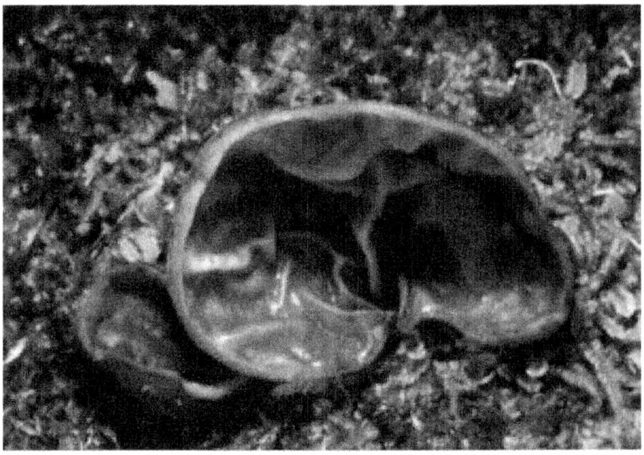

Auricularia.

Known as the jelly fungi. Basidiomata often gelatinous to leathery, phragmobasidia, basidiospores ballistosporic, septa with dolipores and either vesicular or imperforate parenthisomes.

Class 7: Unnamed

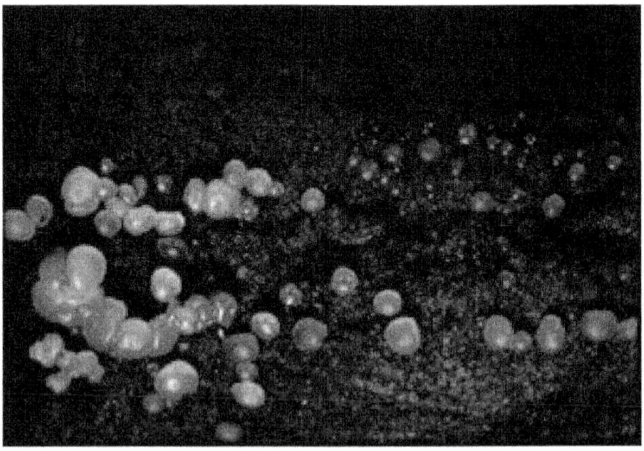

Dacrymyces on Wood.

Known as the holobasidial jelly fungi. Basidiomata not erumpent when parasitic, septa simple with dolipores, imperforate parenthisome, undivided basidia have stout sterigma, basidia may germinate by repetition, conidia or germ tube in some taxa.

Class 8: Euholobasidiomycetes

Known as the true basidiomycetes, include mushrooms, etc, have dolipores with imperforate parenthisomes. Basidia a single, undivided unit, have fine sterigma. Basidiospores germinate by germ tube.

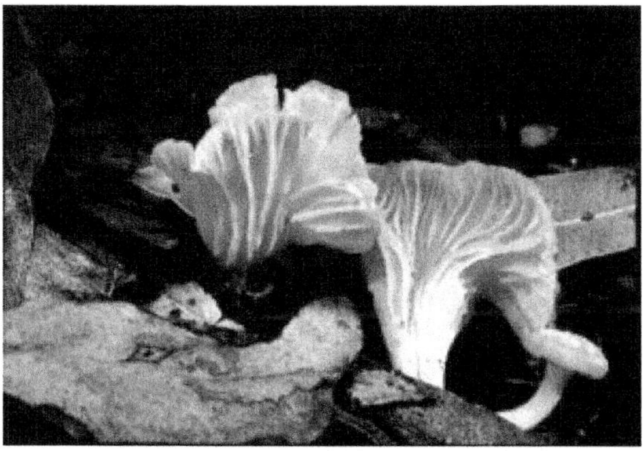

***Chantarelle*. Image from E and R Kearney**

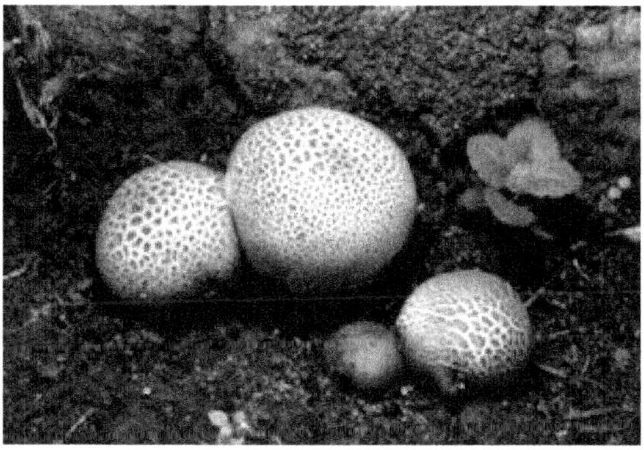

Puff Balls.

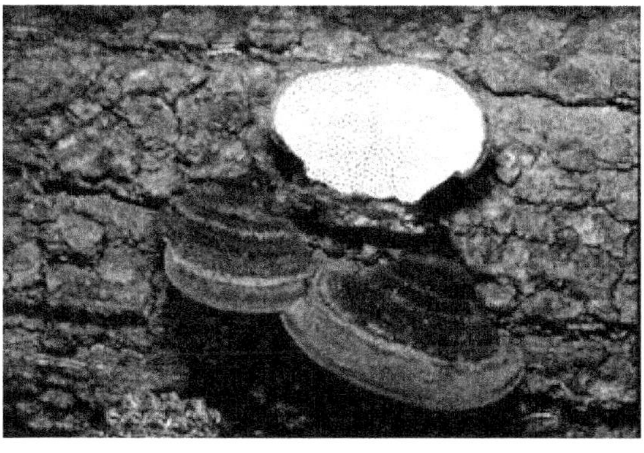

***Stereum*. Image from E and R Kearney**

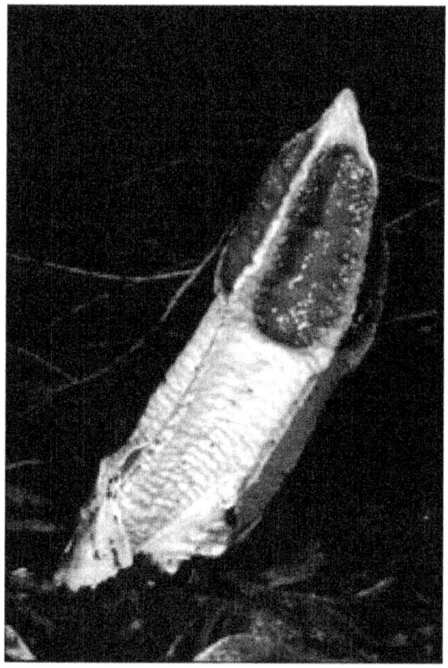

Euholobasidiomycete. Image from E and R Kearney

Cultural Conditions

The division contains a huge diversity of fungi. Some can be cultured on simple substrates. However, many require specific cultural conditions and have the capacity to digest and utilise complex organic compounds such as lignin and wax. Others, such as the rust fungi, are obligate *biotrophs* of specific plants and have a relationship only with specific races of a single plant species. Many of the fungi grow slowly in culture and take a relatively long time to form fruiting structures.

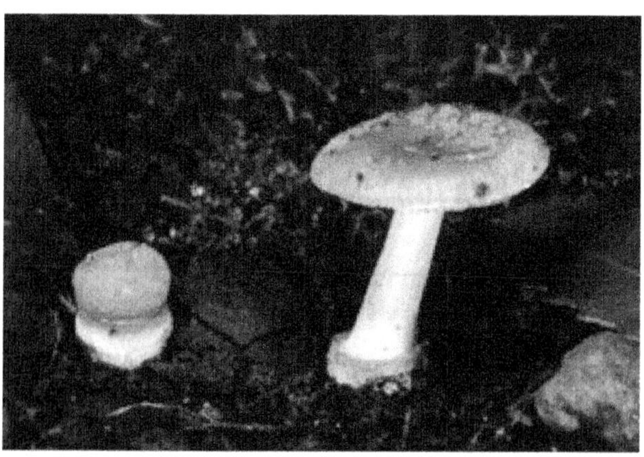

***Amanita*. Image from E and R Kearney**

The conditions under which these fungi are grown also influences the secondary metabolites produced. For instance, toxin production may change from negligible to extreme with changing N concentration in the medium or temperature of growth. Generalisations cannot be made about the cultural conditions for growth.

Ecology

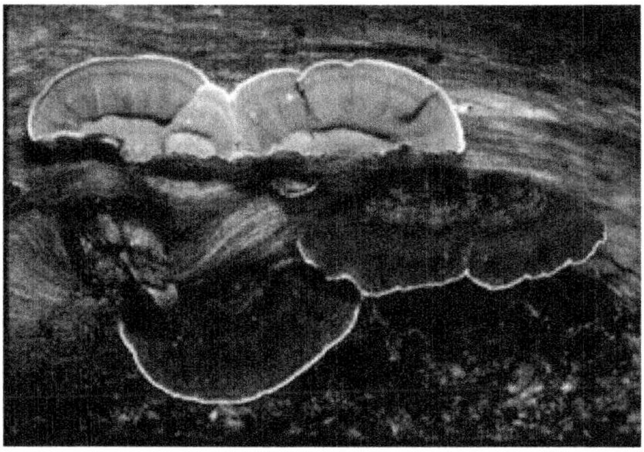

Pathogens of forest trees.Image from E and R Kearney

Most Basidiomycota are found in terrestrial habitats. A few are marine. Within the division are important *saprophytes*, plant *pathogens*, animal pathogens, edible and ecto*mycorrhizal* fungi, most of the fungi that form large, conspicuous fruiting bodies and many of the *yeast*-like fungi. Among the saprophytes are the degradative fungi involved in breakdown of wood and litter in forests and timber products, most edible fungi, many halucinogenic and toxigenic fungi. Plant pathogens include the rusts and smuts, pathogens of forest trees and soil pathogens of crop plants. Animal pathogens are less important, though Filobasidiella (Cryptococcus) neoformans has become important in immuno-compromised people (due to organ transplants or HIV). A huge number of Euholobasidiomycetes form *ectomycorrhiza*. Not only do they improve mineral uptake of the host, they are also responsible to tight cycling of minerals in un-disturbed ecosystems due to their capacity to extract organic N and P from the litter. A relatively small number of fungi are eaten directly. Of these *A. bisporus*, the mushroom, is by far the most economically important.

The fungi are well known for their secondary metabolites. These com-pounds include odours, colours, flavours and toxins. The odours may be unpleasant as in the case of the stinkhorns. Odours can be used by animals to locate the fungal structure. Unfortunately, though odours are used to assist identification, cultural conditions influence which compounds areproduced.

The fungi tend to be long-lived and over their lifecycle produce huge numbers of spores. However, dilution techniques, such as from soil, intending to isolate all fungi usually fail to isolate basidiomycota. This is because their spores, if present, require specific triggers to break dormancy and because they are slow to respond to triggers. Also many of these fungi survive periods of stress as hyphae rather than spores.

Some of the fungi also form *rhizomorphs* enabling access to fresh sources of organic nutrients. The wood rot fungi, in particular, tend to form rhizomorphs, or mycelial cords, as organic nutrient concentrations decline. When concentrations increase, hyphae fan out, penetrating the substrate more thoroughly.

Overall, the fungi tend to be slow growing producers of a wide range of secondary metabolites, utilising a variety of complex substrates and habitats.

Hyphal Characteristics

Hyphal walls have many layers and the cells may contain one, two or many nuclei. If one, the nuceus may be haploid or diploid. Mixtures of haploid and diploid nuclei in cells have also been recorded. The hypha arising from germination of a spore commonly contains a single haploid nucleus and is a *monokaryon*. This hypha will continue to grow until it meets a hypha of a compatible mating type. These mycelia fuse, maybe at several points, and nuclei from one mating type pass into the second mycelium. Nuclei of the migrating type multiply independently of mitosis and nuclei migrate through the mycelium until the mycelium is binucleate (and *dikaryotic*). The hyphae within the cell may fuse in *karyogamy* with the resulting mycelium becoming diploid, though most commonly, the mycelium remains dikaryotic.

A few fungi have a multinucleate state. More than 2 nuclei are present in each cell from germination of spores. It is assumed that the nuclei are haploid replicas of one genotype.

Changes in nuclear state are accompanied by formation of *clamp* connections in many species of Euholobasidiomycetes. It is assumed that the clamp enables multiplication of both nuclei in the cell. However, some species lack clamps and in others formation of clamps seems to be influenced by environmental conditions.

While a single cell is rather small, the entire mycelium may be quite large. Genets of the plant pathogen *Armillaria ostoyae* have been found throughout 15 ha of a single forest suggesting that the fungus is the largest organism ever documented. While the entire mycelium may be spatially separated into units, because basidiomycota anastomose readily, it is likely that the entire mycelium will have cytoplasmic continuity from one end to the other. Further, basidiomycota commonly form rhizomorphs or mycelial cords. The aggregations aid connections between spatially separated zones of nutrient concentrations.

Pores

Any fungus that contains a *dolipore* septum is definitely a member of the Basidiomycota. However, other pore types are seen in members of the division.

Sexual Reproduction

The characteristic of this division is the formation of *basidiospores* from basidia. Within the division we find considerable variation in the structure of the *basidium*. Holobasidia are single cells from which arise *sterigma*. The remaining fungi have basidia that are either septate or isolated teliospores that germinate to form a septate germ tube on which basidiospores emerge. Basidia may be vertically or horizontally septate, with a sterigma attached to each.

Typically, the diploid nucleus goes through meiosis within the basidium and the resulting haploid spores emerge through the sterigma and are held on the tips. Sometimes more than one nucleus passes into a spore and sometimes mitotic division results in the spore being multinucleate.

In most Euholobasidiomycetes, the basidiospore is held on the sterigma obliquely and is forcibly discharged from the sterigma.

Spore are released from structures that are used to characterise the fungi. These are the mushroom, brackets, earth stars, stinkhorns, puffballs, rusts, smuts, bunts, jelly fungi and coral fungi. The fruiting bodies usually collapse following spore release. One group, the bracket fungi, may form perennating fruit bodies and these can release huge numbers (up to 1000,000,000) of spores each year.

Homothallism is rare in this division. Where sexual compatibility has been investigated, most have either single factor (plus/minus) or multifactor (multiple alleles at two or more loci) *heterothallism*.

Division Chytridiomycota

Asexual Spores, Sexual Spores and Life Cycle

The division is in a state of flux. Recent molecular evidence indicates the loss of zoospores on more than four occasions, and that zoospores are found on fungi in at least four quite different groups. For convenience, the presence of zoospores will be used as the basis of the division.

The Chytrid fungi are characterised by their asexual state, a *zoospore* (motile) with a single whiplash *flagellum* oriented and located posteriorly. The zoospore is formed in a *sporangium*. Zoospores are released through an opening in the wall, and their release usually indicates the death of the *thallus*. Note that zoospores with a single tinsel flagellum are found in diverse Protists, though the morphology of some of these microbes is otherwise simialr to some chytrids (*e.g., Hyphochytrium*).

In the few species where it has been determined, zoospores of opposite mating type meet, fuse (*plasmogamy* immediately followed by*karyogamy*) and go on to form a single resting sporangium. After meiosis, the sporangium germinates to release further zoospores, which may be asexual, in which case, the single zoospore attaches (encysts) to and colonises a *substrate*. The cyst may be a resting structure or it may grow to form a sporangium. The sporangium may continue growing, it form a resting stage. The thallus may be located on or in the substrate.

In some species, resting sporangia may take the form of a multicelled, thick walled sorus. The sorus is commonly found within parasitised tissue such as a root, where it remains until the surrounding tissue has rotted away. The resting sporangium may arise from haploid or diploid zoospores.

Classes

The Chytridiomycota is the most ancient group of fungi, and is a polyphyletic taxon. Of the fungi, they share most characteristics with the animal kingdom. At present Chytridiomycota consists of a single class, Chytridiomycetes, containing five orders which are separated on ultrastructural and molecular characteristics. The five orders are:

Chytridiales, Monoblepharidales, Neocallimastigales, Rhizophydialies and Spizellomycetales. The differences in morphology of the flagellar roots enable separation of isolates down to the level of genus, and are particularly important for determing the order.

Blastocladiales, formerly part of this Division, remains to be formally placed in its own division. The plant parasite *Olpidium* also remains to be formally allocated a new division with the Entomophthorales. *Rozella,* a parasite of the Chytrids, is also awating formal separation from the Chytrids.

Cultural Conditions

Saprophytic chytrids are commonly trapped from soil or water using pollen, cellulose, chitin, keratin or other complex organic materials. Some can then be cultured on nutrient media containing simple organic energy and protein. Obligate parasites are cultured using their host, inoculation being achieved by washing zoospores onto the host surface. Gut fungi are commonly isolated by dilution, cultures then being maintained in strict anaerobic conditions in a nutrient broth containing cellulose and its breakdown products.

Ecology

Chytrids require a water film in which zoospores disperse. Thus the fungi can be isolated from habitats that are moist for at least short periods. Chytrids are readily trapped from water and soil, where they may be saprotrophs. Other chytrids may be parasites of plants, animals, algae and other fungi.

Saprotrophic Chytrids appear to use complex carbohydrate as their source of energy. Especially in marine habitats, zoospores are attracted by chemicals released by the host or substrate. Zospores settle on the substrate, and encyst. In some cases, the life cycle may be completed within three days. There appears to be specificity between substrate and fungus, with, for example, *Rhizophlyctis rosea* and close relations only being trapped on cellulose.

An important group are anaerobic chytrids are found in the gut of herbivores. These fungi are believed to function synergistically in the degradation of carbohydrates in the diet of ruminants. Their attraction to plant materials, capacity to penetrate cell walls and their broad range of degradative enzymes has led to the suggestion that they play an important role in the initial attack of structural carbohydrates. Many gut fungi are obligate anaerobes.

THE REST: The plant parasite *Olpidium* is unusual among fungi in that the sporangium in the plant cell lacks a host membrane around the fungal protoplast. The fungus also vectors up to 10 viruses, possibly by carrying the virions in fungal cytoplasm. The lack of a host barrier would reduce the capacity of the plant to prevent ingress by the virus. Apart from carrying viruses, few chytrids cause serious plant disease.

The Thallus

Chytrids lack a true mycelium. Each fungus may be in one of three forms. The simplest is *monocentric*; a single *holocarpic* cell. The monocentric cell may be *eucarpic*, a single cell with or without rhizoids growing into the substrate. The rhizoids lack nuclei. The fungus may form several sporangia on a single rhizomycelium (cells contain nuclei), and the structure is known as polycentric. Each sporangium may release zoospores of different mating types, and release them at different times.

Division Glomeromycota

Asexual Spores

All species in this Division are known to reproduce asexually. The fungi form *blastospores* usually at the terminus of a hypha. The primary separation of groups within the division is associated with the development of blastospores. Minor morphological variation exists between blastospores of each species within each family. Minor and continous molecular separation is found between isolates making identification of species particularly problematic.

Classes

Class: Glomeromycetes

All but one member of the class form arbuscules in cortical cells of the roots of angiosperm plants. The taxon, *Geosiphon*, forms an interesting symbiosis with algae. The taxonomy of this group is undergoing massive change. We will incorporate these changes as we come to understand their implications.

Orders: Archaeosporales, Diversisporales, Glomales, Paraglomales.

Cultural Conditions

The division is united in having an association with photosynthetic hosts. All but one taxon of those examined so far form arbuscular *mycorrhiza*. They have only been cultured with a living plant growing in adequate light intensities or in axenic conditions on transformed roots accessing sucrose from the medium.

Ecology

Glomeromycota are *symbionts* found only in terrestrial habitats. Those that form arbuscular mycorrhiza (AM) with plants constitute at least half the living fungal biomass in soils. They are at a competitive advantage over other soil-borne fungi in that they have access to a constant source of organic nutrients supplied by their host while living in a largely *oligotrophic* medium. The fungi benefit their host by functioning as an extension of the root system. In addition, they ramify through soil, where they release glycoprotein, including glomalin, which may function in gluing together fine soil fragments. Hyphae also physically aggregate soil.

Hyphal Characteristics

Mycelia of AM fungi are characteristically *aseptate* with, septa only being found to delimit spores or dying portions of hyphae. The hyphae can*anastomose*, thereby forming interconnected mycelia. The walls of hyphae contain *chitin, chitosan* and polyglucuronic acid.

Physiology

In general, AM fungi produce enzymes that enable them to colonise plant roots. These enzymes can be recovered from spores or hyphae. However, enzymes to enable digestion of complex carbohydrates are unknown. Energy from the host plant is transferred probably as glucose at the interface between the symbionts, and then transformed into fungal sugars thereby preventing its return to the host. Carbon is moved about the fungal thallus, and may be stored in the root of plants, though only minor transfer into plant tissues seems possible (mycoheterotrophic plants being the exception).

AM fungi also access large quantities of dissolved minerals from soil. Minerals are also transferred around the thallus, but they can be transferred to the host via an as yet unknown mechanism. Transfer of minerals such as phosphate, copper and zinc appear to be essential to the function of the host in soils where these minerals are deficient. While the fungi also access nitrates and other small minerals, their transfer to the host is rarely greater than that supplied by simple diffusion.

Because of their mutual benefit, it seems clear that some form of recognition is necessary for the initiation of each symbiosis. The genes involved include some well characterised in the formation of nitrogen fixing nodules in legumes.

Pores

While the fungi are known to be *aseptate* not a great deal is known about the formation of pores between separate compartments. Multiperforate septa are found in some members. It was assumed that these would be found in all members of the division. However, as the Paraglomaceae is deeply divided phylogenetically from the other orders, further examination is warranted.

Sexual Spores

Zygospores have been found in only one species of *Gigaspora*. The fungus was shown to be *heterothallic*. Molecular studies indicate that members of *Glomus* are asexual. Further studies are needed to clarify this situation.

Division Zygomycota

Asexual Spores

Reproduce asexually by *sporangiospores* in most orders of the Zygomycetes, *conidia* in the Entomophthorales and *arthrospores* in some Trichomycetes. Sporangia may contain from 50 to 100,000 spores. Sporangiola are smaller and contain from one to as many as 30 spores. *Sporangiola* have considerable morphological variation. When spores are found in a single row, they are known as uniseriate merosporangia. Single spored sporangiola found in some Mucorales have been referred to as conidia, but as they form within a sporangial wall, the basal septum is not part of the wall of the spore and they are sporangiospores.

Classes

Class: Zygomycetes

Characterised by well developed mycelium, which may be immersed in *substrate* or host tissue. Please note that the taxonomy of this group is undergoing massive change. We will incorporate these changes as we come to understand their implications.

Orders: Dimargaritales, Endogonales, Kickxellales, Mucorales, Zoopagales.

Note that the Entomophthorales are now placed in a new division which remains to be formally descibed

Class: Trichomycetes

Characterised by a much reduced thallus, with a *holdfast* that attaches to the host exoskeleton. The mycelium is usually simple though may be branched.

Orders: Amoebidiales, Asellariales, Eccrinales, Harpellales.

Cultural Conditions

The division contains a variety of fungi. The members of the Mucorales are commonly referred to as sugar fungi, meaning that *saprophytes* respond rapidly to and grow rapidly on simple sugars in artificial media. Often they are the fungi found to be fruiting first in natural substrates such as dung, presumably because of their rapid response to readily available organic nutrients. Culture of Trichomycetes appears complicated because of the obligate association with arthropod hosts. The order Endogonales contains culturable fungi, some are saprophytes and others form *ectomycorrhiza*.

Yeast forms have been induced in some species by the use of media containing high concentrations of nutrients and anaerobic conditions.

Ecology

Zygomycetes are commonly saprotrophs, *symbionts* or *parasites* in terrestrial habitats. Members of the Mucorales are easily isolated from soil, humus and dung where they grow saprobically.

Trichomycetes are obligate associates of arthropods, including insects, millipedes and crustaceans. The host may be an adult or larva, in terrestrial or aquatic habitats. The fungi are usually found attached to the *chitinous* gut lining of the host by a holdfast. The precise relationship is difficult to determine in most cases. However, one pathogen is known and in another example, the fungus has been found to provide its host with sterols and vitamins. The fungus is assumed to gain nutrients from the gut of the host.

Hyphal Characteristics

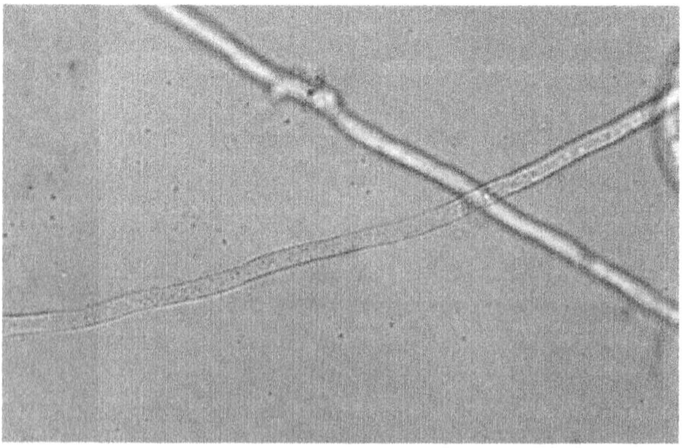

Aseptate hyphae.

Mycelia are characteristically *aseptate* with septa only being found to delimit *sporophores* or dying portions of hyphae. The hyphae rarely *anastomose*, so do not form interconnected mycelia common in the higher fungi. The walls of hyphae contain chitin, *chitosan* and polyglucuronic acid.

Physiology

In general, members of the Zygomycetes do not produce enzymes that enable them to utilise complex polysaccharides. While some fungi possess cellulases and pectinases, these appear to enable the fungi to colonise plants rather than digest their component parts. Only a few plant pathogens are found in this group, and many of these penetrate the plant through existing openings such as wounds, nectaries and stomates.

Some members of the Mucorales are commonly used to ferment foods and produce important industrial products such as lactic acid, amylases, rennin and organic acids. The fermentation of foods is nutritionally important because the fungi contribute substantially to the nutritional value of the foods.

Pores

While the fungi are known to be aseptate in the vegetative mycelium, secondary septa form during separation of sporophores from the *mycelium*. At least four different types of septa have been described. Multiperforate septa are found in members of the Mucorales and Endogonales. Another common type of septum remains open with the edges of the pores retaining a lenticular cavity containing a plug. The lenticular pore is found in members of the Kickxellales, Dimargaritales, Harpellales and Asellariales. Septa of the Zoopagales remain to be examined, though some are known to be perforate.

Note that the Entomophthorales have a complete, bilayered septum lacking perforations.

Sexual Spores

The division is characterised by formation of *zygospores*. The formation of zygospores has been followed in several species of Mucorales and these details follow. In *heterothallic* fungi, presence of hyphae of each of the two mating types induces the formation of specialised sites called gametangia.

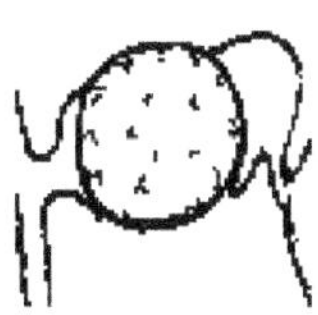

Formation of zygospores.

Different pheromones are produced by each mating type. The gametangia grow towards one another and the tips fuse (*plasmogamy*). Specialised cells at the tips, called progametangia, are walled off and the fusion zone dissolves. The resulting gametangium becomes the zygosporangium and the attached hyphae, the suspensors. Haploid nuclei in the *zygosporangium* fuse (*karyogamy*) and a wall forms around the now diploid zygospore inside the zygosporangium. The zygosporangium wall thickens and often becomes ornamented and melanised. Meiosis takes place prior to germination inside the zygospore. The *germ* tube is always haploid.

Homothallic fungi are also known in the Mucorales. In these fungi, both mating types are found in one mycelium and the isolates are self-mating.

Germination of zygospores is nearly always preceded by a period of *dormancy*. It is assumed that the sexual spore is a structure that enables long-term survival of the fungi.

Sexual reproduction has not yet been demonstrated in the Trichomycetes, though several genera are known the form thick-walled bi-conical spores following conjugation.

Chapter 4

MOLECULAR METHODS FOR IDENTIFICATION OF PLANT PATHOGENIC FUNGI

BIOTROPHIC PLANT PATHOGENS

Quite a lot of plant-pathogenic fungi establish a long-term feeding relationship with the living cells of their hosts, rather than killing the host cells as part of the infection process. These pathogens are termed biotrophic.

Typically, these fungi grow between the host cells and invade only a few of the cells to produce nutrient-absorbing structures termed haustoria. By their feeding acitivities, they create a nutrient sink to the infection site, so that the host is disadvantaged but is not killed. This type of parasitism can result in serious economic losses of crop plants, and in natural environments it can reduce the competitive abilities of the host; indeed, a few biotrophic pathogens have been used successfully as biological control agents of agricultural weeds.

In many ways, this type of parasitism is very sophisticated - keeping the host alive as a long-term source of food. This has led some people to suggest that biotrophic parasitism is evolutionarily advanced. But this is clearly not the case *in general*, because an almost identical type of parasitism is found in the arbuscular mycorrhizal fungi which are thought to have developed on the earliest land plants.

Here we consider the two most important groups of biotrophic plant pathogens:

- The rust fungi (Basidiomycota)
- The powdery mildew fungi (Ascomycota).

A parallel can be made between the behaviour of these fungi and the biotrophic mycoparasites

Powdery Mildew Fungi

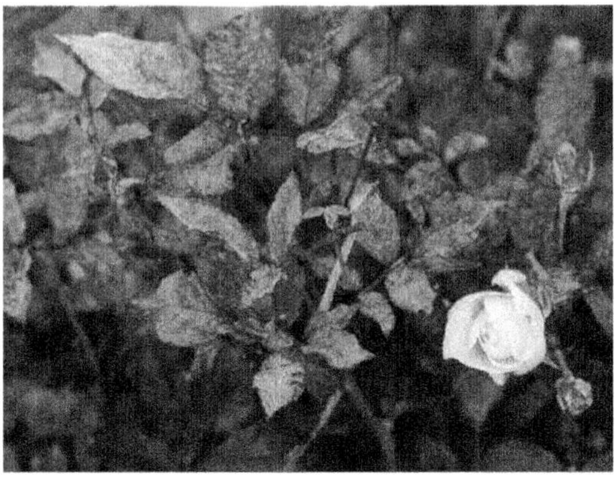

Powdery mildew of roses, caused by the fungus *Sphaerotheca pannosa.*

This is a very common disease, familiar to most gardeners, and it is typical of many powdery mildews, where the fungus forms a powdery coating of white spores on the leaf surface. Other common examples in Britain are powdery mildew of hawthorn (*Podosphaera oxyacanthae*), gooseberry (*Sphaerotheca mors-uvae*), and cereals and grasses (*Erysiphe graminis*).

All these fungi grow superficially on the host, only penetrating the leaf epidermis. But they extract considerable amounts of plant nutrients through their haustoria, and these nutrients are used for sporulation, leading to rapid epidemic spread of these diseases.

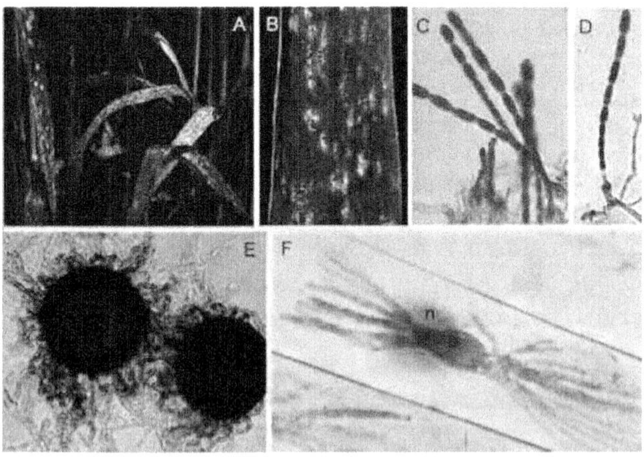

The powdery mildew pathogens are in the fungal group Ascomycota (ascus-forming fungi). They produce chains of asexual spores (conidia) for aerial dispersal, and at the end of the growing season they can produce small fruiting bodies (ascocarps) containing the sexual spores (ascospores) that serve for dormant survival.

Figure A (above) shows many separate, localised lesions of *Erysiphe graminis* on wheat leaves. The conidia from these lesions (C, D, stained with trypan blue) are produced continuously in chains, maturing at the tip of the chain and being wind-dispersed. They are large enough to impact onto cereal leaves at normal wind speeds (typically 1-2 metres per second) in field conditions. Figure B shows similar lesions near the end of the growing season. The small black flecks are the ascocarps. The ascocarps of a different fungus (*Podosphaera*, the powdery mildew fungus of hawthorn) are seen at higher magnification in Figure E. This type of ascocarp is termed a cleistothecium - a closed body containing one or more asci (each with 8 ascospores inside it). The ascospores are released when the cleistothecium wall is ruptured.

The haustorium of *Erysiphe graminis* is highly distinctive , consisting of a rounded body with finger-like projections in a wheat epidermal cell. The fungus in this Figure was stained with trypan blue, which also shows the host cell nucleus (n). The haustoria of *E. graminis*, like those of all biotrophic fungi, are not in direct contact with the host cell contents, because they are surrounded by a membrane - the extrahaustorial membrane - which represents a modified form of the host cell membrane.

G

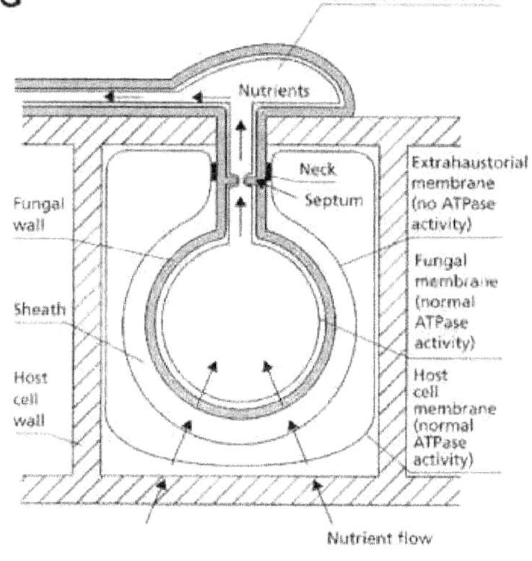

From Deacon (1997) *Modern Mycology*

By digesting the host cell walls with enzymes, it has been possible to isolate "haustorial complexes" consisting of the haustorium and its encasing membrane. Experimental studies on these haustorial complexes of pea powdery mildew have shown that the extrahaustorial membrane lacks ATPase activity and thus lacks the ability to control the movement of nutrients across this membrane. In contrast, both the haustorial membrane (of the fungus) and the plant cell membrane have normal ATPase activity for driving nutrient uptake.

The consequence is that the fungus can take up nutrients from the host cell, with little or no resistance, while the infected host cell can take up nutrients from its neighbours. So there is a one-way flow of nutrients into the haustorium, and from there to the fungal hyphae on the plant surface, where the fungus uses the nutrients for spore production.

Rust Fungi

The infection behaviour of rust fungi is broadly similar to that of the powdery mildews, involving nutrient absorption by haustoria to support abundant sporulation for epidemic spread. These fungi also get their name from the characteristic sporing stage - in this case the (usually) rust-coloured uredospores which develop in pustules where the fungus erupts through the plant surface.

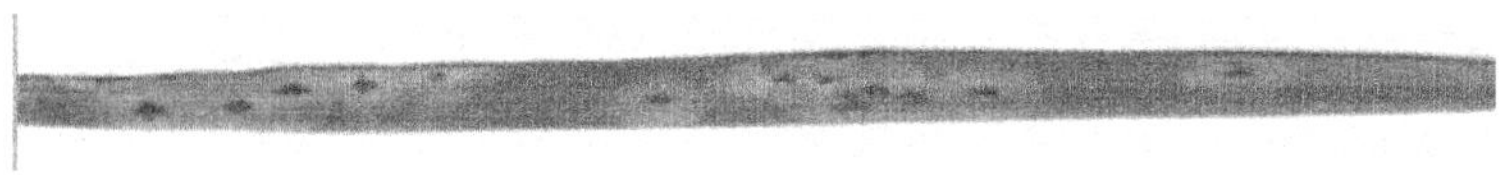

Figure H. Wheat leaf infected by the rust fungus, *Puccinia graminis* var *tritic*i, showing individual lesions (light coloured haloes on the leaf) with pustules of uredospores in their centres. [Image taken by placing an infected leaf on a flat-bed scanner]

The life cycle of rust fungi (basidiomycota, related to the toadstool-producing fungi) is often more complex than that of powdery mildews, because some rust fungi need two different types of host to complete their cycle. These hosts are termed the main host and the alternate host.

For example, *Puccinia graminis* var. *tritici* has wheat as its main host and barberry plants (*Berberis* species) as its alternate host. There is a correspondingly large number of sporing stages - up to 5 in some cases, as shown below.

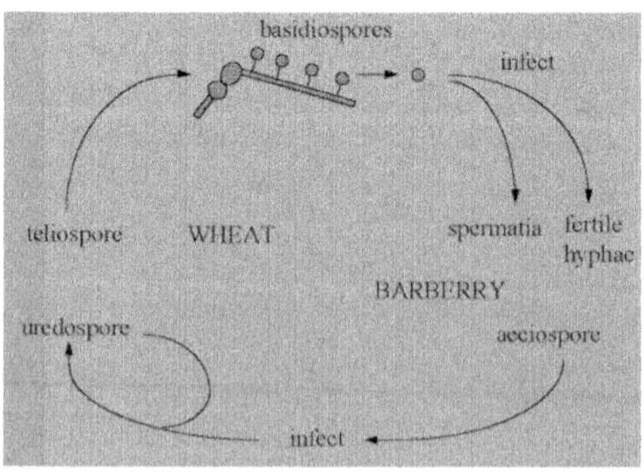

Figure I. Life cycle of *Puccinia graminis* var *tritici.*

On wheat:

- *P. graminis* produces uredospores from a bed of tissue that erupts through the leaf or stem surface. These uredospores can reinfect another wheat plant , leading to multiple cycles of infection during the cropping season. They are binucleate spores, containing nuclei of different mating types, and they germinate to produce hyphae that have 2 nuclei in each hyphal cell. In this condition, the fungus is termed a dikaryon (*i.e.* with two nuclear types).

- Near the end of the growing season, the same pustules produce a different type of spore - the teliospore, which consists of two cells with heavily thickened and darkly pigmented walls. The teliospores also are dikaryons, with two nuclei in each cell.
- The teliospores overwinter, and in spring the nuclear pairs fuse to form diploid nuclei. This is followed immediately by meiosis, then the spore germinates from each cell to form a short hypha that produces 4 uninucleate, haploid basidiospores.

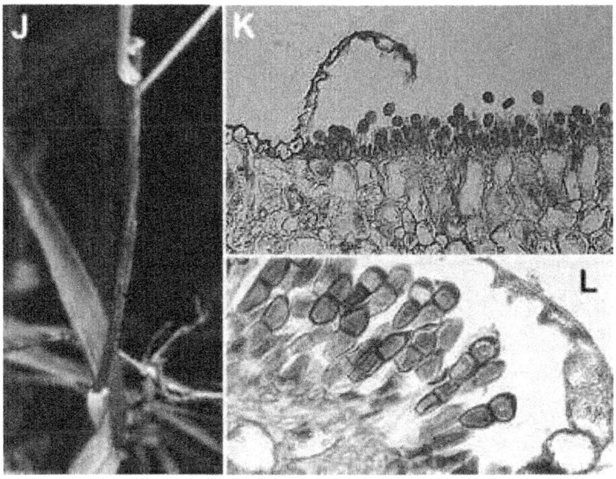

Figures J-L. *Puccinia graminis* on the cereal host. (J) Pustules of uredospores on a cereal stem. (K) Section of a leaf showing eruption of uredospores through the leaf epidermis (stained with safranin). (L) Section of a leaf later in the season, showing teliospores in place of the uredospores that were produced earlier.

On barberry:

- The basidiospores can only infect a barberry plant. They give rise to haploid hyphae of different mating types, which grow through the barberry leaf. These hyphae produce flask-shaped sexual structures termed spermogonia on the upper surface of the barberry leaf. Small "male" sexual spores (spermatia) are formed within the spermogonia, and "female" flexuous hyphae project from the neck of the spermogonium, among the stiffer hairs (arrowhead in Figure N).
- Fertilisation of flexuous hyphae by spermatia of a different mating type is brought about by insects. Then the nuclei pair in the hyphae, forming a dikaryon which gives rise to sporing pustules on the lower surface of the barberry leaf.

- The spores from these pustules are termed aeciospores. They can only infect a cereal host, thereby completing the life cycle.

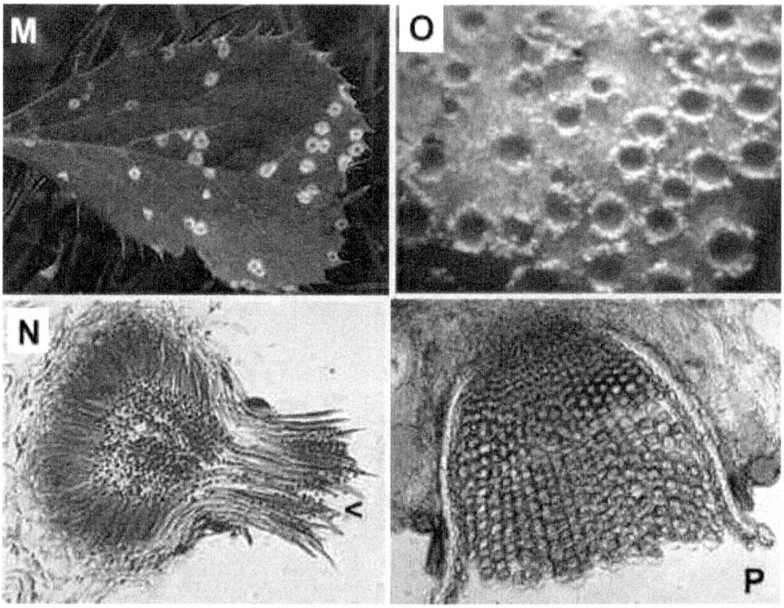

Figures M-P. *Puccinia graminis* on the alternate host, barberry. (M) Small lesions on the upper surface of a barberry leaf, with spermogonia in their centres. (N) Section of a spermogonium, showing the minute spermatia (male sexual cells) and the position (arrowhead) where flexuous (female) hyphae arise. (O) Close-up of lower surface of the leaf, showing cup-shaped pustules of aeciospores. (P) Cross section of a leaf showing the aeciospores developing in tightly packed chains from a pad of fungal tissue.

Some Common Rust Fungi

Rust fungi are remarkably common on both crop plants and wild, native plants. On crops they cause serious economic damage, necessitating the use of fungicides. Although *Puccinia graminis* (black stem rust of cereals) is most important in the USA, *Puccinia striiformis* (yellow rust) and *P. recondita* (brown rust) are more important on cereals in Britain.

Several other rusts are common in Britain.

- *Phragmidium violaceum* produces pustules of violet teliospores on the leaves of blackberry bushes (*Rubus fruticosus*). The stalked teliospores of this fungus are highly distinctive (R). There is no alternate host in this case, only the main host.
- *Puccinia punctiformis* (thistle rust) is also commonly seen. It grows systemically in the thistle *Cirsium arvense*, overwintering as mycelium in the rootstock, and producing chocolate-brown aeciospores. This fungus also has no alternate hosts.

- Another common species is birch rust, *Melampsoridium betulinum*, which forms abundant uredospores and aeciospores on birch leaves. Larch trees are the alternate host of this fungus.
- Further common species include mint rust, groundsel rust (*Coleosporium tussilaginis*; Figures U, V), dandelion rust, hollyhock (mallow) rust and snapdragon (*Antirrhinum*) rust.

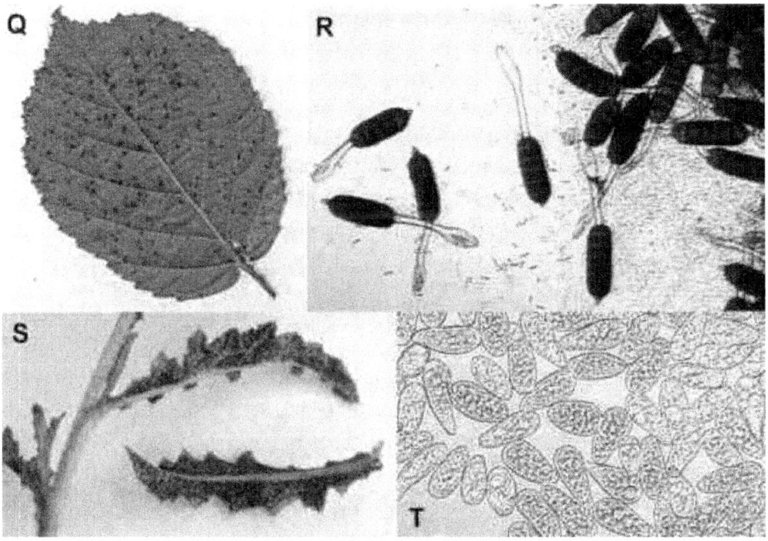

Figure Q-R. Blackberry rust, showing pustules of aeciospores on the leaf surface (Q) and the stalked, multicellular aeciospores under a microscope (R). Figure S.Thistle rust. Figure T. A mass of uredospores of birch rust, each about 30 micrometers long and easily impacted onto leaf surfaces during wind-dispersal.

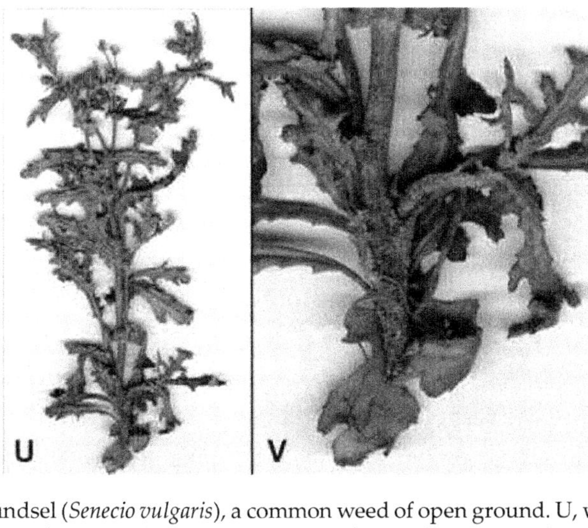

Figures U,V. Groundsel (*Senecio vulgaris*), a common weed of open ground. U, whole plant (about 15 cm tall) with rust infection at the base; V, close-up of base, showing uredospore pustules of *Coleosporium tussilaginis* on the stem and leaves.

SOIL-BORNE PLANT PATHOGENS

Soils contain diverse communities of microscopic organisms that are capable of damaging plants. A detrimental interaction between a soil organism and a plant is often highly specific. For example, a fungus that causes root-rot of wheat may have no effect on the roots of another plant growing in the same soil. Highly specialised interactions between soil organisms and plants can kill seedlings and even adult trees. Many organisms target younger plants but others appear as problems at later stages in the life of the plant. Other pathogens are able to cause disease in many different plant species.

The soil organisms that have the potential to be plant pathogens include fungi, bacteria, viruses, nematodes and protozoa. Some pathogens of the above ground parts of plants (leaves, stems) survive in the soil at various stages in their life cycles. Therefore, a soil phase of a plant pathogen may be important, even if the organism does not infect roots.

In spite of the potential for severe damage to be inflicted on plants by soil organisms, most plants do not display serious symptoms of disease. Disease usually occurs when conditions are particularly unfavourable, or when a soil organism is accidentally introduced where a highly susceptible plant species occurs. Intensive production in agriculture, horticulture or forestry increases the opportunities for diseases to develop compared with undisturbed natural ecosystems. Planting of similar plant species together in monoculture increases the probability of disease outbreak. In contrast, the damage caused by the fungus *Phytophthora cinnamomi* in many different plant species in diverse natural ecosystems in Australia demonstrates the damage that can be caused by a pathogen that infects the roots of many unrelated plants.

Some plant pathogens depend on their host plant for survival and are unable to complete their life cycle without infecting their host plant. Biotrophic organisms of this type are often difficult to grow in laboratory media.

Disease-causing microorganisms and soil animals are a natural component of the soil community. The organisms are normally present in relatively low numbers. An outbreak of disease commonly follows either an increase in the abundance of the pathogen or a change in the susceptibility of the host to the pathogen. Important questions are:

- Why do outbreaks of disease occur?
- What are the soil conditions that allow disease to develop or that suppress disease?
- What are the soil conditions that suppress disease?
- How do management practices influence the soil as a habitat for plant pathogens?
- How do plant pathogens survive in the absence of their specific host plant?

What is Plant Disease?

Plant disease is the result of a complex interaction between the host plant, the pathogen and the environmental conditions including those in the soil. Soil conditions that influence the development of disease are pH, moisture, oxygen, and nutrients. In addition, other soil organisms interact with the pathogen. The level of this interaction is also influenced by soil conditions, which affects the growth and activity of the non-pathogenic organisms. Consequently, whenever the soil environment changes, there is potential for either a positive or negative influence on soil-borne plant pathogens.

Plant disease reduces plant survival and productivity, but the symptoms of plant disease are often difficult to recognise. One reason for this is that they are not necessarily the same in plants of different ages. The amount of a potential pathogen in a soil and the environmental conditions can also alter the symptoms displayed by a plant. Furthermore, symptoms may be related to infection by more than one organism. Interactions between organisms can produce symptoms that are not the same as those that occur when each type of organism is present alone.

Similar symptoms can be caused by different pathogens and furthermore, the symptoms of some pathogens resemble those of various nutrient deficiencies. Therefore, it is not always easy to identify a pathogen responsible for a disease simply from the symptoms on the plant. A great deal of experience and knowledge of the range of symptoms produced by different pathogens and knowledge of symptoms of nutrient deficiency is essential for an accurate identification of most pathogens.

Examples of Major Soil-borne Plant Diseases:

Disease Caused by Phytophthora Cinnamomi

Phytophthora cinnamomi causes a serious disease that threatens forests and other ecosystems especially in south-eastern and south-western Australia. Hundreds of different plant species are killed by this introduced pathogen. Vehicles are often responsible for the widespread distribution of the pathogen by disturbing and transporting infected soil. There is no simple solution to *Phytophthora*disease in the forest. Quarantine methods have been introduced to limit the spread of the fungus and cleaning of vehicles is mandatory.

P. cinnamomi and related fungi also cause disease of horticultural plants such as azalea, pineapple and avocado. In horticultural systems, organic mulches have been used to stimulate the community of soil organisms and reduce the negative impact of the pathogens.

Take-all Disease of Wheat

Take-all disease is caused by the fungus *Gaeumannomyces graminis* var. *tritici*. This pathogen infects the vascular tissue of wheat roots and restricts the transport of water and nutrients within the plant. Severely infected plants have stunted root systems. In addition to root rot, a severe symptom is 'white heads' which occurs if plants survive seedling damage and grow to maturity. Such plants form seed heads with poor grain development that are characteristically white. The fungus survives in the soil on decaying plant material and relies on this material as a carbon source to sustain it until it is able to infect new roots in the following wheat crop or alternate hosts.

The take-all fungus is a major root disease world-wide and is estimated to cause millions of dollars in lost wheat yield each year. The main method for control is by crop rotation with non-host plants and removal of weeds that may act as alternate hosts. There has been little success in breeding for resistance in wheat to the fungus *Gaeumannomyces graminis* var. *tritici*. Under some conditions, soils become naturally suppressive to the fungus. This has been observed in long-term monocultures of wheat. Reduced disease is likely to be due to changes in activity of other soil microorganisms.

Crown Gall

Crown gall occurs on many genera of plants and is characterised by the formation of root tumours caused by the bacterium *Agrobacterium tumefaciens*. The bacteria infect the root and induce plant cells to divide; a tumour-like swelling is formed that contains infected cells around its outer surface. Bacterial DNA is transferred to the host plant.

Root Knot Nematode Disease

Root knot nematodes cause disease on hundreds of plant species, especially horticultural species, in warmer climatic zones. Species of nematodes in the genus *Meloidogone* induce the formation of numerous galls throughout the root system. The damaged roots also have malfunctioning root tips which reduce root growth, resulting in considerable yield losses.

How is the Source of Plant Disease Identified?

The fundamental approach to identifying the organism or organisms responsible for a disease is to follow the protocol known as Koch's Postulates :

Step 1. Visual inspection of the plant to identify signs of the presence of a pathogen and symptoms on the host (observed without the aid of a microscope.

Step 2. Observation of the diseased portion of the plant using a microscope.

Step 3. Isolation and purification of the pathogen (growth of the organism on artificial media).

Step 4. Inoculation of plants of the same type with organisms isolated from the diseased plant to observe symptoms of disease, for comparison with the original observations in Step 1.

The purpose of using Koch's Postulates is to identify the organism responsible for disease in a rigorous manner. This is the first step in working out how to overcome the problem. Not all organisms can be isolated from plants and grown in culture and for many diseases it is difficult to identify the responsible organism.

Once a potential pathogen has been identified, a variety of procedures can be used to determine the conditions under which the disease is likely to be most severe. An example of this approach was used to identify the causal organism of brown spot and root rot of lupin in Western Australia as described below.

Case Study: Pleiochaeta Root Rot of Lupin

The pathogen responsible for a serious root disease of lupin in Western Australia was not immediately identified when it was first detected. In fact, several species of fungi were commonly reported from diseased lupin roots, including species of*Fusarium*, *Pythium*, *Rhizoctonia* and *Pleiochaeta* further hindering the control of the disease.

The protocol used by Sweetingham to identify the causal pathogen was to

- *Observe and describe* the disease symptoms and occurrence in the field.
- *Identify* the fungi isolated from diseased roots. More than 10 species of fungi were isolated, but none were clearly identified as the possible pathogen.
- *Perform* pathogenicity tests to identify the degree of disease caused by each fungus when it was present by itself on lupin roots. Some fungi were more pathogenic than others, but none of the fungi, when present alone, was distinctly more pathogenic than any other.
- *Test* the effect of selected fungicides on the presence of different fungi and on the extent of disease using field soil. This technique was most successful in identifying the likely cause of the root rot.

The use of fungicides that targeted the major fungi present in soil where disease was observed demonstrated that *Pleiochaeta* was the most likely pathogen. The fungicides Ridomil, Benlate, Rovral and a mixture of all three were used. Ridomil eliminated *Pythium* without causing a major decline in the disease rating of roots. Benlate had little effect on *Pythium*, but reduced *Pleiochaeta* and *Fusarium* in the roots leading to a reduced disease rating. Rovral eliminated *Pleiochaeta* but left *Furarium* and *Pythium* and these roots had a much reduced disease rating. This set of fungicide treatments was successful in discriminating among the fungi that were present in the diseased roots.

This case study illustrates the complexity involved in identifying the causal agent of root disease, especially when more than one fungus may be involved. It cannot be presumed that the presence of an organism implicates it as a causal agent of disease. Non-pathogenic fungi often enter damaged roots and therefore a number of organisms may be present in diseased roots without being responsible for the disease.

How are Disease-causing Organisms Dispersed?

Disease occurs under the following conditions:

(i) if there are enough infective hyphae, spores or other fungal structures such as sclerotia of a pathogen present in the soil,

(ii) if there is a suitable environment for growth of the organisms, and

(iii) if the plant is susceptible to infection.

If the organism is present in low numbers, it may have little impact on the plant. If soil conditions are suitable for the multiplication of the pathogen, there will be a greater chance of disease if other factors are also appropriate for disease development. In general, the more organisms present, the more severe the disease , but plant and soil conditions are also important.

The hyphae of some fungal pathogens can grow long distances through the soil. For example, hyphae of the fungus *Armillaria* extend many metres horizontally in soil and simultaneously infect the roots of several plants. Some fungi, for example species of *Phytophthora,* form narrow and short hyphae that do not spread the fungus through the soil. Often these fungi are disseminated in water or in soil during erosion and by other mechanisms including transport of soil on vehicles or footwear.

Spores of fungi that are pathogens of leaves and stems may have a soil phase during which spores remain dormant. Once the dormancy period ceases, spores survive in the soil until conditions are appropriate for their germination and infection of a new plant. Growth of some pathogens is stimulated by the presence of roots which release exudates or emit other signals.

Spores of some soil-borne fungi are dispersed by wind. Dispersal of bacterial plant pathogens is common when soil is transported from one place to another. Some disease-causing organisms are dispersed on plant debris. *Pleiochaeta setosa* spores are dispersed in soil via dead organic material from the previous crop.

How Do Pathogens Enter Roots?

Pathogenic invasion of plant roots involves several steps. The first of these is recognition between the pathogen and the plant, which involves complex molecular mechanisms. Recognition occurs in the diffuse interface between the root and the soil created by root exudates and sloughed off root cells. This interface between soil and root may be a very suitable habitat for growth of the pathogen before it enters the root.

This initial interaction can also be the stimulus for the fungus to penetrate the plant cell wall. Some organisms produce enzymes that dissolve cell wall layers, allowing the hyphae to enter. Pathogens can also enter roots at points damaged by soil animals or other disturbances. A plant may react to the presence of a pathogen by producing chemical or physical barriers that prevent entry or confine potential pathogens to a small part of the root.

Once inside the root, pathogens occupy specific cell types. Some fungi enter the cells that conduct nutrients, carbon and water throughout the plant. If these cells are damaged, important processes that control the life of the plant are disrupted. Other fungi predominantly colonise the cortical root cells that are important in nutrient uptake from the soil.

How Do Roots React to Pathogens?

Plants differ in their susceptibility to plant pathogens. Even cultivars of one plant species can be more or less resistant to a pathogen. Similarly, species or strains of pathogens differ in their capacity to infect a plant. Differences in the degree of disease can therefore be related to:

(i) the susceptibility of the plant;

(ii) the virulence of the pathogen;

(iii) the stage of development of the plant;

(iv) the abundance of the pathogen; and

(v) the environmental conditions.

How Does Plant Nutrition Affect Disease?

The nutritional status of a plant influences many of its physiological processes and this can alter the response that plants have to the presence of pathogens. Changes in the concentration of nutrients such as phosphorus in a plant can alter root and shoot growth as well as the movement of carbohydrate and other molecules within the plant. These changes have the potential to influence the development of a disease.

Fertilisers can also alter plant susceptibility to disease. In this study, in the absence of manganese fertiliser, barley grew less well if nematodes were present in the soil than when manganese was added at sufficient levels to meet the requirements for growth of the plant This occurred even when the numbers of nematodes in both treatments were similar.

What is 'Biological Control' of Plant Disease?

The control of pathogens and prevention of plant disease is a natural soil biological process. Indeed, in most situations, plant disease is not strongly evident even when potentially pathogenic fungi are present in a soil. This suppressive characteristic of soil is common in natural ecosystems, but becomes less apparent with high levels of disturbance and the introduction of plant monocultures. Disturbance creates soil conditions and a high density of susceptible roots that encourages the multiplication of pathogens. Once potentially damaging organisms become present in high numbers in a soil, they may be difficult to eradicate. Management practices are required that create conditions in the soil that are not favourable to pathogens so that their growth is limitedand therefore, disease it restricted.

There are many demonstrations of the potential for soil organisms to reduce disease. As demonstrated in this study, when apple trees were replanted in a soil previously prone to apple replant disease, increased tree growth occurred when various strains of bacteria were added to the soil, indicating a beneficial effect. Similar effects were observed at two sites, but because different quantities of the bacteria were added in the inoculum it is difficult to compare the effectiveness of each organism.

In some environments, take-all disease of wheat has declined naturally. These areas of natural suppression of disease have been investigated extensively to determine how the pathogen is suppressed. Specific bacteria that are capable of limiting the growth of the fungus, including *Pseudomonas* species, have been isolated from disease suppressive soils. But it is still unclear whether these bacteria are responsible for the decline in take-all disease.

PLANT PATHOGEN CONTROL

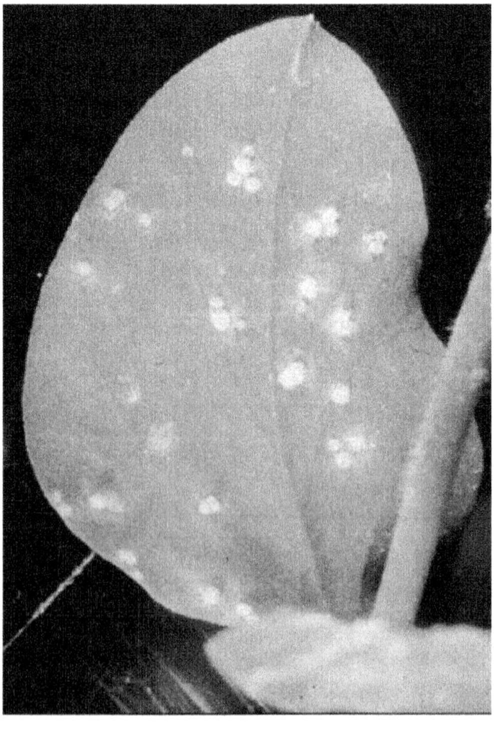

Rust fungi.

Plant disease is caused by abiotic and biotic factors. While other chapters deal with the way fungi improve plant health directly, this chapter will ask the question "Can fungi be used to reduce the effects of biotic causes of plant disease?" As most biotic causes are fungi, we will concentrate on fungus to fungus interactions. The topic of fungus to fungus interactions has been reviewed in depth on many occasions from many different points of view. This chapter will not attempt to summarise this work. Rather we will dip into some of the stories.

The interactions between fungi can be manipulated to reduce the damage caused by one fungus on the plant. Essentially, these interactions include:

- Direct hyphal parasitism
- Antagonistic interactions
- Competition for resources

The direct mechanisms are not mutually exclusive and all may be evident within one interaction at the same or different times.

Direct Hyphal Parasitism

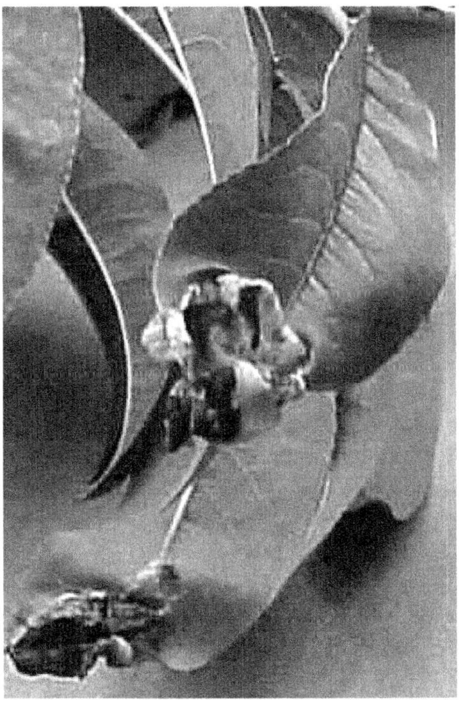

Leaf curl caused by fungi.

Mycoparasitism is where one fungus derives its nutrition from another without any benefit in return. The interaction can be where the parasite is biotrophic or necrotrophic. Biotrophs tend to have highly specialised relations with specific hosts. The parasite is extremely difficult to manipulate, and as a consequence few have ever been used in biocontrol.

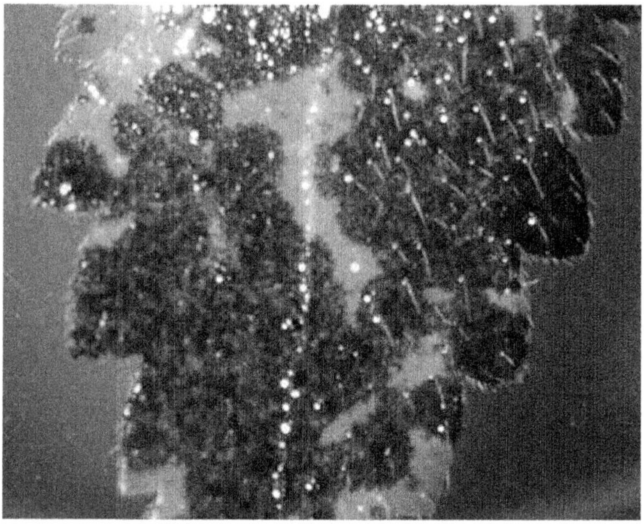

Meloline fungi.

Necrotrophs tend to kill the cell or tissue from which they feed. They tend to have a broad host range, and their activity commonly includes excretion of antibiotics. For instance, the necrotroph*Trichoderma* has the capacity to parasitise a great number of soil fungi, including pathogens *Rhizoctonia, Scerotinia, Fusarium* and *Verticillium,* and the Straminipiles *Pythium* and *Phytophthora. Trichoderma* grows from the host where it is an endophyte, towards hyphae of other fungi where it binds following branching. *Trichoderma* releases hydrolytic enzymes that digest the walls of the host prior to penetration. *Trichoderma* is also thought to release a range of toxins that reduce any response from the host to invasion.

For biocontrol to be successful, the parasite must be placed in an environment favourable for its competitive success. Most soils are oligotrophic. Thus, the introduced fungus is unlikely to be favoured against isolates that are adapted to the local conditions. Biocontrol agents that colonise epidermal or cortical cells of the root (as endophytes) have access to abundant nutrients, placing them at an advantage over the plant pathogen. The judicious increase of organic amendments at inoculation, biocontrol can be attained followed by a decline in the population size of the parasite as the amendment degrades.

Antibiotic Interactions

Antibiosis can be defined as the inhibition of one organism by metabolites of another. Metabolites that kill another organism tend to be called toxins. Thus, dilute solutions of toxins may inhibit and high concentrations of antibiotics may kill.

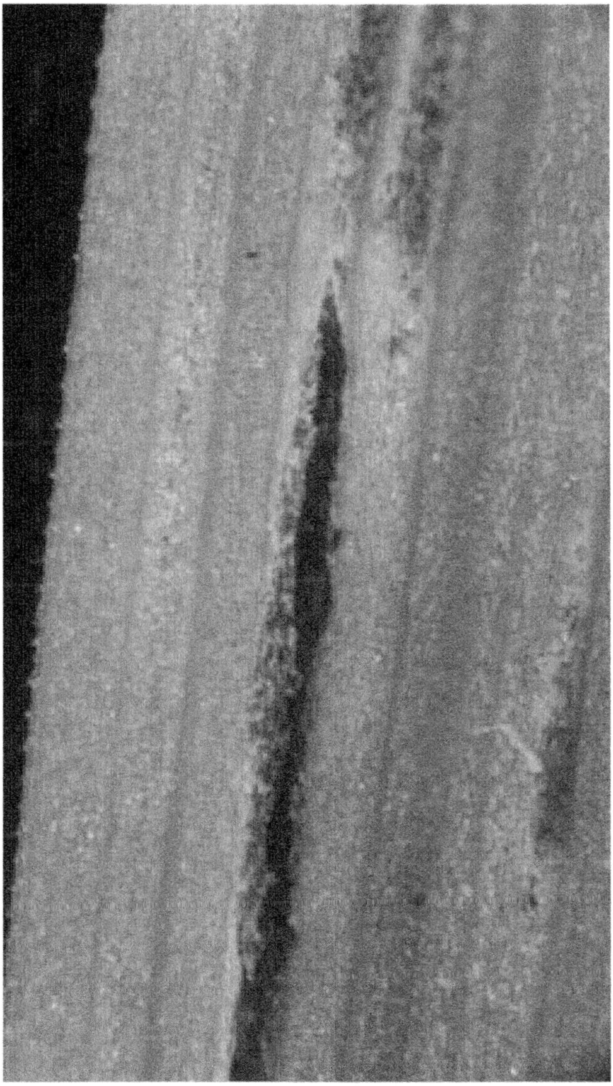

Smut fungi.

Most fungi produce inhibitory metabolites. For instance, *Gliocladium* produces a diketopiperazine that kills *Pythium* probably because of coagulation of proteins in the cytoplasm. Use of the fungus on seed appears to reduce seedling damping-off as much as a common seed treatment. Volatile pyrones produced by *Trichoderma* appear to reduce damping off caused by *Rhizoctonia*. The potential to exploit these metabolites in biocontrol appears feasible.

The role of antibiotics in biocontrol is still unclear. Some spectacular experiments, when repeated in the field, have resulted in enormously variable outcomes indicating that other mechanisms may be important.

Competition for Resources

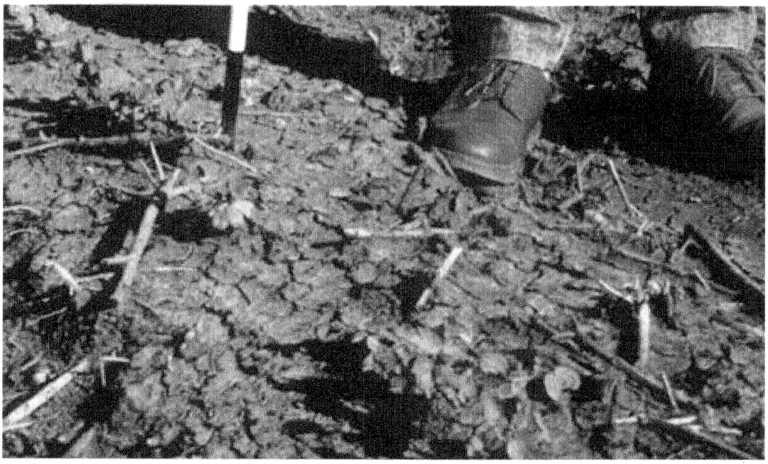

Seedling disease.

Competition is the active requirement for resources in excess of those immediately available to two or more organisms. The resource most commonly considered to be important for fungi is organic energy. Thus much competition exists for access to energy for growth and maintenance. However, competition between fungi is extremely difficult to quantify. The mycelium is constantly changing: growth conditions are dynamic. Resources within the hyphae are shifted between different parts of the thallus, with consequent changes in what is being (experimentally) assessed over very short periods of time.

Competition for nutrients can take place between survival propagules, germinating units and mycelium in any arrangement. The result of competition is stasis of the less successful competitor. Inoculum density does not decline and the potential to initiate disease remains. Thus the application of competition by itself to biocontrol appears limited.

Success

Some examples of fungi used against fungal pathogens with some success:

Agent	Pathogen
Chaetomium	*Venturia*
Coniothyrium	*Scerotinia*
Cladosporium	*Botrytis*
Gliocladium	*Rhizoctonia*
Trichoderma, Penicillium	*Rhizoctonia*
Scerotinia, Fusarium, Verticilium	*Puccinia*

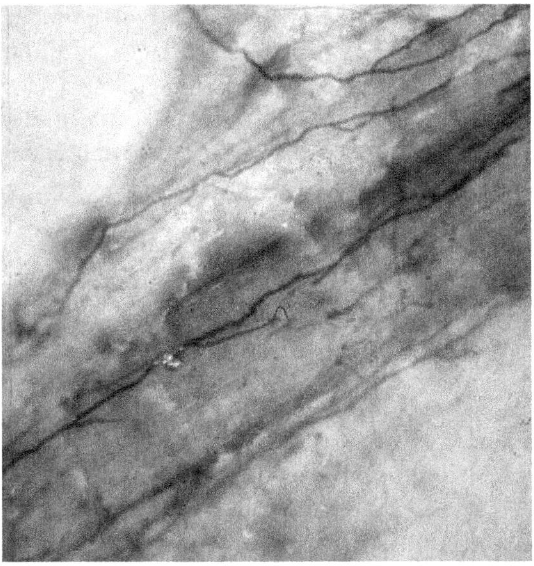

Fungi in roots.

Trichoderma viride, Trichoderma harzianum, Trichoderma polysporum and *Trichoderma hamatum* (all Ascomycota) have all been formulated for commercial use against soil fungi. These fungi are common inhabitants of heated soil where the almost sterile medium is conducive to their explosive colonisation following germination of ascospores of the fungi. Over time other fungi colonise the soil, competitively excluding the species of *Trichoderma*. Thus their obvious presence in soil may be short-lived. They colonise the soil, interact with the pathogen but then generally decline. For instance, *Trichoderma* can be used in potting media to reduce seedling diseases.

For the fungus to be successful in field soil, it must establish an endophytic association with the roots of the crop plant. Given the benefit of a constant source of energy, the biocontrol agent may explore beyond the internal plant tissues interacting with pathogens. Success in field soils is less common than the theory indicates. Many fungi are commercially available but we still have much to understand before reliable control of pathogens can be expected.

While laboratory and glasshouse experiments indicate fungi have a role in biocontrol of plant pathogenic fungi, the results of field application of a single fungus have been discouraging. While it appears that some agents may have a place in integrated approaches to reducing plant pathology, we do not appear to understand the basis of interactions between fungi. Until this understanding is improved, use of fungi in biocontrol is likely to remain a dream.

New theoretical developments argue that release of multiple organisms will have greater potential to regulate populations of ALL microbes, both pathogens and saprotrophs. The real issue is whether the diversity effect is due to taxonomic diversity or functional attributes of the diverse inoculants. If this impact is due to functional diversity, incorporation of organic amendments inocuated with fungi selected to suppress deleterious microbes and other pests may be a sustainable mechanism to reduce reduce reliance on chemical treatment of pests and pathogens. Treatment would be a regular management process, rather than a reaction to a problem.

RESISTANCE OF PLANT PATHOGENS TO PESTICIDES

Resistance of plant pathogens to pesticides was not a widespread concern until the early 1970s. Prior to that time most of the fungicides used were broad spectrum protectants that affected multiple sites within the target fungi. However, in the late 1960s, fungicides were developed that had systemic activity and could control some fungi after infection occurred. These fungicides had specific sites of action and resistant populations of fungi developed, causing diseases on many crop plants. The selection and subsequent increase of resistant forms of the pathogens occurred similarly to that described previously for arthropods.

So far, resistance has become a problem in the United States to only two pathogens that affect apples during the growing season: *Venturia inaequalis,* cause of apple scab, and *Erwinia amylovora,* cause of fire blight.

Resistance of *V. inaequalis* to the benzimidazole fungicides became widespread in the mid and late 1970s throughout the Southeast. Consequently, the benzimidazole fungicides are no longer recommended for apple scab control in most southeastern states. Resistance of *V. inaequalis* to dodine has been reported from several northern apple producing states but has not been reported from the Southeast. Similarly, there have been reports from other countries of the world concerning resistance of *V. inaequalis* to the demelthylation inhibiting fungicides (DMI fungicides), but resistance has not been reported in the United States. Resistance of the fire blight bacterium, *E. amylovora,* to streptomycin has been reported from California, Michigan, and several other states, but not in the Southeast.

Several strategies are available for minimizing the likelihood of resistance development. These strategies involve reducing the selection pressure on the pathogen in one of several ways: by minimizing the use of a particular pesticide or by combining it with another pesticide, which has a different mode of action and broad-spectrum activity.

To avoid the development of dodine resistance, limit dodine use to two or three applications a year. In areas of the United States where dodine resistance is a problem, dodine was often used ten to twelve times through the growing season. Historically in the Southeast, dodine has been used only two to four times, early in the season. A similar strategy should be used to avoid resistance to streptomycin. Applications of streptomycin should be made only when needed during periods favorable for infections.

Chapter 5

THE APPLICATION OF MOLECULAR MARKERS IN THE EPIDEMIOLOGY OF PLANT PATHOGENIC FUNGI

MOLECULAR MARKER

A molecular marker is a molecule contained within a sample taken from an organism (biological markers) or other matter. It can be used to reveal certain characteristics about the respective source. DNA, for example, is a molecular marker containing information about genetic disorders, genealogy and the evolutionary history of life. Specific regions of the DNA (genetic markers) are used to diagnose the autosomal recessive genetic disorder cystic fibrosis, taxonomic affinity (phylogenetics) and identity (DNA Barcoding). Further, life forms are known to shed unique chemicals, including DNA, into the environment as evidence of their presence in a particular location. Other biological markers, like proteins, are used in diagnostic tests for complex neurodegenerative disorders, such as Alzheimer's disease. Non-biological molecular markers are also used, for example, in environmental studies.

Genetic Markers

In genetics, a molecular marker (identified as genetic marker) is a fragment of DNA that is associated with a certain location within the genome. Molecular markers are used in molecular biology and biotechnology to identify a particular sequence of DNA in a pool of unknown DNA.

Application

It has 5 applications in fisheries and aquaculture:

- Species Identification
- Genetic variation and population structure study in natural populations

- Comparison between wild and hatchery populations
- Assessment of demographic bottleneck in natural population
- Propagation assisted rehabilitation programmes. However,there still exists some limitations.

MOLECULAR SYSTEMATICS OF FUNGI

The research areas for this group are fungal molecular systematics, fungal diversity and ecology, exploitation of medicinal and edible fungi, and conservation and sustainable use of fungal resources. Combination of morphology and molecular approaches, incorporated with ecological, bio-geographic and genomic information, is emphasized in the research on fungal phylogeny and evolution. Efforts are also made to develop methods of molecular detection for various fungi and to investigate their implication in the study of ecology and bio-resources of fungi.

Background and Significance

The current research of this group focuses on economically important macro-fungi in China, including *Cordyceps, Ganoderma, Tricholoma, Termitomyces, Boletus, Pleurotus, Tuber* and ectomyrrhizal fungi. These are important medicinal and edible fungi, with great economical values. They are widely explored to meet the needs of domestic markets and also exported to many countries. In addition, all these fungi are playing important roles in the ecosystem of the world.

Major Achievements

During the past four years, more than 4700 fungal specimens have been collected all over China and 361 were received as gifts from overseas herbaria. Over 1100 living strains have been isolated and gathered. More than 6200 fungal specimens have been studied in the laboratory and over 1800 DNA samples extracted, with more than 600 sequences of various fragments were determined.

The fungal resources of *Tuber, Tricholoma, Pleurotus, Termitomyces* and *Boletus* in China have been investigated, with over 620 fungal names clarified. Ten new species and 4 Chinese new records were found and 4 new combinations proposed. The application of molecular markers, *e.g.* ITS, IGS, nLSU, β-tubulin, Cytb, mtLSU and mtSSU, in the molecular systematic study of various fungal groups was evaluated. Based on sequences of these markers, molecular trees of *Tuber, Ganoderma, Pleurotus, Termitomyces, Tricholoma* and *Boletus* were reconstructed to elucidate the phylogeny and evolution of these genera and associated groups. As a part of the results of this research, the molecular identification systems for these groups have been established to serve the need of the public. After DNA cloning, the intra-strain ITS heterogeneity was first found in *Ganoderma* species. The molecular variations of *Cordyceps sinensis* on the Tibetan Plateau and co-evolution between fungi and insects (*e.g. Cordyceps sinensis* v. moths and *Termitomyces* v. termites) are being studied. Investigation on nutritional requirements and bioactive compounds of *Cordyceps sinensis* has also been carried out to stimulate further exploitation and utilization of this important medicinal fungus.

Since 2001, 41 papers (including 20 in SCI cited journals) and one Chinese translation of text book were published by this group, and one patent has been submitted for approval.

Future Research Plan

1. Research on biodiversity of medicinal and edible fungi, and conservation and sustainable use of these fungal resources with special focus on *Cordyceps sinensis, Tricholoma matsutake* and *Ganoderma* spp.
2. Study of molecular systematics, phylogeny, biogeography, co-evolution with other living organisms of economically important fungi, mainly *Cordyceps, Tuber, Ganoderma, Pleurotus, Termitomyces, Boletus* and *Tricholomataceae,* with further development of the molecular identification system.
3. Exploitation of medicinal and edible fungi with focus on isolation, analysis and immunological effects of bioactive compounds from *Cordyceps sinensis, C. militaris, Dictyophora indusiata* and *Ganoderma* spp.

PLANT DISEASE EPIDEMIOLOGY

Plant disease epidemiology is the study of disease in plant populations. Much like diseases of humans and other animals, plant diseases occur due to pathogens such asbacteria, viruses, fungi, oomycetes, nematodes, phytoplasmas, protozoa, and parasitic plants. Plant disease epidemiologists strive for an understanding of the cause and effects of disease and develop strategies to intervene in situations where crop losses may occur. Typically successful intervention will lead to a low enough level of disease to be acceptable, depending upon the value of the crop.

Plant disease epidemiology is often looked at from a multi-disciplinary approach, requiring biological, statistical, agronomic and ecological perspectives. Biology is necessary for understanding the pathogen and its life cycle. It is also necessary for understanding the physiology of the crop and how the pathogen is adversely affecting it. Agronomic practices often influence disease incidence for better or for worse. Ecological influences are numerous. Native species of plants may serve as reservoirs for pathogens that cause disease in crops. Statistical models are often applied in order to summarize and describe the complexity of plant disease epidemiology, so that disease processes can be more readily understood. For example, comparisons between patterns of disease progress for different diseases, cultivars, management strategies, or environmental settings can help in determining how plant diseases may best be managed. Policy can be influential in the occurrence of diseases, through actions such as restrictions on imports from sources where a disease occurs.

In 1963 J. E. van der Plank published "Plant Diseases: Epidemics and Control", a seminal work that created a theoretical framework for the study of the epidemiology of plant diseases. This book provides a theoretical framework based on experiments in many different host pathogen systems and moved the study of plant disease epidemiology forward rapidly, especially for fungal foliar pathogens. Using this framework we can now model and determine thresholds for epidemics that take place in a homogeneous environment such as a mono-cultural crop field.

Elements of an Epidemic

Disease epidemics in plants can cause huge losses in yield of crops as well threatening to wipe out an entire species such as was the case with Dutch Elm Disease and could occur with Sudden Oak Death. An epidemic of potato late blight, caused by *Phytophthora infestans*, led to the Great Irish Famine and the loss of many lives.

Commonly the elements of an epidemic are referred to as the "disease triangle": a susceptible host, pathogen, and conducive environment. For disease to occur all three of these must be present. Below is an illustration of this point. Where all three items meet there is disease. The fourth element missing from this illustration for an epidemic to occur, is time. As long as all three of these elements are present disease can initiate, an epidemic will only ensue if all three continue to be present. Any one of the three might be removed from the equation though. The host might out-grow susceptibility as with high-temperature adult-plant resistance, the environment changes and is not conducive for the pathogen to cause disease, or the pathogen is controlled through a fungicide application for instance.

Sometimes a fourth factor of time is added as the time at which a particular infection occurs, and the length of time conditions remain viable for that infection, can also play an important role in epidemics. The age of the plant species can also play a role, as certain species change in their levels of disease resistance as they mature; a process known as ontogenic resistance.

If all of the criteria are not met, such as a susceptible host and pathogen are present but the environment is not conducive to the pathogen infecting and causing disease, disease cannot occur. For example, corn is planted into a field with corn residue that has the fungus *Cercospora zea-maydis*, the causal agent of Grey leaf spot of corn, but if the weather is too dry and there is no leaf wetness the spores of the fungus in the residue cannot germinate and initiate infection.

Likewise, it stands to reason if the host is susceptible and the environment favours the development of disease but the pathogen is not present there is no disease. Taking the example above, the corn is planted into a ploughed field where there is no corn residue with the fungus *Cercospora zea-maydis*, the causal agent of Grey leaf spot of corn, present but the weather means long periods of leaf wetness, there is no infection initiated.

When a pathogen requires a vector to be spread then for an epidemic to occur the vector must be plentiful and active.

Types of Epidemics

Monocyclic epidemics are caused by pathogens with a low birth rate and death rate, meaning they only have one infection cycle per season. They are typical of soil-borne diseases such as Fusarium wilt of flax. Polycyclic epidemics are caused by pathogens capable of several infection cycles a season. They are most often caused by airborne diseases such as powdery mildew. Bimodal polycyclic epidemics can also occur. For example, in brown rot of stone fruits the blossoms and the fruits are infected at different times.

For some diseases it is important to consider the disease occurrence over several growing seasons, especially if growing the crops in monoculture year after year or growing perennial plants. Such conditions can mean that the inoculum produced in one season can be carried over to the next leading to a build of an inoculum over the years. In the tropics there are no clear-cut breaks between growing seasons as there are in temperate regions and this can lead to accumulation of inoculum.

Epidemics that occur under these conditions are referred to as *polyetic* epidemics and can be caused by both monocyclic and polycyclic pathogens. Apple powdery mildew is an example of a polyetic epidemic caused by a polycyclic pathogen and Dutch Elm disease a polyetic epidemic caused by a monocyclic pathogen.

Employment

Plant disease epidemiologists are typically employed as researchers by universities, or governmental institutions such as the USDA. However, private companies in agricultural fields also employ epidemiologists.

APPLICATIONS OF MOLECULAR MARKERS; GERMPLASM CHARACTERIZATION, GENOTYPE IDENTIFICATION AND PATHOGEN DIVERSITY

Germplasm Characterization

The objective of *ex situ* germplasm collections used for the conservation of the genetic cacao resources is to ensure that a maximum of the genetic diversity is preserved and used in a rational manner. A maximum of genetic information must be obtained from the collections to allow its use for the improvement and conservation of the genetic resources. Molecular markers present diverse applications to efficiently obtain useful genetic information in diverse stages of the collection.

The estimate of the genetic similarity between individuals of a given accession or between accessions of the collection permits information to be obtained on the identity of this accession, identify duplicates, and complement the characterization with morphological and isoenzymatic traits. The partition of the marker-estimated variability among individuals, accessions, populations, types, and geographic origin permits the identification of populations and/or geographic regions that present greater diversity or that are underrepresented in the collection, supporting strategies and the planning of new collections and acquisitions and the development of nuclear collections (core collections) that represent the diversity of the collection. Molecular markers have the potential to help in the identification of genes/alleles of interest in the collection.

This strategy of characterization is being followed in the germplasm collection of Trinidad. Marita (1998) evaluated the diversity and genetic relation of the sample composed of 270 genotypes of the Active Germplasm Bank (AGB) of CEPEC by RAPD markers with the objective of furnishing recommendations for the establishment of nuclear collections and to maximize the genetic diversity between genotypes with witches' broom-resistance. This was the most comprehensive study on genotypes of a single collection, since genotypes had previously been provided by various collections. Thirty-eight primers produced 133 RAPD loci and the estimates of genetic diversity were used to generate a subgroup representing the maximum diversity that would serve as a nuclear collection. The inclusion of samples could be simulated and evaluated based on the availability of molecular data. The frequency of the RAPD loci associated with witches' broom-resistant genotypes was used to indicate possible associations between the markers and the trait. Similar studies could be realized for other traits such as black pod-resistance and fat content evaluated earlier in this germplasm in order to accelerate the identification of associated markers.

The cacao germplasm bank of ERJOH and CEPLAC in the State of Pará is result of the systematic collection of representative samples of the genetic variability of the populations existing in the Brazilian Amazon. The relations and genetic diversity between cacao genotypes contrasting in field resistance against witches' broom (20 resistant; 10 susceptible), collected originally in various watersheds in the Amazon, besides two external susceptible controls (Bahia, Brazil, and Costa Rica) and two genotypes considered resistant ('SCA 6' and 'SCA 12' of Peru), were initially determined by RAPD. Of the 26 tested primers, 19 amplified 163 fragments of good resolution and reproducibility. The mean genetic similarity among all genotypes was 85.3%.

All analyzed genotypes differed significantly from each other. The genotypes had a tendency to group according to their geographic origins; with genotypes collected in proximity presenting a greater similarity among themselves.

The phenogram could be arbitrarily subdivided into two large groups. The first contained 11 resistant genotypes, including all genotypes collected originally in Acre (rivers Chandless, Acre, and Iaco) with one exception, besides two genotypes from the river Solimões and of the actual SCA 6 and SCA 12 clones. This group presented the greatest genetic diversity. The second group contained 16 resistant and 7 susceptible genotypes, including principally "domesticated" genotypes from Belém and Alenquer, PA, from Bahia and from Costa Rica. The results demonstrated the genetic diversity between the various potential sources of resistance to witches' broom. The structuring of the genotypes according to origin and geographic proximity suggests the potential of the use of this technique for an identification of genotypes and the establishment of core collections.

Genotype Identification

The various collections of cacao germplasm existing in the world embrace a large number of accessions with at least 16,300 clones identified by distinct acronyms in publications, representing close to 12,700 individual clones listed in the International Cacao Germplasm Database (ICGD). Various genotypes have more than one name or acronym of identification (synonymy) and some names refer to more than one genotype (homonymy). The transference of results and recommendations between the various improvement programmes is hampered by the lack of agreement between the genotypes of the various germplasm collections. Few geneticists have sufficient experience to recognize individual genotypes and the identification of duplications and errors of identification only became possible and measurable with the advent of molecular markers. The correct identification of trees of progenies from controlled crossings is also necessary in programmes of genomic mapping.

The presence of errors in the identification of genotypes must be more common than expected in the diverse collections when they compared the identity of commonly used clones in improvement by means of RAPD analysis. Based on a sample of only 11 genotypes at least two of their clones (ICS 6 and SIAL 93) or 18.2% did not show the same identity in two important collections (CEPLAC, Brazil, and BAL Plantations, Malaysia) in spite of bearing the same name (homonymy).

If the two 'AMAZON 2-1' accessions were also considered distinct at both sites based on the low similarity found, the error percentage would reach 27.3%. The results obtained by the RAPD analysis were corroborated by the aspect of the fruits. Clone SIAL 93 from the Malaysian collection is derived from an accession of Mayaguez, Puerto Rico where a branch of the rootstock used might have grown and substituted the 'SIAL 93' crown. This accession with an identification error was detected in Mayaguez and was reintroduced in Bahia, Brazil, where it was renamed 'CEPEC 533' (B.G.D. Bartley, personal communication). The error in the identification of 'SIAL 93' was later confirmed with ISSR analysis.

Another demonstration of the applicability of molecular markers in the identification of genotypes refers to the study realized in Costa Rica to confirm the hybrid origin and paternity of the progeny resulting from the crossing of 'Catongo x Pound 12'. By means of the RFLP analysis it was demonstrated that of the 120 trees that formed the trial, only 55 (45.8%) are true 'Catongo x Pound 12' hybrids, 29 (24.2%) are trees originated from self-pollination of 'Pound 12', three (2.5%) corresponded to 'Catongo' self-pollination, and the remaining 33 (27.5%) trees were of unknown origin.

The low rate of confirmation of hybrid trees casts doubts on the results obtained in the various hybrid trials conducted in the country, besides indicating the fragility of the self-incompatibility system of cacao, considering that 'Pound 12' is self-incompatible. Similar results were observed in biclonal orchards of natural hybrid seed production installed in Togo, Ivory Coast, and Cameroon where self-fecundation rates of up to 97% were observed in self-compatible genotypes based on isoenzymatic markers.

The breaking up of the incompatibility is caused by the phenomenon denominated mentor pollen.

Another application of genotyping with an apparent potential is the identification of the most probable parents for selections of witches' broom-resistant cacao (VB selections) realized at plantation level in south Bahia. To carry out this heritability test the DNAs of plants of the VB selections and of the possible parents used in the hybrid generation programme of the period in question (in the 60s) were included in the RAPD analysis. The individuals were grouped based on their genetic similarity pattern. This technique allows the identification of the different resistance sources that contributed to the formation of the actual resistant materials. In the future it will allow orientation of the composition of materials on each farm with a view to amplifying its genetic base.

The need for preservation of cacao genetic resources in active germplasm banks demands a great volume of resources and area due to the seed recalcitrance and the absence of efficient *in vitro* conservation and cryopreservation methods, besides exposing the collections to environmental and biotic risks. However, the diversity of names and acronyms seems to substitute the genetic diversity established in the collections, and the occurrence of homonyms and synonyms diminishes the efficiency in the conservation of the genetic resources. Besides, administration problems in collections, errors of tagging, branch growth from the rootstock etc reinforce the problems of correct identification. The International Cocoa Genebank, Trinidad (ICGT) is being systematically characterized to optimize the conservation of genetic resources and facilitate the improvement.

The genotypes are being characterized using morphological and agronomical traits, isoenzymes and RAPD with the objective of identifying duplications, errors of identification, and selecting clones with a potential for improvement. Christopher *et al.*, 1999 described the use of markers of the RAPD type together with botanic descriptors to verify the identity of accessions. In the identification of 117 accessions, 23 RAPD loci were analysed and 35 accessions containing mixtures identified and of these thirteen did not have any tree with the original identity.

Within the scope of the international project "Conservation and Utilization of Cacao Germplasm", financed by the partnership CFC/ICCO/IPGRI, Risterucci *et al.*, 2000b, evaluated 148 individuals of 28 clones that were sampled in various national collections using 10 microsatellite loci and identified nearly 30% of the individuals with errors of identification.

For the standardization of genotype identification through molecular markers the development of low cost and simple technology, stable systems with precision and reproducibility under the diverse conditions found in the countries that own germplasm and quarantine collections is also necessary. The molecular identifications should be deposited in data banks. Attempts in this sense were initiated by BCCCA, which conducted a test ("ringtest") to evaluate the reproducibility of the RAPD technique for genotype identification at five laboratories localized in Brazil,

Trinidad, France, England and the United States. The results demonstrated that the RAPD analysis was precise for the identification of duplicated genotypes when evaluated in the same laboratory but the reproducibility was low among laboratories. Alternatively, molecular markers of the ISSR type are being evaluated at the University of Reading to characterize cacao genotypes, due to the greater stability and reproducibility. The existence of stable molecular markers that could be used by the various programmes in the world would make the identification of synonymy and homonymy among the accessions easier.

A recent initiative of the US Department of Agriculture (USDA) proposes the molecular characterization of all the important cacao germplasm collections of America using 15 microsatellite loci.

Diversity in Pathogens

The fungus *Crinipellis perniciosa* (Stahel) Singer, etiological agent of witches' broom, endemic in the Amazon region, was detected for the first time in south Bahia in May 1989, in the municipality of Uruçuca. It was followed by the second observation six months later in the municipality of Camacã, 120 km south from Uruçuca. The area with infected plants in Camacã was greater however than that found in Uruçuca, suggesting that this introduction had occurred before the one in Uruçuca. The appearance of the disease at two distant and distinct sites, well in the middle of the cacao region suggests that more than one introduction of the pathogen had occurred, probably from the Amazon region and possibly by human intervention. South Bahia comprises the largest contiguous cacao-producing region in the world. The predominance of the cacao cultivar 'Comum', considered susceptible and presenting low genetic variability contributed to a rapid expansion of the disease, causing progressive loss in the cacao yield.

The most effective method of witches' broom control, although long term, consists in the use of resistant genotypes. Clone SCA 6 has been to date the most used resistance source, in spite of its susceptible performance in the conditions of Ecuador, Colombia and Bolivia. Studies that were based on inoculations of the fungus onto 'SCA 6' progenies, on somatic incompatibility, morphological variability, and on biochemical reactions revealed genetic diversity among the isolates. However these techniques only allow the evaluation of a small number of loci as opposed to the molecular RAPD markers that permit a rapid evaluation of a large number of loci of the fungus, being useful in the determination of the genetic diversity of pathogens. The use of these molecular markers is increasing in other pathosystems. The determination of the genetic variability of the pathogen is an important aspect in the selection process of resistant genotypes that will be used in improvement programmes.

The great majority of the fungi do not fit into the classical model of population genetics. They have varied methods of colonization, dispersion, and survival and their populations can overlap with the passing of time. However, *C. perniciosa* multiplies uniquely by means of basidiospores which in turn have a mean lifespan of a few hours.The genetic analysis of isolates of the freshly collected fungus can reflect the genetic variability of the population at that moment with high confidence.

In a pilot study realized by Cascardo & Figueira, 1995 in the cacao-producing region of south Bahia, infected branches of cacao (dry brooms) were collected during the year of 1994. These branches were placed within infected brooms to induce the reproductive structure of the fungus. Multisporic cultures were cultivated in liquid LB medium for the later extraction of mycelial DNA and posterior analysis via RAPD. In this study, 156 loci RAPD were amplified using 10 primers. The genetic similarity was calculated using the DICE coefficient and a phenogram was constructed based on the UPGMA algorithm. The mean genetic similarity of the fungus isolates was 87%, varying from 60% to 100%. The isolates were grouped into two categories, according to the geographic origin and related to the supposed areas of independent introduction.

An epidemiological monitoring of the disease done in the subsequent years after its discovery revealed that two foci concentrated radially around the sites of introduction of Uruçuca and Camacã in 1991. The disease was later disseminated from the initial sites; an overlapping of the two foci occurred in the region that encompasses the cities of Itabuna and Ilhéus. Comparing the geographic origin of isolates collected in these areas, a clear tendency of grouping of the same in relation to the presence or absence of determined polymorphic bands was observed, noticing also that some isolates of the intersection region presented bands in common.

These results suggest that two independent introductions may have occurred and that genetic recombination could be occurring in the intersection area. This hypothesis was confirmed since in other trials where the DNA of monosporic cultures originated from a single basidiocarp was analyzed, a high variability in the electrophoretic pattern of the RAPD bands was observed, suggesting genetic segregation. Multisporic cultures were therefore chosen to work with. It is known that in the PCR reaction the more frequent alleles are preferentially amplified in detriment of the less frequent ones. Consequently, a DNA bulk of a multisporic cultures would be more representative for the predominant alleles in that population. The other measure taken was to separate the cultures of the basidiocarps individually since in a same broom there may be basidiocarps from different origins.

In a later study, when comparing these isolates together with other isolates from the Brazilian Amazon region it was confirmed that the infection foci were originated from two distinct introductions and the possible origin of the same was traced. In this study, specific polymorphic bands were also observed for a particular isolate group. Some isolates of the south were grouped with isolates of the north. This indicates that the attempt of eradication of the initial focus in the region of Uruçuca may have led to a reduction of the inoculum level of the same, contrary to the focus of Camacã that spread out quickly. The phenogram shows that this tendency was also observed in the study realized by Cascardo & Figueira, 1995 where some isolates of the southern region grouped together with isolates from the north.

These results demonstrate the potential of the use of molecular markers to evaluate the genetic diversity and pathogen gene flow. The studies, together with the work of identifying molecular markers of the pathogen associated with its virulence can contribute significantly to improvement programmes. This relation between the pathogen and its resistance response in cacao has already been demonstrated. For future perspectives, the RAPD analysis permits an evaluation of the gene flow and the genetic diversity of the pathogen throughout the years.

Other cacao pathosystems were also studied using molecular markers. Nyassé *et al.*, 2000, evaluated the genetic diversity of *Phytophthora megakarya* that infected cacao in Central and West Africa. A separation between two large groups, that of West and the other of Central Africa was found. In India, AFLP markers and ITS sequences were successfully used to differentiate various *Phytophtora* species found in intercropped cacao and coconut plantations.

Chapter 6

MOLECULAR BIOLOGY FOR CONTROL OF MYCOTOXIGENIC FUNGI

MYCOTOXIN

A mycotoxin is a toxic secondary metabolite produced by organisms of the fungi kingdom, commonly known as molds. The term 'mycotoxin' is usually reserved for the toxic chemical products produced by fungi that readily colonize crops. One mold species may produce many different mycotoxins, and the same mycotoxin may be produced by several species.

Overview

Most fungi are aerobic (use oxygen) and are found almost everywhere in extremely small quantities due to the minute size of their spores. They consume organic matter wherever humidity and temperature are sufficient. Where conditions are right, fungi proliferate into colonies and mycotoxin levels become high. The reason for the production of mycotoxins is not yet known; they are not necessary for the growth or the development of the fungi. Because mycotoxins weaken the receiving host, the fungus may use them as a strategy to better the environment for further fungal proliferation.

The production of toxins depends on the surrounding intrinsic and extrinsic environments and the toxins vary greatly in their severity, depending on the organism infected and its susceptibility, metabolism, and defense mechanisms. Some of the health effects found in animals and humans include death, identifiable diseases or health problems, weakened immune systems without specificity to a toxin, and as allergens or irritants. Some mycotoxins are harmful to other micro-organisms such as other fungi or even bacteria; penicillin is one example. It has been suggested that mycotoxins in stored animal feed are the cause of rare phenotypical sex changes in hens that causes them to look and act male.

Mycotoxins can appear in the food chain as a result of fungal infection of crops, either by being eaten directly by humans or by being used as livestock feed. Mycotoxins greatly resist decomposition or being broken down in digestion, so they remain in the food chain in meat and dairy products. Even temperature treatments, such as cooking and freezing, do not destroy some mycotoxins.

Although various wild mushrooms contain an assortment of poisons that are definitely fungal metabolites causing noteworthy health problems for humans, they are rather arbitrarily excluded from discussions of mycotoxicology. In such cases the distinction is based on the size of the producing fungus and human intention. Mycotoxin exposure is almost always accidental whereas with mushrooms improper identification and ingestion causing mushroom poisoning is commonly the case. Ingestion of misidentified mushrooms containing mycotoxins may result in hallucinations. The cyclopeptide-producing *Amanita phalloides* is well known for its toxic potential and is responsible for approximately 90% of all mushroom fatalities. The other primary mycotoxin groups found in mushrooms include: orellanine, monomethylhydrazine, disulfiram-like, hallucinogenic indoles, muscarinic, isoxazole, and gastrointestinal (GI)-specific irritants. The bulk of this article is about mycotoxins that are found in microfungi other than poisons from mushrooms or macroscopic fungi.

Many international agencies are trying to achieve universal standardization of regulatory limits for mycotoxins. Currently, over 100 countries have regulations regarding mycotoxins in the feed industry, in which 13 mycotoxins or groups of mycotoxins are of concern. The process of assessing a need for mycotoxin regulation includes a wide array of in-laboratory testing that includes extracting, clean-up and separation techniques. Most official regulations and control methods are based on high-performance liquid techniques (*e.g.*, HPLC) through international bodies. It is implied that any regulations regarding these toxins will be in co-ordinance with any other countries with which a trade agreement exists. Many of the standards for the method performance analysis for mycotoxins is set by the European Committee for Standardization (CEN). However, one must take note that scientific risk assessment is commonly influenced by culture and politics, which, in turn, will affect trade regulations of mycotoxins.

Food-based mycotoxins were studied extensively worldwide throughout the 20th century. In Europe, statutory levels of a range of mycotoxins permitted in food and animal feed are set by a range of European directives and Commission regulations. The U.S. Food and Drug Administration has regulated and enforced limits on concentrations of mycotoxins in foods and feed industries since 1985. It is through various compliance programs that the FDA monitors these industries to guarantee that mycotoxins are kept at a practical level. These compliance programs sample food products including peanuts and peanut products, tree nuts, corn and corn products, cottonseed, and milk. There is still a lack of sufficient surveillance data on some mycotoxins that occur in the U.S., which is due largely to the lack of reliable analytical methods.

Major Groups

Aflatoxins are a type of mycotoxin produced by *Aspergillus* species of fungi, such as *A. flavus* and *A. parasiticus*. The umbrella term aflatoxin refers to four different types of mycotoxins produced, which are B_1, B_2, G_1, and G_2. Aflatoxin B_1, the most toxic, is a potent carcinogen and has been directly correlated to adverse health effects, such asliver cancer, in many animal species. Aflatoxins are largely associated with commodities produced in the tropics and subtropics, such as cotton, peanuts, spices, pistachios and maize.

Ochratoxin is a mycotoxin that comes in three secondary metabolite forms, A, B, and C. All are produced by *Penicillium* and *Aspergillus* species. The three forms differ in that Ochratoxin B (OTB) is a nonchlorinated form of Ochratoxin A (OTA) and that Ochratoxin C (OTC) is an ethyl ester form Ochratoxin A. *Aspergillus ochraceus* is found as acontaminant of a wide range of commodities including beverages such as beer and wine. *Aspergillus carbonarius* is the main species found on vine fruit, which releases its toxin during the juice making process. OTA has been labeled as a carcinogen and a nephrotoxin, and has been linked to tumors in the human urinary tract, although research in humans is limited by confounding factors.

Citrinin is a toxin that was first isolated from *Penicillium citrinum*, but has been identified in over a dozen species of *Penicillium* and several species of *Aspergillus*. Some of these species are used to produce human foodstuffs such as cheese (*Penicillium camemberti*), sake, miso, and soy sauce. Citrinin is associated with yellowed ricedisease in Japan and acts as a nephrotoxin in all animal species tested. Although it is associated with many human foods (wheat, rice, corn, barley, oats, rye, and food colored with Monascus pigment) its full significance for human health is unknown. Citrinin can also act synergistically with Ochratoxin A to depress RNA synthesis in murine kidneys.

Ergot Alkaloids are compounds produced as a toxic mixture of alkaloids in the sclerotia of species of *Claviceps*, which are common pathogens of various grass species. The ingestion of ergot sclerotia from infected cereals, commonly in the form of bread produced from contaminated flour, cause ergotism the human disease historically known as St. Anthony's Fire. There are two forms of ergotism: gangrenous, affecting blood supply to extremities, and convulsive, affecting the central nervous system. Modern methods of grain cleaning have significantly reduced ergotism as a human disease, however it is still an important veterinary problem. Ergot alkaloids have been used pharmaceutically.

Patulin is a toxin produced by the *P. expansum*, *Aspergillus*, *Penicillium*, and *Paecilomyces* fungal species. *P. expansum* is especially associated with a range of moldy fruits and vegetables, in particular rotting apples and figs. It is destroyed by the fermentation process and so is not found in apple beverages, such as cider. Although patulin has not been shown to be carcinogenic, it has been reported to damage the immune system in animals. In 2004, the European

Community set limits to the concentrations of patulin in food products. They currently stand at 50 μg/kg in all fruit juice concentrations, at 25 μg/kg in solid apple products used for direct consumption, and at 10 μg/kg for children's apple products, including apple juice.

Fusarium toxins are produced by over 50 species of *Fusarium* and have a history of infecting the grain of developing cereals such as wheat and maize. They include a range of mycotoxins, such as: the fumonisins, which affect the nervous systems of horses and may cause cancer in rodents; the trichothecenes, which are most strongly associated with chronic and fatal toxic effects in animals and humans; and zearalenone, which is not correlated to any fatal toxic effects in animals or humans. Some of the other major types of *Fusarium* toxins include: beauvercin and enniatins, butenolide, equisetin, and fusarins.

Binding Agents and Deactivators

In the feed and food industry it has become common practice to add mycotoxin binding agents such as Montmorillonite or bentonite clay in order to affectively adsorb the mycotoxins. To reverse the adverse effects of mycotoxins, the following criteria are used to evaluate the functionality of any binding additive:

- Efficacy of active component verified by scientific data
- A low effective inclusion rate
- Stability over a wide pH range
- High capacity to absorb high concentrations of mycotoxins
- High affinity to absorb low concentrations of mycotoxins
- Affirmation of chemical interaction between mycotoxin and adsorbent
- Proven *in vivo* data with all major mycotoxins
- Non-toxic, environmentally friendly component

Since not all mycotoxins can be bound to such agents, the latest approach to mycotoxin control is mycotoxin deactivation. By means of enzymes (esterase, de-epoxidase), yeast(*Trichosporon mycotoxinvorans*) or bacterial strains , mycotoxins can be reduced during pre-harvesting contamination. Other removal methods include physical separation, washing, milling, heat-treatment, radiation, extraction with solvents, and the use of chemical or biological agents. Irradiation methods have proven to be effective treatment against mold growth and toxin production.

In the Indoor Environment

Buildings are another source of mycotoxins and people living or working in areas with mold increase their chances of adverse health effects. Molds growing in buildings can be divided into three groups — primary, secondary, and tertiary colonizers. Each group is categorized by the ability to grow at a certain water activity requirement. It has become difficult to identify mycotoxin production by indoor molds for many variables, such as:

(i) they may be masked as derivatives;

(ii) they are poorly documented; and

(iii) the fact that they are likely to produce different metabolites on building materials.

Some of the mycotoxins in the indoor environment are produced by *Alternaria, Aspergillus* (multiple forms),*Penicillium,* and *Stachybotrys. Stachybotrys chartarum* contains a higher number of mycotoxins than other molds grown in the indoor environment and has been associated with allergies and respiratory inflammation. The infestation of *S. chartarum* in buildings containing gypsum board, as well as on ceiling tiles, is very common and has recently become a more recognized problem. When gypsum board has been repeatedly introduced to moisture *S. chartarum* grows readily on its cellulose face. This stresses the importance of moisture controls and ventilation within residential homes and other buildings. The negative health effects of mycotoxins are a function of the concentration, the duration of exposure and the subject's sensitivities. The concentrations experienced in a normal home, office or school are often too low to trigger a health response in occupants.

In the 1990s, public concern over mycotoxins increased following multimillion-dollar toxic mold settlements. The lawsuits took place after the Center for Disease Control (CDC) did a study in Cleveland, Ohio and claimed that there was an association between mycotoxins from *Stachybotrys* spores and pulmonary hemorrhage in infants. However, in 2000, based on internal and external reviews of their data, the CDC concluded that because of flaws in their methods, the association was not proven. *Stachybotrys* spores in animal studies have been shown to cause lung hemorrhaging, but only at very high concentrations.

One study by the Center of Integrative Toxicology at Michigan State University investigated the causes of Damp Building Related Illness (DBRI). They found that *Stachybotrys* is possibly an important contributing factor to DBRI. So far animal models indicate that airway exposure to *S. chartarum* can evoke allergic sensitization, inflammation, and cytotoxicity in the upper and lower respiratory tracts. Trichothecene toxicity appears to be an underlying cause of many of these adverse effects. Recent findings indicate that lower doses (studies usually involve high doses) can cause these symptoms.

Some toxicologists have used the Concentration of No Toxicological Concern (CoNTC) measure to represent the airborne concentration of mycotoxins that are expected to cause no hazard to humans (exposed continuously throughout a 70–yr lifetime). The resulting data of several studies have thus far demonstrated that common exposures to airborne mycotoxins in the built indoor environment are below the CoNTC, however agricultural environments have potential to produce levels greater than the CoNTC.

Human Health Effects

Mycotoxicosis is the term used for poisoning associated with exposures to mycotoxins. The symptoms of mycotoxicosis depend on the type of mycotoxin; the concentration and length of exposure; as well as age, health, and sex of the exposed individual. The synergistic effects associated with several other factors such as genetics, diet, and interactions with other toxins have been poorly studied. Therefore, it is possible that vitamin deficiency, caloric deprivation, alcohol abuse, and infectious disease status can all have confounding effects with mycotoxins. In turn, mycotoxins have the potential for both acute and chronic health effects via ingestion, skin contact, and inhalation. These toxins can enter the blood stream and lymphatic system; they inhibit protein synthesis, damage macrophage systems, inhibit particle clearance of the lung, and increase sensitivity to bacterial endotoxin.

In 2004 in Kenya, 125 people died and nearly 200 others were treated after eating aflatoxin-contaminated maize. The deaths were mainly associated with homegrown maize that had not been treated with fungicides or properly dried before storage. Due to food shortages at the time, farmers may have been harvesting maize earlier than normal to prevent thefts from their fields, so that the grain had not fully matured and was more susceptible to infection.

COMMON MYCOTOXIGENIC SPECIES OF FUSARIUM

Fusarium

Fusarium is one of the most important genera of plant pathogenic fungi on earth, with a record of devastating infections in many kinds of economically important plants. Fusarium species are responsible for wilts, blights, root rots and cankers in legumes, coffee, pine trees, wheat, corn, carnations and grasses. The importance of Fusarium species in the current context is that infection may sometimes occur in developing seeds, especially in cereals, and also in maturing fruits and vegetables. An immediate potential for toxin production in foods is apparent.

The very important role of Fusaria as mycotoxin producers appears to have remained largely unsuspected until the 1970s. Research has now unequivocally established the role of Fusaria as the cause of alimentary toxic aleukia (ATA). This was the previously mentioned human mycotoxicosis epidemic in the USSR which killed an estimated 100,000 people between 1942 and 1948. ATA is also known to have occurred in Russia in 1932 and 1913, and there is little doubt that outbreaks occurred in earlier years as well. Matossian (1981) has argued persuasively that ATA occurred in other countries, including England, in the 16th to 18th centuries at least.

Research since 1970 has shown that Fusarium species are capable of producing a bewildering array of mycotoxins. Foremost among these are the trichothecenes, of which at least 50 are known: the majority are produced by Fusaria. The most notorious is T-2 toxin, which was responsible for ATA. Other Fusarium mycotoxins are known to be highly toxic to animals, and are suspected to be responsible for acute and chronic human diseases also.

Taxonomy

The taxonomy of Fusarium was in disarray until recently, with several competing taxonomic schemes, recognising from 9 to 60 species in the genus. A determined attack on this problem by an international collaborative group has resolved most of the conflict, and the taxonomy of Nelson *et al.* (1983) has met with widespread approval. Nelson *et al.* (1983) accepted 30 species.

A direct consequence of confusion in taxonomy is confusion over species mycotoxin associations. Fusarium isolates producing a particular toxin have been given different names as a result of the different taxonomic systems used, or simply as a result of misidentification. Marasas *et al.* (1984) intensively studied more than 200 toxigenic Fusarium isolates, and provided accurate information on species identifications and the corresponding toxins produced. They listed 24 Fusarium species with confirmed toxigenicity. The four species judged to be most important from the viewpoint of human health, F. sporotrichioides, F. equiseti, F. graminerarum and F. moniliforme, are discussed below.

Enumeration

Growth of Fusarium species is favoured by dilute, high aw (water acturty) media. Enumeration of fusaria can be effectively carried out on PDA provided chloramphenicol or other broad spectrum antibiotics are added to suppress bacteria. Howerver, acidified PDA, a frequently used antibacterial medium, is not recommended because it may inhibit sensitive cells (King *et al.*, 1986). DCPA is an effective enumeration and isolation medium for most food-borne Fusarium species. Recognition of Fusarium colonies on any of these media requires careful observation and experience. Presumptive identification to genus level can usually be made from colony appearance: low to floccose colonies, coloured white, pink or purple, with pale to red or purple reverses, are indicative of Fusarium. Confirmation requires microscopical examination, where the crescent-shaped macroconidia characteristic of the genus should be observed. These are not always produced on enumeration media, especially PDA, however. Differentiation of some species on enumeration media is possible, but also requires experience.

Identification

Identification of Fusarium species is ideally carried from growth on carnation leaf agar, an effective medium for macroconidium production , but this medium is not readily available to the nonspecialist. Pitt and Hocking (1985a) provided keys and descriptions enabling identification of the species of interest in foods after growth on the readily available media CYA, MEA and PDA. Hocking and Andrews (1987) reported that DCPA, a medium which encourages production of macroconidia, is a practical alternative medium to carnation leaves for identifying most food-borne species.

As noted above, the current definitive taxonomy is that of Nelson *et al.* (1983). Burgess *et al.* (1988) have published a very useful and up to date guide to most species. Pitt and Hocking (1985a) give keys and descriptions of common food-borne species.

Toxins, Toxicity and Symptoms: General

Trichothecenes, the principal toxins produced by Fusaria, are sesquiterpenes with a basic 12,13-epoxytrichothec-9-ene ring system. Trichothecenes are often produced in mixtures even under pure culture conditions, and are very difficult to separate, so the toxicity of many of these compounds remains uncertain. Some are known to be highly toxic: none appear to be benign.

Because of the variable quality of the available data, trichothecene toxicity will not be considered in detail. However, as an example, the acute LD50 values for T-2 toxin are of the order of 8-4 mg/kg in rats, pigs and mice. LD50 values for the much less toxic deoxynivalenol have been reported as at least 70 mg/kh in mice; however, only 5 mg/kg in feed causes vomiting in pigs.

The biochemical basis of trichothecene toxicity is noncompetitive inhibition of protein synthesis. Points of attack at the molecular level differ: some trichothecenes attack initiation of protein synthesis, others inhibit elongation or termination. Differences in toxicity and symptoms result.

It is difficult indeed to catalogue the symptoms of trichothecene poisoning. Vomiting, diarrhoea, anorexia and gastro-intestinal inflammation are rapid responses which sometimes occur, but less immediate effects such as skin necrosis, leukopenia, ataxia, haemorrhaging of muscular tissue, and degeneration of nerve cells are all known. Mortalities may result not only from injection or oral ingestion, but sometimes also from topical application. With the possible exception of the hepatocarcinogenicity of the aflatoxins, trichothecenes constitute the greatest known mycotoxin threat to human and animal health, the more insidious because the symptoms are so variable.

Some other mycotoxins are also produced by Fusarium species: zearalenone, in fact an oestrogen rather than a true mycotoxin; moniliformin, which has a unique four carbon ring structure; fusaric acid, better known as a phytotoxin involved in plant pathogenicity; and, from obviously toxic isolates, toxic principles which have defied isolation or characterisation. The most notable of these, produced by F. moniliforme, eluded investigators for 15 years. It has recently been isolated, characterised, and named fumoinisin B.

Most Fusarium toxins have been shown to possess only acute toxicity. However, strong circumstantial evidence suggests that some may be involved in human cancer. Fumonis B. which is a bizarre molecule believed to cause leucoencephalomalasia in horses, is also reported to induce liver cancer in rats. Involvement of this compound in oesophageal cancer in southern Africa appears likely.

Finally, the possible implication of trichothecenes in the controversial "yellow rain" episode in Laos must be mentioned. The discussions of this controversy show strong political bias and the facts about the alleged use of mycotoxins as a biological warfare agent remain obscure. However, the fact that trichothecene toxins could have caused many of the reported symptoms is beyond dispute.

Distribution in Nature and in Foods

Fusarium species are primarily plant pathogens, and occur mostly in association with plants and cultivated soils. In many cases, particular plant species associations are known or can be predicted. Such associations will be described under the individual species which follow.

Unlike most Aspergillus and Penicillium species, Fusaria grow in crops before harvest, and grow only at high aw levels. Mycotoxins are therefore usually only produced before or immediately after harvest.

F. Sporotrichloides

In the years 1942-1948, at least 100,000 Russian people died from a mysterious epidemic. Illness and death occurred mostly but not exclusively in the Orenburg district near the Caspian Sea. In some localities, up to 60% of the population were affected, and up to 10% died. The disease, now called Alimentary Toxic Aleukia (ATA), has since been shown to have occurred in Russia twice previously in this century, in 1932 and 1913, and perhaps elsewhere in Europe, including England, in earlier centuries.

ATA is an exceptionally unpleasant disease. Symptoms include fever, haemorrhagic rash, bleeding from nose, throat and gums, necrotic angina, extreme leucopenia, agranulocytosis, sepsis and exhaustion of the bone marrow. These symptoms more closely resemble those of radiation sickness than bacterial or other fungal toxicoses.

The direct cause of ATA in Russia in the 1940s was consumption of bread and other cereal products made from grains which were left in fields over winter due to wartime labour shortages. This became clear about 1950. However the fact that ATA was a mycotoxicosis was not finally established until the mid1970s, when it was proved that T-2 toxin was produced by Fusarium sporotrichioides and the closely related species F. pose during growth in freezing and thawing cycles.

F. sporotrichioides has also been implicated in a variety of very serious animal diseases, including scabby grain intoxication and bean hull poisoning in Japan, mouldy corn toxicosis in the USA, and fescue foot in the USA, Australia and New Zealand.

Taxonomy

Fusarium sporotrichioides, classified in Fusarium Section Sporotrichiella, is now a well circumscribed species. However, many literature reports have misidentified this species as F. tricinctum sensu Snyder and no boldface currently accepted species. F. tricinctum sensu stricto is in fact a species of low toxicity. Much of this confusion has been rectified by Marasas *et al.* (1984).

Identification

Colonies of F. sporotrichioides grow rapidly on CYA, MEA, PDA or DCPA, are deep and floccose, with mycelium coloured pale pink or salmon, and reverse on PDA greyish rose to burgundy. Macroconidia and microconidia are abundant on DCPA: microconidia are both fusiform and pear-shaped, and are borne from phialides with more than one fertile pore.

Toxins and Toxicity

As well as T-2 toxin, some isolates of F sporotrichioides are known to produce butenolide, fusarenon-X, neosolaniol and nivalenol. Zearalenone, deoxynivalenol and some less well characterised trichothecenes occur less frequently.

The patterns of toxicity shown by isolates of this species depend on the relative production of these various toxins. All animal species studied are affected by them, in various ways.

Symptoms

T-2 toxin, produced by F. sporotrichioides, is the most important human food poison which can result from ingestion of mouldy grain. The symptoms of this intoxication have been described above.

Distribution in Nature and Food

Fortunately, Fusarium sporotrichioides is not a commonly occurring species. It is found mainly in temperate regions on cereal crops, although it has also been isolated from peanuts and soy beans. Indications are that T-2 production is favoured by growth at low temperatures, but experimental evidence remains incomplete.

F. equiseti

A broad spectrum plant pathogen and soil saprophyte of widespread distribution, F. equiseti has a long history of association with animal disease, and a possible implication in human leukaemia. Conclusive evidence of the latter is lacking. However, given its widespread distribution and the long list of mycotoxins produced , the potential for this species to cause human and animal disease cannot be ignored. Confusion over the name of this species in earlier years precludes any detailed discussion of its history.

Taxonomy

F. equiseti is classified by Nelson *et al.* (1983) in Fusarium Section Gibbosum. It has a teleomorph (sexual state) called Gibberella intricans, which has been recorded from nature quite rarely. Consequently, use of the Fusarium name for this species is to be preferred.

Much of the earlier literature on F. equiseti is located under the names F. roseum sensu Snyder and Hansen (1940), F. roseum 'Gibbosum', F. roseum 'Avenaceum', F. roseum 'Culmorum', or F. roseum 'equiseti' and several other distinct species as well. Hence it is now frequently impossible to determine which of the names used by earlier authors actually refers to F. equiseti.

Identification

Colonies of F. equiseti on CYA, MEA and PDA usually cover the whole Petri dish, while those on DCPA are smaller. Mycelium and colony reverses are not strongly coloured, but white or pale salmon to brown. Macroconidia are distinctly curved ("hunch backed"); microconidia are not produced.

Toxins and Toxicity

Mycotoxins produced by F. equiseti include nivalenol, fusarenon X, T-2, diacetoxyscirpenol, butenolide, zearalenone and several others less well characterised: an impressive list. Several of these may be produced simultaneously. Suzuki *et al.* (1980) reported production of nivalenol, diacetoxyscirpenol and fusarenon-X by 16 of 25 isolates of F. equiseti in Japan. T-2, zearalenone and butenolide occur less commonly.

Association of F. equiseti with human leukaemia has been reported, and diacetoxyscirpenol is suggested as a possible cause. F. equiseti was isolated from dust in a house in the United States where two people had developed leukaemia, and the isolate was capable of depressing the immune response of guinea pigs. Previous work by the same authors had suggested a Fusarium species was the cause of leukaemia in a U.S. house where four cases had occurred.

Associations have been reported linking F. equiseti with animal diseases such as degnala, a disease of buffalo eating rice straw in Pakistan and India, bean hull poisoning of horses in Japan, and tibial dyschondroplasia, a bone disease of poultry. Clear evidence of causation remains elusive in each case

Symptoms

As with other intoxications caused by Fusaria, symptoms of poisoning caused by F equiseti are very varied, reflecting the insidious nature of trichothecene toxicoses, the range of toxins produced by a single species and the proportion of each toxin formed under the influence of substrate, temperature and water activity.

F. equiseti has been reported from several types of grains, and no doubt toxins produced by this species are consumed in human food in many places from time to time. The wide range of diseases produced in animals, as indicated above, provide little clue to the recognition of symptoms in humans.

Distribution in Nature and Foods

A cosmopolitan soil fungus, F equiseti has a distribution extending from Alaska to the tropics. It has also been isolated from a wide variety of plants, where it causes stem and root rots in particular. F equiseti has been reported from a variety of cereal grains, especially maize and barley but relatively from other foods.

F. Graminearum

As with most Fusarium species, the history and toxicological importance of F graminearum is obscured by the confusion over its identity. Now that the taxonomic problems have been clarified, it is recognised that F. graminearum causes oestrogenic syndromes, feed refusal and emetic syndromes in pigs and sometimes other animals, and is very likely to be the cause of human Akakabi-byo (scabby grain intoxication) in Japan.

Taxonomy

In the classification of Nelson *et al.* (1983), F graminearum is placed in the Section Discolor. This species has a well recognised teleomorph (sexual state), Gibberella zeae. Literature references to G. zeae are frequent, and usually correct. Choice of the Fusarium or Gibberella name for this species ultimately depends on the individual author's preference, but should also reflect the predominant state being isolated or studied.

Two quite distinct biotypes of this species have been encountered in Australia. F. graminearum Group I, the cause of crown rot of wheat, produces the Gibberella state only when isolates are mated in compatible pairs. However, F. graminearum Group II, which causes diseases in aerial parts and grains of cereals, readily forms the Gibberella state in single isolate (and single spore) culture.

As with F. equiseti, the name F. roseum sensu Snyder and Hansen (1940) has caused great confusion in the literature related to F. graminearum.

Identification

Colonies of F. graminearum fill the Petri dish when grown on CYA, MEA or PDA for 7 days; colonies on DCPA are somewhat smaller. On CYA and MEA, colonies are floccose, in muted or pastel shades of greyish red or yellow. On PDA, colonies are usually highly coloured, with dense to floccose greyish rose to golden brown mycelium and a dark ruby reverse, while on DCPA, colony appearance is dominated by salmon to orange clusters of macroconidia, often in concentric rings. Macroconidia are relatively straight and thick walled, with a foot-shaped basal cell. Microconidia are not produced.

Toxins, Toxicity and Symptoms

The principal toxins produced by F. graminearum are well defined: deoxynivalenol (DON; also known as vomitoxin), nivalenol ad zearalenone. Reports of the occurrence of diacetoxyscirpenol, fusarenon-X and butenolide are accepted by Marasas *et al.* (1984), but the frequency of production appears to be much lower. Production of T-2 remains equivocal. Some minor toxins are also produced.

The toxicity of the major F: graminearum toxins is undoubted, but until recently the picture has been clouded by the fact that some isolates identified as this species can produce T-2. Even low levels of this very toxic compound, difficult to separate out or often even to detect, can cause symptoms which have been wrongly attributed to the other toxins. The recent production of gram quantities of pure DON will shortly result in much more accurate toxicity studies.

One point is clear: DON causes vomiting and feed refusal in pigs at levels near 5 mg/kg of feed. Although very low limits have been set for DON in human foods in the USA, Canada and Japan, its toxicity to species other than pigs remains to be defined, and appears unlikely to be high.

The oestrogenic effect of zearalenone in animals is a well defined syndrome. Corn, barley and wheat grains infected with F. graminearum and producing zearalenone cause genital problems in domestic animals, especially pigs. Symptoms include hyperemia and edematous swelling of the vulva in prepubertal gilts, or in more severe cases prolapse of the vagina and rectum. Reproductive disorders in sows include infertility, foetal resorption or mummification, abortions, reduced litter size and small piglets. Male pigs are also affected: atrophy of testes, decreased libido and hypertrophy of the mammary glands are all well documented.

At present, F. graminearum isolates which produce nivalenol are known only from Japan, and the significance of this mycotoxin in the environment is not clear. However, in Japan, sporadic epiphytotics of "akakabi-byo" (red mould disease) occur, most probably due to the common occurrence of F. graminearum on wheat, barley, oats, rye and rice in Japan. Symptoms include anorexia, nausea, vomiting, headache, abdominal pain, diarrhoea, chills, giddiness and convulsions. Deoxynivalenol and zearalenone appear unlikely to be the prime causes of this range of symptoms: nivalenol is a more likely candidate.

Metabolites of F. graminearum enter the human diet through cereal consumption in other parts of the world also. Deoxynivalenol, nivalenol and zearalenone have all been reported from corn, corn meal and other corn products, wheat and breakfast cereals in the USA, Canada and Africa. Possible effects in humans remain undefined.

Distribution in Nature and Food

F. graminearum is primarily a pathogen of gramineous plants, particularly wheat, causing crown rot at the base of the stem, and head scab in developing grain. It also causes cob rot of corn in many countries, including the wetter areas of Europe, North America, Africa and Australia It is uncommon in other situations, or other foods.

F. Moniliforme

Fusarium moniliforme Sheldon, also known as verticillioides (Sacc.) Nirenberg, was described more than a century ago as a species occurring on corn. Reports of its possible involvement in human or animal disease date back almost as far, coming from Italy, Russia and the United States by 1904.

The only animal disease for which the causal role of F. moniliforme has been established beyond doubt is the disease of horses and related animals known as equine leukoencephalomalacia (LEM). This disease was known as early as 1850 in the corn belts in the United States, with epidemics involving hundreds or thousands of horses in 1900, the 1930s and as recently as 197879. It also occurs in other parts of the world, including Argentina, China, Egypt, New Caledonia and South Africa. However, F. moniliforme was not positively identified as the cause of LEM until 1971.

F. moniliforme has been suggested to be the cause of a variety of other animal diseases, including bean hull poisoning of horses in Japan, abnormal bone development or rickets in chickens and pigs in the U.S.A., France and Germany, and a toxicosis due to mouldy sweet potatoes in the U.S.A..

The high rate of human oesophageal cancer which occurs in some parts of Transkei in southern Africa appears to be associated with corn consumption, ad perhaps, therefore, with F. moniliforme.

Taxonomy

Unlike other species considered here, F. moniliforme has been recognised as a distinct entity for many years. Disagreement still exists over the correct name: the name F. verticillioides (Sacc.) Nirenberg undoubtedly has nomenclatural priority, but has not been accepted by Nelson *et al.* (1983), their grounds being that the provisions of the International Code of Botanical Nomenclature are difficult to apply to Fusarium species in the absence of type material. Eventual conservation of F. moniliforme seems likely.

Identification

Colonies of F. moniliforme on PDA are white, sometimes tinged with purple. Macroconidia vary from slightly sickle shaped to almost straight. Microconidia are abundant, and are usually single celled, ellipsoidal to clavate with a flattened base, and formed in long chains.

Toxins and Toxicity

Despite early warnings that F. moniliforme was a highly toxic fungus, the road to understanding of the toxins responsible has been long and difficult. The most important toxin produced by this species is undoubtedly fumonisin B. a mycotoxin only characterised very recently. Fumonisin B is a bizarre molecule, consisting of a 20 carbon aliphatic chain with two ester-linked hydrophilic side chains.

F. moniliforme growing in corn is known to be responsible for LEM, a brain disease of horses. LEM has been as a serious problem in the United States corn belt for more than 100 years, and has caused the deaths of thousands of horses. Recent work indicates that fumonisins are the toxins most likely to be responsible for LEM.

The effect of fumonisins on humans is not known, but the fact that fumonisin B can induce cancer in rats suggests that this toxin may have a role in human oesophageal cancer. Corn is the major staple food in areas of the Transkei where oesophageal cancer is endemic, and the most striking difference between areas of low and high incidence was the much greater infection of corn by F. moniliforme in the high incidence areas. Kriek et al (1981) showed that several isolates of F. moniliforme from high incidence areas were acutely toxic to ducklings, but did not produce other known toxins such as moniliformin. The discovery of the fumonisins should help in the elucidation of the role of moniliforme in human oesophageal cancer.

Because of the intense interest over the past two decades in the toxicity of F. moniliforme which has recently culminated in the discovery of the fumonisins, other toxins produced by this species have been carefully studied. A few isolates of moniliforme produces moniliformin, a compound which is known to be toxic, but which lacks a known disease role. Other known compounds of greater or less toxicity include fusaric acid, fusarins and fusariocins. The production of T-2 toxin, diacetoxyscirpenol and zearalenone by F. moniliforme have been reported, but are regarded as unlikely by Marasas *et al.* (1984).

The known toxicity of F. moniliforme to a wide variety of animals, and its probable role in human oesophageal cancer, may well result from the production of the newly discovered fumonisins. However, the diversity of the demonstrated toxicity of authentic isolates of F. moniliforme to a wide variety of animals is such that the possibility of other potent toxins cannot be ruled out.

Symptoms

Symptoms of F. moniliforme poisoning vary widelywith animal type, dosage and toxigenic fungal isolate. The best defined disease produced by F. moniliforme, LEM, is characterised by liquefactive necrotic lesions in the white matter of the cerebral hemispheres of horses and other equine species. Marked neurotoxicity is evident, with aimless walking and loss of muscle control followed by death, which usually occurs about 2 weeks after toxin ingestion. In baboons, F. moniliforme toxicity has been shown to lead to heart failure. Chickens and ducklings are sensitive to feed containing F. moniliforme, and there is some evidence that the toxin responsible is moniliformin.

The toxicity of one F. moniliforme isolate to rats was characterised by cirrhosis and hyperplasia in the liver, and thrombosis in the heart and other organs. Of 20 other isolates, 15 showed mortalities and some symptoms, while 5 were nontoxic.

F. moniliforme may also be involved in abnormal bone development and diseases similar to rickets in chickens, sometimes with high mortalities.

The principal human disease with which F. moniliforme may be associated is oesophageal cancer, which has an abnormally high prevalence in the Transkei and in Henam Province, China. However, direct evidence that fumonisins or other known F. moniliforme toxins are causally related to this disease remain lacking.

Distribution in Nature and Foods

Like other Fusarium species, F. moniliforme is primarily a plant pathogen, causing both stalk and cob rot of corn, and diseases in rice, sorghum, sugar cane and other Graminae. It is much more common in the tropics than temperate zones. By far its most common source in foods is corn, both under field and storage conditions, but it has also been isolated from nuts, yams and occasionally other commodities.

BIOLOGICAL CONTROL OF PATHOGENIC FUNGI AND OOMYCETES

Rest in biological control of pathogens by antagonistic bacteria has been stimulated in recent years by trends in agriculture and horticulture towards greater sustainability and public concern about the use of hazardous pesticides. The overall aim of this research topic is to identify antagonistic bacteria in diverse agro-ecosystems and aquaculture to exploit their potential to control economically important pathogens of plants and fish.

In a project funded by the Vietnamese Government (MOET), we are developing new methods to control soil-borne diseases of groundnut. The project specifically focuses on biological control of *Sclerotium rolfsii*, causal agent of root and pod rot. Beneficial bacteria indigenous to the groundnut agro-ecosystem are isolated, characterized and tested for their activity against *S. rolfsii*, both in small-scale greenhouse assays as well as field trials.

Black pod disease of cacao (*Theobroma cacao*) causes substantial yield worldwide and it is the main limiting factor in cocoa production. The aim of this project is to study the genetic diversity of *Phytophthora*, causal agent of the disease, and to explore and exploit the diversity of beneficial microorganisms in the cocoa ecosystem to develop novel strategies to manage black pod disease. The goal is to develop an integrated control strategy by combining selected antagonistic microorganisms, phytosanitary measures and fungicides.

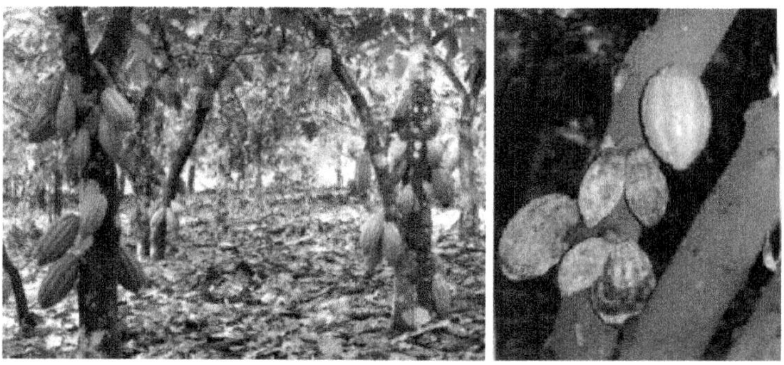

Fig. : Cocoa field in Ghana. Healthy cocoa pods (left), Phytophthora megakarya infected prods (right).

In a project funded by the European Union (EU), the research focus is to develop novel and sustainable measures to control *Saprolegnia parasitica*, an important oomycete pathogen of fish. In this project (www.abdn.ac.uk/oomycetes/SAPRO.html), the ecological, molecular, immunological and biochemical factors/traits underlying pathogen-host interactions are investigated. The objectives of our sub-project are to:

i) isolate and characterize beneficial bacteria from fish or fish eggs, and

ii) identify bacterial metabolites with activity against Saprolegnia.

The goal is to develop sustainable control strategies for *Saprolegniosis* that will be tested and validated by the industrial partners within the proposed time frame of the program. In a VIDI-project with the Phytophthora Group, the phenotypic and molecular responses of the late blight pathogen *Phytophthora infestans*to metabolites produced by beneficial bacteria are investigated. Microarray-based analyses of *Phytophthora* exposed to cyclic lipopeptide surfactants has resulted in the identification of a number of differentially regulated genes. The role of these genes in the biology and interaction of *Phytophthora* with the host plant are being analysed to identify new molecular targets for pathogen control.

BIOLOGY AND CONTROL OF TAKE-ALL DISEASE

The take-all fungus, *Gaeumannomyces graminis* (ascomycota), is a major root-rot pathogen of cereals and grasses. It is most damaging to intensively grown wheat and barley crops, when the same crop is grown year after year in a site. It survives in the infected residues of one crop, then invades the roots of thc following crop, progressively destroying the root system. In exceptional cases it can kill the whole crop; hence the name "take-all".

G. graminis also causes a patch disease of turf grasses, especially of *Agrostis* species (the "bent grasses") which are used in high-quality turf such as on bowling greens and golf-course greens.

Here we review briefly the biology of this pathogen, but focus mainly on two remarkable biological control systems that operate in natural field conditions and which can be exploited commercially. One of these involves the actions of antibiotic-producing fluorescent pseudomonads; the other involves competition from weakly parasitic fungi.

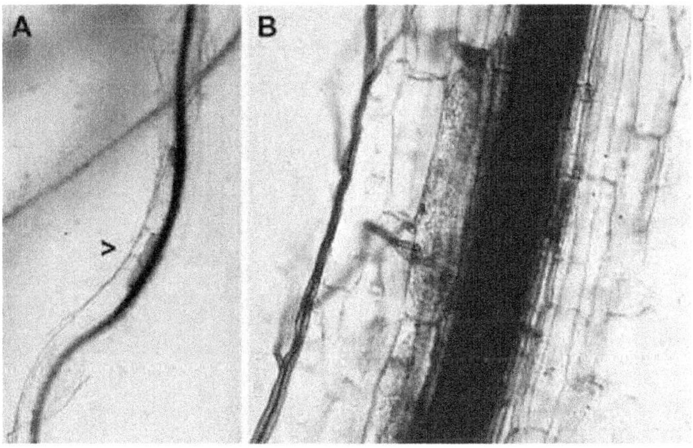

Figure A, low power micrograph of a wheat root infected by *G. graminis*, showing conspicuous vascular discoloration and dark runner hyphae of the fungus on the root surface. Figure B, close-up of the region shown as arrowhead in Fig. A, showing invasion of the root cortex, some browning reactions in the inner cortex and phloem, and intense blockage of the xylem by darkly pigmented vascular gels.

The Fungus and the Infection Process

The species *Gaeumannomyces graminis* is sub-divided into three varieties, which differ in minor morphological features but, more importantly, differ in pathogenicity:

- *G. graminis* var. ***tritici*** is an aggressive pathogen of wheat and barley crops

- *G. graminis* var. ***avenae*** is a pathogen of oats (*Avena* species) and of turf grasses
- *G. graminis* var. ***graminis*** is usually a weak parasite, found on maize and rhizomatous or stoloniferous grasses.

All these fungi are specialised parasites of the grass family (Gramineae), and they grow along the surface of roots as darkly pigmented runner hyphae. Branches that arise from these runner hyphae penetrate the root cortex at intervals, but their ability to infect depends on two factors - first, the nutrient reserves of the fungus (*e.g.* size of the organic food base from which it infects) and, second, the degree of host tissue resistance. In a successful infection by a pathogenic strain , the invading hyphae grow through the root cortex, invade and destroy the root phloem (sugar-transport system) and then invade the xylem (water-transport system), causing discoloration and blockage of the xylem vessels. The root is effectively cut off and decays below this point. The fungus then spreads progressively upwards towards the stem base, by producing runner hyphae and invading from these. Eventually it can colonise the stem base and kill the plant. Then the fungus survives as a saprotroph in the infected roots and stem bases or, in appropriate conditions, it can produce its sexual stage - a flask-shaped perithecium from which many elongated ascospores are ejected forcibly to disperse the fungus.

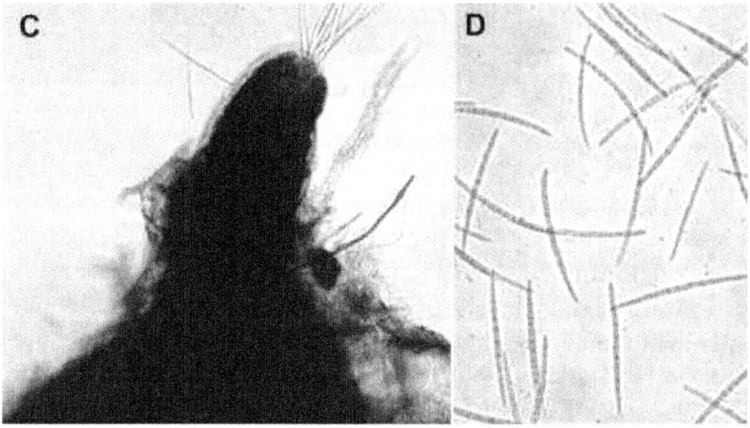

Figure C, the neck region of the perithecium of *G. graminis*, showing needle-shaped ascospores emerging from the apical pore where the perithecium was crushed.Figure D, individual ascospores, each about 100 micrometres long and with several septa (cross walls).

Take-all Patch Disease of Turf Grasses

Take-all patch disease of turf grasses is caused by *G. graminis* var. *avenae*. However, most grass species are quite resistant to take-all infection, and only the "bent grasses" (*Agrostis* species) are seriously affected. Even these grasses are seldom diseased in field sites, although they are extremely susceptible to attack in glasshouse tests. The reason is that they are protected by a natural biocontrol system, and disease occurs only when the natural biocontrol agents are eliminated by management practices.

Specifically, the take-all patch disease is seen when:

- grass is resown in sites that have been fumigated to destroy weeds, pathogens or insect pests,
- the existing turf is limed heavily to correct for excessive acidity (pH 4.5 or less) caused by repeated applications of ammonium fertilisers (H^+ ions are released when ammonium is taken up by roots, leading to progressive lowering of the soil pH),
- grass is sown in sites with a low resident microbial population - for example, in the "sand carpet" method of producing golf-course greens, where grass is sown on a 15 cm depth of sand over gravel.

In all these circumstances, the pathogen infects from air-borne ascospores, which become washed into the soil and initiate infection of roots. In all normal soils the ascospores do not have sufficient nutrient resources to infect roots.

This infection from air-borne ascospores explains why the disease occurs in patches, each representing a separate focus of infection. But even these disease patches only spread for two or three years before they become quiescent, probably because the normal biocontrol agents of grass roots become re-established and halt the spread of disease.

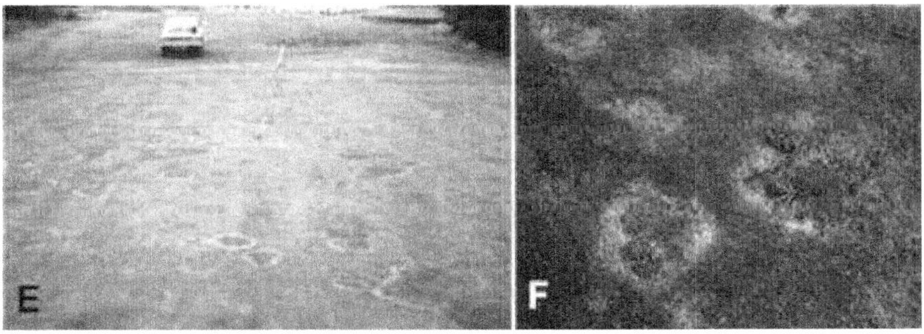

Take-all patch disease of bent-grass turf (*Agrostis* species). Figure E shows part of a golf-course green with numerous spreading disease patches. The grass has been killed at the advancing (white) margins of these patches. Figure F shows a close-up of disease patches, where the fine turf (*Agrostis* spp.) has been killed and the centres of the patches have been invaded by coarser-leaved grasses and weeds that affect the playing qualities of the turf.

The Natural Control Agents

Several fungi and bacteria might contribute to biocontrol of take-all patch disease, but most attention has focused on the weakly parasitic fungi of grass roots. One of these fungi, *Phialophora graminicola,* is extremely common on the roots of turf grasses in Britain and continental Europe, and in glasshouse conditions it is a highly effective biocontrol agent of take-all. It grows on the surface of roots as dark runner hyphae , but it is only a weak parasite, unable to invade the living root cortical cells and, crucially, unable to penetrate the root endodermis - the layer of suberised cells that surrounds the phloem and xylem. Thus, this fungus does not enter the vascular system and does not cause disease. When its attempted penetration of living root cells is halted by host resistance factors, *Phialophora*produces characteristic groups of swollen, pigmented cells which provide a simple way of identifying its presence in roots.

G. graminis var *graminis* (another weak parasite) plays a similar role in controlling the take-all pathogen on grasses in Australia and other warm countries.

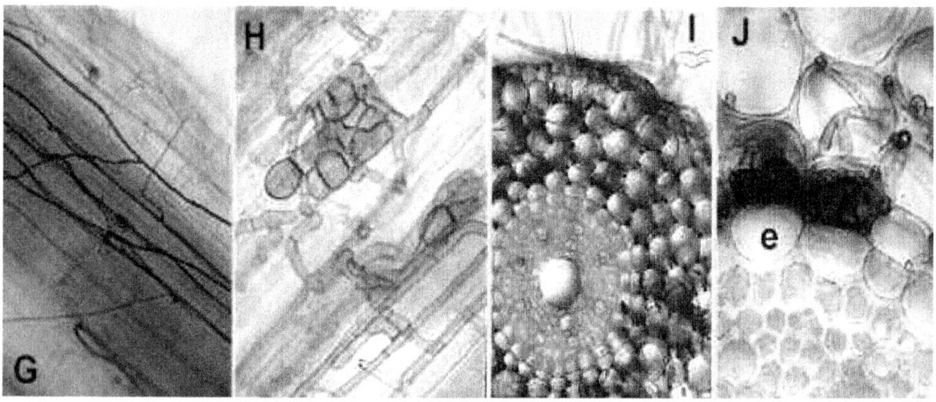

Figures G-J. *Phialophora graminicola,* a weak parasite of cereals and turf grasses. G, dark runner hyphae on the surface of a cereal or grass root. H, groups of darkly pigmented cells produced by the fungus when its attempted penetration of host cells is halted. I, cross section of an older root, in which the root cortical cells have died and been colonised by *P. graminicola,* but the fungus has been stopped at the root endodermis. J, close-up of part of Fig I, showing *Phialophora* hyphae in the root cortex (upper part of the Figure), but no invasion of the endodermis (e).

The involvement of these fungi helps to explain the specific conditions in which take-all patch disease occurs. For example, the table below shows that *Phialophoragraminicola* is abundant in turf of relatively high pH (pH 6 and above), but repeated applications of ammonium fertiliser can lower the turf pH to a level (about pH 4) where *Phialophora* is eliminated. At such low pH levels the grass itself begins to suffer, and when it is limed heavily to correct for this the take-all fungus can infect from ascospores. Eventually, *Phialophora* becomes re-established and prevents the disease patches from spreading further.

In short, take-all patch disease is caused by mismanagement of turf pH, and by other practices which fail to conserve the natural biocontrol agents.

Occurrence of *Phialophora graminicola* in cores of turf taken from golf-course greens of different pH

A wheat seed was placed in each core of turf, then the wheat seedling roots were examined microscopically for presence of *Phialophora* after 3 weeks in a glasshouse. Data from Deacon (1973).

Turf pH	% of cores with *Phialophora*	% of wheat assay roots with *Phialophora*
< 4.0	3	<1
4.0 - 4.5	6	3
4.5 - 5.0	27	21
5.0 - 5.5	75	25
5.5 - 6.0	67	48
> 6.0	100	100

Mode of Action of P. Graminicola

Phialophora and other weak parasites like *G. graminis* var *graminis* have no effect on the take-all fungus in laboratory culture - they do not damage or destroy the pathogen hyphae (compare this with the effects of *Pythium oligandrum* or *Phlebiopsis gigantea*) and they do not produce inhibitory compounds (compare this with*Agrobacterium radiobacter*). So there are only two possible mechanisms to explain their roles in biocontrol: they might activate the host defences against the pathogen or they might compete for resources - a form of biocontrol termed competitive niche exclusion.

There is strong evidence that competition plays at least some role. The weak parasites grow in the same way as the pathogen on the root surface (compare Figures A or B with Figure G) and, presumably, compete for the same resources. The key to understanding this came from the discovery that cereal and grass roots show an early natural senescence of the root cortex. It begins in the root epidermis soon after the root hairs die, and it progresses inwards with increasing distance (and age of the root region) behind the root tip. Meanwhile, the root tips continue to extend into fresh zones of soil, and the older root regions (with senescent cortex) have a predominantly transport function - the vascular tissues remain alive, protected by the living root endodermis.

Phialophora and other weak parasites are specifically adapted to exploit this natural root cortical death. They rapidly colonise the root cortical cells as these begin to senesce, and before the cells can be colonised by soil saprotrophs. In doing so, these fungi can prevent the invasion of *G. graminis* from ascospores, which have low nutrient reserves, because *G. graminis* then depends on the senescing cortical cells to build up its nutrient reserves for infection.

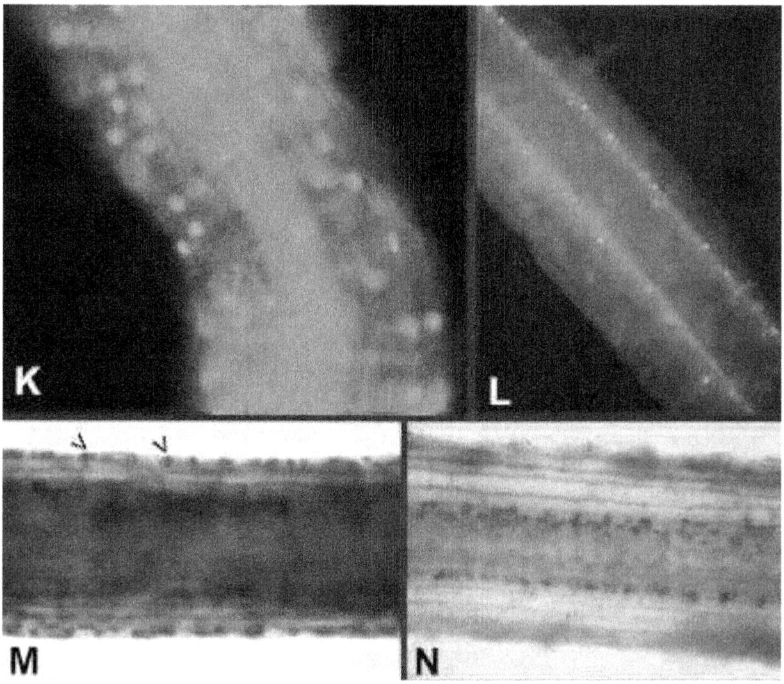

Figures K-M. Natural root cortical senescence of cereals and grasses. This natural senescence can be shown in several ways, including nuclear staining with acridine orange (K, L) and cytoplasmic staining with neutral red (M, N). Figure K. Part of a wheat root about 1 week old, stained with acridine orange and viewed by fluorescence microscopy. Nuclei are seen in all cell layers of the cortex, evidenced by fluorescence of acridine orange intercalated in the DNA. Fig. L. Part of a wheat root in a region 3 weeks old, showing nuclei in the inner cortex, but the outer cortical cell layers are anucleate and (right-hand side of root) have been shed. Fig. M. Young region of a wheat root treated with neutral red and strong salt solution. Neutral red stains the cytoplasm, while the salt solution plasmolyses the living cells, which are seen as contracted, densely stained protoplasts (arrowheads). Fig. N. Older region of a similar root, showing that living cells are present only in the inner cortex.

Take-all of Cereal Crops

A different type of biocontrol has been identified in continuously grown cereal crops. It is termed take-all decline or take-all suppressiveness. As shown in Figure O, the level of take-all infection builds up progressively in the first few years of cereal monocropping, because the inoculum in the residues of one crop serves as a food base for infection of the next. But the disease level reaches a peak in the third to fifth year and then declines spontaneously to a level that gives economically acceptable yields. Fluorescent pseudomonads are thought to be the main organisms responsible for this dramatic disease decline. Recent work implicates two specific sub-groups of fluorescent pseudomonad - those that produce phenazine-type antibiotics (Phe) and those that produce 2,4-diacetylphloroglucinol (Phl).

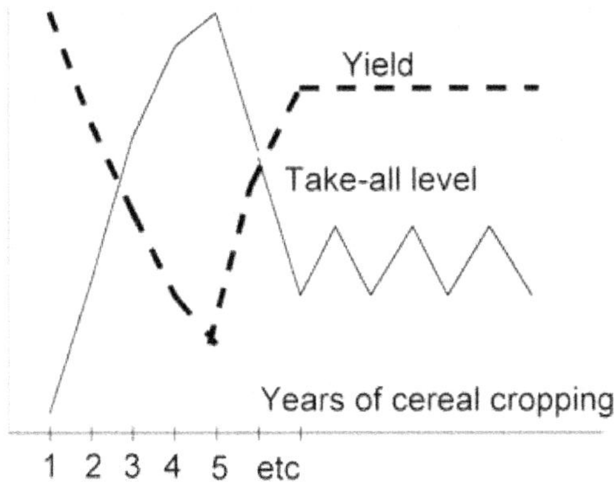

Figure O. Diagram to illustrate take-all decline in continuous cereal cropping. The disease level builds to a peak in the third to fifth year (depending on climatic and field conditions) then spontaneously declines to a lower, relatively stable level. Grain yields respond accordingly.

The populations of Phl strains, in particular, are markedly higher in suppressive than in non-suppressive soils, suggesting that these bacteria are favoured by the build-up of disease, which they subsequently help to control. The Phl strains have been grown in laboratory conditions and applied to wheat seeds which are then sown in soil. They colonise the emerging wheat roots, attaining high population levels, and the Phl antibiotic can be detected in the root zone by high-performance liquid chromatography. Also of interest, naturally occurring Phl-producing pseudomonads have been implicated in suppression of other root diseases in various crops in Europe and the USA. Consistent with this, 2,4-diacetylphloroglucinol is a broad-spectrum antibiotic (also toxic to plant roots if it is over-produced in the root zone). The gene coding for biosynthesis of this antibiotic has been identified in *Pseudomonas fluorescens* and used as a probe to show that it is a highly conserved gene among fluorescent *Pseudomonas* strains from many parts of the world.

The hope is that these antibiotic-producing bacteria might be applied commercially as seed coatings, to prevent or reduce take-all or other root diseases.

PREVENTION AND CONTROL OF MYCOTOXINS

Serveral mycotoxins in agricultural products cause health hazards to people and animals and economical problem. Dangerous mycotoxins are naturally present in foods, feeds and our environment. They are pathologically classified as hepatotoxins, nephrotoxins, vomitoxin and neuro-musculotoxin, some of which are potentially carcinogenic and mutagenic. Aflatoxin, for example, is the most potent hepatocarcinogen and mutagen among mycotoxins. Therefore, the contamination of mycotoxins should be minimized by designing a series of measures of prevention and control.

Strategies for Prevention and Control of Mycotoxins

To design strategies for the reduction or elimination of mycotoxins, knowledge about their fungal sources are needed. The growth of fungi in crops and agricultural products is the main cause of toxin formation and related to the concentration of the toxic substances. Many factors are involved in enhancing the formation of mycotoxins. They are plant susceptibility to fungi infestation, suitability of fungal substrate, temperate climate, moisture content and physical damage of seeds due to insects and pests.

Toxin-producing fungi may invade at pre-harvesting period, harvest-time, during post-harvest handling and in storage. According to the site where fungi infest grains, toxinogenic fungi can be divided into three groups:

(a) field fungi;

(b) storage fungi; and

(c) advanced deterioration fungi.

The first category includes species of plant pathogenic fungi, namely, genus Fusarium, *e.g.* F. moniliforme, F roseus, F. tricinctum and F. nivale. The "storage fungi" are principally the general Aspergillius and Penicillium, *e.g.* A. flavus and A. parasiticus. The "advanced deterioration fungi" normally do not infest intact grains but easily attack damaged ones and require high moisure content. The examples of the third group are A. clavatus, A. fumigatus, Chaetomium, Scopulariopsis, Rhizopus, Mucor, and Absidia.

The prevention of mycotoxins in our environment is a big task. In general, prevention of the contamination of fungi and their mycotoxins in agricultural commodities can be divided into these following three levels.

Primary Prevention

The step of prevention should be initially carried out before the fungal infestation and mycotoxin contamination. This level of prevention is the most important and effective plan for reducing fungal growth and mycotoxin production. Several practices have been recommended to keep the conditions unfavorable for any fungal growth. These include:

1. development of fungal resistant varieties of growing plants;
2. control field infection by fungi of planting crops;
3. making schedule for suitable pre-harvest, harvest and post-harvest;
4. lowering moisture content of plant seeds, after post harvesting and during storage;
5. Store commodities at low temperature whenever possible;
6. Using fungicides and preservatives against fungal growth;
7. Control insect infestation in stored bulk grains with approved insecticides.

Table : Some mycotoxins, their sources and potential toxicities.

Toxins	Producing fungi	Toxicities
Aflatoxin	*Aspergillus flavus*	Hepatocarcinogen
	Aspergillus parasiticus	and fatty liver
Citreoviridin	*Penicillium viridicatum*	Cardiac beri-beri
Citrinin	*Penicillium vindicatum*	Nephrotoxin
	Penicillium citrinum	
Cyclochlorotine	*Penicillium islandicum*	Hepatotoxin
Cytochalasin E	*Aspergillus clavatus*	Cytotoxicity
Maltoryzine	*Aspergillus oryzae*	
Ochratoxins	*Aspergillus ochraceus*	Hepatotoxin
Patulin	*Penicilliumc-expansum*	Brain & lung hemorrhage
	Penicillium patulum	and carcinogenicity
PR Toxin	*Penicillium requeforti*	
Rubratoxin	*Penicillium rubrum*	Liver hemmorrhage and fatty infiltration
Rugulosin	*Penicillium islandicum*	Nephrosis & liver damage
Sterigmatocystin	*Aspergillus flavus*	Hepatocarcinogen
	Aspergillus versicolor	
Tremorgens	*Penicillium and Aspergillus*	
Trichothecenes	*Fusarium graminearum*	Cytotoxicity
Vomitoxin (Deoxynivalenol)	*Fusarium graminearum*	Vomiting
Zearalenone	*Fusarium*	Hyper-estrogenic effect

Secondary Prevention

If the invasion of some fungi begins in commodities at early phase, this level of prevention will then be required. The existing toxigenic-fungi should be eliminated or its growth to be stopped to prevent further deterioration and mycotoxin contamination. Several measures are suggested as follows:

- Stop growth of infested fungi by re-drying the products;
- Removal of contaminated seeds;
- Inactivation or detoxification of mycotoxins contaminated;
- Protect stored products from any conditions which favour continuing fungal growth.

Tertiary Prevention

Once the products are heavily infested by toxic fungi, the primary and secondary preventions would not be then feasible. Any action would not be as effective as the practices mentioned above, since it will be quite late to completely stop toxic fungi and reduce their toxin formation. However, some measures should be done to prevent the transfer of fungi and their health hazardous toxins highly contaminated in products into our daily foods and environment. For example, peanut oil extracted from poor-graded peanut seeds always contains very high levels of aflatoxins and the oil-soluble toxin has to be eliminated by absorption and alkalinization during oilrefining process. Only a few practices are recommended:

- Complete destruction of the contaminated products;
- Detoxification or destruction of mycotoxins to the minimal level.

Since aflatoxin is the most well-known mycotoxin ever throughly studied and its prevention and control has been most successfully practiced in various countries, therefore, this paper will focus on such practices in certain detail for the prevention and control of aflatoxins mycotoxin contamination. Successful development will bring a great impact for the increased production of crops and safe and nutritious foods around the world. A number of researchers have been working on A. flavus-resistant or tolerant varieties of corn (2-3) and peanut (4-6)

It has been clear that the fungal-resistance of each variety is genotypic. However, the resistance to invasion of A. flavus has been attributed to several biochemical, environmental and physical factors. Uncontrollable factors could bring the failure in the utilization of selected fungal-resistant variety, as shown by laboratory screening, in the field.

Davis and his co-workers reported the survey and comparison of aflatoxin contamination in upto 215 corn hybrids grown in Alabama, USA during 1976-81. Unfortunately, they could not find any hybrid tested resistant to aflatoxin formation. They were convinced that significant aflatoxin levels generally accompanied stress caused by high temperature, low rainfall, low moisture-holding capacity of sandy soils and insect infestation.

A differential pathogenic capacity of various toxigenic strains of A. flavus have been observed. Some strains would require physical damage for their infestation and others would not. The association of mycotoxin production and physical damage to grain and drought during grain ripening indicates that Aspergillus spp. are weak pathogens. During long grain storage, the biochemical activity of grain is much reduced, while invasion of storage fungi and mycotoxin contamination would increase. More data is needed on the biochemistry and pathogenesis of toxigenic fungi to understand and evaluate their genotype.

The germination and viability of maize seeds could be affected by attack of Aspergillus and Penicillium species and their fungal infestation have been found to be different among maize genotypes.

Similarly, genotypes of peanut and biochemical properties of its seed such as tannin content , thin pericarp , small amount of cuticular wax and chemical composition of the pericarps and embryos have been shown to inhibit fungal invasion by A. flavus and aflatoxin formation.

Recently, antifungal enzymes, chitinase and B-1, 3-glucanase , found in a number of plant seeds, may act as defense against pathogenic fungi, since chitin and glucan are major polymeric components of many fungal cell walls. Such polysaccharides in fungal cell wall could be enzymically hydrolysed into smaller products resulting the damage or killing of fungal mycelia or spores. The role of these enzymes for genotype evaluation is now being studied. It is foreseen that seeds rich in such antifungal enzymes likely resist the infestation of fungi. If so, the seeds for breeding would be easily screened out and used a stock one.

Even there are many technical problems in searching for the "super" plant against pathogenicity, the development of fungal-resistant plant varieties utilizing genetic resistance to mycotoxin contamination is still possible and encouraged.

Fungal Growth Inhibition

How to prevent growth and invasion of pathogenic fungi in agricultural commodities is very important in preventing mycotoxin contamination. The inhibition of fungal growth can be achieved by physical, chemical and biological treatments.

- Physical treatment. After the crops have been harvested, drying and proper storage and suitable transportation of the commodities are of prime importance. Several flavourable factors contribute to the growth of fungi and aflatoxin production, namely high moisture content, humid climate, warm temperature (2540°C), insect infestation and pes damage. Many means and measures to prevention of fungal contamination have been emphasized and practically done.
- Drying seeds and commodities to the safe moisture levels (<9% for peanut kernel, and < 13.5% for corn).
- maintenance of the container or warehouse at low temperature and humidity.
- keep out insects and pests from the storage
- Gamma-irradiation of large-scale commodities.
- Chemical treatment with synthetic fungicides
- organic acids: acetic acid , propionic acid and butyric acid , malonic acid , benzoic acid , sorbic acid , lactic acid , citric acid and their sodium salts
- sodium chloride
- Benzoic acid derivatives : Onitrobenzoate, O-aminobenzoate, paminobenzoate, benzocain (ethly aminobenzoate), ethyl benzoate, methyl benzoate and aspirin (O-acetoxy benzoic acid)
- potassium sulfite and potassium fluoride
- dichorvos

- fumigant: ammonia and phosphine.
- treatment with natural products from plants or herbs.
- allicin and related substances from garlic and onion extracts
- chitosan or derivative of chitin isolated from crustacean shells
- cinnamon extract: trans-cinnamic acid, trans-cinnamaldehyde, and ferulic acid (phydroxy-3-methyl cinnamic acid)
- clove oil
- other herbs: thyme, star anise seeds , black and white peper. plumbago indica.

Decontamination of Mycotoxins

Contaminated mycotoxins in foods and feeds should be removed, inactivated or detoxified by physical, chemical and biological means depending on the conditions. However, the treatment has its own limitations, since the treated products should be healthsafe from the chemicals used and their essential nutritive value should not be deteriorated. The following methods are suggested to be applied for effective decontamination of some mycotoxins.

Physically, fungi-contaminated seeds can be removed by hand picking or photoelectric detecting machines. The method would consume time and Iabor or expensive.

Organic solvents (chloroform, acetone, hexane and methanol) have been used to extract aflatoxins for agricultural products, but mainly in vegetable oil refining process.

Heating and cooking under pressure can destroy nearly 70% of aflatoxin in rice compared to under atmospheric pressure only 50% destroyed. Dry and oil roastings can reduce about 50-70% of aflatoxin B1. We could show that only about 10% of total 1242 ppb of aflatoxin B. decreased in naturally contaminated peanut by heating at upto 100°C. Since aflatoxin resists to higher temperature upto 260°C, long-time cooking and overheating would destruct essential vitamins and amino acids in treated foods.

Ionizing radiation such as gamma-rays can stop growth of food spoilage organisms, including bacteria, molds and yeasts. It also inactivates pathogenic organisms including parasitic worms and insect pests. It has been reported that gammairradiation (5-10 M-rad) caused reduction of aflatoxin. The irradiation, however, could not completely destroy the toxin and its mutagenicity. In our laboratory, only about 30% of total 600 ppb at aflatoxin B1, either pure toxin or in contaminated peanut, was destroyed by 1 and 5 Mrad or gamma irradiation. The treatment combination of gamma irradiation and ammoniation should be therefore attempted for more aflatoxin decontamination.

Chemical treatment has been used as the most effective means for the removal of mycotoxins from contaminated commodities. The method should be sure that the detoxification system is capable of converting the toxin to a nontoxic derivative (s) without deleterious change in the raw product. Mutagenicity of the treated products should be assessed. The toxicity may be checked by feeding animals including bouts, egg embryos, chicken, ducklings and rats. Many common chemicals have been brought to test the effectiveness in detoxification of aflatoxin. These chemicals include the followings:

- acetic acid (C_2H_5OH)
- ammonia gas (NH_3) or NH_4OH or ammonium salts, 3-5%
- calcium hydroxide ($Ca(OH)_2$)
- formaldehyde
- hydrogen peroxide (H_2O_2)
- methylamine (CH_3-NH_2)
- ozone gas (O_3)
- phosphoric acid (H_3PO_4)
- phosphine gas (PH_3), very highly toxic!
- sodium bicarbonate ($NaHCO_3$)
- sodium bisulfite ($NaHSO_3$)
- sodium bisulfite (NaOH)
- sodium hypochlorite (NaOCI).

The chemical reactions of detoxification of aflatoxin are primary addition of the double bond of the furan ring and oxidation involving phenol formation and opening of the lactone ring. In the presence of acid, aflatoxins B. and G. will be converted into their 2-hydroxy derivatives, aflatoxins B2a respectively.

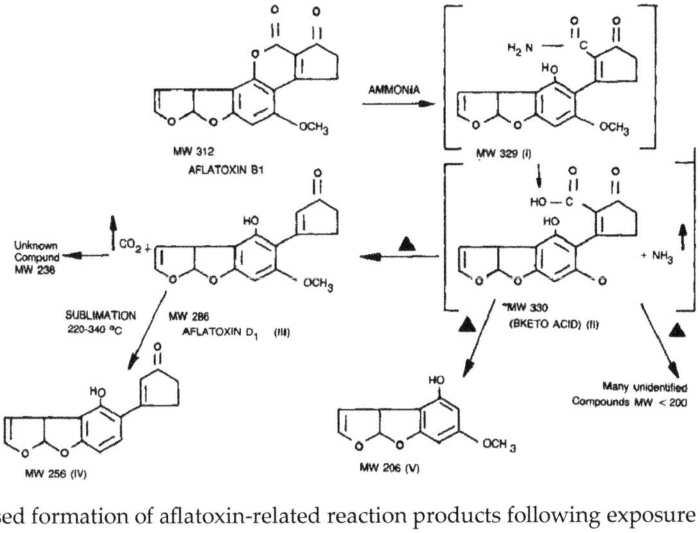

Fig. : Proposed formation of aflatoxin-related reaction products following exposure to ammonia

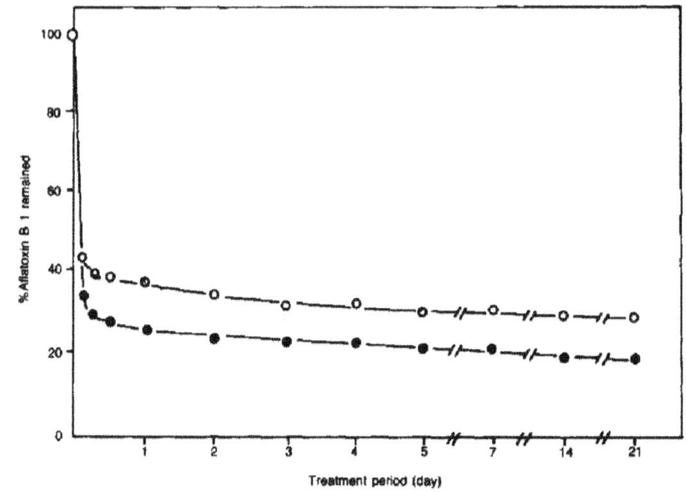

Figure Comparison of detoxification efficiency of 3% ammonium bicarbonate and 3% ammonium carbonate at different length of periods
● : NH_4HCO_3
○ : $(NH_4)_2CO_3$

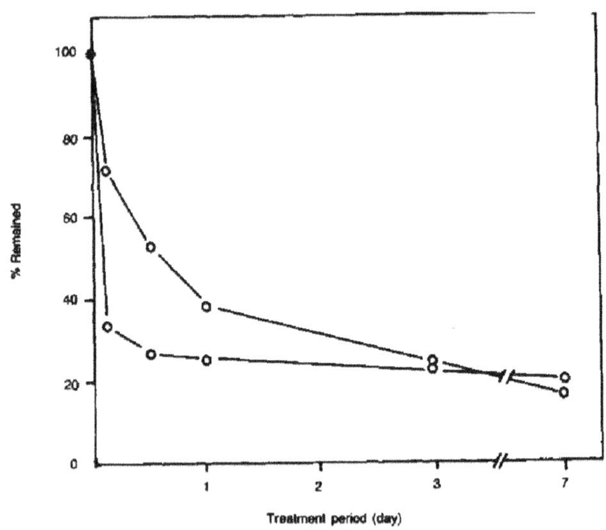

Figure Comparison of reduction of AFB_1 content in different treatment periods and their mutagenicities
● : Quantitative determination
○ : Mutagenicity test

Other mycotoxins which are like aflatoxin and have a lactone grouping in the molecule can be similarly destroyed by alkaline condition using ammonia, sodium hydroxide and sodium bicarbonate. These toxins are patulin, penicillin acid, citreoviridin, citrinin, cyclochlorotin, ochratoxin A, rubratoxin, trichothecenes and zearalenone.

Certain conditions such as moisture content, heat, ultraviolet or gamma irradiation, sunlight and pressure at different treatment-periods have been simultaneously combined with the chemicals for the enhancement of detoxification.

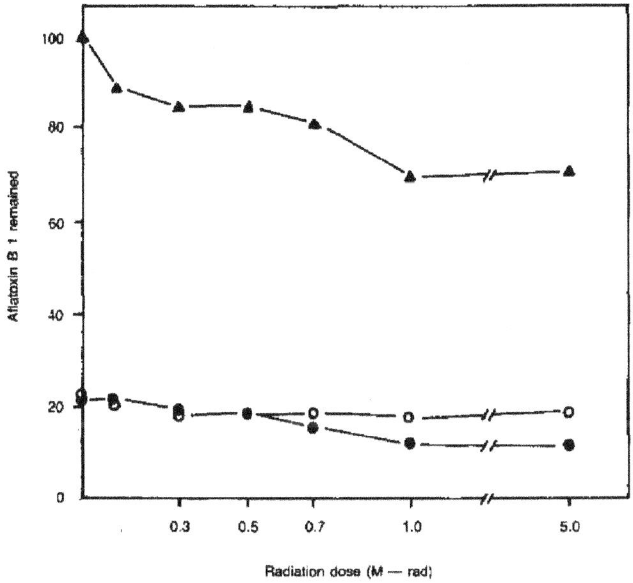

Figure Effect of gamma irradiation with and without ammonium bicdrbondte (3%) on AFB_1 in peanut.
▲ : Peanut without NH_4HCO_3 treated
○ : Peanut contdining 782 ppb of AFB_1 treated with NH_4HCO_3
● : Peanut contdining 1415 ppb of AFB_1 treated with NH_4HCO_3

Inactivation methods can be achieved by mixing, packing, fumigation and immersion with the chemical used.

Control of Mycotoxins

Careful control of mycotoxins should be started and administered by the government of each country through ministries and organizations such as the Ministry of Health, the Ministry of Agriculture, Food and Drug Administration, National Environment Committee Board and Consumer Protection Committee Board. The control program may be set up by a special administrative committee and the legislative body who regulate the national policy of food safety and the maximum tolerance limits for mycotoxins. Farmers, middlemen, food and feed factories and exporters will be well educated about mycotoxins and encouraged to prevent and control the contamination of microflora and their health-hazardous mycotoxins in their commodities as much as possible.

International cooperation for the mycotoxin regulation in trading products or commodities is also needed. The countries should establish quality control limits for certain commodities intended for export or import. The producer countries would be stimulated to be aware of mycotoxin contamination in their exported susceptible commodities. For example, the European Economic Community (EEC) has already established certain maximum tolerance limits of aflatoxins for animal feeds, *i.e.* not more than 20 ppb for complete feed for gigs, poultry and feed supplements for dairy animal; 50 ppb for produce to be processed into mixed feed and complete feed for cattle, sheep and goats.

International organizations such as FAO, WHO and UNEP in the UN system are engaged in providing essential information on various aspects of prevention and control of mycotoxin problems to all the countries. Guidelines for international trade include: a) procedure of sampling and analysis, b) surveillance and food control inspection systems, c) use of contaminated produce in feeding of different animals, d) protocols for detoxification and the quality control of the products. Conferences, symposiums, trainings and workshops on current informations of mycotoxins should be promoted. Low-cost technology for assessment, prevention and control of environmental mycotoxins could be then transferred from developed countries to developing ones.

MYCOTOXINS IN FOOD AND FEED

Mycotoxins are harmful substances produced by fungi in various foods and animal feeds. They are carcinogenic substances and may carry over through a number of links in the feed and food chains. They may occur in cereals, grains and pasture grasses in Europe and in countries producing food and feed imported into Europe. The EU has established a "European Mycotoxin Awareness Network" (EMAN), using an interactive website to provide information on mycotoxins to the food and animal feed sectors.

Mycotoxins may occur in anything between 30% and 75% of human food and animal feed The EU Scientific Steering Committee have issued a warning about the increased occurrence of mycotoxins in European cereals, grains and pasture grass and stated that "mycotoxins pose a notifiable risk to human and animal health". Mycotoxins occur regularly in "Food Alerts" issued by the Food Safety Authority Ireland (FSAI), showing how important they are as a food safety issue.

So what are mycotoxins and why should we be concerned about them? Mycotoxins, from the Greek *mukés*, meaning mushroom (fungus), are produced by fungi which may occur on plants, such as cereals, during growth and particularly on grain, or other plant materials, during storage. The occurrence of fungi on plant materials during growth and storage is influenced by conditions of heat and moisture (humidity).

Potential Risks to the Food Chain

A major concern about some mycotoxins, such as the aflatoxins (named from the fungus,*Aspergillus flavus*), is that they are carcinogenic, while others may be responsible for kidney and liver damage (ochratoxin A), affect reproductive function (zearalenone), and interfere with the body's immune system (trichothecenes). These negative effects may occur both in animals and humans and what is of most concern is that the effects may be chronic, and therefore not easily detected.

Farm to Fork Issue

Given that farming and food manufacturing are integrated in a total "farm to fork" system, the occurrence of mycotoxins in animal feed or in food ingredients is of concern to all involved in food production. For example, aflatoxins or ochratoxin A in animal feed fed to dairy cows or pigs may arise as aflatoxin M1 in milk and dairy products or as ochratoxin A in pork. In the case of food manufacturing, ingredients such as cereals, nuts and spices are important potential sources of mycotoxins in food products. Poor performance of animals fed mouldy hay/silage or mycotoxin-containing compound feed is an additional economic concern for farmers.

European Mycotoxin Awareness Network.

Mycotoxins are a serious issue for anyone involved in feed manufacturing, farming, or food product manufacturing. Quality of raw materials, prevention of the occurrence of mycotoxins, and control and testing systems are all essential to reducing the exposure of humans and animals to mycotoxins. An important initiative by the European Commission has been to establish the "European Mycotoxin Awareness Network" EMAN. EMAN is a dissemination activity with the objective of providing information on mycotoxins through an interactive website.

The material on the website is provided by seven institutions in Europe with particular expertise on various aspects of mycotoxins in food and feed. Fact Sheets give information on the different types of mycotoxins, how and in which plants and plant materials they occur, their toxicity, stability and persistence, general issues on sampling and analysis and legislation and control. Another area covered is the whole issue of prevention of the occurrence of mycotoxins and the application of HACCP (Hazard Analysis and Critical Control Points) approaches both before harvest and during storage. Decontamination systems for agricultural and food commodities containing mycotoxins are explained. The effect of various food processing procedures, such as grain cleaning and cooking, on reducing mycotoxins in food is described. The EMAN site deals with Sampling and Surveillance for mycotoxins in food/feed; the sampling techniques used are very important because the occurrence of mycotoxins normally is not evenly spread throughout a bulk lot of food/feed material. In another section of the website, testing for mycotoxins in food/feed is covered, including rapid screening methods, confirmatory analyses and quality assurance of testing. Finally, the important areas of Control and Legislation are described.

Access to the website is free to all and there is the possibility to seek additional information by way of on-line queries.

Table: Effect of mould contamination on nutritional value of stored maize

	Metabolisable Energy (ME) (Kcal/kg)	Crude Protein (%)	Fat (%)
Good corn	3,410	8.9	4.0
Mouldy corn	3,252	8.3	1.5

Table: Mycotoxin decontamination methods

Physical methods	Mechanical separation/sorting Washing under pressure Thermal inactivation Adsorbent materials
Chemical methods	Ammonia treatment Ozonisation
Microbiological methods	Yeast, bacterial inactivation

Table: Maximum Residue Limits (MRLs) for mycotoxins in food

	MRL (µg/kg, ppb)		
	Aflatoxin B_1	Aflatoxin M_1	Ochratoxin A
Nuts and dried fruit	2-8		10
Cereals	2		3-5
Spices	5		
Milk		0.05	

Information Dissemination

This information is disseminated through an interactive web-site - EMAN - through newsletters, through on-line training courses, and through workshops throughout Europe.

CURRENT TRENDS IN MYCOTOXIN ANALYSIS

Fungal toxins (mycotoxins) comprise one of the major sources of food poisoning in humans, agricultural animals and wildlife. The impacts of these naturally occurring toxins range from acute disease and deaths to chronic disease with production of tumors and significant immunosuppression underlying the occurrence of other infectious bacterial and fungal diseases. Mycotoxins are formed frequently in field crops during the growing season and are unavoidable as a result of our inability to control the environmental and crop susceptibility factors that allow for their production by toxigenic, plant pathogenic fungi.

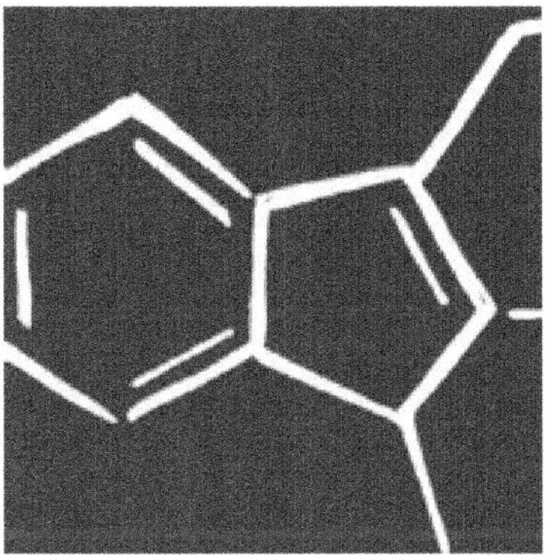

The risks associated with these mycotoxins occurring in field crops, foods, feeds and animal products impact both food safety and economic aspects of the marketing of grain for food and feed. Thus, food and feed manufacturers have a great concern for the occurrence of mycotoxins in the commodities used in their products and resort to testing for mycotoxins to provide a healthful and non-toxic product. However, among these manufacturers there is need to consider the amount of testing, the mycotoxins of concern and whether to test in-house or rely on testing of commodities and products by a commercial laboratory. For this reason, we have developed this report as an aid to understanding the analysis of mycotoxins and to assist in choosing appropriate testing measures.

The analysis of mycotoxins is the final part of overall mycotoxin testing. Sampling, subsampling and sample preparation are the first steps in the testing scheme. These first steps are where the greatest variability occurs in the overall testing scheme.[1] In this discussion, aimed at describing the function and use of the specific tests, we will assume that an adequate sample has been collected and it has been subsampled and prepared for analysis by a suitable extraction methodology for the respective mycotoxin(s)analysis. Also, we will address those methods that are most applicable in a well-developed analytical laboratory using reference methodology and discuss methodology that is generally applied to test kits that are used by feed and food manufacturers and other personnel who desire a rather rapid testing procedure. Finally, we will briefly mention some new technology that is being evaluated for analysis of mycotoxins.

ANALYTICAL METHODS MOST APPLICABLE TO AGRI-FOOD LABORATORIES

The most common analytical procedures used for mycotoxin analysis in labs include thin-layer chromatography (TLC); high-performance liquid chromatography (HPLC); gas chromatography (GC); and mass spectrometry (MS).

Thin-Layer Chromatography

TLC is the separation of compounds in a mixture that is spotted near one end of a plate (commonly glass is used as the plate) that is coated with a thin layer of an adsorbent matrix such as silica gel. Separation occurs when the end of the plate nearest the spotted mixture is placed in a solvent system in the bottom of a closed vessel (glass tank or jar) and the solvent is allowed to migrate through the adsorbent matrix and moves toward the top of the plate. As this occurs the mixture of compounds separate based on their interactions between the solvent system (mobile phase) and the matrix (stationary phase). Because these properties differ for the various compounds in a mixture, they will migrate on the plate at different rates from their starting point. If the analyte (mycotoxin in this case) of interest is present in the mixture it will be identified by comparison with standard spots of this mycotoxin that are necessarily spotted on the plate in a similar manner as the mixture.

TLC often is used as a screening technique for mycotoxins, because although it is a very powerful tool for the determination of the presence of one or more mycotoxins present in a sample, it does not allow for critical quantitation that may be required. Often, unknown mycotoxins have been found on a TLC plate that may not have been very evident using other more quantitative methods. Confirmation of the identity of a specific mycotoxin may be carried out directly on a TLC plate.

Densitometric quantitation can be conducted on a TLC plate and is more accurate when the compounds are colored or fluorescent. Thus, the analyst does not have to spray or dip the plate in a solution for visualization of the spots. TLC can often be used with little or no cleanup prior to the analysis. In some cases, two-dimensional TLC can be utilized to find mycotoxins in extremely dirty samples.

High-Performance Liquid Chromatography

HPLC is similar to TLC in principle, but is a much more sophisticated system. Basically, a small portion of a sample to be analyzed is injected into a stream of solvent being pumped through a column of an adsorptive matrix. When the sample reaches the column from the injection port it is, of course, still a mixture of compounds. As the sample moves through the column the compounds are separated by the same basic principles as described for TLC. The sample components elute off of the column as separate (it is hoped) entities, and the flow of solvent with the respective compounds passes through some sort of detector to measure the response of the specific compound. The measurements are determined based on detector response (peak height or area under the peak), which are compared to selected concentrations of standard that were injected into the instrument as part of the analysis sequence. The time for the peak appearance under the conditions of operation of the HPLC is of importance in identifying the compound of interest.

HPLC currently is the most frequently and widely used method for the analysis for mycotoxins. Methods have been developed for most of the major mycotoxins and have reasonably low levels of detection. Thus, HPLC usually is a good quantitative tool. The time for analysis of most mycotoxins is less than 15 minutes, but the methods require substantial cleanup of the extracts before injection into the system. The equipment is considerably more expensive than that required for TLC.

Gas Chromatography

GC uses sophisticated equipment in which compounds (often derivatized) are separated by a gas flowing through a heated glass column coated with a stationary nonvolatile liquid. Similar to HPLC, samples injected into the system are separated into the specific components on the column, and the separated analytes coming off of the column are detected by some chemical or physical detection system.

GC is used in more highly technical laboratories for some of the mycotoxins and is used for the analysis of the Type A trichothecenes, which do not render themselves readily amenable to HPLC analysis. This is a highly sensitive technique and the time of analysis varies with the compound of interest, but is usually less than 15 minutes. Again, the equipment is expensive and cleanup of the extract is necessary.

Mass Spectrometry

Both HPLC and GC utilize multiple, simple two-dimension detection systems that generate a chromatogram generally consisting of the concentration of an analyte as the X-axis and total detector response as the Y-axis. MS is a detection technique that can be coupled to HPLC and GC in a benchtop instrument system, and has gained significant popularity over the past decade since three dimensions of data can be acquired for an analysis. In addition to the typical chromatogram, a third dimension of information is obtained from the analytes entering the detector in the form of the mass spectrum. A mass spectrum is the "fingerprint" of a compound based on the mass to charge (m/z) ratio of the molecular ion and molecular fragments created through the ionization process. The mass spectrum is very useful in unknown identification or confirmation of the identity of peaks in a chromatogram. Mass spectrometers can also be set up to be very compound-specific detectors using selected ion monitoring in which only specific mass ions are monitored that are characteristic of the compound of interest under the conditions of analysis.

There are several types of mass spectrometers, but this article will only address those commonly used for mycotoxin analysis. The mass spectrometers used with HPLC are significantly different than those used for GC. In general, a mass spectrometer consists of an ionization source, a mass separator and an ion detection system. Ionization sources is the point at which the compounds eluting from the chromatographic column are ionized and commonly some fragmentation of the parent molecules occur. After ionization, the charged molecules and fragments are filtered by a mass separator, which allows separation of the ions based on their m/z ratio, which are then detected.

For liquid chromatography/mass spectrometry (LC-MS), the ionization sources typically are atmospheric pressure chemical ionization (APCI) or electrospray ionization (ESI), both of which are referred to as "soft" ionization techniques. The molecular ion and some molecular fragments from the parent compound are then separated by a mass separator. LC-MS is a very versatile tool in that a very wide number of compounds are suitable for detection, but a common limitation is that different compounds require different ionization conditions making general screening of samples very difficult.

For GC-MS, the historic ionization source for both identification and quantitation is electron impact (EI), which is called a "hard" ionization technique. As the compound elutes from the CC column, the molecules are bombarded with high-energy electrons. The charge induced on the molecules commonly causes them to fragment significantly. This fragmentation leads to the characteristic EI mass spectra which is very useful in the identification of unknowns either by mass spectral interpretation, or by comparison of the fragmentation pattern with commercial spectral libraries. Methods that utilize another ionization technique, negative chemical ionization (NCI), are becoming more common in the literature due to remarkable sensitivity and low background noise. After ionization, the fragments are separated based on their m/z ratio and detected. GC-MS is a very versatile tool for screening samples for unknowns, as well as for the quantitation of compounds suitable for CC analysis.

METHODS USED PRIMARILY FOR TEST KITS

The methods used in mycotoxin test kits are primarily designed to achieve a rapid analysis of mycotoxins occurring in commodities and some products derived from them. The primary types of procedures used are enzyme-linked immunosorbent assay (ELISA), affinity column chromatography and fluorometry.

Enzyme-linked immunosorbent assay. This is one of the most popular immunologically based methods used in test kits for the analysis of mycotoxins in foods and feeds. Most methods employed for mycotoxins use monoclonal antibodies specific for the mycotoxin in the assay. Often, ELISA methods will give quantitative answers greater than that by HPLC, because the antibodies may cross-react with other compounds of similar structure in the sample. Once a test has been developed, the method is quite easy to use and the equipment necessary is relatively inexpensive.

The technique for conducting an ELISA may use one of several possible approaches, but a commonly used approach is a competitive assay in which a known amount of labeled (enzyme) toxin competes with any possible toxin in the sample for the specific antibodies attached to the reaction vessel (usually a microtiter well or a test tube). Any unbound toxin is washed from the vessel. The quantitation is dependent upon the amount of enzyme-labeled toxin remaining in the vessel to react with the substrate for the enzyme. This reaction results in a colored product that can be measured optically. Therefore, a highly colored reaction would indicate the absence of the specific mycotoxin in the sample.

Affinity Column Chromatography

Affinity column chromatography has been used for quite a few years in the analysis of mycotoxins. Essentially, the technique is used to clean up extracts prior to an analytical procedure used to measure the mycotoxin. In these columns, monoclonal (usually) antibodies specific for a given toxin are attached to the column matrix. When an extract is placed on the column and the column is washed, only the specific toxin remains on the column as it has bound to the antibody. The attached toxin is then eluted from the column by changing the solvent system to 90% methanol in water; thus, the eluate contains rather pure toxin, which can be analyzed using any technique such as fluorometry or HPLC.

It is possible to apply more than one antibody to the column for use in simultaneous isolation of more than one mycotoxin. This has been applied to fumonisin B1 and its hydrolysis product prior to HPLC analysis. This potential exists for other mycotoxins, as well.

Fluorometry

The basis of fluorometry is the quantification of compounds by measuring their fluorescence using a fluorometer. In some cases, the compounds may be innately fluorescent, while in others, the compounds are rendered fluorescent through some chemical derivatization. The tests for compounds using fluorometry includes the extraction of the compound from the specific matrix, followed by a cleanup process using either affinity columns or solid phase cleanup columns, and then derivatization (if necessary) and measurement of the fluorescence. The equipment for fluorometry is rather inexpensive, since the fluorometer is a one-time expense and the test kits usually contain most of the necessary materials to run the test.

CURRENT TECHNOLOGY IN DEVELOPMENT

One of the newer technologies applied to the analysis of mycotoxins is capillary electrophoresis (CE). Using this technique, a mixture of compounds having dissimilar charges can be separated in a capillary column using electrophoresis. Because of the use of capillary columns, small sample volumes can be used for analysis and thus eliminates the use of large quantities of hazardous solvents. This method is not as easy to perform as HPLC, requires considerable clean up, requires longer derivatization time, at least for some mycotoxins, and the equipment is quite expensive. However, it is rapid, the capillary columns are easily regenerated with rinsing and only small amounts of reagents are required for the test.

Cell-based assays are often employed in the analysis for mycotoxins because they reflect both toxin interaction with the receptor on the cells and the subsequent intracellular responses of the cell that result from the toxin/receptor interaction. Toxin class specificity can be achieved, at least to some extent, by choosing a particular cell line known to express a given response to a particular toxin or class of toxins. Endpoints in these assays include cytotoxicity or quantifiable changes in cell responses.

Table : Emerging technologies used in mycotoxin testing and analysis.

Technology	Brief Test Concept
Aptamer	Uses oligonucleotides specific for certain analytes. These can be used instead of antibodies in a variety of formats.
Array Biosensor	Uses capture antibodies in lines on a surface. Sample is applied in lines at 90 degrees. Apply a second fluorescent-labeled antibody for visualization.
Fiber-Optic Immunosensor	Specific antibody is coated on fiber optic whisker which has fluorescent labeled antigen attached. When whisker is dipped in extract containing analyte there is a diminished fluorescence (measuring the evanescent wave) in a quantitative manner. Can be used in a direct format if the compound of interest is fluorescent. Has been applied to fumonisins and aflatoxins in maize.
Fluorescence Polarization	Specific antibody and fluorescent ligand has a certain polarization of the emitted fluorescence. When the analyte is present the polarization changes relative to the antigen-antibody reaction and it is concentration dependent.
Immunoaffinity Fluorometric Biosensor	An extract is put through an immunoaffinity column into an inline cuvette and the fluorescence is measured with a fluorometer. Developed as a single, handheld device. Has been applied to aflatoxins.
Surface Plasmon Resonance Immunosensor	Antibodies are adsorbed on a gold surface. Diode emitting planar light passing through prism causes excitation of surface plasmon resonance at gold surface. When a sample containing the analyte is added to a cell on outside of gold film, the angular profile of light intensity shifts. Changes in resonance angle and beam intensity are proportional to analyte concentration.
Microarrays	This is a name given to any kind of small "dot type" array that one would desire using in a format with antibodies or perhaps aptamers. Most likely format is a silicon chip with microwells in which the reaction takes place.

Currently, a developmental strategy for many natural products and pesticides is the use of biosensors. These biosensors utilize diverse technological advances such as monoclonal antibodies, optics, microchips and recombinant genetic materials put together in combinations to achieve rapid, specific and quantitative measurement of the desired analytes. Table describes some of the emerging techniques in the field of mycotoxin analysis that are in research status with no test kits or only limited applications for mycotoxins available commercially.

Our ability to control and manage the contamination of mycotoxins in feeds and foods is dependent upon the development of procedures for analyzing mycotoxins in them. These methods must be valid, efficient, sensitive and accurate to make the world's food safer.

INSPECTION, SAMPLING AND ANALYSIS OF MAIZE AND GROUNDNUTS FOR AFLATOXIN

Farm Level

The effect of environmental and cultural factors on aflatoxin contamination in maize have been studied and have identified the 5 major factors that affect the development of toxin-producing fungi; inoculum level, moisture, temperature, time, and damaged kernels, Each of these variables play a complex role in toxinproducing capability of fungi. From the farmer prospective the following practices are major contributors to the overall high aflatoxin level in maize.

- *Moisture:* Harvesting before maturity where the moisture of the grain is high (greater than 25%).
- *Damage:* Shelling grain with high moisture content causes severe hernel damage.
- *Time:* Temporary storage of moist grain (very common) (Known as bulking up in the maize trade).
- *Inoculum Source:* Storage of grain in mould infested bin (last year mould spores will infect this year crop). A bottle of bleach and water will reduce the inoculum level in many storage bins and it is inexpensive.

In most cases this will require more than one sample and one analysis. Because of the variability associated with test methods for aflatoxin in maize, analyses of samples from good lots may indicate that the lots are bad (low level of aflatoxin); (exporters' risk) and analyses of samples from bad lots (high level of aflatoxin) may indicate that the lots are good (importer's risk). Testing program for aflatoxin has been developed for maize in the United States. This testing program involves taking a 4.5 kg. sample from a lot, grinding the entire sample and subdividing into five 50 9. samples. The aflatoxin is solvent-extracted from the maize meal by the AOAC CB method. If the analysis of one sample is less than 20 PPM, the lot is accepted. Otherwise, two more samples are analyzed. If the difference between each of the 3 analyses and their average (A) is less than 0.3 A. The lot is accepted when a is less than 25 PPB and the lot is rejected when the average is greater than 25 PPB. When any one of the 3 analyses is not within the range of (A ± 0.3 A) PPB. the fourth and fifth samples are analyzed. The lot is accepted when the average of the 5 anlyses is greater than 25 PPB.

Even with this elaborate test program, the exporter still has a 1% probability of making a wrong decision based on the results of 5 analyses, and the importer has a 7.3% probability of making a wrong decision.

Decisions based on mini-column method, that is not qualitative, will expose the exporter to a great risk and likelihood of being wrong with respect to the amount of aflatoxin in particular lot of maize being exported. Also, the importer has the same increase risk of being wrong using a non-quantitative method.

Special Precaution and Decontamination Procedures

Handle aflatoxins as very toxic substance, use gloves when handling standards. Perform manipulations under hood whenever possible. Take special precautions with toxins in dry form because of electrostatic nature and tendency to disperse in work area. Decontaminate all volumetric flasks used in making standard solutions and all vials containing concentrated sample extract or reference standard solutions by the following procedure. Prepare decontamination solution by adding 3 ml of sodium hypochlorite solution (household bleach) per microgram of each aflatoxin.

After mixture is left for several hours or overnight, an equal volume of 5% aqueous acetone is added. After 30 minutes, the mixture may be safely discarded. Oils and solids extracts should be shaken well with the sodium hypochlorite solution. Large volume of organic solvent should be evaporated and then add the sodium hypochlorite to residue. It is very important to add the acetone to sodium hypochlorite solution. One of the products of treatment of waste containing aflatoxin b1, -2, 3 dichloride, which is also carcinogenic. Rinsings of glassware and TLC plates should be done in sink in washer room to minimize bleach use in area where aflatoxin analysis is ongoing.

Use dust mask when grinding grains, feeds or any other products generating dust into atmosphere during preparation. Sample preparation area should be specially cleaned after preparation of sample containing high levels of mycotoxin.

Extraction and clean-up steps should be performed under hood whenever possible. Since aflatoxin is destroyed by alkali and strong acids, do not handle samples in same area where vapors of strong acid/base may be present. Spot TLC plates under hood and avoid exposure of spotted plates to reactive vapors such as O3, SO2 and HC1, which effect stability of aflatoxins on plates and which can also effect the absorbents. Dry plates thoroughly before exposure to UV light since presence of solvent enhances degradation of aflatoxins expased to UV light including fivorescent Light

Sampling Procedures

Samples may be taken from crops growing in the field, during handling, during storage, and at other points in the production, marketing, and processing system.

Stream Sampling

The most effective sampling method is to take small portions from a moving stream at periodic intervals and combine the portions into a sample. The stream should be sampled frequently but the amount taken at each interval should be small to avoid accumulating too large a sample. The samples must be taken from the stream throughout the time the lot is moved.

Probe Sampling

Probe sampling is probably adequate for lots which have relently been blended by harvesting lots which have relently been blended by harvesting or handling operations. Recommended methods for taking probe samples are published by the American Oil Chemists Society Official Methods of Analysis.

Field Sampling

Diagram each field and divide the field into approximately N equal sections. If the field is rectangular, a count of rows for width and a count of paces for length can be helpful in dividing and identifying N equal field sections. Devise a method to randomly (table of random number) select one ear from each section. Shell and composite the N ears in a mesh bag as a single sample.

Sample Size

Sampling must yield a minimum of one 4.54 kg. sample per unit sample. If more than 4.54 kg. is taken, reduce by dividing or by blending and quartering.

Note on Sampling Methods

Stream sampling is preferable to probe sampling which in turn is preferable to field sampling of ears of corn. For example, the estimated coefficients of variation (%) among aflatoxin test when the indicated number of ears of maize is randomly selected from a field with an average of 100 ppm aflatoxin a 133% for 36 ears harvested and shelled, 83% for 100 ears, 62% for 200 ears, 48% for 400 ears, and 39% for 800 ears. However, if the same entire field is harvested, shell and a stream sample is taken the estimated coefficient of variation is only 29%

Sample Collection Report

A sample collection report should have a sample number and give the date sample was collected, where collected, observation at collection site (insects, rodents), description of sampling method used to collect sample, size of lot sample was drawn from, where sample is stored in laboratory and a description of method used to prevent aflatoxin accumulation during period of sample collection and sample analysis.

Handling and Safe Storage

Each sample should be handled in a manner that minimizes the post-collection production of aflatoxin. Moist samples should be held in cloth or paper bags, cooled if feasible and transported to drying facilities as soon as possible. Samples should not be placed in plastic bags unless seeds are dead and dry, or any other confined area where humidity and temperature can increase around them.

Significant mould growth and aflatoxin production can occur in just a few hours and this must be prevented. As soon as possible after collection, the samples should be dried at approximately 80-90 °C for 3 hours or more to reduce the grain moisture to about 12-13% (where moulds are to be studied, 60°C for a longer time is recommended). If samples were refrigerated before drying, they should be dried immediately and kept dry until analysed.

Sub-Sampling

The entire 4.54 kg. sample should be ground to pass a No. 14 sieve, thoroughly blended, and properly sub-divided to a 1 kg. sample. The entire 1 kg. sample should be ground to pass a No. 20 sieve, thoroughly blended, and properly subdivided to 50 9 analytical samples. When feasible, the samples may be ground and sub-sampled immediately after collection. The subsamples can be analyzed immediately, stored under refrigeration, or dried for storage.

Methods of Analysis

The method for analysis should be selected in advance and used uniformly throughout the survey. Official methods that have been collaboratively studied should be used as the method of analysis for aflatoxin. Unofficial methods or methods reported in literature should be validated with respect to the method accuracy and variability before survey is started. The selected method should be retested during the survey by periodic analyses of blank, standards and "spiked" samples. Analytical procedures once set must be followed to the letter. No alternative procedures are permitted unless expressly authorised in operations manual or by project leader.

Laboratory Report

A standard format for reporting data should be adopted and used throughout the survery. Laboratory report form should include sample number, date analysed, method used, commodity, analyst's name, all calculations of results, source of standards and concentration of standard.

Preparation and Storage of Aflatoxin Standards

Aflatoxin standards should be prepared and stored as per AOAC Official Book of Method 14 edition.

Survey Sample

A survey summary should be used to capture data generated by the survey. This form should be kept by the project leader of the survey and can be used to summarize the progress of the survey. The form should include sample number, date collected, date analysed, analyst's name, analytical result, and space for reporting miscellaneous observations, such as recovery from spiked sample and/or duplicate samples analyzed at different times by different analysts.

Recommendations

Inspection/Sampling/Analysis

- Develop protocols for surverys, sampling, postcollection handling, and analysis of grain sample involved in the aflatoxins.
- Maize grading standard should be standardized throughout maize marketing system.
- Develop a standard operating procedure for aflatoxin analysis. Standardization of operation procedure will aid in the assessment of the true aflatoxin concentration.
- Develop the necessary mechanism or an agreement with the largest importer of Thai maize to obtain information about the aflatoxin contamination levels found at time of import. This project could be a very cost-effective way in obtaining contamination levels in Thai maize and will aid in resolving the problems associated with level of aflatoxin found at time of export and the level of aflatoxin found at time of import.
- Develop and evaluate a testing program for aflatoxin in Thai maize.
- Establish aflatoxin certification laboratory. This laboratory should be subjected to periodic proficiency testing.
- Establish a centralized aflatoxin standard distribution system for Thailand.
- Develop a national aflatoxin check sample program.
- Centralize data collection from both government and private sector.
- Maintain a high level of qualified analysts in aflatoxin area in government and private sector. Training should be given periodically in aflatoxin survey techniques, quality control, sampling and analytical techniques.

Prevention and Control

The development of reliable aflatoxin prevention programs requires considerable scientific knowledge about when and where in the marketing system that aflatoxin contamination occurs. Guidelines based on lesser scientific knowledge most often are costly and ineffective. The environmental conditions that Thai maize is subjected to during post-harvest handling will determine the contamination level at the exporter's silo. Reducing moisture content of maize to below the safe level (13-14%) before any mould growth occurs is at present the safest and most reliable method of preventing aflatoxin contamination. However, reducing moisture content of very large volume of maize in a short period of time in high humidity environment is costly. Other methods should be investigated (particularly aeration, on cob drying on plant, effects of insects, and inoculum levels in bin) to determine if control of these environmental conditions will retard the accumulation of aflatoxin during post-harvest processing. Optimization of these parameters may increase the time moist maize may be safely stored before final drying and final storage.

Guildelines for Mould Control and Aflatoxin Prevention in High-Moisture Maize :

1. Reduce the moisture and temperature of shelled maize. Recognize that high levels of moisture and temperature, together with damage to kernels, are the chief reasons for the growth of molds in shelled maize. Mould growth is negligible when maize moisture content is below 13%.
2. Never exceed the maximum storage time for a given combination of moisture and maize temperature. Remember that the maximum time begins with harvest.
3. Dry wet maize to 12 to 14 percent moisture immediately after harvest. This is the surest method of limiting mould growth.
4. If you must hold wet maize at moisture levels above 20 percent -without even a small amount of drying, maintain maize temperatures as low as posible under some form of aeration or maintain maximum ventilation as possible.
5. Follow those cultural practices most likely to produce a vigorous maize crop and make it fully mature before harvesting. If maize is fully mature, there is less infection and mould growth while ears are still on the plant.
6. Keep damage to kernels to a minimum. Causes of damage include: high-moisture shelling, high speed operation of sheller, drying process, loading and unloading, equipment and Insect damage in field and in storage.
7. Use caution in blending lots of maize that differ substantially in either quality or moisture content. Do not blend maize of 20 percent moisture with 10 percent maize in the belief that the mixed batch will equalize overall at 15 percent. It will not unless the mixing is unusually thorough.
8. After the maize has been dried to a moisture level adequate for storage, provide sufficient air flow to bring the maize to a uniform temperature.
9. Check maize regularly for moisture, heat, mold, insects or off odors.

10. Follow good sanitation practices, good insect control, and disease control practices.

BIOTECHNOLOGICAL POTENTIAL OF ERGOT ALKALOIDS ERGOT ALKALOIDS

The ergot alkaloids are mycotoxins produced by several species of fungi in the genus *Claviceps*. The name *ergot* means 'spur' in French, since grains colonized with *Claviceps* spp. often resemble the spurs on the legs of a rooster. There are four main groups of ergot alkaloids: the clavines, the lysergic acids, the lysergic acid amides, and the ergopeptides.

Ergot poisoning in humans and domestic animals is known as ergotism. This disease may cause strange hallucinations, the feeling of itchy and burning skin, gangrene, loss of hands and feet, and even death. Ergotism is one of the oldest known human diseases caused by mycotoxins. In the Middle Ages, humans suffering from a disease called St. Anthony's fire reportedly experienced symptoms that suggested ergot poisoning such as burning skin, insects crawling under their skin, and the loss of hands and feet. In the 1690s, ergot alkaloids may have been the cause of symptoms associated with witchraft surrounding the Salem Witch Trials in Salem, Massachusetts. Ergotamine is one of the building blocks of the psychoactive drug lysergic acid diethylamide (LSD). Today, ergot alkaloids are being exploited for medicinal purposes including the treatment of migraines, the induction of child birth, and the control of post-partum bleeding.

Many different grasses may become contaminated with ergot alkaloids. *Claviceps purpurea* may infect the reproductive structures of cereal plants during cool, wet weather, with rye being the main economic host. Specialized fungal structures known as sclerotia (the ergots) develop from these reproductive structures and may be harvested with the grain. If the sclerotia are processed into flour, high levels of contamination with ergot alkaloids may result. *Claviceps africana* and *C. sorghi* may contaminate sorghum with ergot alkaloids worldwide, but the relative toxicity of mycotoxins produced by these fungi has not been evaluated in detail. Common pasture grasses may be contaminated with alkaloids similar to those produced by ergot fungi, but these fungi belong to different genera and do not produce sclerotia

ERGOT AND ITS ALKALOIDS

History of Ergot

Students at the University of Pittsburgh receive an introduction to pharmacognosy and natural products during their first-professional year in an introductory course in *Drug Development*. The role of natural products as both historical and continuing sources of drugs, as well as sources of precursors for semisynthetic modification and sources of probes for yet undiscovered drug moieties, is emphasized. In addition, students are continually exposed to the concept that complex natural products are a result of secondary metabolism, and as such are produced via the unique combination of a relatively limited number of structurally unsophisticated primary metabolites. As a consequence, secondary metabolites have a more limited distribution in nature, and their occurrence is an expression of the individuality of the parent species. This curricular dialog with pharmacognosy and various bioactive natural products continues in courses in the second-professional year and third-professional year.

Early History

Although it has been speculated that the 4,000-year old Eleusinian Mysteries of ancient Greece were connected with ergot-induced hallucinations, the earliest authenticated reports of the effects of ergot occurred in Chinese writings in approximately 1100 BC, when the substance was used in obstetrics. A magic spell found in a small temple in Mesopotamia dating to 1900-1700 BC referred to abnormally infested grain as*mehru*, while Sumerian clay tablets of the same period described the reddening of damp grain as *samona*. The Assyrians of this era were sufficiently knowledgeable to differentiate between different diseases affecting grain and by 600 BC writings on an Assyrian tablet alluded to a "noxious pustule in the ear of grain." References to grain diseases have also been found in various books of the Bible in the Old Testament (850-550 BC).

In 550 BC the Hearst Papyrus of Egypt described a particular preparation in which a mixture of ergot, oil, and honey was recommended as a treatment for hair growth. In 370 BC Hippocrates furnished a description of corn blight and subsequently described ergot as *melanthion*, noting its use to halt postpartum hemorrhage. Around 350 BC the Parsi wrote of "noxious grasses that cause pregnant women to drop the womb and die in childbed," while in 322 BC Aristotle postulated that grain rust was caused by warm vapors. Around 286 BC the Greeks concluded that barley was more susceptible than wheat to rust infections, and that windy fields had less rust than damp, shady low-lying ones.

Middle Ages to Twentieth Century

The first documented epidemic of *ergotism* likely occurred in 944-945 AD, when some 20,000 people of the Aquitane region of France (about half of the population) died of the effects of ergot poisoning. Some 50 years later, about 40,000 people reportedly died because of the "holy fire."

Up through the 18th Century botanists persisted in considering ergot to be a "super" rye, possessing an enlarged kernel. Finally, in 1764, ergot was recognized as a fungus by von Munchhausen. Epidemics of ergot poisoning, often termed *ergotism*, continued to ravage continental Europe through the Middle Ages and outbreaks of ergotism occurred in Germany in 1581, 1587, and 1596. These epidemics were due, in part, to the fact that rye was grown in larger quantities in medieval times, and many people (particularly those less wealthy) ingested contaminated rye flour.

These outbreaks were characterized by the production of 2 distinct forms of toxic reactions, with these reactions now being understood to be attributable to the effects of the alkaloids produced by the parasitic ergot fungus which was contained within the ergot fungal body (sclerotium). The fungus-contaminated grain crops along with their fungal metabolites (ergot alkaloids) were ingested with flour prepared from the grain. The first gangrenous form (*Ergotismus gangraenosus*), commonly known as "holy fire," "infernal fire," or "St. Anthony's Fire," was more common in France and its effects were characterized by pronounced peripheral vasoconstriction of the extremities (limbs).

Hands, feet, and whole limbs would swell, producing a violent, burning pain that ultimately culminated in the separation of a dry gangrenous limb (usually a foot) at a joint, without pain or loss of blood. St. Anthony's Fire was so named because the Order of St. Anthony traditionally cared for sufferers in the Middle Ages, and the condition was characterized by severe burning pain ("fire"). The second form of ergotism, also known as the convulsive form (*Ergotismus convulsivus*), was particularly common in Germany and was typically characterized by the development of delirium and hallucinations, accompanied by rigid, extremely painful flexed limbs, muscle spasms, convulsions, and severe diarrhea.

The term *ignis sacer* (holy fire) was commonly employed for epidemics of ergotism, but numerous other terms, mainly of Latin derivation, were coined, including: *Ignis judicalis, Ignis occultus, Morbus hic tabificus, Mortifer ardor, Pestilens ille morbus, Pestis ignaria, Plaga ignis, Plaga illa, and Plaga invisibilis*. Many were quite naturally translated into various European languages, including French (*Feu sacré*) and German (*heilige Feuer*). Other terms for ergotism included names of regions (*Mal de Cologne*) and of saints to whom an appeal for help was made (St. Anthony, St. Martin, St. Martial).

In 1597 scientists at Marburg University observed that signs of ergotism often appeared after blighted rye grains were consumed, and that blight was promoted by cold, damp growing seasons. In 1630 it was observed that feeding of blighted grain to animals produced an illness similar to human ergotism and by the end of the 18th century poisoning was demonstrated in animals. Although the source of ergotism was linked to the consumption of infected rye by 1676, it was the large outbreaks of ergotism in Europe in 1770 and 1777 that prompted the introduction of legislation in France and elsewhere. Epidemics of gangrenous ergotism were recorded from the Middle Ages to the nineteenth century, while that of convulsive ergotism were documented between 1581 and 1928.

Twentieth Century

From 1926 to 1927, some 11,319 cases of ergotism were reported in a population of 506,000 in the vicinity of Sarapol, near the Ural mountains. In 1928, 200 Jewish refugees in Manchester, England, were sickened when they consumed rye bread that had been prepared from rye grown in South Yorkshire. In mid-August of 1951, 230 villagers of the popular French tourist town of Pont Saint-Esprit on the Rhone River were sickened after ingesting contaminated goods from a local baker. They became violently ill with symptoms of severe gastrointestinal upset, dramatic reduction in body temperature, hallucinations, euphoria, and suicidal ideation. Within days, some became extremely delirious and others complained of excruciating burning pains in the extremities, culminating in the development of gangrene in some patients.

Early Medicinal Uses

In 1582 a preparation of ergot that was employed in small doses by midwives to produce strong uterine contractions was described by Adam Lonicer in his *Kreuterbuch*. The use of ergot as an oxytocic in childbirth became very popular in France, Germany, and the United States. The first use of the drug in official medicine was described by the American physician John Stearns in 1808, when he reported on the uterine contractile actions of a preparation of ergot obtained from blackened granary rye as a remedy for "quickening childbirth." However, shortly thereafter the number of stillborn neonates rose to a point that the Medical Society of New York initiated an investigation. As a result of this enquiry, it was recommended in 1824 that ergot only be used in the control of postpartum hemorrhage. Ergot was introduced into the first edition of the *United States Pharmacopeia* in 1820 and into the *London Pharmacopeia* in 1836.

Livestock Poisoning

Ergot alkaloid contamination of livestock feeds has long been known and has been described in various places. Although reports of ergotism in livestock vary from year-to-year in the United States according to rainfall and temperature, this mycotoxicosis occurs in the grain-producing areas of the Northern plains in most years. Three syndromes have been described in animals: nervous ergotism, gangrenous ergotism, and agalactia.

Biology and Chemistry

Biology and Lifecycle

The genus *Claviceps* is a group of phytopathogenic ascomycetes that is composed of approximately 36 different species of filamentous fungi. These species are known to parasitize over 600 monocotyledonous plants of the families Poaceae, Juncaceae and Cyperaceae, including forage grasses, corn, wheat, barley, oats, millet, sorghum, rice, and rye. Ergot was first recognized as a fungus in 1711 but its lifecycle was not described in the form of a general outline until 1853 by Tulasne. The term ergot or *Secale cornutum* derives from the French word *argot* (a spur) and represents the dark brown, horn-shaped pegs that project from ripening ears of rye in place of rye grains. These tuberous projections are collected before and during harvesting or are separated from the threshed rye. In a histologic sense, these bodies consist of compactly interwoven hyphae of the filamentous fungus *Claviceps purpurea* (Fries) Tulasne but biologically these compact grains are designated as sclerotia, the form in which the fungus passes the winter.

The parasitic life cycle of the ergot fungi begins in the spring, with windborne ascopores landing on susceptible host plants. Hyphae invade and colonize the ovary, producing masses of anamorphic spores that are exuded into a syrupy fluid (honeydew). Insect vectors, rainsplash, or head-to-head contact transfer this honeydew to other blooming florets, allowing the spread of the ergot fungi in a field. When the sclerotia begin to form, production of honeydew and condiation cease, and the sclerotia mature in about 5 weeks. The number and size of sclerotia produced on each spike of cereal by *C. purpurea* varies according to grain, with rye usually bearing a considerable number, while wheat has relatively few. The sclerotia are considered as the early stage of sexual differentiation of *Claviceps*. In autumn, the ripe pigmented sclerotium leaves the spike and falls to the ground, ultimately producing asci and nonseptate ascospores, thereby completing the cycle.

Sources of Ergot Alkaloids

The industrial production of the ergot alkaloids began in 1918 when Arthur Stoll patented the isolation of ergotamine tartrate, which was subsequently marketed by Sandoz in 1921. Sandoz dominated the world industrial market in ergot alkaloid production up until the 1950s, when other competitors begin to appear. Today Novartis (the successor to Sandoz) still retains leadership in the world production of ergot alkaloids. Some other major producers of these alkaloids market their products as bulk pharmaceutical chemicals, including: Boehringer Ingelheim (Germany), Galena (Czech Republic), Gedeon Richter (Hungary), Lek (Slovenia), and Poli (Italy). Others active in the marketplace include Eli Lilly and Farmitalia. Annual world production of ergot alkaloids has been estimated at 5,000-8,000 kg of all ergopeptines (peptidic ergot alkaloids) and 10,000-15,000 kg of lysergic acid, the latter being mainly used in the manufacture of semisynthetic derivatives. The greater part of this production occurs as a result of fermentations (around 60%) while field cultivation of tritcale (a hybrid of wheat and rye) accounts for the balance.

Chemistry of Ergot Alkaloids

The ergot alkaloids are indole compounds that are biosynthetically derived from L-tryptophan and represent the largest group of nitrogenous fungal metabolites found in nature. Over 80 different ergot alkaloids have been isolated, mainly from various *Claviceps* species (over 70 alkaloids), but also from other fungi and from higher plants. Ergot sclerotia contain about 0.15%-0.5% alkaloids, with the medicinally useful compounds separated into 2 classes: the water-soluble amino alcohol derivatives (about 20% of the total alkaloid mixture) and the water-insoluble peptide derivatives (up to 80% of the total alkaloids). A common portion of the ergot alkaloids is a tetracyclic ring system assigned the trivial name of *ergoline,* which is a partially hydrogenated indole[4,3-f,g] quinoline. These alkaloids may be conveniently divided into 3 major structural groups: clavines, lysergic acid amides (paspalic acid amides), and peptides (sometimes designated ergopeptides or ergopeptines).

The nomenclature of this group of alkaloids is quite complex, with the naturally occurring compounds commonly being assigned a trivial name by their discoverer(s). Systematic names tend to be used only for semisynthetic derivatives or to ascribe an exact chemical description of the molecule. Many of the trivial names of these alkaloids are derived from the botanical names of the host plant or producer, as ergosecaline (*Secale spp*).

Still others are a product of special circumstances of their discovery, such as ergokryptine, an alkaloid that remained elusive (*cryptic; kryptos* [Gr]) and obscured for many years. Ergobasine was so named because of its basic properties, while lysergic acid received its name because it was a product of the lysis of various ergot alkaloids. Some alkaloids bear nomenclature that reflects specific pharmacological properties, such as ergometrine for its actions on the uterus (endometrium uteri). Still other alkaloids have been named to reflect some personal attachment, such as ergocristine for Cristine Stoll, daughter of the scientist Arthur Stoll who isolated ergotamine and later was President of Sandoz AG in Basel, Switzerland.

There have been 3 forms of systematic nomenclature reflecting different chemistry that are employed for the alkaloids of this group. The first utilizes a system found in chemical abstracts that employs ergoline as the name for the tetracyclic system present in most ergot alkaloids and ergotaman for the heptacyclic system occurring in most of the peptidic alkaloids. The second type employs the name ergopeptine, and is used only for the full heptacyclic peptidic alkaloid system. The final type utilizes the IUPAC system, and as such is the most rigorous and rational, albeit complicated. In this variation, the ergoline system is designated as 7-methyl-4,6,6a,7,8,9,10,10a-octahydro-indolo[4,3-f,g]quinoline.

Ergot alkaloids contain several centers of chirality of varying configuration, but the R- chirality at C-5 is constant and nonvariable, reflecting the derivation of these alkaloids from L-tryptophan (the amino acid precursor of the indole ring) as well as the C-4, C-5, and N-6 atoms. The medicinally useful ergot alkaloids are all C-8 amide/peptide derivatives of (+)-lysergic acid, a compound bearing the R-chirality at C-8.

Clavine Ergot Alkaloids

The clavines are substituted 6,8-dimethylergolines but include a few members, such as the chanoclavines, that possess a 6,7-seco D-ring. Although at least 35 alkaloids of this type have been isolated and characterized, none of the group is used medicinally.

Lysergic Acid Amide Ergot Alkaloids

Amidation of the C-8 carboxy-group of lysergic acid results in the formation of 2 types of compounds: the simple nonpeptidic amides that bear relatively short carbon chains and the peptidic amides that commonly exist as tripeptides. Common 5R,8R-nonpeptidic amides found in ergot include ergonovine (ergometrine, ergobasine), lysergic acid 2-hydroxyethylamide, lysergic acid amide (ergine), and paspalic acid. Lysergic acid-derived amides are highly active pharmacologically, and ergonovine and its semisynthetic derivatives methylergonovine and methysergide are used medicinally.

As a consequence of the adjacency of the C-8 chiral carbon of lysergic acid to a carbonyl, the configuration at this center may be changed as a result of heat- or base-catalyzed enolization (particularly in polar solvents) proceeding through a symmetric intermediate to afford the pharmacologically-inactive epimeric (+)-isolysergic acid and its derivatives. In the bioactive lysergic acid derivatives, the amide group is present in the 8-equatorial position, while in the inactive iso-forms, the group is axial. The lysergic acid derivatives commonly end in the suffix "-ine", while the their epimeric counterparts (the isolysergic acid derivatives) are assigned the suffix "inine."

Peptide Ergot Alkaloids (Ergopeptines)

The peptide alkaloids may be biosynthetically envisioned as tetrapeptides containing lysergic acid as the first member of the peptide chain. The other 3 classical amino acids are variable which accounts for the great diversity of the peptidic group of these alkaloids. Ergotamine and the other ergopeptines are composed of (+)-lysergic acid and a L-proline-containing complex tripeptide moiety. A unique structural feature not found in other naturally-occurring molecules is the cyclol part of the molecule, which results from the reaction of an α-hydroxyamino acid adjacent to lysergic acid with the carboxyl group of proline.

Other commonly occurring amino acids present in the tripeptide portion of the ergopeptines include L-alanine, L-phenylalanine, L-valine, L-leucine, and L-isoleucine, as well as 2-aminobutyric acid. Ergotamine is the only naturally occurring ergopeptine to be used medicinally in the United States; however, useful semi-synthetic derivatives of peptide alkaloids include dihydroergotamine, bromocriptine (brominated ergocryptine), and dihydrogenated ergot alkaloids (ergoloid).

Ergot Alkaloid Biosynthesis

The fundamental building blocks for the lysergic acid skeleton of the ergot alkaloids are the amino acid L-tryptophan and the isoprene dimethylallyl diphosphate, the latter deriving from 3R-mevalonic acid. Alkylation of L-tryptophan with dimethylallyldiphosphate affords 4-dimethylallyl-L-tryptophan which is N-methylated with S-adenosyl-L-methionine. Oxidative ring closure followed by decarboxylation, reduction, cyclization, oxidation, and allylic isomerization yields D-(+)-lysergic acid. The simple alkylamide derivatives of D-(+)-lysergic acid are readily formed, including ergine [D-(+)-lysergic acid amide] and ergonovine (ergometrine).

The somewhat more complex peptides (ergopeptines) are formed via the sequential addition of activated amino acid residues (ATP-mediated) to thioester-bound lysergic acid, affording a linear lysergyl-tripeptide that is attached to the enzyme complex via covalent linkage as an enzyme complex. The cyclized tripeptide residue of ergotamine and other ergopeptines undergoes lactam formation that releases the product from the enzyme, followed by formation of a heimketal-like linkage.

Ergot Alkaloid Pharmacodynamics

The pharmacological effects of the ergot alkaloids as a group tend to be complex and variable, with the net result of their actions being a sum of the effects of partial agonism or antagonism at adrenergic, dopaminergic, and serotonergic receptors. Variables relating to these effects are influenced by the agent, dosage, species, tissue, physiological, and endocrinological state, and experimental conditions.

Therapeutically Significant Lysergic Acid Amide Alkaloids

Ergonovine (Ergometrine, Ergobasine)

Ergonovine was discovered in 4 different laboratories almost simultaneously, with 4 different names (ergometrine, ergotocine, ergosterine, and ergobasine) being assigned to the alkaloid. The names ergometrine and ergobasine have persisted in Europe, while ergonovine was adopted in the United States. The structure of ergonovine was elucidated in 1935 when it was shown that hydrolysis of the alkaloid afforded (+)-lysergic acid and (+)-2-aminopropanol. Ergonovine was introduced into world commerce in 1936 and first synthesized in 1938 via amidation of (+)-lysergic acid with (+)-2-aminopropanol. This represented the first synthesis of an ergot alkaloid. Ergonovine is a light-sensitive, water soluble compound that is commercially marketed as its water soluble maleate salt. The compound is presently obtained from 3 different sources: isolation from field ergot as a minor byproduct, isolation from fermentation broth, and synthesis from (+)-lysergic acid and L-(+)-2-aminopropanol using variable coupling reagents.

Ergonovine is a selective and moderately potent tryptaminergic receptor antagonist in various smooth muscles, being only a partially agonistic or antagonistic at tryptaminergic receptors in the central nervous system. In blood vessels the alkaloid is only weakly antagonistic of dopaminergic receptors and partitally agonistic of α-adrenergic receptors. The most pronounced effect of ergonovine is one of direct stimulation of the uterine smooth musculature, resulting in increased muscular tone and an enhancement of the rate and force of rhythmical contractions.

This stimulant effect seems to be most closely associated with agonist or partial agonist effects at 5-HT2 receptors. Food and Drug Administration (FDA) approved indications are for the treatment and prophylaxis of abortion complicated by delayed and/or excessive hemorrhage, and in the treatment and prophylaxis of postpartum hemorrhage due to uterine atony or subinvolution. The drug is administered after the expulsion of the placenta because prior administration may result in placental entrapment. Ergonovine maleate is typically administered intramuscularly or intravenously, with the intravenous route being reserved for emergency use.

The drug is contraindicated for use in the induction of labor because it may jeopardize placental blood flow and fetal oxygen supply, and in cases of threatened spontaneous abortion, or in pregnancy (FDA Pregnancy Category X). Adverse reactions are generally gastrointestinal and are limited to nausea and vomiting (1%-10%). Ergonovine derivatives are substrates of CYP3A4 metabolism, and as such are contraindicated for concomitant use with compounds established as strong CYP3A4 inhibitors (including protease inhibitors, some macrolide antibiotics, quinolones, azole antifungals) because of the production of acute ergot toxicity.

Finally, the drug has an off-label indication for use as a diagnositic test for Prinzmetal's angina (variant angina, vasospastic angina) in which it has been used successfully for noninvasive diagnosis of coronary vasospasm as a cause of chest pain. The drug is marketed as *Ergotrate* Eli Lilly and Co., Indianapolis, Ind) and is supplied as an intravenous solution (0.2 mg/mL) and oral tablets (0.2 mg).

Methylergonovine

Methylergonovine does not occur naturally in ergot, but was first introduced as a synthetic product for medical use in 1946. The drug is currently prepared via reaction of (+)-lysergic acid with L-(+)-aminobutanol, using different coupling reagents. Unlike its homologue ergonovine, methylergonovine has a low water solubility and as such is marketed as its water soluble maleate salt.

It is used therapeutically much in the same manner as ergonovine, with the FDA approving the compound for the routine management of postpartum uterine atony and hemorrhage, after the delivery of the placenta. Adverse reactions and contraindications are similar to ergonovine. The drug is marketed as *Methergine* (Sandoz Pharmaceuticals, Princeton, NJ) and is administered via the same route and dosage as ergonovine.

Methysergide

Methysergide is another alkaloid-derivative that does not occur naturally in ergot, but was first introduced as a synthetic product into medical use in 1960. The drug is currently prepared via synthesis from (+)-lysergic acid or by methylation of ergonovine (ergometrine, ergobasine) and is marketed as its water soluble maleate salt. This semisynthetic compound is a potent 5-HT2 receptor antagonist that is postulated to stabilize neurotransmission in the trigeminovascular system to block the development of neurogenic inflammation. The FDA-labeled indication is for the prophylaxis of vascular headache, but therapy tends to be limited to those patients who suffer frequent and/or severe and uncontrollable vascular headaches that do not respond to other prophylactic measures.

The compound is not indicated in the symptomatic treatment of acute headaches. The compound is known to produce retroperitoneal, pleuorpulmonary, and endocardial fibrosis on long-term administration in a small number of patients (0.02%). Methysergide is administered as an oral tablet (2 mg) and should not be taken continuously for longer than 6 months, with appropriate therapy requiring a 3-4 week treatment interruption, prior to resumption of therapy. The dosage of the medication needs to be gradually reduced over 2-3 weeks prior to discontinuation in order to avoid rebound headaches.

Some gastrointestinal upset is a common adverse effect. Patients should be advised to discontinue the medication and consult their physician immediately in case of chest pain, cold feeling or bluish discoloration in the extremities, or pain in the calf. The drug is assigned as a FDA Pregnancy Category X compound. Methysergide has been marketed as *Sansert*.

Therapeutically Significant Peptide Alkaloids

Ergotamine

The isolation and naming of ergotamine by Stoll occurred in 1925 but the complete elucidation of structure was not achieved until 1951, with synthesis following some 10 years later. Current sources of ergotamine include the isolation from field ergot and fermentation broth, as well as synthesis via coupling of (+)-lysergic acid with the appropriate synthetic peptidic moiety. Ergotamine was introduced into world commerce in 1921, and is currently marketed as its water soluble tartrate salt.

Ergotamine is a partial agonist at various tryptaminergic receptors (including the serotonin receptor [5-HT2]) and at various α-adrenergic receptors in blood vessels and various smooth muscles. It is likely that the major activity of ergotamine and related alkaloids is one of agonism at the 5-HT1B/1D receptors, just as with the "triptan" antimigraine compounds. FDA-labeled indications for ergotamine tartrate are in the abortion or prevention of vascular headaches, such as migraine, migraine variant, cluster headache, and histaminic cephalalgia. The alkaloid is considered useful in the therapy of moderate to severe migraine attacks in which it acts to constrict intracranial blood vessels and inhibit the development of neurogenic inflammation in the trigeminovascular system.

Both venous and arterial constriction occur at therapeutic dosage. Ergotamine is most effective when administered early in the migraine attack, preferably at the first indication of an impending event. Nausea and vomiting are the most common adverse effects of therapy due to stimulation of the medullary chemoreceptor trigger zone. Other common adverse effects include muscle weakness, fatigue, paresthesias, tightness in the chest, and diarrhea. Ergotamine tartrate is available as both oral (1 mg, with caffeine [100 mg]) and sublingual (2 mg) tablets (*Ergomar*), as well as rectal suppositories (2 mg, with caffeine [100 mg]). Caffeine is added to these preparations to enhance absorption and potentiate analgesia. Oral dosage is 2 mg at the onset of the attack, followed by 1-2 mg every 30 minutes as needed, with maximum dosage being 6 mg/day or 10 mg/week. Rectal dosage is 1-2 mg at onset, followed 1-2 mg every hour, as needed, with a maximum dosage of 4 mg/day or 10 mg/week.

Dosage requirements to achieve an effective therapeutic subnauseating endpoint should be established and adhered to for future attacks. Prolonged administration or excessive dosage may result in severe peripheral vasoconstriction, ergotism, gangrene, or various fibrotic complications (cardiac valvular, retroperitoneal, pleuropulmonary). Ergotamine derivatives are contraindicated in patients with peripheral vascular disease, hepatic or renal disease, coronary artery disease, hypertension, sepsis, or pregnancy (FDA Pregnancy Category X). Ergotamine derivatives are substrates of CYP3A4 metabolism and as such are contraindicated for concomitant use with medications established as strong CYP3A4 inhibitors (protease inhibitors, some macrolide antibiotics, azole antifungals) because of the production of acute ergot toxicity.

Dihydroergotamine

Dihydroergotamine, which does not occur naturally in ergot, was first introduced as a semi-synthetic product in 1946. The drug is currently prepared either by hydrogenation of ergotamine isolated from field ergot/fermentation broth or via synthesis from dihydrolysergic acid and the appropriate synthetic tripeptide. Dihydroergotamine has a very low water solubility and is marketed for parenteral use as its water soluble mesylate (methanesulfonate) salt. FDA-labeled indications for the use of dihydroergotamine mesylate are for the acute treatment of the symptoms of migraine headache (w/wo aura) and for the acute treatment of cluster headache.

The compound is poorly and erratically absorbed from the gastrointestinal tract and for this reason is available for intranasal and parenteral (intravenous, intramuscular, subcutaneous) administration. Following intranasal administration, the relative bioavailability is 30%-40%, with peak plasma concentrations reached in 30-60 minutes. The adverse effect profile resembles that of ergotamine, although dihydroergotamine is rarely associated with serious toxicities. Contraindications, precautions, and drug interactions are similar to those of ergotamine.

Although parenteral administration was once mainly viewed as therapy reserved for inpatient or emergency department treatment for moderate to severe migraine, patients have been trained to administer the drug intramuscularly or subcutaneously. The administration of dihydroergotamine mesylate is not generally associated with the production of rebound headache, but dosage restrictions cited for ergotamine tartrate should be strictly observed to prevent this phenomenon. Dihydroergotamine mesylate is commonly available under the trade names of *D.H.E. 45* injection and *Migranal* nasal spray.

Bromocriptine

Bromocriptine (2-bromo-α-ergocriptine) is not a naturally occurring ergot alkaloid, but is a semisynthetic derivative of the naturally occurring peptide alkaloid α-ergocriptine (α-ergocryptine, α-ergokriptine, α-ergokryptine) that has been isolated from field ergot/fermentation broth or prepared via synthesis from (+)-lysergic acid. Bromocriptine is prepared via the bromination of α-ergocriptine with different brominating agents and is marketed as its water soluble mesylate (methanesulfonate) salt. The compound was introduced onto the world market in 1975. Although bromocriptine exhibits only weakly antagonistic interactions with tryptaminergic and α-adrenergic receptors, the drug is a potent dopamine receptor agonist via activation of certain central dopamine D2 receptors. Bromocriptine is known to stimulate both pre- and post-synaptic sites, promoting the release of dopamine and inhibiting dopamine uptake. The net effect is a decrease in the turnover rate of dopamine without significant changes in concentration.

FDA-labeled indications for bromocriptine are for the therapy of hyperprolactinemia and pituitary prolactinoma. Hyperprolactinemia is a state of persistent elevation of serum prolactin levels and may result in infertility and amenorrhea in females, and galactorrhea in both males and females. Prolactin is a hormone produced by the anterior pituitary, with increased serum levels of prolactin being associated with both benign secreting tumors of the gland (prolactinomas [prolactin-secreting adenomas]) and the use of centrally acting dopamine antagonists. Bromocriptine inhibits the release of prolactin by direct stimulation of postsynaptic dopamine receptors in the hypothalamus.

Other FDA-labeled indications for bromocriptine are in the therapy of acromegaly and as an adjunct to L-dopa therapy in patients with Parkinson's disease who are experiencing a deteriorating response to L-dopa or who are undergoing fluctuations in response to the drug. Other patients who may benefit from bromocriptine therapy include those with limited clinical response to L-dopa owing to an inability to tolerate higher doses. Dopamine agonist drug therapy is associated with an L-dopa-sparing effect and with a decrease in the frequency of off periods. Adverse effects of therapy may be dose-limiting, tending to occur more frequently at the onset of therapy, and being more likely with higher dosage or rapid escalation of dosage.

Gastrointestinal upset (particularly nausea) is common and may also be accompanied by asymptomatic postural hypotension, sedation, light-headedness, and vivid dreams. Daytime sedation, confusion, and hallucinations are psychic effects that are often dose limiting. Ergot-like fibrotic complications are rare but may occur. Bromocriptine is marketed as a tablet (2.5 mg) and a capsule (5 mg) under the name of *Parlodel*. (Sandoz Pharmaceuticals, Princeton, NJ).

Closely related, orally administered synthetic drugs include pergolide mesylate (*Permax*) and cabergoline (*Dostinex*). Pergolide, a potent dopamine receptor agonist at both D1 and D2 receptor sites, has an FDA-labeled indication for use in the therapy of Parkinson's disease and non-FDA labeled indications for the therapy of hyperprolactinemia and Tourette's disorder (syndrome). Cabergoline, a potent dopamine D2 receptor agonist with a low affinity for dopamine D1 receptors, is represented by an FDA labeled indication for use in the therapy of hyperprolactinemia, and non-FDA labeled indications for the therapy of Parkinson's disease, and suppression of puerperal lactation.

Dihydrogenated Ergot Alkaloids

The peptide alkaloid ergotoxine was originally believed to be a single compound, but was subsequently shown to be a mixture of 3 peptide ergot alkaloids: ergocristine (30.0%-36.5%), ergocriptine (ergocryptine) (30.0%-36.5%), and ergocornine (30.0%-36.5%). Ergocristine was first isolated from Iberian ergot in 1937 and later found to be a constituent of ergotoxine in 1943. Both ergocryptine and ergocornine were first discovered as members of the ergotoxine complex in 1943. Later, ergocriptine was shown to be a mixture of 2 compounds α-ergocriptine and β-ergocriptine, in a ratio of 1.5:1.0 – 2.5-1.0.

Each of the ergotoxine alkaloids contains a tripeptide ring, with 2 of the amino acids (valine and proline) being in common in all of the compounds, while the third differs. Dihydroergotoxine is prepared via isolation of the individual alkaloids or their mixtures from field ergot or fermentation broths, followed by catalytic hydrogenation and preparation of the methanesulfonate (mesylate) salts. Alternatively, dihydroergotoxine may be prepared via synthesis of the individual dihydroergotoxines by coupling dihydrolysergic acid with the corresponding synthetic peptidic portions, adjustment to the required ratio of individual components, and preparation of the mesylate salts.

Dihydroergotoxine mesylates, commonly known as ergoloid mesylates, has an FDA-labeled indication for use in the symptomatic therapy of age-related dementia, being employed in individuals over 60 years of age who manifest signs and symptoms of an idiopathic decline in mental capacity. Those who respond are thought to be ones suffering from some process related to aging or have some underlying condition of dementia, such as primary progressive dementia, Alzheimer's dementia, senile onset dementia, or multi-infarct dementia. The mechanism of action of dihydroergotoxine (ergoloid) mesylates in geriatric senility has not been determined.

Although the drug has some α-blocking activity, it has no vasoconstrictor or oxytocic properties and controversy continues to exist as to whether there is any effect on cerebral arteriosclerosis, cerebrovascular insufficiency, or cerebral blood flow. Systemic bioavailability is higher with the liquid capsule than the tablet. Common adverse effects include gastrointestinal upset and headache. The drug is available in a liquid filled oral capsule (1 mg), oral tablet (1 mg), oral solution (1 mg/mL), and sublingual tablet (0.5 mg, 1 mg). Alleviation of symptoms may not be noticeable for 3-4 weeks and up to 6 months of therapy may be necessary to determine efficacy. The drug is a major substrate of CYP3A4, and drug interactions may be anticipated.

Hallucinogenic Ergot Alkaloid Derivatives

Lysergic Acid Diethylamide

Lysergic acid diethylamide (German term: *lysergsaurediethylamid*) was first synthesized in 1938 in a screening of compounds for oxytocic activity and was the 25th semisynthetic ergot that Dr. Albert Hofmann of Sandoz AG in Basel, Switzerland, prepared by combining (+)-lysergic acid with different amines. The compound was tested in comparison with ergonovine and found to have less oxytocic activity and thus Sandoz pharmacologists ceased any further development of this particular formulation. According to company policy, LSD-25 should thereafter have been discounted and never resynthesized. However, on Friday, April 16, 1943, Dr. Hofmann, acting under a "peculiar presentiment" that the compound possessed as yet undiscovered properties, resynthesized the compound. He was interested in further investigating the analeptic activity of LSD-25 in comparison to nikethamide because of the partial structural similarities of the 2 compounds.

Unknown to him, Hofmann accidentally ingested some of the product during the purification and crystallization of lysergic acid diethylamide tartrate, and was compelled to go home in midafternoon and lay down. He experienced a dazed and dream-like state, in which he sensed a distortion of time and was flooded with vivid, highly colored, kaleidoscopic images of extreme plasticity and unusual dimension. Several hours later he returned to normal and inherently understanding that these reactions had to be a result of something that he had come in contact with in the laboratory, was prompted to undertake a series of self-experiments ingesting the smallest quantity of an ergot alkaloid (0.25 mg) that at the time might be expected to produce a biological effect.

On Monday, April 19, he separately ingested every product (in a dosage of 0.25 mg) with which he had worked the previous Friday. About 40 minutes after ingesting a solution of lysergic acid diethylamide tartrate the hallucinations and emotions that he had previously experienced returned. He spoke and wrote only with difficulty, and requested that his laboratory assistant accompany him home. As they rode their bicycles to his home (no automobiles used during WWII, except for designated individuals), Hoffmann's field of vision alternately wavered and became distorted, and he had the sensation of not being able to move from place to place, even though his assistant later confirmed that they had traveled rapidly.

On arriving home, he asked his assistant to call for the family physician and also requested milk from his neighbors. He recovered in about 14 hours and his colleagues at Sandoz were astounded at his description of the event. A few of them ingested small doses of lysergic acid diethylamide (LSD) and confirmed Hofmann's description and experience. Since it is currently believed that marked behavioral changes occur in dosages of 25-50 nanograms (ng = 10^{-9} g), it can be estimated that Hofmann ingested roughly a 10,000-fold overdose of the hallucinogen. This unique man, who made such a staggering discovery in his seminal research with LSD, just recently reached his 100th birthday. Sandoz tried to find a therapeutic, commercial utility for LSD, even providing free samples and financial support to many investigators, but ultimately researchers observed that the experiences of a numerous variety of patients (psychiatric, alcoholic, cancer, and others) included as many adverse reactions ("bad trips") as beneficial ones.

LSD and related hallucinogens are known to interact with brain 5-HT receptors to produce agonist or partial antagonist effects on serotonin activity. This includes both the presynaptic 5-HT1A and 5-HT1B receptors, as well as the postsynaptic 5-HT2 receptors. Theoretically, LSD and other hallucinogens promote glutamate release in thalamocortical terminals, thus producing a dissociation between sensory relay centers and cortical output. It is not known whether the agonist property of these hallucinogens at 5-HT2C receptors contributes to behavioral alterations. LSD has also been demonstrated to interact with many other 5-HT receptors, including cloned receptors whose functions have not yet been determined. In any event, the precise mechanism by which hallucinogens exert their effects remains unsolved.

The onset of action of LSD is within 30-60 minutes, with effects peaking over 1-6 hours and dissipating in 8-12 hours. Acute toxicity includes gastrointestinal upset, chills, hyperglycemia, hypertension, mydriasis, tachycardia and panic, while chronic effects include flashbacks, and exacerbation of latent mental disorders, particularly schizophrenia. Overdosage most commonly results in "bad trips" that are characterized by intense anxiety, combativeness, confusion, and panic. Physical dependence and a withdrawal syndrome are absent and psychological dependence is low. Individuals experiencing a "bad trip" may usually be "talked down" or treated with a benzodiazepine. Because of its relatively high therapeutic index, no deaths have been directly attributed to LSD use alone.

LSD is sold on the illicit market in a variety of forms, including tablets, capsules, and a liquid form. Drops of solution are added to blotter paper, gelatin wraps, gelatin squares, candies, and sugar cubes. Stamped blotter paper containing common or popular cartoon characters is particularly common. When found in a solid form, the substance is a powder or crystal that is incorporated into capsules or tablets.

THE ORIGIN OF DRUGS IN CURRENT USE

Throughout history man has turned to nature for medicines. Microorganisms, such as fungi, offer a huge source of pharmaceutically useful molecules. For the fungi the relatively recent discovery of penicillin marked a new era. Of the top 20 selling prescription medicines in 1995, six of these contained compounds derived from fungi.

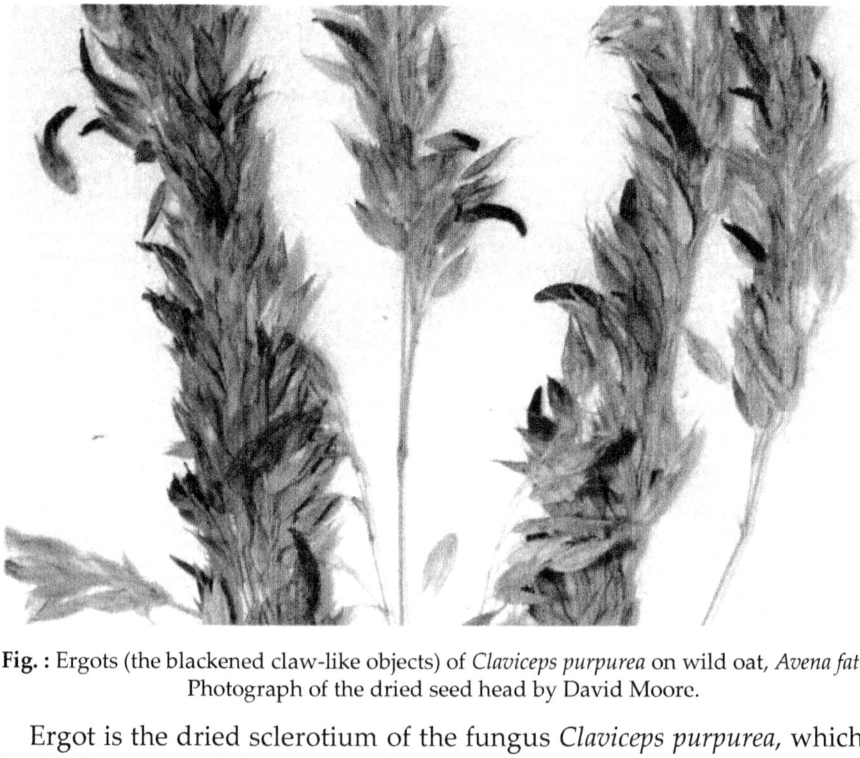

Fig. : Ergots (the blackened claw-like objects) of *Claviceps purpurea* on wild oat, *Avena fatua*. Photograph of the dried seed head by David Moore.

Ergot is the dried sclerotium of the fungus *Claviceps purpurea*, which can arise in the ovary of the rye *Secale cereale*. These sclerotia can replace one or more of the kernels in a mature grain head with a hard, dark coloured, horn-like mass. Although ergot most readily attacks rye it can also infect other wild and cultivated cereals (barley, wheat, oats), and other grasses. These sclerotia contain the fungal toxins and are harvested along with the rest of the grain so that the ergots contaminate the food chain. Widespread contamination by ergot, known as ergotism, occurred frequently during the Middle Ages. Eating rye bread contaminated with C. *purpurea* resulted in gangrene of limbs, disruption in functions of the CNS and even death. Ergot does however have its uses in medicine, with midwives as far back as 1582 using it to reduce haemorrhage after childbirth. Ergotamine, an alkaloid of ergot, has been found to be useful in the treatment of migraine headaches.

History

It is possible that ergot infected grasses produced in the first agricultural settlements of Mesopotamia around 9000 BC, but ergot is thought to have first been mentioned around 600 BC by the Assyrians. The Roman historian Lucretius (98-55 BC) referred to ergotism as *'Ignis sacer'*, meaning Holy Fire, which was the name given to ergotism during the Middle Ages, and it was during these times that ergotism occurred frequently.

The word ergot is actually derived from an old French word argot, meaning the cock's spur. The violet or black sclerotia formed by ergot consist of a mass of hyphae, and may be two to ten times the length of a normal kernel. When bread was prepared from grain contaminated by the black spurs, epidemics of ergotism occurred. Poor people mainly ate rye bread, especially during famines, when even the spurs were collected because of hunger. The first mention of gangrenous ergotism was in Germany in 857 AD, and the first epidemic of convulsive ergotism occurred in 945 in Paris. These are the two distinct types of ergotism, with the gangrenous type seen mostly in France and the convulsive one in Germany. The name Holy Fire or St. Anthony's Fire given to ergot epidemics during the Middle Ages referred to the burning sensation in the limbs experienced by sufferers.

Ergot was probably first used in medicine as an oxytocic drug. In 1582 Adam Loncier in Germany made the first note of ergot stimulating uterine contractions of labour by administering three sclerotia. It was the most effective drug for this purpose at the time resulting in a rapid and sudden termination of labour, with a delivery time lasting less than three hours. But ergot was eventually deemed unsuitable for this purpose as the dosage could not be given accurately due to large variations in the active ingredients. Ergot also caused severe adverse effects, such as violent nausea and vomiting. And it was in 1822 when Hosack from New York stated that many stillbirths were due to uterine rupture resulting in maternal death, and the use of ergot as an oxytocic was virtually abandoned by the end of the 19th century. It was only during the 20th century that ergot was shown to be useful in the treatment of attacks of migraine. This would mainly involve the alkaloid of ergot, ergotamine.

Life Cycle

As already mentioned, ergot is derived from the fungus *Claviceps purpurea* and is formed in the body of the sclerotium of the fungus that grows mainly in the seed of the rye plant and various other cereal grasses. There are 32 recognized species of ergot. The sclerotia of the fungus will remain dormant through periods of winter or drought and will only germinate under favourable, moist conditions to give rise to small club-shaped fruiting bodies. The heads of these fruiting bodies are shiny and spherical and produce long, filamentous ascospores, which are the sexual spores of the fungus.

The ascospores are ejected from the fruit body of the fungus and are blown by the wind. If they land on a healthy stigma of a grass flower or rye plant, they germinate and their hyphae enter, as do pollen tubes, into the plant's ovary and begin to form a fine mycelial network. Soon after infection, a special structure called the sphacelium develops between the ovary and the base of the floral cavity. The sphacelium produces hundreds of thousands of conidia. These conidia are the asexual spores of the fungus. They are embedded in a sugary matrix called honeydew, a yellowish saccharine secretion produced by the sphacelia, which will reduce Fehling's solution. The honeydew is very sticky and its strong, fetid smell attracts insects to the plant to feed on these sugars. When the insects move on to another plant, conidia stuck to their legs and mouthparts infect the healthy ovaries of other plants. After several weeks, the sclerotium develops between the sphacelium and the flower base, bringing us back to the beginning of the cycle and completing the life cycle.

Composition of Ergot

The ergot sclerotium contains up to 30-40% of fatty oils and up to 2% of alkaloids. The other components of the sclerotium are free amino acids, ergosterin, choline, acetylcholine, betaine, ergothionine, uracil, guanidine, free aromatic and heterocyclic amines (tyramine, histamine, agmatine) and alkylamines (the natural representatives of which were originally found in ergot). The outer shell of sclerotium contains acid pigments belonging to anthraquinolinic acid derivatives, including orange- red (endocrinin, clavorubin) and light yellow (ergochromes, ergochrysins). These pigments form part of the sclerotium shell giving rise to the greyish-brown-violet colour. Albino ergot strains also exist which are incapable of producing pigments. Naturally growing*Claviceps purpurea* species has several geographic types differing in both qualitive and quantitative composition of alkaloids. So ergot strains are specially selected that are capable of producing predominantly a single alkaloid or a certain group of alkaloids: ergotamine, ergotoxine, ergocristine, *etc.*

The ergot alkaloids have a high biological activity and a broad spectrum of pharmacological effects, hence they are of considerable importance to medicine. They have adrenoblocking, antiserotonin and dopaminomimetic properties. Ergot alkaloids have a therapeutic effect on some forms of migraine, post-partum haemorrhages, mastopathy, and a sedative effect on the central nervous system. These compounds are now obtained both by methods of artificial parasitic cultivation on rye and by techniques using *in vitro* culture.

Naturally occurring ergot consists of two types of alkaloids. The first, the clavine-type alkaloids, are derivatives of 6,8-dimethylergoline. The second type comprises the lysergic acid derivatives, which are peptide alkaloids. All ergot alkaloids can be considered as derivatives of the tetracyclic compound 6-methylergoline. It is the lysergic acid derivatives, or peptide alkaloids, that are the pharmacologically active alkaloids. Each active alkaloid occurs with an inactive isomer involving isolysergic acid. These alkaloids have been studied over many years and were not easy to characterize. Since its isolation in 1906 (by Barger and Carr and independently by Kraft), ergotoxine had been accepted as a pure substance, and in the form of ergotoxine ethanosulphate was formerly used as a standard. It was shown to be a mixture of the three alkaloids ergocristine, ergocornine and ergocryptine.

Six pairs of alkaloids predominate in the sclerotium and fall into the water-soluble ergometrine group or the water-insoluble ergotamine and ergotoxine groups. Alkaloids of groups 2 and 3 are polypeptides in which lysergic acid or isolysergic acid is linked to other amino acids. In the ergometrine alkaloids lysergic acid or its isomer is linked to an amino alcohol. Ergometrine was isolated in 1932 by Dudley & Moir, and was synthesized by Stoll & Hoffmann in 1943.

Table: Alkaloids of the ergot sclerotium

Group	Alkaloid	Formula
1.Ergometrine group	Ergotmetrine	$C_{19}H_{22}O_2N_3$
	Ergotmetrinine	
	Ergotamine	$C_{33}H_{35}O_5N_5$
2.Ergotamine group	Ergotaminine	
	Ergosine	$C_{30}H_{37}O_5N_5$
	Ergosinine	
	Ergocristine	$C_{35}H_{39}O_5N_5$
3.Ergotoxine group	Ergocristinine	
	Ergocryptine	$C_{32}H_{41}O_5N_5$
	Ergocryptinine	
	Ergocornine	$C_{31}H_{39}O_5N_5$
	Ergocorninine	

Alkaloid Production in Artificial Cultures

Ergot alkaloids can be produced in submerged fermentation by *Claviceps* or *Penicillium* species, which are used for their industrial production. Initial work in Japan showed that submerged cultures did not produce the typical alkaloids associated with the sclerotium but instead produced a series of new non-peptide bases (clavines) which did not posses any significant pharmacological action. Attempts were made by many workers to influence alkaloid production by modification of the culture medium and the fungus strain.

The first pure ergot alkaloid, ergotamine, was obtained by Stoll in 1920. Moir reported the discovery of the "water soluble uterotonic principle of ergot" in 1932. This was subsequently determined to be ergonovine (also called ergometrine). As a result of successful experiments in 1960, the commercial manufacture of simple lysergic acid derivatives by fermentative growth of a strain of *Claviceps paspali* became feasible.

The final fermentation broth contains a complex mixture of alkaloids, salts, polysaccharides, fats, solids, *etc.*; and organic solvent extraction is commonly used for the separation of ergot alkaloids from the broth. To avoid the formation of emulsions during the extraction, a separation process taking advantage of solid-liquid adsorption of alkaloids on activated carbon bentonite and other silicate sorbents was developed in 1986. In 1988 selective adsorption of polycarboxylic ester resin XAD-7 to isolate alkaloids from plant cell cultures was used.

In 1990 the Kawaken Fine Chemicals Company submitted two patents for isolation and purification of ergot alkaloids from *Claviceps* species fermentation broth by adsorption on porous, synthetic polymer, ion-exchanging adsorbents such as styrene divinylbenzene copolymer or phenolic resin. The fermentation broth was passed through a packed bed column and subsequently washed by water and eluted by methanol. Methanol was removed under reduced pressure. The residue was extracted with ethyl acetate and dried with a yield of 68% and purity of 94%. In more recent studies, the simultaneous uptake of agroclavine, elymoclavine, chanoclavine and chanoclavine aldehyde by inorganic sorbents has been investigated.

Chemistry

The naturally occurring alkaloids contain a substituent in the β configuration at position 8 and a double bond in ring D. The natural alkaloids of therapeutic interest are amide derivatives of δ-lysergic acid; these compounds contain a double bond between C9 and C10 and thus belong to the family of 9-ergolene compounds. Many alkaloids, containing either a methyl or a hydroxymethyl group at position 8, are present in ergot in small quantities. These are the clavine alkaloids and consist mainly of both 9-ergolenes (*e.g.* lysergol) and 8-ergolenes (*e.g.* elymoclavine, the 8-ergolene isomer of lysergol).

Amine alkaloids and congeners

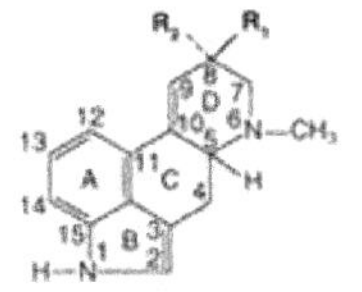

Amino acid alkaloids

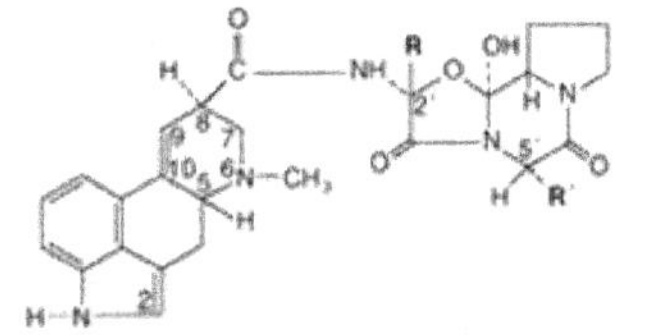

Alkaloid	R_1	R_2
d-Lysergic acid	$-COOH$	$-H$
d-Isolysergic acid	$-H$	$-COOH$
d-Lysergic acid diethylamide (LSD)	$-C(=O)-N(CH_2CH_3)_2$	$-H$
Ergometrine	$-C(=O)-NH-CH(CH_3)CH_2OH$	$-H$
		$-H$
Methylergometrine	$-C(=O)-NH-CH(CH_2CH_3)(CH_2OH)$	$-H$
Methysergide[a]	$-C(=O)-NH-CH(CH_2CH_3)(CH_2OH)$	$-H$
Lisuride	$-H$	$-NH-C(=O)-N(CH_2CH_3)_2$
Lysergol	$-CH_2OH$	$-H$
Lergotrile[b,c]	$-CH_2CN$	$-H$
Metergoline[a,b]	$-CH_2-NH-C(=O)-O-CH_2-$phenyl	$-H$

Alkaloid[d]	R(2′)	R′(5′)
Ergotamine	$-CH_3$	$-CH_2-$phenyl
Ergosine	$-CH_3$	$-CH_2CH(CH_3)_2$
Ergostine	$-CH_2CH_3$	$-CH_2-$phenyl
Ergotoxine group		
ergocornine	$-CH(CH_3)_2$	$-CH(CH_3)_2$
ergocristine	$-CH(CH_3)_2$	$-CH_2-$phenyl
α-ergocryptine	$-CH(CH_3)_2$	$-CH_2CH(CH_3)_2$
β-ergocryptine	$-CH(CH_3)_2$	$-CH(CH_3)CH_2CH_3$
Bromocriptine[e]	$-CH(CH_3)_2$	$-CH_2CH(CH_3)_2$

a Contains methyl substitution at N1
b Contains hydrogen atoms at C9 and C10
c Contains chlorine atom at C2
d Dihydro derivatives contain hydrogen atoms at C9 and C10
e Contains bromine atom at C2

Fig. : An abbreviated catalogue of naturally occurring ergot alkaloids.

Optical isomerism exhibited by the ergot alkaloids is due to the two asymmetrical carbon atoms (position 5 and 8) in the lysergic part of the molecule. Derivatives of *l*-lysergic acid (the epimer at position 5) and of δ-isolysergic acid (the epimer at position 8) display relatively little biological activity. Upon hydrolysis, ergonovine and its derivatives yield lysergic acid and an amine; consequently they are called amine alkaloids. The alkaloids of higher molecular weight yield lysergic acid, ammonia, pyruvic acid (or a derivative of), proline, and one other amino acid (either phenylaline, leucine, isoleucine, or valine) and so are known as amino acid alkaloids.

Numerous semisynthetic derivatives of the ergot alkaloids have been prepared, and several are of therapeutic interest. The earliest derivatives were prepared by the catalytic hydrogenation of the natural alkaloids, yielding a series of compounds that are saturated in ring D of lysergic acid. These have been designated dihydroergotamine, dihydroergocristine, and so on, and possess different pharmacological properties than do the parent alkaloids. Another ergopeptine derivative is bromocriptine (2-bromo-α-ergocryptine). It is also possible to prepare different amides of lysergic acid. Two products of this series, lysergic acid diethylamide (LSD) and lysergic acid hydroxybutylamide (methylergonovine) are of pharmacological interest. Methylation of the indole nitrogen of the latter compound yields 1-methylmethylergonovine (methysergide). A large number of related compounds that are not derivatives of lysergic acid have also been prepared. These include lisuride, lergotrile

(2-chloro-6-methyl-8β-cyanomethyl-ergoline), and metergoline (1,6-dimethyl-8β-carbobenzoxyaminomethyl-ergoline).

Pharmacological Properties

In general, the effects of all the ergot alkaloids appear to results from their actions as partial agonists or antagonists at adrenergic, dopaminergic, and tryptaminergic receptors. The spectrum of effects depends on the agent, dosage, species, tissue, and experimental or physiological conditions. However, some of the actions of the ergot alkaloids are not entirely compatible with this view:

(1) while agonistic effects are generally apparent only at concentrations that are lower than those required to observe antagonism, this is not always the case (*e.g.* the action of methysergide on cerebral blood vessels);

(2) the effects of full agonists (*e.g.* norepinephrine) are usually augmented by low concentrations of ergot alkaloids, even those with weak efficacy as partial agonists (*e.g.* the action ergonovine on arterioles); and

(3) the contractile responses to other agents, such as acetylcholine or angiotensin, are sometimes also augmented by adrenergic or tryptaminergic blocking agents. These and other observations emphasize the importance of the physiological or pathophysiological state in determining the spectrum and intensity of effects produced in animals or patients.

Apart from the stereochemical considerations mentioned above, few rules governing structure-activity relationships have emerged. In general, small amide derivatives of lysergic acid are potent and relatively selective antagonists of 5-HT, while the amino acid alkaloids are usually less selective and show similar affinities as blocking agents at α-adrenergic and tryptaminergic receptors. Dihydrogenated derivatives usually have fewer and less intense agonistic actions than do parent alkaloids. Finally, insertion of a methyl group at position 1 usually results in compounds with less affinity for receptors for catecholamines and with more selective ability to block tryptaminergic receptors.

All of the natural alkaloids of ergot significantly increase the motor activity of the uterus. After small doses contractions are increased in force or frequency, or both, but are followed by a normal degree of relaxation. As the dose is increased, contractions become more forceful and prolonged, resting tonus is markedly increased, and sustained contracture can result. Although this characteristic precludes their use for induction or facilitation of labour, it is quite compatible with their use postpartum or after abortion to control bleeding and maintain uterine contraction. The gravid uterus is very sensitive, and small doses of ergot alkaloids can be given immediately postpartum to obtain a marked uterine response, usually without significant side effects.

Effects on the Uterus

Although all natural ergot alkaloids have qualitatively the same effect on the uterus, ergonovine is most active and also less toxic than ergotamine. For these reasons ergonovine and its semi-synthetic derivative methylergonovine have replaced other ergot preparations as uterine-stimulating agents in obstetrics. Methylergonovine differs little from ergonovine in its uterine actions. The dihydrogenated alkaloids do not have the uterine-stimulating properties of the parent alkaloids when tested in experimental animals; however they are capable of exerting a marked uterine-stimulating action on the pregnant human uterus at term.

Ergonovine and methylergonovine are rapidly and virtually completely absorbed after oral administration and reach peak concentrations in plasma within 60 to 90 minutes that are ten times those achieved with an equivalent dose of ergotamine. A uterotonic effect can be observed within 10 minutes after oral administration of 0.2 mg of ergonovine to women postpartum. Judging from the relative duration of action, ergonovine is metabolised and/or eliminated more rapidly than ergotamine. The half-life of methylergonovine in plasma ranges between 0.5 and 2 hours.

Effects on the Cardiovascular System

The natural amino acid alkaloids, particularly ergotamine, constrict both arteries and veins. While dihydroergotamine retains appreciable vasoconstrictor activity, it is far more effective on capacitance than on resistance vessels. This property is the basis for investigation of its usefulness in the treatment of postural hypotension. The dihydrogenated derivatives of the ergotoxine group are considerably less active and usually produce hypotension because of effects on the CNS.

In doses used in the treatment of migraine, the rectal administration of ergotamine produces little change in blood pressure but does cause a slowly progressing increase in peripheral vascular resistance that persists for up to 24 hours. This could reflect constriction of arteries mediated by stimulation of trypaminergic receptors. Dihydroergotamine has relatively little capacity to produce such effects in human beings.

At the higher plasma concentrations achieved by intravenous administration, both ergotamine and dihydroergotamine cause a rapid increase in blood pressure that dissipates in a few hours. This is thought to reflect action on arterioles, which in rats results from stimulation of $\alpha 2$-adrenergic receptors. With the notable exception of the cerebrum, the prolonged increase in vascular resistance is accompanied by decreased blood flow in various organs. This results in part from reduced flow through non-nutritive arteriovenous anastomoses.

While less potent than ergotamine, the amine alkaloids can also raise blood pressure slightly and decrease blood flow in the extremities when administered in therapeutic doses. The intensity of pressor effects is greater when the blood pressure is raised. Ergot alkaloids that produce peripheral vasoconstriction can also damage the capillary endothelium. The mechanism of this toxic action is poorly understood. Vascular stasis, thrombosis, and gangrene are prominent features of ergot poisoning. The natural tendency of these alkaloids to cause gangrene appears to parallel their vasoconstrictor activity.

Pharmacological Properties Related to Treatment of Migraine

Ergot derivatives were first found to be effective anti-migraine agents in the 1920s, and they continue to be a major class of therapeutic agents for the acute relief of moderate or severe migraine. However, ergot alkaloids are non-selective pharmacological agents in that they interact with numerous neurotransmitter receptors, including all known 5-HT1 and 5-HT2 receptors, as well as adrenergic and dopaminergic receptors. As shown in Table 2, the ergot alkaloid dihydroergotamine (DHE) can compete for radioligands for binding to a variety of receptor subpopulations.

DHE is potent at all known 5-HT1 receptors as well as at a number of other biogenic amine receptors, such as 5-HT2A, 5-HT2B, D2 dopamine, and α1- and α2-adrenergic receptors. The multiple pharmacological effects of ergot alkaloids have complicated the determination of their precise mechanism of action in the acute treatment of migraine. These findings should be compared with those for the selective 5-HT receptor agonist, sumatriptan, which has its highest affinity for 5-HT1D and 5-HT1B receptors, with lower affinity for 5-HT1A and 5-HT1E receptors.

Table 2. Ergot alkaloid reactivity with neurotransmitter receptors

Receptor	Ki Values (nM)	
	Dihydroergotamine	**Sumatriptan**
Serotonergic		
5-HT1D	0.55	6.7
5-HT1A	0.83	120
5-HT1B	6.2	35
5-HT1E	8.8	920
5-HT2C	39	>10,000
5-HT2A	78	>10,000
5-HT3	>10,000	>10,000
Adrenergic		
a1	6.6	>10,000
a2	3.4	>10,000
b	960	>10,000
Dopaminergic		
D1	700	>10,000
D2	98	>10,000
Other sites		
Muscarinic	>10,000	>10,000
Benzodiazepine	>10,000	>10,000

Absorption and Excretion

The oral administration of ergotamine by itself results in undetectable systemic drug concentrations because of extensive first-pass metabolism. Bioavailability after sublingual administration also is poor and is often inadequate for therapeutic purposes. Although the con-current administration of caffeine improves both the rate and extent of absorption, the bioavailability of ergotamine is probably still less than 1%. The bioavailability after administration of rectal suppositories is greater, and maximal plasma concentrations of ergotamine of over 400 pg/ml can be achieved following a 2-mg dose, compared to peak plasma concentrations of approximately 20pg/ml in plasma 70 minutes after a 2-mg dose taken orally.

Ergotamine is metabolised in the liver by largely undefined pathways, and 90% of the metabolites are excreted in the bile. Only traces of unmetabolized drug can be found in urine and faeces. Ergotamine produces vasoconstriction that lasts for 24 hours or longer, despite a plasma half-life of about 2 hours. Dihydroergotamine is much less completely absorbed and is eliminated more rapidly than ergotamine, presumably due to its rapid hepatic clearance.

Ergot Derivatives and Parkinson's Disease

Recent studies involving ergot derivatives in the treatment of Parkinson's disease have investigated their effects on dopamine receptors. Most dopamine agonists employed at present are ergot derivatives. The similarity of the ergoline ring structure to the endogenous monoamines explains the action of these compounds on dopaminergic, serotonergic and adrenergic receptors. Bromocriptine, lisuride and pergolide are the three oral ergot dopamine agonists presently available.

Bromocriptine, which has potent D2 agonist properties and a weak D1 receptor antagonistic effect, was the first ergoline to be successfully used in Parkinson's disease. The synthetic ergoline derivative pergolide is a long acting agonist at D2 dopamine receptors and has a mild D1 receptor agonistic effect. Both bromocriptine and pergolide are effective in relieving the symptoms of Parkinson's disease and can achieve a reduction of on-off fluctuations. Comparative studies of both drugs have shown that they have similar efficacy and adverse effects. The efficacy of a long acting, slow release formulation of bromocriptine has been shown to bc equivalent to standard treatment with bromocriptine, but patients needed fewer doses daily and had fewer adverse effects.

Another ergot derivative, lisuride, stimulates postsynaptic striatal D2 receptors and is a mild D1 receptor agonist. The antiparkinsonian efficacy of this drug is equivalent to that of bromocriptine and pergolide. Lisuride is highly water-soluble and can therefore be used for parenteral therapy via ambulatory infusion pumps. This compound has been shown to be very effective in controlling motor fluctuations in Parkinson's disease when administered by continuous infusion. However, long-term studies of lisuride have shown that its parenteral use is complicated by a high incidence of psychiatric adverse effects, possibly because of its serotonergic properties.

MIND-ALTERING PLANT ALKALOIDS

The chapter will discuss plants containing different types of mind-altering alkaloids. It will contain photos of some of the plants, illustrations of their potent alkaloids, simplified explanations of the mechanisms of these poisons, and anecdotal tidbits not usually found in this type of subject matter.

Some of the plants discussed in this chapter contain very poisonous alkaloids which can be lethal if ingested in sufficient quantities. Native people developed time-tested religious rituals using these plants that were passed down through countless generations.

This chapter is mostly about the adaptive advantage of alkaloids in plants, and some remarkable examples of psychoactive (hallucinogenic) alkaloids found in some of the most amazing plants on earth. Stimulant (or tranquilizer) alkaloids such as nicotine from tobacco (Nicotiana tabacum), caffeine from coffee (Coffea arabica), and cocaine from coca (Erythroxylum coca) are highly addictive and clearly alter one's mental and physical state, but are beyond the scope of this chapter. Likewise, narcotic, pain-relieving alkaloids from the opium poppy (Papaver somniferum), including codeine and morphine, are excluded from this chapter. Although marijuana (Cannabis sativa) is often discussed in books about mind-altering plants, its active chemical tetrahydrocannabinol (THC) is not an alkaloid. It does not contain nitrogen and consists of both terpene and phenolic subunits--hence, it is called a terpenophenolic compound.

During the past 130 million years, flowering plants have colonized practically every habitat on earth, from arid deserts, boggy meadows and windswept alpine summits, to sun-baked grasslands, lush rain forests and wave-battered rocky shores. They have replaced most of the ancient ferns and seed plants that dinosaurs subsisted on, and developed a complex and fascinating relationship with insects and mammals. During these countless centuries of time, flowering plants have gradually evolved all sorts of ingenious protective devices to discourage hungry herbivorous animals. Leaves and stems have developed a variety of vicious spines and stinging hairs (trichomes).

In some plants, the dense covering of silvery hairs may also provide other ecological advantages such as solar reflection and insulation in arid environments. But mechanical defenses, such as spines and trichomes, are of limited value and probably would not deter all hungry herbivores, particularly the chewing and sucking insects. Therefore, plants have developed a "chemical warfare," a defense strategy based on a vast arsenal of chemicals which are toxic or distasteful to animals.

According to Daniel Janzen, noted authority of the Costa Rican rain forest, developing seeds of the tropical liana Mucuna (including M. urens and M. pruriens) are nearly free from seed predators. [These vines produce the distinctive hamburger-shaped "sea beans" that drift ashore on tropical beaches throughout the world.] In addition to a dense covering of stinging trichomes, the pods of Mucuna are rich in the potentially toxic amino acid L-dopa. [L-dopa, precursor of the brain neurotransmitter dopamine, is given to patients suffering from Parkinson's Disease.] Just as genetic variability and time allow agricultural pests to tolerate pesticides, so can some herbivores circumvent a plant's natural chemical defenses.

This "predatory pressure" has resulted in the evolution of an endless array of complex plant molecules, from gums and terpenes to alkaloids and phenolic compounds. For example, in nettles (Urtica species) the sophisticated defense chemicals acetylcholine and histamine are employed in an ingenious system of "injection hairs" strategically placed throughout the plant. When we touch these plants, we may be accidental casualties in a chemical warfare between plants and herbivores that has waged through countless millennia.

One of the largest groups of chemical arsenals produced by plants are the alkaloids. Many of these metabolic by-products are derived from amino acids and include an enormous number of bitter, nitrogenous compounds. According to R.F. Raffauf, more than 10,000 different alkaloids have been discovered in species from over 300 plant families. Alkaloids often contain one or more rings of carbon atoms, usually with a nitrogen atom in the ring. The position of the nitrogen atom in the carbon ring varies with different alkaloids and with different plant families.

In some alkaloids, such as mescaline, the nitrogen atom is not within a carbon ring. In fact, it is the precise position of the nitrogen atom that effects the properties of these alkaloids. Although they undoubtedly existed long before humans, some alkaloids have remarkable structural similarities with neurotransmitters in the central nervous system of humans, including dopamine, serotonin and acetylcholine. The amazing effect of these alkaloids on humans has led to the development of powerful pain-killer medications, spiritual drugs, and serious addictions by people who are ignorant of the properties of these powerful chemicals.

A Tree Protected by Ants Instead of Alkaloids

Thousands of different alkaloids have been discovered from throughout the plant kingdom, but there are some species that do not contain any of these bitter, nitrogenous compounds. One of the most remarkable examples of the "competitive disadvantage" of a tree unable to produce alkaloids is the bullhorn acacia (Acacia cornigera) of Central America. The common name of "bullhorn" refers to the large, swollen thorns (technically called stipular spines) that occur in pairs at the base of leaves, and superficially resemble the horns of a steer. In fact, the thorns of A. cornigera, and a related species A. collinsii, are so striking that they are often strung into unusual necklaces and belts. In El Salvador the horn-shaped thorns provide the legs for small ballerina seed dolls which are worn as decorative pins.

According to the famous botanist G.L. Stebbins, other species ofAcacia growing in the same region of Central America (including A. chiapensis and A. macrantha) produce alkaloids, but grow much more slowly and live in drier areas than A. cornigera. Apparently these other species of Acacia pay the price for their alkaloid defense system in terms of a slower growth rate, which enables them to compete only in the drier regions where the growth of competing vegetation is also slower and less vigorous. The ability of A. cornigera to enter more moist, lush vegetation areas is apparently related to its evolution of another defense system that depends on a little hymenopteran helpmate rather than the production of bitter alkaloids.

The "swollen-thorn" acacias of Central America are truly remarkable trees. In the wild, their enlarged, hollowed-out stipular spines are occupied by fiercely biting-stinging ants that protect them from browsing herbivores and epiphytic plants that might shade them out. The swollen thorns are not galls, they are not produced in response to chemical or physical stimuli from invasive insects imbedded in their tissues. The amazing thorns are genetically-programmed structures that are formed with or without the presence of symbiotic ants. On wild trees they are inhabited by colonies of Pseudomyrmex ferruginea, very aggressive ants with a painful sting. Disturbed ants release an alarm pheromone and rush out of their thorn "barracks" in great numbers. According to Daniel Janzen, livestock can apparently smell the pheromone and avoid these acacias day and night.

Getting stung in the mouth and tongue is an effective deterrent to browsing on the tender foliage. In addition to protecting A. collinsii from leaf-cutting ants and other unwanted herbivores, the ants also clear away invasive seedlings around the base of the tree that might overgrow it and block out vital sunlight. According to Daniel Janzen, the Central American swollen-thorn acacias lack the chemical defenses of most other acacias to discourage ravaging insect predators and competition, and symbiotic ants have taken over this vital role. The acacias reward their ant helpmates with thorn "condos" to live in, carbohydrate-rich nectar from glands on the leaf stalks, and nourishing, protein-lipid morsels called Beltian bodies on the leaflet tips. There is no known function for Beltian bodies, except to provide food for symbiotic ants.

Prior to settling on a thorn acacia, the winged virgin queen ant goes on a mating flight to the highest treetop or nearby hill. Here she gets inseminated by a winged male and then hunts for an acacia in which to lay her eggs. [This "hilltopping" phenomenon also occurs in San Diego County, California, where sexually mature adult male and female butterflies fly upslope to a rendezvous point at the summit!] The queen ant cuts an entrance hole into a green thorn, hollows it out, and then deposits her eggs. Subsequent entrance holes are cut by the new generations of worker ants. All this astonishing, complicated behavior has evolved in a tree that for some reason doesn't synthesize alkaloids.

An Acacia Protected By Ants Instead of Alkaloids

Chile peppers in the wild may also benefit from a naturally-occurring alkaloid present in their fruits. The active ingredient causing the intense burning pain when you chew into the walls of these fruits, especially the placental region where the seeds are attached, is the alkaloid capsaicin. So potent is this alkaloid that one millionth of a drop can be detected by the human tongue. Capsaicin is not broken down during the digestion process--this is why you often get burned several hours later after dining on chile peppers. Like other alkaloids in the chemical arsenal of plants, capsaicin may serve to discourage mammalian fruit predators. Botanists believe that birds are immune to the burning sensation of capsaicin, and may serve to disperse the seeds. Capsaicin may prevent hungry mammals from devouring the fruits, so that they can be eaten by fruit-eating birds who are attracted to bright red fruits. Passing through the bird's digestive tract relatively unharmed, the small seeds are dispersed to other favorable regions. According to D. Dewitt and P. W. Bosland, there are 5 species of Capsicum peppers native to the New World: C. pubescens, C. baccatum, C. annuum, C. frutescens and C. chinense. The hottest chile peppers belong the C. chinensegroup, including the notorious habanero. Although this species is named "chinense," it is not from China. Actually, its center of origin is thought to be the Amazon Basin of South America.

The Deadly Datura

There are approximately 25 different species of Datura throughout the world, including Europe, Africa, southeast Asia, Central and South America, Mexico and the United States. They are often called jimsonweed or "thornapple." The latter name refers to the spiny seed-bearing capsules. Most of the species are low, shrubby or sprawling annuals or perennials, but some tree-like forms may reach 11 meters (36 feet) in height. The tree-like forms are occasionally cultivated as ornamentals and are usually placed in the genus Brugmansia. They generally have the same characteristic trumpet-shaped flowers as Datura with striking color variations of red, pink and yellow. Both genera belong to the Nightshade Family (Solanaceae), along with the deadly nightshade (Solanum dulcamara), tomato (Lycopersicum esculentum), eggplant (Solanum melongena) and potato (S. tuberosum). Many species of Solanum (called nightshades) are quite poisonous to humans if eaten raw. Incidentally, the new sprouts of potato tubers contain the toxic alkaloids solanine and solanidine. Until about a century ago the tomato was thought to be poisonous because of its toxic relatives.

The common native Datura of the western United States, with rank-smelling foliage and large, white flowers up to 20 centimeters (8 inches) long is D. wrightii. Sometimes the flowers are tinged with purple and resemble oversized petunia or morning glory blossoms. It is a sprawling perennial with an enormous taproot that may extend more than 60 centimeters (2 feet) into the ground. D. wrightii has a large geographic range, including California, Utah, Arizona, New Mexico and Texas. It is often listed in older references as D. meteloides, a name that actually applies to a similar Mexican species, D. inoxia ssp. inoxia. Another species with smaller flowers, D. discolor, is native to desert washes and riverbeds of southeastern California and Arizona.

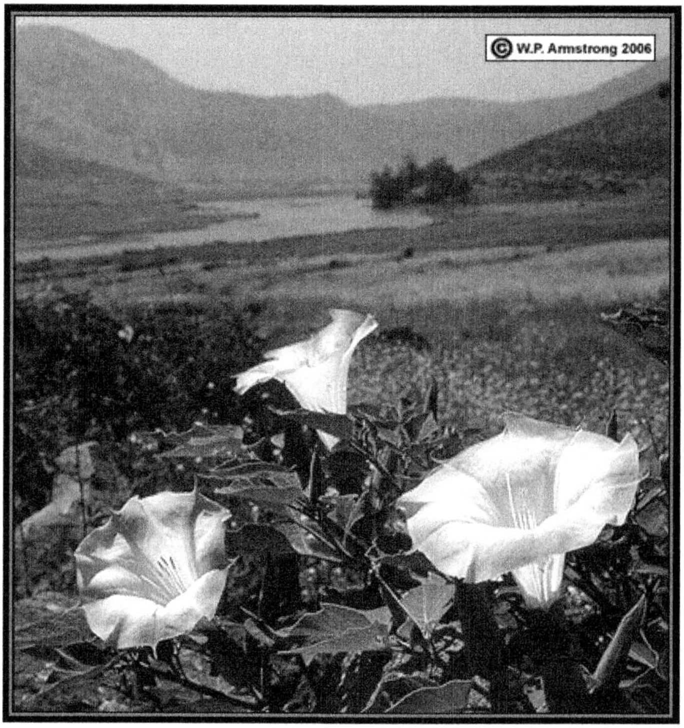

The large, trumpet-shaped flowers of jimsonweed (Datura wrightii) are sometimes tinged with purple and resemble giant morning glory blossoms. It actually belongs to the tomato family (Solanaceae) along with petunias, potatoes and tomatoes. This remarkable plant is common along roadsides and riverbeds throughout the western United States, and is one of the largest and most striking of all native wildflowers.

Athird species, Datura stramonium, is naturalized throughout the United States. Most older references list D. stramonium as an Old World species naturalized as an weed in North America. According to Dr. Robert Bye of the Universidad Nacional Autonoma de Mexico , D. stramonium probably originated in the New World and migrated to the Old World during pre-Columbian time.

It can readily be distinguished from D. wrightii by its more erect habit, smaller flowers and distinctive angular or prismatic calyx. It is a cosmopolitan annual weed and prolific seed-producer. A dozen viable seeds planted in fertile soil in spring may give rise to nearly 50,000 seeds by late summer, an increase of more than 400,000 percent.

Although jimsonweed is synonymous with several species of Datura, this unusual common name is actually derived from D. stramonium. In 1676 British soldiers stationed in Jamestown, Virginia became intoxicated by D. stramonium when it was inadvertently included in their salads by the regimental cooks. The episode was widely publicized and the plant culprit became known as "Jamestown weed", and later as jimsonweed.

The leaves, stem, root and fruits of Datura contain a battery of tropane alkaloids, the most potent of which are atropine, hyoscyamine and scopolamine. These alkaloids affect the central nervous system, including nerve cells of the brain and spinal cord which control many direct body functions and the behavior of men and women. They may also affect the autonomic nervous system, which includes the regulation of internal organs, heartbeat, circulation and breathing. One autonomic response of atropine is the dilation of pupils, once considered to be a beautiful and mysterious look in Italian women. In fact, belladonna means "beautiful lady," so named because sap from the closely related belladonna plant (Atropa belladonna) was used as eye drops to dilate the pupils.

The action of tropane alkaloids at the cellular level is complex. It is summarized by R.E. Schultes and A. Hofmann in the The Botany and Chemistry of Hallucinogens (1973) and in Medical Botany by W.H. Lewis and M.P.F. Elvin-Lewis (1977). Tropane alkaloids are found in many other poisonous plants, including henbane (Hyoscyamus niger), pituri (Duboisia spp.), and mandrake (Mandragora officinarum), all of which were used extensively in witches' brews and folk medicines.

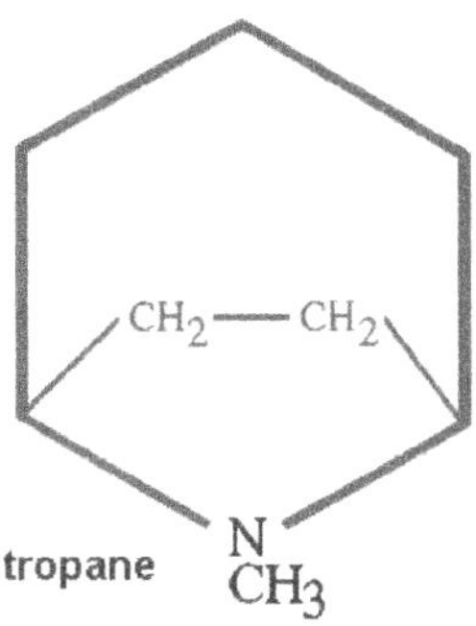

tropane

A common property of tropane alkaloids is a methylated nitrogen atom N-CH_3 at one end of the molecule. This chemical structure is also found in the neurotransmitter acetylcholine, which transmits impulses between nerves in the brain and neuromuscular junctions.

The anesthetic properties of tropane alkaloids may relate to their interference with acetylcholine, perhaps by competing with it at the synaptic junctions, thus blocking or inhibiting nerve impulses. It is interesting to note that the infamous tropane alkaloid, cocaine, is also a local anesthetic when injected into skin or muscle tissue. This property led to the discovery and synthesis of the more potent compound, novocain, widely used in dentistry.

Without getting into complicated anatomy and physiology, one nerve cell (neuron) connects to an adjacent neuron by a long extension called an axon. The axon branches into axonal endings, each of which attaches to the adjacent neuron at a synaptic knob filled with acetylcholine. The minute gap or synaptic cleft within this knob is only about 0.02 micrometers. As a nerve impulse (wave of depolarization or action potential) reaches this gap, acetylcholine diffuses across the synaptic cleft and activates the adjacent neuron. Acetylcholine in the synaptic cleft is deactivated or broken down by the enzyme acetylcholinesterase, thus shutting off the action potential. Organophosphate insecticides, such as malathion and parathion, bind to active sites on this enzyme, thus preventing the normal shut down of nerve impulses and destroying the nervous control of insects. Nerve gasses developed during World War II have a similar effect on the nervous system. Gulf War soldiers carried an atropine syrette to counter the possible effects of nerve gas.

Depending on the dosages, several tropane alkaloids of Datura (when absorbed together) may have synergistic properties resulting in extreme hallucinations, delirium and death. Since the alkaloids are fat soluble they are readily absorbed through the skin and mucous membranes. Volumes have been written about the uses and properties of Datura in the Middle Ages. Most of the uses involved the consumption of potions or concoctions made from various parts of the plant.

During ancient religious rituals in India, seeds were eaten by priests to induce hallucinogenic, prophetic and oracular states. European priests apparently drank Datura for the same reason. Some authorities believe the intoxicating smoke inhaled by Greek priests over 2,000 years ago at the Oracle of Delphi was Datura. Thieves in India and Europe used Datura for centuries as "knock-out drops" to rob their stupefied victims. The plant was also known in China, where a law prohibited mixing it with wine and other drinks.

In the East Indies, women fed Datura leaves to beetles, and then fed the poisonous dung to faithless lovers. Prostitutes in India added the seeds to their patron's drinks to induce sexual excitement. In fact, the use of Datura as an aphrodisiac spread throughout India, the Far East and Europe, and was an important ingredient in love potions and witches' brews. Specially prepared salves and ointments were also applied to various parts of the body. The famous seventeenth century Dutch artist, David Teniers the Younger, made several paintings of witches preparing for their demonic orgy or sabbat. The scenes frequently depicted a nude witch being anointed while she straddled a broom.

According to M.J. Harner, writing in Hallucinogens and Shamanism (1973), the use of a broom or staff was undoubtedly more than a symbolic Freudian act, for it served to apply the salves to sensitive vaginal membranes.

Clay tablets from Babylonian and Assyrian ruins indicate that Datura was used medically in ancient civilizations several thousand years ago. Greek and Roman physicians used Datura mixed with opium as a sedative and general anesthetic during surgery. In fact, the use of scopolamine (one of the alkaloids in Datura) plus morphine as an effective pain reliever and sleep inducer was common practice until the nineteenth century.

One of the most interesting medical cases of the 20th century involved a German physician, Dr. Carl Gauss, and the Lutheran Church. In 1905, Dr. Gauss used extracts from Datura and morphine to induce twilight sleep treatment for women experiencing difficult child birth. The church elders denounced Dr. Gauss because the Old Testament said that women were to bring forth in pain (Genesis 3:16). The combination of scopolamine (one of the active alkaloids found in Datura) plus morphine was used for years as an effective pain reliever and sleep inducer; however, it is generally considered unsatisfactory for women in labor because of hallucinogenic side effects on the mother, and it may repress breathing of the newborn.

Datura had a number of uses among Indian tribes of the United States, as well as Mexico and South America. In fact, it is difficult to find a tribe that didn't use a species of Datura from their region in one way or another. The hallucinogenic uses involved drinking an infusion made from the crushed roots or sometimes the crushed seeds fermented in water. Some aboriginal Indians in South America gave a Datura-alcohol beverage to wives and slaves of dead warriors and chieftains. The powerful brew induced stupor before they were buried alive to accompany their dead husbands and masters on the long journey to heaven. The high priests of some tribes took Datura in order to communicate with spirits of the dead and with their gods. The brilliant red-flowered tree Datura (Brugmansia sanguinea) is used today by herbal healers called "curanderos" in several countries of South America.

Probably the best known use of Datura by several North American Indian tribes was the puberty ceremonial dances involving the drinking of a "toloache" (Datura) infusion by young boys preparing to enter manhood. The ceremonies generally involved wild erratic dancing, varied hallucinations, and finally unconsciousness.

Datura alkaloids even have modern medicinal uses. The fat solubility of tropane alkaloids is employed in a remarkable remedy for seasick mariners called Transderm Scop®. A small, medicated disc resembling a circular Band-Aid® releases small amounts of scopolamine into the bloodstream for up to 3 days. The anti-nausea effects of scopolamine may relate to its anticholinergic action on the central nervous system, which results in inhibition of the vomiting reflex.

The Jimsonweed Junkie Moth

The nocturnal blossoms and pollination ecology of Datura wrightii are among the most ingenious and unusual of all wildflowers of western North America. The entire corolla is neatly folded or pleated (plicate) and twisted (convolute) in the bud forming a compact cylinder. Each day at dusk during the summer months the buds begin to gradually unfold. Each corolla slowly unfurls and then suddenly snaps open as the intertwined lobes come lose from one another. At this instant a powerful fragrance is emitted.

Five long stamen filaments are attached to the funnelform corolla at the throat. The filaments extend as ridges down the inner corolla tube, forming 5 narrow canals to the base of the ovary where the disc-shaped nectary is located. The nectar canals can best be seen by examining a corolla tube in cross section. The curious name "revolver flower" refers to nectar canals which resemble the cylinder chambers of a 5-shot revolver. At night the nectar oozes along the length of the canals within the corolla tube.

During mid and late summer the white, fragrant blossoms are frequently visited by large nocturnal hawk moths (family Sphingidae). They are sometimes called sphinx moths because the alarm posture of the larva resembles the Egyptian sphinx. Several species of hawk moths are known to visit blossoms of Datura wrightii, but two of the most common are Manduca quinquemaculata and M. sexta. The larval forms of both are better known as tomato and tobacco hornworms.

Since Datura, tomatoes and tobacco all belong to the Nightshade family (Solanaceae), the larvae are apparently content to feed on whichever plant is available to them. The larvae are remarkably camouflaged with green markings and are difficult to spot as they rapidly devour your tomato plants. After feasting on Datura (or your tomato plants) all summer, the robust, ravenous caterpillars crawl to the ground and burrow into the soil where they undergo pupation. Unlike many other moth larvae they do not spin a cocoon. Probably every tomato gardener has unearthed the large, carmel-colored pupa with its peculiar "jug handle" appendage, which is actually a case for the developing proboscis of the adult moth.

In warm regions there may be two generations per year: summer pupae produce adults after only a week or two; fall pupae remain in the ground until the following late spring or summer. The adult moth emerges from the pupal case and pushes out of the soil, eventually flying off into the night. After mating the gravid (pregnant) female lays her eggs on a convenient Datura plant--or on your prize-winning tomato plants while you sleep.

Adult moths have remarkable proboscides (tongues) up to 12 centimeters long (over 4 inches), long enough to reach the base of a 10-11 centimeter (4 inch) corolla tube. When the moth is not feeding the remarkable tongue rolls up into a neat, compact coil. The common white-lined sphinx moth (Hyles lineata) of the American southwest, with a medium proboscis of 3-5 centimeters

(2 inches), can only reach nectar in the upper part of the nectar canals. Unlike the Manduca moths, its pupa lacks the characteristic "jug handle" appendage.

Drs. Verne and Karen Grant at the University of Texas have made detailed observations of hawk moth behavior. The hawk moth approaches the Datura flower with its proboscis extended, lands on the blossom, crawls onto the corolla throat, and inserts its proboscis into the nectar canal. This is somewhat unusual because on most types of flowers hawk moths feed from a hovering position. While feeding they often flap their wings against the corolla throat. The proboscis becomes moist and sticky after insertion into the nectar canal, and pollen from anthers adheres to it readily. Pollen also gets on the head and other furry body parts. Thus pollen is transferred from one flower to another as the moth rubs against sticky receptive stigmas at the tips of long exerted styles. According to the Grants, hawk moths are the only flower-visiting insect with mouthparts that fit the long, slender nectar canals of Datura wrightii.

Another fascinating aspect of the Grant's research concerns a rather unconventional type of floral reward for hawk moths visitingDatura blossoms. Several intoxicating alkaloids are known to occur in Datura, but heretofore have not been correlated with pollination. Apparently Datura nectar is "spiked" with alkaloids and the hawk moths seem to like it and come back for more. Sometimes they arrive early and hover around the flowers, impatiently waiting for the blossoms to "pop" open. Outside the windows at the WAYNE'S WORD® headquarters we have observed what appear to be intoxicated moths flying erratically around D. wrightii, clumsily landing on blossoms and crashing into leaves or falling upon the ground. With a flashlight their eyes glow bright red in the darkness.

Although the coincidence of proboscis and floral lengths in hawk moths and Datura is quite remarkable, these coadapted partners have the ability to obtain nectar and set seeds in the absence of each other. According to Verne Grant, hawk moths may feed on a variety of bee flowers and hummingbird flowers, and pollen-collecting bees may visit Daturablossoms in late afternoon and early morning. The hippo staff at WAYNE'S WORD® has observed large carpenter bees (Xylocopa) and bumble bees (Bombus) on the blossoms of D. wrightii, busily collecting pollen from the anthers.

Flowers of some Datura species are also homogamous (stamens and pistils mature at the same time) and are self-compatible and partially self-pollinating. It is thus possible for one partner to migrate beyond the range of the other partner for a time. According to Verne Grant, flexibility in the pollination system permits migrational and evolutionary change. During times of general flower scarcity, hawk moths are more dependent on Datura and related sphingophilous (hawk moth-loving) blossoms because of the reliable nectar source which is unavailable to other animals.

There are many other examples of large, white blossoms pollinated by nocturnal hawk moths, including the fragrant gardenias of French Polynesia and South Africa. In fact, to reach the nectar of a Madagascar orchid (Macroplectrum sesquipedale) requires a 30 centimeter (eleven inch) proboscis. This remarkable orchid is also known as Angraecum sesquipedale. Long before it was found in nature, both Charles Darwin and Alfred Wallace (founding fathers of the principles of evolution) predicted that it would be a hawk moth.

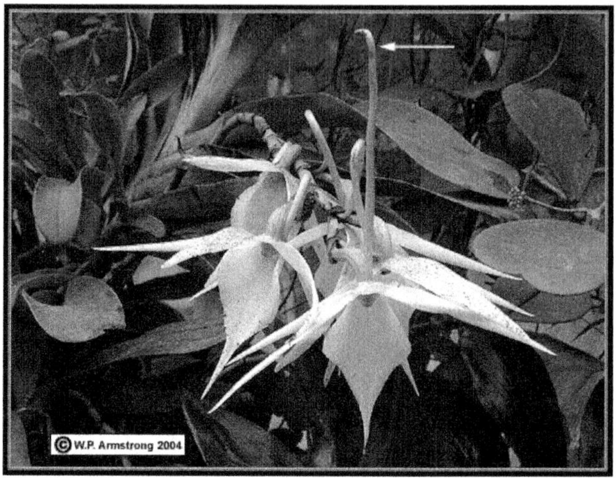

This beautiful orchid native to Madagascar is often called the "Darwin orchid" (Macroplectrum sesquipedale), also known as Angraecum sesquipedale. In fact, to reach the nectar of this orchid requires a 30 centimeter (eleven inch) proboscis to penetrate the long nectar spur (white arrow). Long before it was found in nature, both Charles Darwin and Alfred Wallace (founding fathers of the principles of evolution) predicted that it would be a long-tongued hawk moth.

Although the potent alkaloids of Datura have produced untold suffering and psychedelic binges among people through countless generations, they may also be an ingenious strategy to insure repeat visits by long-tongued hawk moths through the medium of drug addiction. When the moths sober up they come straight back for more nectar. Only recently are scientists beginning to solve the mysteries of plant toxins and how they may serve as chemical defenses or attractants, rather than simply nonadaptive by-products of plant evolution.

Old and New World Hallucinogenic Mushrooms

Most fungi are not deadly to humans and many are perfectly edible (and are quite delicious); however, some species are poisonous and contain potent neurotoxins. Placing a silver coin in a pan of cooking mushrooms to see if it turns black is not a reliable method of testing poisonous mushrooms. Unless you understand fungal terminology and know how to use a good taxonomic key, the staff at WAYNE'S WORD® does not encourage self indulgence on wild mushrooms.

The beautiful, red, fly agaric mushroom (Amanita muscaria) is unmistakable with its bright red cap covered with white scales. It contains the toxic alkaloid, muscimol, which is derived from ibotenic acid--an amino acid. In Europe the mushrooms were reportedly left in open dishes to kill flies; however, according to some authorities, the flies are merely stunned or stupefied by the toxin, and may even regain control and fly away. Although it is poisonous to humans, there are other species of Amanita that are much more dangerous and are potentially lethal if ingested. Some of these dangerously poisonous species are death cap (A. phalloides), death angel (A. ocreata), and panther amanita (A. pantherina).

Fortunately these latter deadly poisonous species are not bright red and are seldom confused with A. muscaria; however, they may be confused with other edible mushrooms by inexperienced gourmets. According to David Arora, this species may be one of the most common causes of mushroom poisoning in the Pacific Northwest. David Isaak (personal communication, 2002) stated in a recent e-mail message that in the Pacific Northwest, a number of people gather "the panther" for culinary purposes; cooking it in several changes of water to remove "most" of the psychactive materials.

When ingested by humans, Amanita muscaria may produce visions and delirium, and it is perhaps one of the oldest known hallucinogens. Recent studies suggest that this mushroom was the mysterious God-narcotic "Divine Soma" of ancient India. Thousands of years ago, Aryan conquerors who swept across India, worshiped soma, drinking it in religious ceremonies. Many hymns in the Indian Rig-Veda are devoted to Soma and describe the mushroom and its effects. According to the Rig-Veda, Soma is without leaves, seeds or branches, with a head and stalk or pillar [the structure of a mushroom]; its dazzling red skin is like the hide of the bull [the red cap]; its dress like that of a sheep, with woolly fragments remaining when the envelop bursts [the outer membranous envelop called the universal veil breaks as the stalk grows upward, leaving white remnants on the red cap]. This is a remarkably accurate description of the fly agaric mushroom *(A. muscaria)*. There are reports of Siberian tribesmen who ingested the mushroom to get intoxicated. Since the active chemical (muscimol) passes through the body relatively unaltered, others would drink the urine from these men to get high. This way a few mushrooms could inebriate many people relatively safely and efficiently. Lapland shamans eat fly agaric mushrooms for enlightenment, and some authors have postulated that this may have given rise to the flying reindeer and the red- and white-costumed Santa Claus legends.

Apparently not everyone agrees that the "Divine Soma" is Amanita muscaria. According to Terrence McKenna , the active alkaloid in fly agaric mushrooms (muscimol) doesn't produce the psychoactive effects described in the Rig Veda and other literature. Although the identity of the Divine Soma was eloquently presented by R.G. Wasson in 1971, Terrence McKenna has continued to question the effects of Amanita muscaria, and suggested other possible candidates. Based on first-hand experience with these hallucinogens, he has suggested that a psilocybin "magic mushroom," such as Stropharia cubensis, is the true Divine Soma. In fact, he also states that the use of mind-altering psilocybin mushrooms by ancient humans in Africa may have been a catalyst in the development of language and religion in primitive cultures.

Amanita muscaria was apparently one of the sacred hallucinogenic mushrooms of the Incas, Mayans and Aztecs. For the Indians of Mexico, Central and South America, partaking of these mushrooms was a deeply religious experience, enabling them to communicate with their gods. Cortez reported a mushroom (resembling Amanita muscaria) being eaten during the coronation of Montezuma, and in Guatemala stone carvings dating back to 1000 BC depict curious figures with umbrella-like tops resembling the caps and stalks of an Amanita mushroom. Mushrooms are also depicted in ancient Peruvian vessels and in the Mexican Codices. One drawing shows an animal-like messenger from god offering the sacred Amanita to a ruler seated on a throne. And a fresco in a Roman Catholic Church in Plaincouralt (Indre), France depicts Adam and Eve on either side of a tree of knowledge that is unequivocally a branched Amanita mushroom.

Some scholars believe that the original story of Alice's Adventures in Wonderland, where Alice speaks to a green caterpillar who is seated on a red- and white-capped mushroom, is actually the interpretation of a mushroom experience by the author, Rev. C.L. Dodgson of Christ Church College in Oxford (better known by his pen name of Lewis Carroll). Another hallucinogenic "high" that is commonly depicted in paintings and children's stories is the infamous, "politically incorrect" picture of a witch flying on a broom--the effects of a potion made from the deadly alkaloids of several solanaceous herbs, including jimsonweed (Datura stramonium).

New World Mushrooms Containing Natural LSD

According to R.E. Schultes and A. Hofmann , a number of other hallucinogenic mushrooms are used by shamans in Mexico, Central and South America, including the four genera Psilocybe, Paneolus, Conocybe and Stropharia. The exact species selected by different shamans is determined partly by personal preference and partly by the purpose of their use. Since mushrooms are often seasonal, their regional availability may also be a factor in the selection process. The Aztecs referred to these mushrooms as "teonanacatl" (flesh of the gods).

When the Spaniards conquered Mexico, they were appalled to find the natives worshipping their deities with the help of inebriating plants. Although the Spanish conquerers hated and attacked the religious use of all hallucinogens, including "peyotyl" (peyote), "ololiuqui" (morning glory), and "toloache" (Datura), they especially resented the psilocybin mushrooms "teonanacatl." In order for the Aztecs to carry on their cultural traditions without persecution, the use of some of these hallucinogens went "underground." According to R.E. Schultes (1976), the Aztecs may have tried to protect their real sacred plant (peyote) by convincing the Spaniards that all their teonanacatl were dried mushrooms, when actually many of the "mushrooms" were really the dried, shriveled crowns of a sacred peyote cactus.

The Spaniards first misidentified peyote as a mushroom in the sixteenth century when they stated that the Aztec substance "teonanacatl" and peyote were the same. It is easy to see how the Spanish authorities could have mistaken the dried crowns or tops (buttons) of peyote cactus for the caps of teonanacatl mushrooms. The Aztecs may have fooled their conquerers into thinking that these religious plants were mushrooms, while the identity of one of their most spiritual and sacred plants (peyote) was a cactus. Even in the 1900s, botanists and anthropologists concluded that teonanacatl and peyote were the same drug. This mistake was perpetuated until 1936 when hallucinogenic mushrooms were rediscovered and definitely linked to early Mexican ceremonies.

The psychoactive alkaloid in the teonanacatl mushrooms is psilocybin, a potent indole alkaloid. Psilocin, a dephosphorylated version of psilocybin, is about 10 times stronger. After ingestion by humans, psilocybin is automatically converted into psilocin. Most psilocybin-containing mushrooms have only a trace of psilocin. According to David Aurora (1986), the common psilocibin mushroom of the Pacific coast of North America, Psilocybe cyanescens has a higher concentration of natural psilocin and is appropriately named "potent psilocybe." Although two of the most famous species of psilocybin mushrooms are Psilocybe mexicana andStropharia cubensis, there are literally dozens of other species in the above 4 genera with similar hallucinogenic properties. In fact, Paul Stamets describes all of the species and includes color photographs. Like so many LBM's (Little Brown Mushrooms), they are difficult to identify unless you are familiar with mushroom structure and spore taxonomy, and have a good compound microscope at your disposal. In fact, two deadly look-alike LBM's (Galerina autumnalis and Pholiotina filaris) resemble certain species of Psilocybe. The small ring on their stems (called an annulus) and rusty brown spores (rather than black spores) are "dead" give aways to avoid these potentially lethal mushrooms.

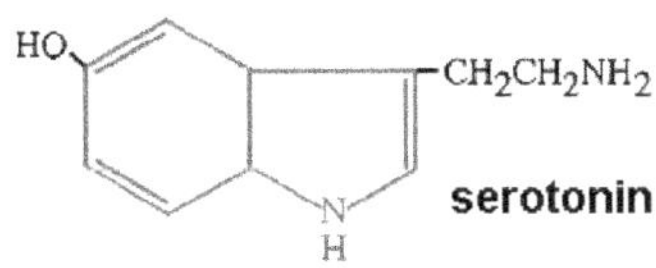

serotonin

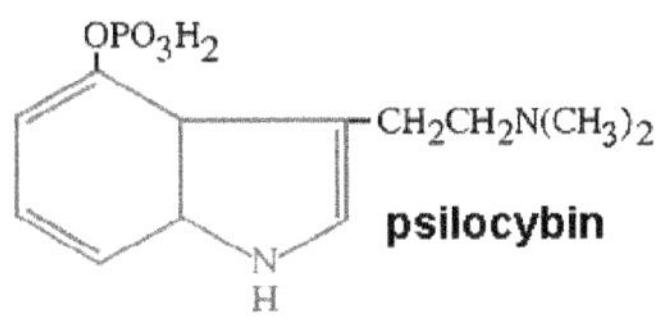

psilocybin

Indole alkaloids contain the indole carbon-nitrogen ring which is also found in the fungal alkaloids ergine and psilocybin, the neurotransmitter serotonin, and the mind-altering drug LSD. These alkaloids may interfere or compete with the action of serotonin in the brain.

Ergot: A Fungus Disease of Rye that Contains LSD

One of the most amazing stories about naturally-occurring alkaloids in fungi concerns ergot (Claviceps purpurea), a fungus that infects grains of rye and related grasses. One of the psychoactive components of ergot fungus is the alkaloid ergine (d-lysergic acid amide), better known as natural LSD. The more potent synthetic LSD, (d-lysergic acid diethylamide), also known as LSD 25, is one of the most powerful psychoactive drugs known. LSD 25 was originally synthesized from natural pshchoactive alkaloids in ergot. According to Lewis and Lewis (1977), it is 4,000 times more powerful than mescaline. Natural LSD (ergine) is also found in the seeds of two species of Mexican morning glory vines which are still ingested by native Indians in an important medicinal and religious ritual.

Ergot forms a dark, compact, fungal mass called a sclerotium where the grain would normally develop. One or several of these pelletlike sclerotia can be seen in an infected grain spike, typically extending out from the bracts (glumes). When separated from the grain spike, the sclerotia superficially resemble rat droppings (rat pellets). The sclerotia are the source of the potent alkaloids in Claviceps purpurea. In late spring, when rye plants are in bloom, the overwintering sclerotia from the previous year's crop produce stalked ascocarps resembling microscopic fungal fruiting bodies. The head of each ascocarp contains many embedded perithecia. The perithecia contain numerous saclike asci, each with eight ascospores. The ascospores infect the young, developing grains (ovaries) of rye plants, eventually replacing them with purplish-black sclerotia. Because it produces ascospores within saclike asci, Claviceps is placed in the fungal Class Ascomycetes.

During the Middle Ages, tens of thousands of people in Europe were afflicted with ergotism, a malady characterized by gangrenous extremities, convulsions, madness and death. They ate rye bread infested with ergot fungus containing several peptide alkaloids of the ergotamine group (including ergotamine, ergosine and ergocristine) that affect blood vessels. Since they are potent vasoconstrictors, these alkaloids can cause gangrene if ingested in sufficient dosages. Known as "St. Anthony's Fire," ergotism was a dreaded disease in Europe. Between 990 and 1129, more than 50,000 people died of this disease in France. The disease became so devastating that in 1093 in southern France the people formed an order to take care of the afflicted, and they chose St. Anthony as their patron saint.

One of the symptons of the disease was an intense burning sensation, hence the name St. Anthony's Fire. It wasn't until 1597 (500 years after the first epidemic of ergotism) that physicians finally associated this horrendous disease with the ergot on rye. Another form of ergot poisoning involves severe hallucinations and madness, caused by pschoactive alkaloids in the sclerotia. [In case you are wondering, another hallucinogenic alkaloid, mescaline, comes from the Mexican peyote cactus (Lophophora williamsii) and the South American San Pedro cactus (Trichocereus pachanoi), and has a chemical structure remarkably similar to the brain neurotransmitter dopamine. Mescaline and the sacred peyote cactus are discussed in a subsequent section.] For more information about these alkaloids please refer to the references below, particularly those by Richard E. Schultes.

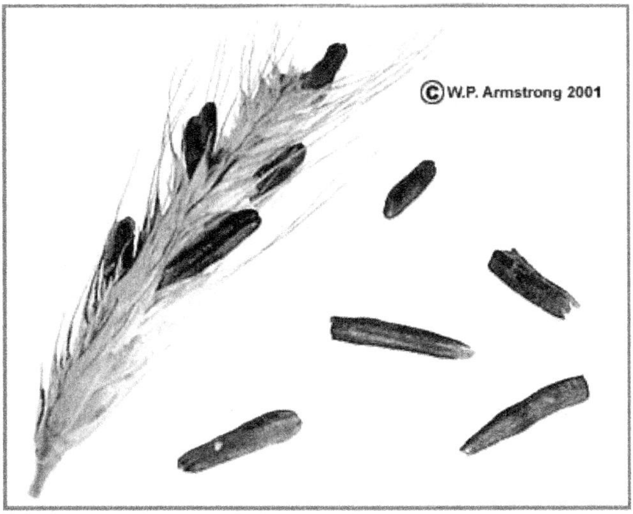

A dried grain spike of rye grass (Secale cereale) infected with ergot (Claviceps purpurea). Some of the grains have been replaced by a dark, compact, fungal mass called a sclerotium. The hardened sclerotia superficially resemble rat pellets (the droppings of a rat). The sclerotia contain valuable vasoconstricting alkaloids of the ergotamine group and lysergic acid alkaloids which are the precursor for LSD 25 (d-lysergic acid diethylamide).

A number of important medical discoveries have come from the study of ergot fungus and ergotism. In 1935 the alkaloid ergonovine was isolated from ergot. Since it causes strong muscular contractions, it has been used to induce labor and to control hemmorrhaging. The alkaloid ergotamine has been used extensively to relieve migraine headaches through the constriction of blood vessels. Thousands of pounds of ergot sclerotia are harvested each year from midwestern rye farms, and are used for various prescription drugs.

In 1943 chemist Albert Hofmann was studying ergot fungus, whose nuclei contain lysergic acid. When he added diethylamide he produced lysergic acid diethylamide, better known as LSD. While working on this new compound, Hoffman discovered that its strong hallucinogenic effects were similar to that of natural lysergic acid alkaloids in the seeds of "ololiuqui," morning glories used by the Aztecs in their religious ceremonies.

A Hallucinogen Alkaloid Found in Seeds and Toads

Another fascinating indole alkaloid called bufotenine occurs in the seeds of Yopo or Paricá (Anadenanthera peregrina), a South American leguminous tree of the Orinoco River basin (not to be confused with the leguminous genus Adenanthera). Indians of this region prepare a powder from the ground seeds which they use as a hallucinogenic snuff. Bufotenine (5-hydroxydimethyltryptamine) is a derivative of the indole alkaloid tryptamine, which is derived from the essential amino acid tryptophan. Tryptophan is one of the 8 (9) essential dietary amino acids in humans (which we cannot synthesize), and is widely distributed in the animal kingdom. Interestingly enough, bufotenine is also present in the skin secretion of certain toads of the genus Bufo, and explains the practice of licking toads by some people. The popular dietary supplement 5-HTP, sold at natural food stores, is 5-hydroxytryptophan. It is made from an extract ofGriffonia simplicifolia seeds from coastal West Africa. 5-HTP is a metabolic precursor of serotonin and is taken as a treatment for depression and sleeping disorders. This herb may conflict with other antidepressant medications, and it would be wise to consult with a knowledgeable physician before taking it.

The Glorious Morning Glories

The morning-glory family (Convolvulaceae) contains at least 50 genera and more than 1000 species, from high-climbing vines and woody lianas of the tropical rain forest to prostrate, trailing perennials. They decorate our fences, trellises and walls with lush green foliage and colorful funnel-shaped blossoms, and form lovely green carpets of dichondra lawn. Several vines of this family provide us with valuable and nutritious root crops, including jicama and sweet potatoes. Although the Morning-Glory Family is usually associated with climbing vines, it also includes erect herbs, shrubs and a few trees. One unusual genus includes the parasitic dodders (Cuscuta) which smother their host with masses of twining, spaghetti-like orange stems. Other morning glories have invaded cultivated fields and have become troublesome weeds. Called "bindweeds," they literally twine themselves over other plants that happen to be in their growth path. Some morning glories, including the infamous *Mary's bean* (Merremia discoidesperma) are excellent seed voyagers and have colonized the distant beaches of tropical islands and atolls.

Seeds of two infamous Mexican morning glories, *Ipomoea tricolor* (syn. I. violacea) and the white-flowered *Turbina corymbosa* (syn. Ipomoea burmanni), were taken in a drink by Aztec priests in order to commune with their gods. The ground seeds were only ingested by experienced persons who understood the proper "spiritual dosage." The seeds from the white-flowered morning gloryTurbina corymbosa are called "ololiuqui" (pronounced o-low-lee-oo-key) by native Indians of Mexico, and to this day provide them with an important medicinal and religious ritual in their cultures. The black, angular seeds from the pink-flowered Ipomoea tricolor are called "tlitliltzin." An intoxicating drink made from the ground seeds of these species is administered by a shaman and is used by a number of different tribes for the devine recovery of illness. This fascinating story is explained in more detail by R.E. Schultes and A. Hofmann (1979) in Plants of the Gods.

Like other sacred, mind-altering plants used by the Aztecs for worshipping their gods, the use of ololiuqui was forbidden by the Spaniards. Spanish attempts to eradicate the use of these plants resulted in secrecy by the Aztecs. As in the teonanacatl mushrooms, the identity of the plants known as ololiuqui was shrouded in mystery. Because of the similar, trumpet-shaped (funnel-shaped) blossoms, the plant was thought to be Datura, a known hallucinogen still used in Mexico. Finally in the 1930s, the seeds were clearly linked to species of morning glories.

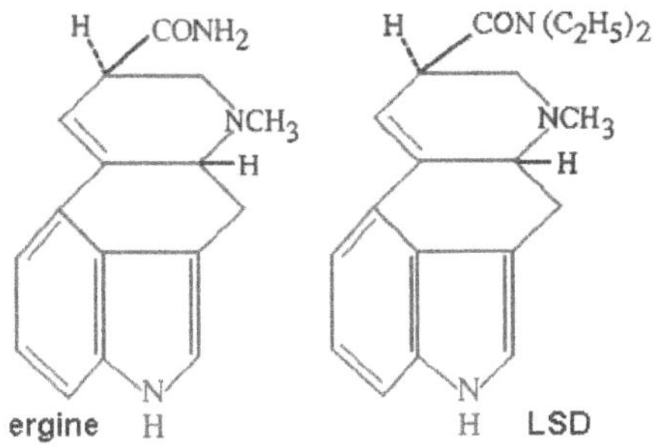

According to R. E. Schultes and A. Hofmann (1979), the morning glory seeds contain a lysergic acid alkaloid called ergine (d-lysergic acid amide), better known as "natural" LSD. The more potent synthetic LSD is d-lysergic acid diethylamide.

Synthetic LSD has two additional ethyl groups (C_2H_5) and is about 100 times more potent. Before it was discovered in morning glories, ergine was only known from ergot (Claviceps purpurea), a rust fungus that infects grains. Psychoactive alkaloids, such as ergine and psilocybin (from the mushrooms Psilocybe, Stropharia, Paneolus and Conocybe) contain the indole structure, a double carbon-nitrogen ring also found in the natural neurotransmitter *serotonin*. These alkaloids may interfere or compete with the action of serotonin in the brain, causing psychedelic visions, delusions and hallucinations.

Locoweeds: Wild and Crazy Plants

Unlike the previous mind-altering plants, the next group to be discussed are not taken as drugs plants. If ingested they can cause serious and permanent damage to the central nervous system. Nonetheless, their poisonous effects on unfortunate herbivores is quite fascinating and may lead to a better understanding of certain human genetic disorders. These plants are most commonly associated with abnormal behavior in livestock and other range animals, and are often referred to as locoweeds. They belong to the large and diverse genus of flowering plants, Astragalus. A few western locoweeds also belong to the closely related genus, Oxytropis. Literally hundreds of different species of locoweeds grow in almost every conceivable habitat, from coastal bluffs, grasslands and deep desert canyons to sun-baked sand dunes and rocky alpine summits. In fact, Astragalus is one of the largest genera of flowering plants with approximately 2,000 different species in the northern hemisphere. Just trying to identify all the different kinds of locoweeds in the western states can be truly overwhelming. Although most North American species are poisonous, several kinds were apparently eaten by native Americans and some are actually good forage plants.

Locoweeds belong to the enormous legume family (Fabaceae), along with beans, peas, clover, alfalfa and more than 15,000 other related species. They are sometimes called milk vetches from the notion that milk secretion in goats was increased when they fed on the common Old World forage species (Astragalus cicer). Like many legumes, the leaves are typically divided into a dozen or more leaflets, and the flowers resemble small pea blossoms. Some species produce inflated seed pods that make a distinct popping sound if you step on them. Locoweeds are also called rattleweed because of the bladder-like, inflated pods, particularly when a gust of wind rattles the seeds inside.

Astragalus is derived from a Greek word meaning anklebone, the plural of which means dice. Perhaps the dice connotation refers to the rattling of seeds inside papery pods, like the sound of dice in a thrower. In anatomy, the astragalus or talus is one of seven bones in the ankle joint. Anklebones were apparently used for dice by ancient Greeks, and to this day, veteran crapshooters in Las Vegas refer to dice as "bones." Since adult astragalus bones are a little too large for dice, some of the smaller, cuboidal or cuneiform anklebones were probably used.

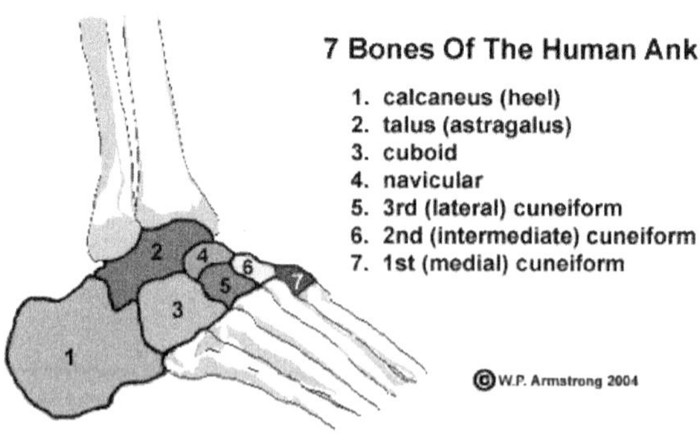

Many species of locoweeds native to the western United States are known to be poisonous to livestock. The actual mechanism of locoweed poisoning may involve:

(1) Toxic levels of selenium absorbed from the soil,

(2) nitrogen-containing sugar compounds called nitroglycosides, and

(3) potent swainsonine alkaloids causing a condition called "locoism."

According to P.R. Cheeke and L.R. Shull, of the 372 species of Astragalusin North America, about 25 species contain toxic levels of selenium, 263 contain poisonous nitroglycosides, and 13 cause locoism. Some of the poisonous species may be placed in more than one of the above three categories. Bee keepers in Nevada have reported serious honey bee losses after working the flowers of spotted locoweed (Astragalus lentiginosus). The toxicity of Astragalus to bees is unusual because relatively few plants produce nectar or pollen which is poisonous to honey bees. The notoriously toxic "diablo locoweed" (A. oxyphysus) is well-known in California and some ranchers have attempted to eradicate it from their property. Apparently it wasn't removed from all grazing land because several years ago the editor of WAYNE'S WORD® observed a deceased steer lying in a pasture containing this locoweed in Kern County.

Some species of Astragalus accumulate toxic levels of selenium and often have a peculiar malodorous foliage. In fact, specimens of A. bisulcatus and the desert species A. crotalariae can produce the dominant scent in herbarium cabinets for many years. Locoweed indicators of selenium-rich soils are sometimes referred to as "poison vetches," and may contain selenium levels of several hundred to 10,000 ppm (parts per million) (P.R. Cheeke and L.R. Shull, 1985). Several names have been applied to selenium poisoning in cattle, including "blind staggers" and "alkali disease."

In severe cases, the animals become lame and emaciated, and often fall into a kneeling position from which they are unable to rise. Toxic levels of selenium also occur in other wild plants of western North America, such as species of Stanleya, colorful shrubby perennials in the mustard family (Brassicaceae). Selenium poisoning of fish and waterfowl from has been well documented in California's Central Valley. In the human body, the trace element selenium is needed for the function of certain enzymes, and may serve as a cellular antioxidant thought to suppress certain tumors. Natural sources of selenium come from the allium vegetables, including onions and garlic.

The majority of locoweed species poisonous to livestock contain toxic nitroglycosides. The extreme toxicity of nitroglycosides, such as miserotoxin, in livestock is caused by the production 3-nitropropionic acid (NPA) through a complex metabolic pathway involving enzymes of rumen microbes and the liver. Another toxic reaction involves the oxidation of hemoglobin by nitrites produced by the hydrolysis of miserotoxin. The oxidized hemoglobin (called methemoglobin) is incapable of carrying oxygen. NPA is lethal to animals because it inhibits the vital enzyme succinate dehydrogenase inside mitochondria, thus blocking ATP synthesis and ultimately causing cellular death.

In some animals with different gastrointestinal microbes, such as rats and rabbits, the miserotoxin is not converted into lethal NPA; however, studies indicate that death in rabbits is due to nitrite poisoning. Symptoms of acute livestock poisoning involve general weakness and loss of neural control, convulsions, blindness, coma, and death. More than 200 species of Astragalus in North America contain nitroglycosides, including A. cibaria, A. falcatus, and A. miser. The latter species has also been known to cause high mortality in foraging honey bees.

One of the most destructive types of livestock poisoning by locoweed ingestion is called "loco disease" or locoism. It is caused by at least two very potent indolizidine alkaloids, swainsonine and swainsonine-N-oxide, and impels horses and cattle to act in a wild and crazy manner. Some authorities only apply the term "locoweed" to species of Astragalus and Oxytropis containing these alkaloids, such as A. lentiginosus, A. mollissimus, A. wootonii, O. lambertii, and O. sericea. Swainsonine was originally isolated from the Australian darling pea (Swainsona) and more recently from the spotted locoweed (Astragalus lentiginosus). Consumption of locoweeds for two weeks to a month is necessary before obvious signs of poisoning are evident. Animals suffering from locoism may exhibit depression, a staggering gait, and general muscular incoordination. They may withdraw from other animals and become solitary. The final stages are characterized by difficulty in eating or drinking, paralysis and death. Afflicted animals often become nervous, or easily aggravated, and exhibit a complete loss of depth perception. There are reports of animals being injured or killed by running through fences or falling into ponds and streams. When plenty of other forage is available, animals tend to avoid locoweed; however, poisoned animals apparently acquire a taste for it and actually seek it out.

The cause of locoism at the cellular level is very complex. The swainsonine alkaloids inhibit or tie up the key enzyme mannosidase resulting in the accumulation of mannose sugar in nerve cells and irreparable damage to brain tissue. This condition is remarkably similar to a genetic deficiency of the same vital enzyme in humans called mannosidosis. Mannosidosis is a genetic disorder called a lysosomal storage disease, in which cells of the central nervous system become filled with cytoplasmic vacuoles of mannose due to the lack of the vital enzyme mannosidase that is essential in breaking down mannose.

The actual vacuoles are swollen organelles called lysosomes where the enzymatic breakdown process normally occurs. Lysosomal storage diseases, such as mannosidosis, are often caused by recessive genes and result in paralysis and death within a few years following birth. Perhaps one of the better known storage diseases is Tay Sachs Disease, in which nerve cells fill up with a lipid called ganglioside or GM2 because they lack the vital enzyme HEX A needed to break down GM2. Studies of the biochemical effects of locoweed poisoning on the central nervous system of cattle may lead to a better understanding of these tragic human neurological disorders.

Native American Indians were apparently aware of the poisonous properties of some locoweeds and avoided them. Other drug plants are occasionally called locoweeds, such as the hallucinogenic jimsonweed (Datura), and some of these were commonly utilized by Indians during ceremonial rituals. In the southwest, Cahuilla Indians utilized the pods of at least one unknown Astragalus locoweed. The pods were pounded up and mixed with beans and other foods, perhaps as a spice or flavor enhancer. An Arizona locoweed (A. ceramicus), with beautiful brown-mottled pods, produces sweet, edible rootstocks which are reportedly eaten in spring by Hopi children. Another locoweed of the prairie and plains states (A. crassicarpus) produces edible fleshy pods. It is known locally as ground plum, Indian pea, or buffalo bean, and the pods were eaten, raw or cooked, by Indians and white settlers. The perennial roots of three species, including A. canadensis, A. caryocarpus, and A. pictus-filifolius, were eaten raw or cooked by Blackfoot Indians of Montana and Hopi of Arizona. Not all locoweeds are poisonous to people; however, because of the difficulty in identifying some species, it would definitely NOT be advisable to try any of them in gourmet dishes. It would actually be easier to select edible mushrooms from the deadly species, especially with all of the pictorial mushroom guides available today.

The Mescal Bean and Peyote Cactus

Probably the most famous New World hallucinogenic plant is peyote (Lophophora williamsii), a small, spineless cactus native to the Rio Grande valley of Texas and the northern and central parts of the Mexican plateau region. Another species (L. diffusa) is native to the Mexican state of Queretaro. The rounded, gray-green stem crown (top) is radially-divided into sections, each bearing a small meristematic region (called an areole) from which arises a tuft of hairs. The crown tapers into a thick carrot-like root that extends into the ground.

Called "peyotyl" by the Aztecs, this was truly a sacred plant used in religious ceremonies where they communicated with their gods. The spiritual, religious usage of peyote persists to this day by the Tarahumara, Huichol and other Mexican Indians, as well as by members of the Native American Church in the United States and western Canada. The Indians cut off the crowns and sun-dry them into brown, mushroom-shaped "mescal buttons" that last for long periods and can be shipped to distant places for use. After the crown (button) dries, the soft, fleshy tissue is reduced in volume. For this reason the tufts of hairs appear much larger and occupy a larger proportion of the button. The Aztec word peyotyl means "caterpillar cocoon," referring to the white, woolly tufts of hairs. When the top is severed, the plant often resprouts with new crowns so that many-crowned peyote plants are common. Except for the woolly hairs, the dried, peyote buttons superficially resemble the cap of a dried mushroom.

Spanish conquerers tried to ban the use of peyote by native Indians. By 1720 the eating of peyote was prohibited throughout Mexico, but peyote was so strongly rooted in native Indian lore that its use actually spread to other tribes. The Spanish thought the peyote buttons used by Aztecs were mushrooms, and it wasn't until the 1930s that researchers finally discovered that the Aztecs were actually using both peyote (peyotyl) and psilocybin mushrooms (teonanacatl) in their religious ceremonies.

The peyote cactus contains more than 50 different alkaloids, but the most active hallucinogen is mescaline. The chemistry, botany and history of peyote is discussed in a fascinating book by E.F. Anderson. Mescaline has a chemical structure similar to the brain neurotransmitter dopamine. It is also structurally similar to the neurohormone norepinephrine (noradrenalin) and to the stimulant amphetamine. In the peyote cactus, mescaline is formed in a complex pathway from the amino acid tyrosine. A similar pathway in humans produces epinephrine (adrenalin) and its demethylated precursor norepinephrine from tyrosine. Dopamine and its precursor L-dopa are also derived from a tyrosine pathway. The following illustration shows the remarkable molecular similarity between mescaline and dopamine:

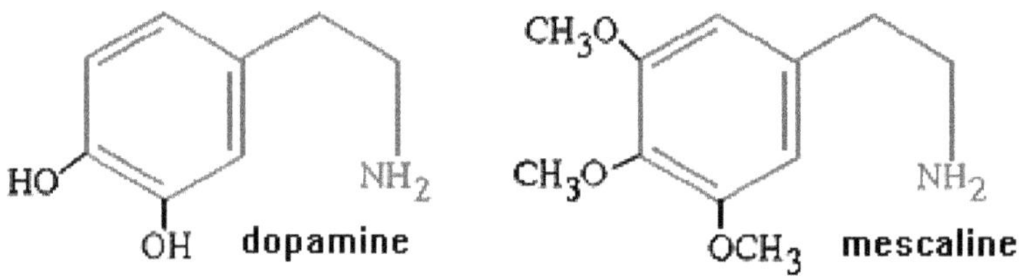

Mescaline also occurs in several other cactus species, including the commonly cultivated, night-blooming, South American San Pedro cactus (Trichocereus pachanoi). In the Andes of Peru, Ecuador and Bolivia the natives call this tall, columnar cactus "aguacolla" or "giganton." An intoxicating drink called "cimora" is made from the boiled stems. According to R.E. Schultes (1976), the drink may be spiked with other potent hallucinogens, including species of Brugmansia (Datura). The cut stems are sometimes seen in market places of the northern Peruvian Andes in neatly stacked piles. There are about 25 species of Trichocereus native to South America. They are typically large, tree-like cacti with cylindrical, ribbed stems and large, nocturnal, white blossoms. Other species within this fascinating genus may also contain mescaline.

Contrary to popular rumors, mescaline is not found in the bright red, poisonous seeds of the beautiful, drought-resistant shrub called mescal bean (Sophora secundiflora); however, they do inhabit a similar range in the arid lands of the southwestern United States and Mexico. Mescaline is also not related to the highly intoxicating beverage called "mescal" or "mezcal," made from the fermented and distilled juices of several North American species of Agave, including A. americana and A. atrovirens. Incidentally, the fermented juice is called pulque, and the highly-alcoholic distilled products include mezcal and tequila.

The seeds of mescal bean contain another potent and dangerously poisonous alkaloid, cytisine, which was ingested by some North American Indian tribes in a vision-seeking "Red Bean Dance" prior to the widespread use of peyote.

According to R.E. Schultes (1976) mescal beans have been discovered in Indian sites dating before AD 1000, and from one site dating back to 1500 BC. In fact, to this day the leader of the peyote ceremony in some of these tribes (called the "roadman") wears a necklace made from bright red mescal beans.

One of the most interesting stories concerning the use of peyote north of Mexico concerns Quanah Parker, son of a Comanche war chief (Nokoni) and Cynthia Parker (a white woman who was captured by the Indians). In 1884, Quanah Parker became seriously ill and was treated by a Mexican curandera. Quanah regained his health and recognized that peyote could be a useful factor in binding his people together. The latter half of the 19th century (post Civil War period) was a time of turmoil and humiliation for the American Indian tribes. Their pristine hunting lands were gradually being taken away from them; they were swindled by federal authorities; they were forced to move to reservations far from their homes; they were forced to attend schools that denied their heritage; they were corrupted by the white man's "firewater" (alcohol); and they had Christianity forced upon them by missionaries.

Quanah fashioned a series of ceremonies, with cultural elements from Comanche, Kiowa, Apache, and parts of Christianity. This was a good time for the peyote religion--to bring native people dignity and hope of survival, and to bring them spiritual sustenance. The peyote religion spread rapidly during the late 1800s and early 1900s. With the assistance of James Mooney, an ethnologist with the Smithsonian Institution, the Native American Church was finally established in 1918. According to Benjamin Capps , Quanah Parker once said to a white listener: "The white man goes into his church house and talks about Jesus; the Indian goes into his teepee and talks to Jesus."

Chapter 7

FUNGI AS PLANT GROWTH PROMOTER AND DISEASE SUPPRESSOR

CONTROL OF DISEASES

Besides stimulation of plant growth, Albit also protects plants through suppressing of development of the broad range of main crop diseases. Albit is not toxic for pathogenic organisms, but it can increase the plant immunity , natural ability of plants to resist diseases.

Chemical plant protection was traditionally based on use of pesticides toxic for pathogenic fungi. Historically, there was three generations of pesticides. Fungicides of the first and second generations (inorganic and organic contact poisonous preparations) are non-specific and toxic for both fungi and plants. This fungicides suppress the development of pathogenic fungi stronger than that of plants just because of plant's more rigid and resistant cell walls protecting from penetration of undesirable substances. This features make troubles for the chemical treatments, because, at the first, the surface of treated plants must be very carefully coated with fungicide (since such fungicides acts only when contacting with parasites), and, at the second, treatments must be multiply repeated.

Fungicides of the third generation are systematic organic compounds that are specific for particular groups of fungi and low-toxic for plants. They penetrate to plant, spread over tissues, and kill pathogenic fungi not only on the plant surface, but also inside of plants, that makes them considerably more convenient than the simple contact fungicides. However, their specificity to targets in metabolic pathways of parasites (for example, in chitin or steroid synthesis), leads to complete loss of toxicity in case of target modification. Thus, continuous use of systemic fungicides in general case leads to loss of their effectiveness due to rising of resistant pathogenic strains. Nowadays, the fort generation of pesticides has come. The pesticides of the forth generation does not kill pathogenic organisms, but decrease their pathogenicity and increase resistance (immunity) of plants.

Due to reinforcement of natural plant protecting mechanisms, Albit acts as a systemic broad fungicide. Owing to immunizing action, Albit have pronounced prophylactic and weak curing effect, but it can not destroy the pathogenic fauna. Thus, according to Prof. Yu. T. Djakov's (Moscow State University) classification, Albit belongs to pesticides of the forth generation.

Albit decreases the infestation of plants by number of diseases: root rots, mealy dew, leaf spots, brown rust, fusarioses, anthracnose, late blight, black leg, white and grey rots, bacteriosis, *etc.*

The basic parameter used for the characterization of fungicidal properties of pesticides is their biological efficiency which is calculated by formula:

$$BE = \frac{R_C - R_E}{R_C} \times 100\%$$, where

R_C — incidence (severity) of disease in the control,

R_E — incidence (severity) of disease in experiment (after fungicide treatment).

Biological efficiency of 100 % means full eradication of disease, 0 % – absence of activity of a pesticide. BE is an analogue of Control Efficacy, which is more common for publications in the USA.

Biological efficiency of Albit against plant diseases is 40-80 % averagely. Albit is most effective against diseases of cereals and leguminous crops, sugar beet, sunflower, vine and flax. Against these diseases, biological efficiency of Albit can reach 90-100 % that is excellent for biostimulant.

By now we have analyzed the results of all 250 field trials with Albit and summarized the data about its efficiency against plant diseases. The results are presented in Table.

Table: Fungicidal activity of Albit (according to results of all field trials 1997–2004). (—) no data.

Crop	Diseases and pathogens	Biological efficiency, %		
		average	min	Max
Winter wheat	Brown and yellow rust *Puccinia recondita Roberge: Desm. f. sp. tritici (Erikss.) C.O. Johnston; Puccinia striiformis West.*	49.1	24	80
	Root rots *Fusarium spp., Bipolaris sorokiniana (Sacc.) Shoemaker, Ophiobolus graminis Sacc.*	81.0	51	100
	Powdery mildew *Blumeria (Erysiphe) graminis (DC.)*	51.0	21	83
	Loose smut *Ustilago tritici Jens.*	39.4	10	60
	Septoria leaf spot *Septoria spp.*	52.1	30	81
	Fusarium head scab *Fusarium graminearum Shwabe, F. avenaceum Sacc.*	35.0	25	50

Spring wheat	Kernel black point *Alternaria tenuis Nees*	50.0	-	-
	Brown rust *Puccinia recondita Roberge: Desm. f. sp. tritici (Eriks.) C.O. Johnston*	46.4	20	84
	Root rots *Fusarium spp., Bipolaris sorokiniana (Sacc.) Shoemaker*	59.2	14	100
	Fusarium head scab *Fusarium graminearum Shwabe, F. oxysporum Schlecht, F. avenaceum Sacc*	43.4	-	-
	Fusarium wilt *Fusarium oxysporum Schlecht, F. culmorum Sacc., F. avenaceum Sacc.*	40.4	10	70
	Helminthosporium rot *Bipolaris sorokiniana Shoemaker*	46.4	14	93
	Powdery mildew *Blumeria (Erysiphe) graminis (DC.)*	81.7	67	90
	Loose smut *Ustilago tritici Jens.*	48.5	10	60
	Stinking smut *Tilletiia caries Tul.*	79.0	-	-
	Septoria leaf spot *Septoria spp.*	51.2	21	100
Spring and winter barley	Brown rust *Puccinia hordeina Lawrov*	50.0	30	91
	Root rots *Bipolaris sorokiniana (Sacc.) Shoemaker, Fusarium spp.*	69.7	26	100
	Powdery mildew *Erysiphe graminis DC. f. hor-dei Em. Marchal*	89.9	72	96
	Ustilaginaceae (smut) diseases *Ustilago nuda Kell.et Swing, U. hordei Kell.et Swing*	52.7	10	70
	Septoria leaf spot *Septoria nodorum Berk.*	45.0	39	55
	Net blotch(helminthosporiose) *Drechslera teres (Sacc.)*	60.4	20	94
	Stem rust *Puccinia graminis Pers. f. hordei Erikss. et P. Henn*	68.6	50	87
	Spot blotch *Bipolaris sorokiniana (Sacc.) Shoemaker*	65.2	62	69
Rice	Root rots *Fusarium oxysporum Schlecht f. oryzae Bilaj, Fusarium nivale (Fr.) Ces.*	73.0	69	77
Sunflower	Watery soft rot *Sclerotinia sclerotiorum DB.*	55.8	14	100
	Head grey mould blight *Botrytis cinerea Pers.*	66.3	26	100
	Verticillium wilt *Verticillium dahliae Kleb.*	6.9	4	10
	Black leg *Phoma helianthi Aleks.*	67.0	62.1	71.4

Sugar beet	Damping-off, black leg *Phoma betae Frank., Pythium debaryanum Hesse, Fusarium spp., Aphanomyces cochlioides Drechs., Rhizoctonia aderholdii Kolosch., Penicillium spp.*	70.3	59	88
	Downy mildew *Peronospora schachtii Fuck.*	48.7	-	-
	Cercospora leaf spot *Cercospora beticola Sacc.*	39.7	38	41
Potato	Early blight *Alternaria solani Ell. et Mart*	44.8	31	59
	Common scab *Streptomyces scabies Waks.et Heur.*	61.8	30	69
	Black scurf, Rhizoctonia disease *Rhizoctonia solani Pr. et Del.*	60.6	50	68
	Late bligh *Phytophthora infestans DB.*	58.5	50	67
Maize	Helminthosporium leaf blight *Helminthosporium turcicum Pass.*	43.5	42	45
	Blister smut *Ustilago maydis (DC.) Cda*	53.0	39	77
Flax	Anthracnose, Seedling blight *Colletotrichum lini Manns & Bolley*	71.8	40	90
	Bacteriosis *Bacillus macerans Schardinger*	84.4	77	91
	Ozoniozis *Fungus sterilis*	83.9	78	90
	Pasmo *Septoria linicola Speg.*	80.8	70	90
Amaranth	Root rots *Fusarium spp.*	97.5	95	100
Pea	Root rots *Fusarium spp., Pythium de-baryanum Hesse, Rhizoctonia sp.*	41.4	17	73
Millet	Head smut *Sphacelotheca panicimiliacei (Pers.) Bub.*	31.0	2	50
Soybean	Fusarium seedling root rot *Fusarium gibbosum Fhh. et. Wr., F. oxysporum Schl*	61.5	27	80
	Septoria brown spot *Septoria glycinea Hemmi*	52.1	48	58
	Ascochyta leaf and pod spot *Ascochyta sojaecola Abramov*	53.3	47	60
White cabbage	Bacterial black rot *Xanthomonas campestris (Pammel) Dowson*	59.1	34	100
	Olpidium root rot *Pythium debaryanum Hesse, Olpidium brassicae Dang.*	30.0	20	40
Vine	Downy mildew *Plasmopara viticola Berl. et Bunk.*	70.0	17	100
	Powdery mildew *Uncinula necator Burril*	74.3	26	100
	Phoma leaf spot *Phomopsis viticola Sacc.*	79.2	67	92

Strawberry	Marssonia leaf scorch *Marssonia potentillae (Desm.) Magn. f. fragariae (Lib.) Ohl.*	34.0	-	-
	Gray mould fruit rot *Botrytis cinerea Pers*	49.0	42	59
Black currant	American gooseberry mildew *Sphaerotheca morsuvae Berk. et Curt.*	71.2	58	89
Apple	Scab *Venturia inaequalis Wint.*	56.0	51	60

The main parameter for describing of fungicide activity of Albit is its average biological efficiency against this or that disease. However, depending on specific agroclimatic and phytosanitary conditions, the real efficiency of pesticide might differs from the average one, that is shown as minimal and maximal values in Table. Averagely, the biological efficiency of Albit against diseases can vary approximately by 30% of average value, presented in table.

For visual presentation of data from Table, there are diagrams of efficiency of Albit against diseases of flax, winter wheat and vine. Winter wheat is the most widespread and economically significant crop which is subjected to treatment with Albit in many regions. Flax and vine, respectively, are the most northern and southern cultures tested with Albit. Thus, presented diagram reflects general fungicidal activity of Albit.

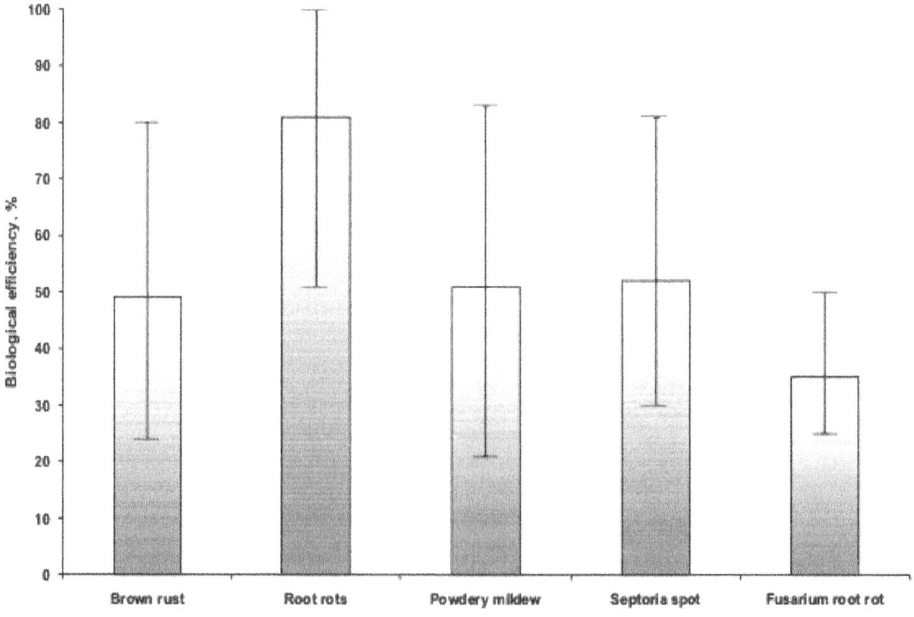

Fig. : Fungicidal activity of Albit against diseases of winter wheat (field trials, 1997-2004). Green bars are the average biological efficiency of Albit, intervals are the minimal and maximal efficiency obtained in trials.

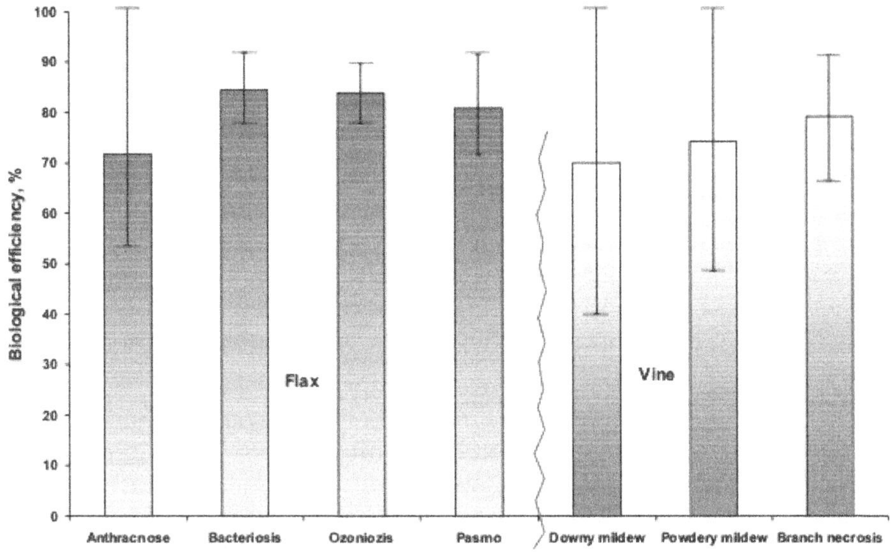

Fig. : Fungicidal activity of Albit against diseases of flax (blue bars) vine (green bars). Bars are the average biological efficiency of Albit, intervals are the minimal and maximal efficiency obtained in trials.

According to decision of Federal Veterinary and Phytosanitation Service (Act Nr 0470, 4 Oct. 2006, State registration Nr 09-00496-0378-1), biostimulant Albit registered as fungicide against:

- Root rots, brown rust, powdery mildew, septoria leaf spot of spring and winter wheat;
- Root rots, net and spot blotch of spring barley;
- Anthracnose, ozoniozis, bacteriosis and pasmo of flax;
- Fusarium root rot of pea;
- Blister smut of maize;
- Watery soft rot and head grey mould blight of sunflower;
- Black leg, cercospora leaf spot and downy mildew of sugar beet;
- Late and early blight of potato;
- Scab of apple;
- American gooseberry mildew of currant;
- Downy and powdery mildew of vine;
- Gray mould fruit rot of strawberry;
- Bacterial black rot of white cabbage.

Universality of immunogenic influence of Albit and new data coming reveal fungicidal activity of Albit against many other diseases.

Summarizing all data of trials, Albit exhibits the most pronounced fungicidal activity against root rots of cereal crops of different etiology - 15-100 %. Biological efficiency of Albit against root rots of spring wheat is 59%, barley -70%, winter wheat - 81%. According to a number of trials carried out by Saratov Regional Plant Protection Station, 2001, Soil institute, 2002, Kirov Regional Plant Protection Station, 2004, *etc.*, Albit suppressed this disease totally (BE 100%). According to All-Russia Institute of Plant Protection RAAS (2004-2005), BE of Albit against Fusarium and Helminthosporium root rots of winter wheat is 57-77 even under conditions of high artificial infectious environment. As far as the effectiveness of majority of recommended chemical protectants against root rots is 40-70%, Albit is effective enough and competitive even among chemical analogues.

The main mechanism of action of Albit is immunization of plants through seed treatment and correction of community of root soil microorganisms through sowing of treated seeds. Consequently, amount of causative agents in rhizosphere gets lower. Chemical protectants surely defend plants against seed root rot infection, but due to stress action, they decrease plantlets resistance to secondary infection from soil. Albit provides more complex and prolonged defense, which is independent on the way of infection soil or seed).

As is well known, the subsidiary phosphoric supplement restricts development of root rots of cereals. Therefore, increase of cereal resistance by Albit can be also partially explained through its ability to increase phosphorus uptake from soil approximately by one third.

Table: Biological efficiency of different means of protection against Fusarium and Helminthosporium root rots of winter wheat var. Ofelia

Set	Application rate, kg(L)/ tonne, kg(L)/ hectare	EC stage 21 (tillering)			EC stage 29 (end of tillering)			EC stage 82-90 (wax stage of ripeness)		
		Prevalence (P), %	Development (R), %	Effectiveness*, %	Prevalence (P), %	Development (R), %	Effectiveness*, %	Prevalence (P), %	Development (R), %	Effectiveness*, %
Albit, seed treatment	0.04	15.7	2.5	66.7	24.9	6,5	67.0	41.5	10.0	56.5
Vitavaks 200 (standard chemical protectant), 200+200 g/L	3.0	14.2	2.0	73.3	23.2	5.8	70.5	38.6	9.6	60.8
Fitosporin-M (standard biological protectant), seed treatment	0.5	19.2	3.5	53.3	29.4	9.5	51.2	50.0	14.0	39.1
Albit, seed treatment + foliar spray	0.04+0.04	14.8	2.5	66.7	28.2	6.4	67.5	21.4	5.4	76.5

Fitosporin-M, seed treatment + foliar spray	0.5+3.0	21.0	3.5	53.3	29.0	9.0	54.3	38.3	10.2	55.6
Dividend star (standard chemical protectant), 30+6,3 г/л	1.0	13.3	1.8	76.0	17.3	4.0	79.7	42.8	11.1	51.7
Dividend star + Albit, (half dose of chemical protectant in combination with Albit	0.5+0.04	13.5	2.0	73.3	23.3	4.8	75.6	40.5	9.8	57.4
Control1 (contamination without treatment with protectants)	-	23.8	7.5	-	42.2	19.7	-	86.5	23.0	-
Control2 (without contamination and treatment with protectants)	-	16.2	4.5	-	35.8	15.2	-	66.3	16.8	-

* Biological effectiveness was calculated basing on disease development (R).

The harmfulness of root rots is increasing every year. For comparison, yield losses due to smut infection and root rots make annually 5-10% and 15-30% respectively. Thus, Albit takes on special significance in agriculture. Though Albit, like the other protectants, is recommended to use only at low and moderate infestation (below 30%), its using is economically sound, since economical harmfulness threshold determined by intensity of disease development is 10-15%. Depending on etiology, Albit in laboratory and field conditions have the highest effectiveness against Fusarium and Helminthosporium root rots. The continuous immunizing activity of Albit provides pronounced plant defense not only against root rots itself, but also against late manifestations caused by the same pathogenic agents (net and spot blotch, helminthosporiose).

Albit is also effective against root rots of other cultures. The average effectiveness of Albit against root rots of: pea, soybean, rice, amaranth is 41%, 62%, 73%, 98% respectively, against black leg of sugar beet is 70%.

Trials of All-Russia Flax Institute revealed the high effectiveness of treatment of flax seeds with Albit against the main diseases of flax. The average effectiveness of Albit against Anthracnose, ozoniozis and bacteriosis was 72–84 % that is inferior to none of the chemical protectants.

Among leaf and stem infections of eared cereals, Albit was most effective against brown rust, powdery mildew and septoria leaf spot. Biological efficiency of Albit against those diseases is 45–90 % averagely, but immunizing activity of Albit is not sufficient for total eradication of those diseases. For full guaranteed defense one can recommend to apply Albit in mixture with half dose of chemical fungicides. The joint treatment with Albit and herbicides in stage of tillering is also high effective, especially for winter crops in spring. In this case, Albit relieves the herbicide-mediated stress which increase probability of the leaf and stem infections.

Foliar spraying with Albit is also high-effective for immunization of plants against: late blight of potatoes, downy and powdery mildew of vine, bacterial black rot of white cabbage, watery soft rot and head grey mould blight of sunflower, black leg, cercospora leaf spot and downy mildew of sugar beet. Regular and well-timed foliar sprayings with Albit allows to decrease plant infestations by 40-70 % or, in case of joint use with chemical fungicides, to eliminate pathogens completely.

Effectiveness of Albit against internal diseases is considerably less pronounced. For example, Albit is not effective against Verticillium wilt of sunflower. Plants should be immunized before infestation, therefore Albit can hardly disinfect seeds, especially mould contaminated ones (Mucor, Penicillium, Aspergillus).Tests of All-Russia Institute of Plant Protection RAAS with winter wheat var. 'Ofelia' demonstrated that seed treatment with Albit did not reduce the infestation of seeds with pathogens and saprophytes, but suppressed the further development of pathogens on plant germs (effectiveness against root rots approx. 67%). Albit itself is unable to disinfect seeds, but it can increase disinfection effectiveness of chemical protectants. For example, tests of All-Russia Flax Institute (2003) demonstrated that use of chemical fungicides (even reduced rates) in mixture with Albit for seed treatment considerably increased the seed disinfection effectiveness. The general contamination of seeds with bunch of infections (anthracnose, ozoniozis, bacteriosis) was 27 % in control, use of full dose of Vital and cut dose of Vital with Albit reduced contamination to 8,0 %, and 4,5 % respectively. Analogously, combination of TMTD with Albit decreased final contamination from 14,4 % (seed treatment with TMTD alone) to 6,0 %.

Albit alone as a fungicidal seed treatment agent also is not effective enough against smut infections. Albit effectiveness decreases in series loose smut stinking smut dwarf smut. Spores of dwarf smut (Tilletia controversa Kuehn.) infect plantlets of cereals generally from soil, spores of stinking smut (Tilletia caries Tul.) remain intact also on the surface of seeds. Spores of loose smut (Ustilago tritici Jens) infect plants in blossoming stage, hyphae of pathogen contaminate seed buds, that make senseless seed treatment with immunizing agents. However, foliar spray in blossoming stage (the second foliar spray) can immunize cereals and decrease loose smut contamination of seeds. Also, joint use with Albit provides dramatic increase of effectiveness of chemical seed treatment agents against smut infections, as it was demonstrated in tests performed by Agricultural Scientific Institute of South-East.

Nowadays, problem of plant diseases of bacterial etiology also become very important for agriculture. Albit effectively suppress chemical fungicide resistant bacteriosis through immunization and correction of plant-associated microbal community. In tests of All-Russia Flax Institute, average biological effectiveness of Albit against bacteriosis was 84,4 %, whereas BE of chemical seed treatment agents was less than 50 %.

In number of trials, Albit acting via immunization was not less effective than direct-acting chemical fungicides. For example, tests of Lipetsk Plant Protection Station with spring barley var. Odesskij-115 demonstrated the same effectiveness against root rots and powdery mildew (91-98 %) of Albit (30 ml/tonne of seeds+2 foliar sprays, 30 ml/hectare) and chemical (Raksil, 1,5 kg/tonne of seeds+foliar spray with Falkon, 0,6 L/hectare) treatments. Analogously, in tests of Tula Plant Protection Station with spring barley var. Zaozerskij-85, effectiveness of presowing treatment with Albit against root rots was 82–85 %, that was not inferior to treatment with standard chemical protectant (Agrosil). In tests of Agricultural Scientific Institute of South-East (2003, 2004), BE of Albit against watery soft rot and head grey mould blight of sunflower was not inferior to BE of standard chemical protectant Fundazol. In tests of North-Caucasian Institute of Horticulture and Viticulture (2004), BE of Albit against downy and powdery mildew of vine was the same (86-93 % and 94-100 % respectively) as BE of chemical pesticides (Ridomil, Falkon, Tiovit, Impakt).

In tests of All-Russia Flax Institute (2002-2004), BE of Albit was 86–90 %, that is approximately the same as BE of chemical standards (Vital, TMTD, Fundazol, Fenoram-super). In tests of Leningrad Plant Protection Station with spring wheat var. Irgina, BE of Albit against septoria leaf spot was 77–100 %, whereas chemical fungicide Fundazol was ineffective. According to conclusion of All-Russia Institute of Plant Protection (2003, 2004), Albit could be used instead of TMTD for coating of sugar beet seeds, because their biological effectiveness (68-88%) are the same. Presented data reveal great plant protective potential of Albit, but it is necessary to note, that fungicidal activity of Albit in general case was 20-30% lower than that of most advanced chemical fungicides.

In summary, as shown in details further, in respect of fungicidal activity Albit was not inferior than majority of standard pesticides in field trials, but in many cases it was less efficient than specialized expensive chemical means of plant protection(protectants and fungicides). Therefore, Albit is useful for plant protection in conditions of low and moderate infestation level and for reduction of application rates of chemical means of plant protection.

According to results of tests, Albit suppresses a broad range infections. However, as in case of other chemical and biological fungicides, the higher infestation level, the lower Albit effectiveness. Fungicidal activity of Albit was detected in experiments only when infestation was less than 30% (low and moderate infestation), and there was no internal infections. At higher infestation level, BE of Albit decreases noticeably.

For example, at disease prevalence of 3,5 %, 35%, 58% and 100% effectiveness of Albit foliar sprays against brown rust was 80 %, 57,1%, 52% and 38.3% respectively (data of Lipetsk Plant Protection Station, 2003, All-Russia Institute of Agrochemistry, 2001, All-Russia Institute of Biological Plant Protection, 2004,and Krasnodar Plant Protection Station, 2004, respectively). Analogously, according to tests of Vladimir Plant Protection Station (2004) with barley, BE of Albit for high (60-100%) and moderate helminthosporiose prevalence was 20-40% and 60,4% respectively. BE of Albit against Bacterial black rot of white cabbage (Xanthomonas campestris Dowson) was 100 and 60% for prevalence of 4 and 13%.

Table: Development of Albit sensitive diseases The average data of all field trials 1997-2004.(–) no data.

Crop	Disease	Prevalence (P), %	Development (R), %
Winter wheat	root rots, brown and yellow rust , powdery mildew, loose smut, septoria leaf spot, fusarium head scab	4-100	2-34
Spring wheat	root rots, kernel black point, brown rust, fusarium head scab and Fusarium wilt, Helminthosporium rot, powdery mildew, loose and stinking smut, Septoria leaf spot	2-100	1-60
Spring and winter barley	root rots, brown rust, powdery mildew, smut infections, septoria leaf spot, net and spot blotch, stem rust	1-90	2-39
Rice	root rots	6-10	22
Millet	smut diseases	-	1-95
Maize	helminthosporium leaf blight, blister smut	-	2-5
Sunflower	watery soft rot and head grey mould blight , verticillium wilt, black leg	21-30	2-10
Sugar beet	black leg, downy mildew, cercospora leaf spot	41-50	7-38
Soybean	fusarium seedling root rot , septoria brown spot	6-22	2-5
Potato	early and late blight, black scurf, common scab	3-79	1-29
Flax	anthracnose, ozoniozis, bacteriosis and pasmo	6-24	-
Pea	root rots	90-100	25-61
Vine	downy mildew, powdery mildew, phoma leaf spot	9-100	1-89
White cabbage	bacterial black rot, olpidium root rot	4-13	-
Apple	scab	2-72	1-35
Strawberry	marssonia leaf scorch, gray mould fruit rot	5-23	7-13
Black currant	american gooseberry mildew	2-8	2-49

Therefore, to get reproducible effect of Albit treatment under more than 30-50 % infestation (10 % infestation with internal diseases), it is necessary to combine Albit with chemical protectants. Albit increases their effectiveness, that might reduce application rates.

Albit, as growth stimulator, accelerate growth and development. Analogously, protective activity of Albit acts through retarding of disease development.

For example, in trials of All-Russia Institute of Agrochemistry and Ryazan Agricultural Academy with spring wheat var. 'Voronezhskaya-10' (2001), brown rust appeared in control and standard series quite fast, in stage of tubering, and infestation has reached 60-65% by stage of ear formation and wax stage of ripeness. In series with Albit treatment, the first signs of disease appeared only in stage of ear formation. Vine var. 'Shardone' treated with Albit exhibited the first signs of powdery mildew only by the end of vegetation. In control sets, there was 36 fold increase of number of infected bunches and 0,05-12,1% increase of disease intensity by the same time, (North-Caucasian Institute of Horticulture and Viticulture (2004). In trials of All-Russia Institute of Viticulture and Wine processing (2003) Albit treatment retarded development of vine diseases (downy mildew and phoma leaf spot) up to 1,5 month.

Similarly to yield increase, contribution of foliar spraying and seed treatment to protection from diseases is different. The most effective immunization against soil and seed infection (*e.g.* root rots) can be reached through seed treatment, against air infections – through foliar spraying. Immunization of vegetables and potato should be performed through foliar spray; reliable defense of apple, vine and currant takes (as shown in tests of All-Russia Institute of Plant Protection, All-Russia Institute of Horticulture, All-Russia Institute of Viticulture and Wine processing and other institutes) at least 2 sets of foliar spray in vegetation season.

In tests of All-Russia Institute of Leguminous and Groat Crops (2003), BE of presowing seed treatment of pea (var. Orlus, Truzhenik and Vega) with Albit against root rots was 35,7–59,1 % (*i.e.* almost the same, as BE of standard TMTD), whereas one or two foliar sprays was completely ineffective. Thus, protection of pea against root rots requires only seed treatment, whereas yield increase can be provided with both seed treatment and foliar spray. For flax (trials of All-Russia Flax Institute), presowing seed treatment with Albit demonstrated high BE (81-90%) against diseases of early growth stages (anthracnose, ozoniozis, bacteriosis), but was useless against pasmo disease in ripening stage. However, combination of Albit seed treatment and foliar spray with allowed to provide BE against pasmo of 70-77%.

According to reports of All-Russia Institute of Plant Protection (2004), effectiveness of Albit (presowing seed treatment) against blister smut of maize was 65,9 %, additional foliar spray with Albit (in stage of 4-6 leaves) increased

BE by 10%, and the second treatment (in stage of blossoming) had no influence on disease development and propagation. As one can see, the earliest treatment was the most effective. The fungicidal effect of Albit against watery soft rot and head grey mould blight of sunflower was almost completely (by 94%) caused by presowing treatment; foliar spray gave only 6% of total BE (Agricultural Scientific Institute of South-East, 2003). In tests of All-Russia Institute of Plant Protection (2002) with sunflower var. Voronezhskij-638, BE of foliar spray with Albit against root and stem form of watery soft rot was approx. 3 times lower than BE of presowing seed treatment. On the other hand, twice repeated foliar spray in the same trial decreased infestation with anthodium (head) form of watery soft rot and head grey mould twice more effectively than seed treatment. The highest BE was provided by combination of seed treatment and double foliar spraying.

ACCELERATION OF PLANT GROWTH

Observed yield increase, which is the final indicator of effectiveness of Albit, depends on stimulation through the whole process of plant growth (biochemical reactions and cellular processes). In result, there might be changes of biometrical characteristics of vegetative growth (germination, development of rootage, bushiness, terms of alteration of phenological phases) and characteristics of yield structure (number of seeds in one ear, weight of 1000 seeds, marketability of potato tubers, weight and size of sunflower head, *etc.*).

Specific mechanisms of Albit influence on different plant organs which determine increased resulting yield are different for different cultures. In trials of All-Russia Institute of Leguminous and Groat Crops and Agricultural Scientific Institute of South-East it was established that increase of buckwheat yield occurs due to increase of amount of seeds (without changes of single seed weight), whereas increase of millet yield occurs mainly due to increase of seed weight. Analogously, according to data of All-Russia Plant Protection Institute and Agricultural Scientific Institute of South-East, Albit do not demonstrates any influence on number of seeds in soybean pod, but increases number of pods per plant, whereas increase of sunflower yield occurs mainly due to increase of number of seeds in head. Important mechanism of yield increase of rice is redistribution of seed formation from lateral panicles to the main one. Increase of yield of eared cereals is formed by combination of several mechanisms: improvement of tillering, increase of single seed weight, increase of number of seeds in ear. Also, Albit helps to reduce ear emptiness by 20% (for rice, analogous characteristic is reduced by 1-6 %).

As is well known, root rots of cereals considerably decrease productive tillering, number of seeds in ear and weight of seeds. Possible, high effectiveness of Albit against root rots also makes contribution to improvement of yield characteristics.

Table: Influence of presowing seed treatment with Albit on germination of different agricultures.

Culture	Increase of yield over control, %
Winter wheat	5-7
Spring wheat	4-16
Spring and winter barley	4-7
Rice	5-8
Millet	6-7
Sunflower	5-11
Potato	3-5
Flax	4-5
Soy bean	2-15
Pea	3-11
Vetch	4-7
Galega	8-13
Lupine	5-10
White cabbage	4-5
Chinese cabbage	1-4
Cucumber	2-12
Sweet pepper	8-15
Tomato	5-10
French bean	5-10

As a rule, stimulating activity of Albit can be detected even in earliest stages of plant growth that is manifested in increased germination. Increase of germination of spring wheat, sunflower, legumes and vegetables might reach of 10-15% due to Albit. Positive influence of the formulation on germination in laboratory and field conditions was detected for the majority of agricultures, and Albit's influence in field conditions is more pronounced. For example, in laboratory trials performed by Albit LLC, Albit increased germination of wheat and barley by 3-4%, whereas increase of germination in field conditions (according to results of long-term trials in different regions of Russia) was 2-15%. Influence of Albit on germination capacity (energy) is exhibited in the same manner. Thus, Albit increases germination capacity of cereals by 1-10% (5.8% averagely). Increase of germination and germination capacity is accompanied by increase of weight of shoots (by 3-5%).

As was revealed in experiments of laboratory of Albit LLC, while treatment of seeds with commercial formulations based on triterpene acids results in analogous stimulation of germination, treatment with Albit also leads to balanced growth stimulation of both over- and underground biomass. Under influence of Albit, weight of shoot roots is increased by 4-7%, length of roots – by 6-11%.

Simultaneous stimulation of germination, growth of over- and underground biomass with Albit helps plants to resist diseases and to assimilate nutrients from very beginning of development. Otherwise, according to our data, stimulation of just vegetative growth which is not accompanied with growth of rootage, make plants more sensitive to diseases, for example, to root rots.

Albit improves development of strong rootage and forming of secondary roots of cereals and other plants. As it was demonstrated in field trials, treatment with Albit increases root length of winter wheat, pea, rice, buckwheat averagely by 11, 11-41, 9, 16-37% respectively. Early acceleration of root growth helps plants to better resist drought and root rots and paves the way for future higher yield.

Besides of increase of germination, on eared cereals Albit increases:

- number of productive stems per square meter (stem density) by 53.3 units averagely,
- number of productive stems per plant (productive tillering) by 25.3% averagely,
- growth of flag leaf (from 1.1 to 1.9 points),
- drought resistance and winter survival.

One important feature of Albit is its ability to visibly accelerate alteration of phenological phases of agricultures. According to obtained data, for cereals this acceleration is 1-12 days.

For example, in trial performed by Tula Regional Plant Protection Station (2003) on spring barley var. Zaozerskiy-85, treatment with Albit (one presowing treatment and two foliar sprayings) accelerated alteration of phenological stages, and the difference with control sets was getting more pronounced with time: coming of EC stage 31 in set with Albit was 3 days earlier than in control, whereas coming of EC stage 83 – even 12 days earlier.

On the contrary, treatment with standard chemical fungicide based on a.i. tebuconazole led to delay of alteration of phonological stages by 1-5 days comparing to control. In trial performed by Voronezh Regional Plant Protection Station (2002), seeds of spring barley treated with Albit germinated 3 days earlier than in control (24th of May instead of 27th). In trial of All-Russia Institute of Plant Protection RAAS on winter wheat (2004), 10 days after sowing the germination of Albit-treated seeds was 60%; in control set germination was 50%; in set treated with chemical standard based on a.i. difenoconazole and cyproconazole, germination was just 10%.

In trial of Lipetsk Regional Plant Protection Station (2003) on winter wheat, coming of EC stage 50-51 was detected after 47-54 days of growth in set treated with Albit, and after 51-62 days in control. In trials performed by the same institution in 2002 with spring wheat var. *Prokhorovka*, coming of EC stage 83 was detected on 78-81 days of growth; in control set it was detected on 85-87 days; in set with application of standard chemical protective treatment (presowing treatment + foliar spraying with chemicals based on a.i. tebuconazole, spiroxamine and triadimenol) – on 90-95 days. The dynamics of alteration of phonological stages is demonstrated on diagram. As you can see, coming of next phases in set with Albit treatment was earlier than in control and in set with chemical treatment throughout all vegetation.

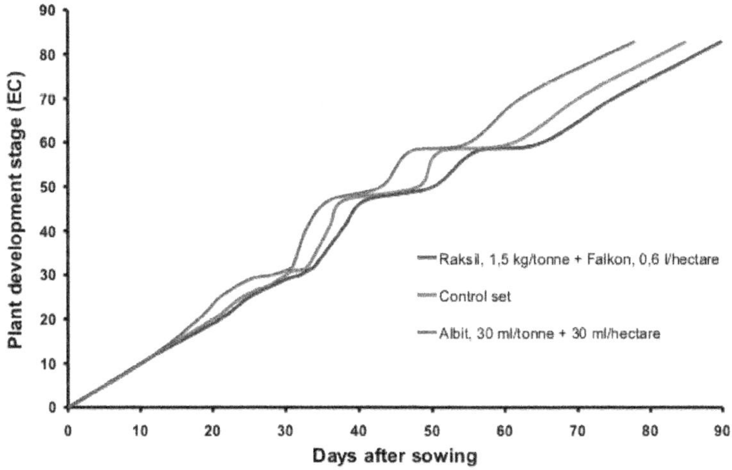

Fig. : Influience of different variants of treatments on alteration of phonological stages of spring wheat.

This growth stimulating feature is even more pronounced for combinations of Albit and chemical fungicides (seed treatment agents). For example, in trial of All-Russia Institute of Biological Plant Protection performed on barley in 2005, plants treated with mix of Albit and decreased rate of chemical fungicide based on a.i. thiabendazole and flutriafol, reached EC stage 50-51 while plants treated with chemical fungicides only were just in EC 30; acceleration of development was clearly visible. In other words, advance of development was about 15 days. Accelerated development of plants at application of combinations of Albit with fungicides is due to not only growth stimulation activity of Albit, but also due to its ability to relieve stress effect of chemical pesticides. In details, it is described here.

Acceleration of alteration of phonological stages due to application of Albit was detected not only for eared cereals, but also for other cultures, for example, for rice and sugar beet. In trial of Tula Regional Plant Protection Station on sugar beet var. Ramonskaya-47, 2 weeks after treatment with Albit plants had 5-6 pairs of real leaves, whereas plants in control set had just 2-4 pairs. In 2003, farmers of Tambov oblast managed to harvest yield of sunflower seeds before autumn rains due to accelerated maturation of sunflower on Albit-treated fields. In collective farm *Kushchevskoe* of Krasnodar krai, physiological ripeness of sunflower seeds on fields treated with Albit came 1 week earlier than in control set.

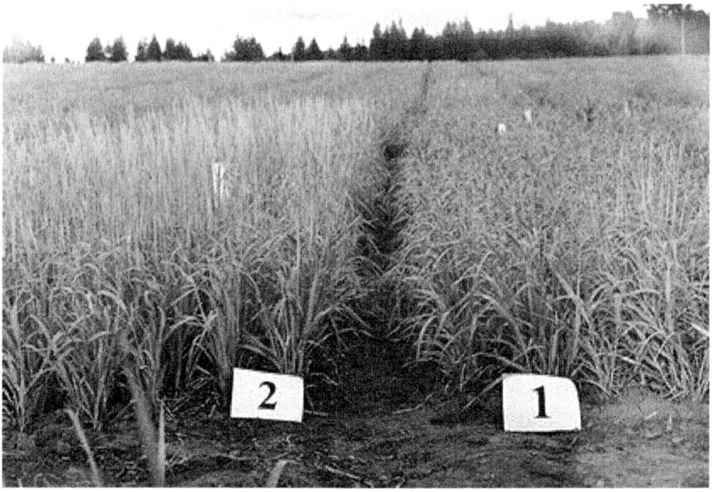

Fig. : Growth of spring barley treated with fungicide based on a.i. thiabendazole and flutriafol (1) and with its combination with Albit (2).

In tests performed by leading potato growing collective farm 'Meristemnye kultury', shoots of potato came up 15-24 days after planting in variant with using of chemical pesticides, whereas shoots of potato treated with Albit came up after 12 days. According to data obtained by collective farm Odoyevskiye zori (2004), application of Albit made possible harvesting of potato 1 week earlier than in control. In trial of All-Russia Flax Institute (2004), flax plants grown on plots treated with mix of Albit and herbicides based on a.i. chlorsulfuron and quizalofop-p-ethyl demonstrated abundant blossoming by 8th of July, whereas plant treated with pure herbicides were just in very beginning of blossoming.

In trials of All-Russia Institute of Vegetable Selection and Seed Breeding (2001-2003) on vegetables (tomato, cucumber, sweet pepper) application of Albit lead to accelerated and more simultaneous ripening; yield was almost completely harvested in 1st and 2nd harvesting; Along with total yield, early yield was also increased. Control set and sets with application of chemical analogues demonstrated more late and 'prolonged' ripening.

Table: Influence of Albit on characteristics of growth structure and yield of eared cereals in field conditions.

Characteristics	Increase of yield over control	Source (Institutions which conducted trials)
Germination, %	6.1	All-Russia Institute of Plant Protection RAAS, All-Russia Institute of Plant Protection (of Ministry of Agriculture), All-Russia Institute of Biological Plant Protection, Far East Institute of Plant Protection, Kurgan Institute of Grain Growing, Agricultural Scientific Institute of South-East , Central Institute of Agrochemical Service, Kursk and Primorye Regional Plant Protection Stations, Ryazan State Agricultural Academy

Germination power, %	5.8	All-Russia Institute of Plant Protection RAAS, All-Russia Institute of Biological Plant Protection, Agricultural Scientific Institute of South-East, Kursk Regional Plant Protection Station
Productive bushiness, %	25.3	Agricultural Scientific Institute of South-East, Central Institute of Agrochemical Service, Vladimir, Krasnodar, Kurgan, Lipetsk, Tula Plant Protection Stations
Stem density, units / m^2	53,3	Soil Institute, Kurgan Institute of Grain Growing, Kursk Scientific Research Institute of Agroindustrial Manufacture, Central Institute of Agrochemical Service, Kemerovo and Kursk Regional Plant Protection Stations
Rootage length, %	8.7	Krasnodar Regional Plant Protection Station, Quality Control Laboratory of Albit LLC
Ear length, cm	1.5	Agricultural Scientific Institute of South-East, Kemerovo, Kurgan, Tula Regional Plant Protection Stations, Lipetsk State Plant Varieties Testing Station
Number of seeds in ear	2.6	Kursk Scientific Research Institute of Agroindustrial Manufacture, Soil Institute, Kurgan Institute of Grain Growing, Kursk, Kemerovo, Kurgan, Lipetsk, Tuva Regional Plant Protection Stations, Lipetsk State Plant Varieties Testing Station, Central Institute of Agrochemical Service, Ryazan State Agricultural Academy
Weight of 100 ears, g	27.2	Klimovo Regional Plant Protection Station of Bryansk oblast
Ear emptiness, %	-20	Central Institute of Agrochemical Service, Ryazan State Agricultural Academy, Penza Regional Plant Protection Station
Weight of 1000 seeds, g	1.8	All-Russia Institute of Plant Protection, Kursk Scientific Research Institute of Agroindustrial Manufacture, Kurgan Institute of Grain Growing , Agricultural Scientific Institute of South-East, Soil Institute, Voronezh, Kursk, Lipetsk, Kurgan, Tula Regional Plant Protection Stations, Central Institute of Agrochemical Service, Ryazan State Agricultural Academy, Aleysk Station of Agrochemical Service
Gluten content, %	2.3	All-Russia Institute of Plant Protection, Soil Institute, Kurgan Institute of Grain Growing, Kursk Scientific Research Institute of Agroindustrial Manufacture, Central Institute of Agrochemical Service, Kemerovo and Kursk Regional Plant Protection Stations, Prikumsk Agricultural Experimental Station, Aleysk Station of Agrochemical Service , Zolotaya Niva LLC of Stavropol krai, Praskoveya Collective Farm of Stavropol krai, Kushchevskoe Collective Farm of Krasnodar krai

Thus, acceleration of growth and ripening of agricultures is universal and characteristic property of Albit. Obviously, this phenomenon depends on high auxin activity of the bioformulation. Besides, Albit promotes accumulation of antioxidants in plant cells that eases accelerated forming of new structural elements of plant organism, and saves metabolites, which can be used for plant growth.

Accelerated plant growth due to Albit might be unfavorable sometimes. For example, accelerated growth of winter cereals might lead to their overgrowth and 'sweatening'. Therefore foliar spraying of plants is recommended to perform after hibernation. But anyway, in most cases this property of Albit is positive for plants. In 2003, farmers of Tambov oblast managed to harvest yield of sunflower seeds before autumn rains due to accelerated maturation of sunflower on Albit-treated fields. Incorporation of Albit into cultivation of sugar beet shortens its vegetation period and therefore makes possible harvesting of good yield even in very north of its cultivation area: in Moscow oblast and in republic of Mordovia. Acceleration of plant growth and alteration of phonological stages is especially important for regions with continental climate (such as Povolzhye and Siberia) which limits vegetation periods.

Intensification of plant growth due to Albit influences the accumulation of metabolites in generative organs, and, as a consequence, the characteristics of the yield. For example, application of Albit on cereals increases:

- weight of 1000 seeds averagely by 1.8 g,
- number of seeds in ear by 2.6,
- ear length by 1.5 cm,
- decreases ear emptiness by 20%.

In trial performed by Soil Institute (2002), treatment of seeds with Albit increased number of seeds in ear of spring wheat by 2, following foliar spraying also added 2 more seeds.

Here is data about influence of Albit on yield characteristics and plant development of other cultures. (besides eared cereals):

Sugar beet: increase of average leaf area by 0,5–3 cm2, acceleration of maturing by 2–7 days, stimulation of metabolite redistribution from leaves into root;

Sunflower: increase of germination by 5–11.2%, acceleration of blossoming and ripening of seeds by 6–7 days, increase of head diameter by 2.2-4.9 cm, increase of oil output by 22.5%, increase of weight of 1000 seeds by 3-7.6 g;

Maize: increase of green biomass yield by 13.4–110 centners/hectare and corn yield by 1.19–6.4%, increase of ear weight by 3.2–6.2 %;

Potato: acceleration of coming-up from several days to one week, acceleration of maturing, improvement of quality and marketability averagely by 6.9 %;

Buckwheat: increase of germination by 4–7 %, rootage development by 3–12 %, weight of 1000 seeds by 1–4 %, weight of seeds obtained from one plant averagely by 33.8 %, drought resistance (yield increase up to 50 % in extremely dry conditions);

Millet: increase of germination by 5.7–7 %, rootage development by 7–20 %, number of seeds from one plant by 9-17, weight of 1000 seeds by 0.21–0.59 g, weight of seeds obtained from one plant by 0.15–0.40 g;

Rice: increase of germination by 5–8%, germination power by 7–9 %, tillering coefficient by 0.7, length of roots by 3.9 cm, stem density by 5.1–8.2%, panicle length by 1.02–1.1 cm, weight of 1000 seeds by 0.4–1.46 g, average number of seeds in one panicle by4.3, weight of seeds from main panicle by 0.16–0.62 g;

Flax: increase of germination by 3–7%, germination power by 3–6%, flax straw yield by 0.8–13 centners/hectare, yield of seeds by 0.4–1.4 centners/hectare, flax fibre quality (average number of long fibre by 0.5–1.3%, percentage-number of flax fibre by 26.1–55%), stem density by 19.7–37%;

Legume groats (peas, chick-pea, soybean, french (kidney) bean, lentil, fodder beans): increase of germination by 3–15%, stem length by 3–36%, rootage development by 11–41%, number of productive nodes by 13.7–16.8%, number of pods per one plant by 12.3–30.9%, number of beans per plant by 11.1–28.9%, weight of beans from one plant by 32.7–36.6%;

Foorage grasses (galega, clover, vetch, alfalfa, sweet clover, lupine, amaranth, canola, *etc.*): increase of germination by 4–157%, drought and winterkill resistance, regrowth after skewing by 14.1–34%, height and density of the grass by 10%, yield of the green biomass by 4,9–91 centners/hectare, yield of seeds by 0,39–1 centners/hectare; in Gramineous forage grasses Albit increases tillering and regrowth by 33.6-68%, increment of green biomass by 7.6–17 %, chlorophyll content by 8.6–9.7%;

Vegetables (tomatoes, cucumber, squash, lettuce, peppers, eggplants carrots, table beet): increase of total and early yields by 11-23%, germination by 2–12%, stimulation of vegetative growth and blossoming, increase of resistance to drought and other adverse factors, increase of ovary number per plant by 1-3, acceleration of blossoming and ripening by 4–5 days, more simultaneous and fast ripening. Specifically, Albit increases length of onion leaves by 12.6-21% and dry weight of onion leaves by 7.4-10.7%;

Cabbage (white and chinese): increase of average weight of cabbage head by 0.2–8.3%, germination and germination power by 3–7%, quality of seedlings (increase of the seedlings weight averagely by 11,5 %, seedlings height by 21,5 %);

Vine: increase of number of ripened grapes by 3.7–13.6%, annual increment of shoots by 28–64%, single shoot increment by 28–56%, ripen shoot increment by 59–64%;

Horticultures (apple, pear, cherry): increase of annual increment of shoots by 12.6–25.6%, photosynthetic activity of the leaves by 6.9–18.6%, assimilation leaf surface by 73–76%;

Berry-bushes (currants, gooseberries, dog rose): faster and more simultaneous ripening, improved marketability of berries, increase of number of flowers on one bush by 9-10%, setting of the berries by 9–13%, berry size by 10–20%, vitamin C content by 5.8–13.8%, chlorophyll content in leaves by 6.2–11.8% and annual increment of shoots by 10–30%, decrease of amount of rotten berries, improvement of physiological state of leaves, increase of resistance to adverse environmental factors (air and soil pollution, dramatic temperature jumps);

Strawberry: increase of yield by 9.3–20 centners/hectare, average number of berries from one plant by 0.4–2, average weight of a berry by 1.3 g, leaf photosynthetic activity(F_v/F_m) by 0.04 units, improvement of stress resistance;

Decorative bushes and forest breeds: increase of annual increment of shoots averagely by 18.2%, chlorophyll content by 6.2-11.8%, leaf assimilation surface by 15-76%. Improvement of physiological state of leaves, acceleration of blossoming by 3-10 days, increase of number of flowers per one bush by 9-10%, increase of resistance to adverse environmental factors (air and soil pollution, temperature jumps). Albit also increases rootage of cuttings by 40-90%, improves initial growth of cuttings, increases chlorophyll content averagely by 7.4%;

Field and greenhouse flowers (roses, gladiolus, pinks, cyclamens, Kaffir lily, amaryllis, violet, primula, *etc.*): earlier blossoming (by 3–15 days), increase of blossoming duration by 4-35 days (up to 1.5-2.5 times), improvement of decorative appearance, increase of number of flowers per plant.

Important features of Albit are blossoming induction, acceleration of opening of flowers, providing of more good and even blossoming, increase of number of flowers per plant. Although this phenomenon was shown for wide range of plant species, it is flowers and decorative cultures that demonstrate the most pronounced and clear response to Albit. It is well known that watering of indoor flowers (geranium, primula, amaryllis, violets, azalea and others) with Albit solution considerably increases longevity and intensity of blossoming. For example, Kaffir lily (*Clivia miniata*) at the head office of Albit LLC blossoms during major portion of winter, spring and summer, that is far out of the typical blossoming period from January to April. Induction of blossoming is most pronounced for flowers, but it is also primary important for all agricultures. Blossoming induced by Albit was detected for most cultures (cereals, legumes, apple, currant, vegetables, forage grasses) and is very important for yield increase.

Table: Influence of Albit on mineral uptake of spring wheat.

Element	*Uptake from soil, mg/pot*	
	into straw	**into seeds**
Nitrogen	151–197	178–229
Phosphorus	15-17	28–37

In trials conducted by All-Russia Institute of Plant Protection and farmers of Krasnodar krai it was shown that foliar spraying with Albit promotes redistribution of metabolites from leaves of sugar beet into root. Analogous redistribution of nutrients from vegetative organs to seeds induced by Albit was detected in trials of Department of Agrochemistry of Moscow State University on spring barley and in trials of Aleysk Station of Agrochemical Service on spring wheat.

In trial of Department of Agrochemistry of Moscow State University, yield increase of straw was twice smaller than that of seeds, whereas in control set the yield of straw and seeds was approximately equal.

In other words, positive influence of Albit was more significant for forming of generative organs. In trial of Aleysk Station of Agrochemical Service, treatment with Albit intensified arrival of nitrogen, phosphorus and potassium to seeds, but decreased that to straw, that indicates redistribution of nutrients to seeds under influence of Albit. Long-term data of many scientific institutions indicates that application of Albit leads to additional seed setting of cereals and legumes, increases setting of fruits of apple, currant and vegetables.

These facts demonstrate that influence of Albit on agricultures is not just stimulation of vegetative growth. Of course, accelerated vegetative growth lets plants accumulate sufficient amount of metabolites. However, with the lapse of time stimulation of growth gets accompanied with stimulation of earlier blossoming, formation of generative organs and redistribution of nutrients to them, that finally provide high yield increases.

YIELD QUALITY

One important feature of Albit is its positive influence on yield quality. Nowadays, one of the main problems in agriculture isgluten content in grains which determines its market price. Treatments with fungicides, herbicides and insecticides causes stress and retardation of plant growth, that leads to considerable decrease of gluten content. Finally, harvested yield has low quality. To relieve stress and increase yield quality, it is necessary to use liquid nitric fertilizers (that is quite expensive) or antistressors.

Using of bioformulatlons as an antistressors in combination with chemical pesticides is a well tested technique to increase yield quality. Being an antistressor, Albit was shown to be very effective in such role. According to data of field trials, Albit increases yield quality (gluten content in wheat by 0.5-5.1%, sugar content in sugar beet by 0.3-2.5%, content of vitamins in vegetables by 11-25 relative %, fraction of commercial yield of potato, vegetables, grape vine (fraction of commercial yield of grape vine (table varieties) is increased by 10-25 %), improves organoleptic and flavouring properties of grape vine, contributes the best storage (losses of flavouring and aroma are decreased by 75 % during storage)). Using of Albit in mixes with chemical pesticides (fungicides, herbicides, and insecticides) provides high yield quality due to both its own stimulating activity and decreasing the stress effect of pesticides.

Presowing seed treatment with Albit influences mostly the productivity of plants; to increase gluten content, using of foliar sprayings is more effective. For example, in trials on winter wheat performed by Kursk Regional Plant Protection Station (2001-2003), presowing seed treatment with Albit increased the average gluten content in grain by 0.8% whereas foliar spraying in EC stages 20-49 increased it by 3.8%. In southern regions of Russia (Rostov oblast, Krasnodar kray), high effectiveness for increase of gluten content was demonstrated by 2nd recommended foliar spraying in EC stages 50-69.

Joint using of Albit with liquid urea and some plant growth stimulators also effectively increases yield quality. As it was shown by Agrochemical Department of Moscow State University (2000), application of Albit increased gluten content in spring wheat by 0.5%, whereas using of Albit/urea combination increased it by 0.8%.

High effectiveness of application of Albt together with urea might be explained through the tact that usual application rates of urea used in field conditions is up to 30 kg/hectare, or 10% solution approximately, whereas concentration of even 1% is considered to be toxic for plant growth. Albit relieves this toxic influence on plant growth while urea nitrogen is maximally used for protein (gluten) synthesis in grain.

Joint using of Albit with humates, as it was shown in trials in Krasnodar kray, increases average gluten content up to 5.3%, though such effect can not be provided by Aibit or humates separately (mutual enforcement of action, aka synergism phenomenon).

It is well known that infestation of grain crops with root rots leads to decrease of gluten content in grain. Thus, the hign efficiency of Albit against these diseases might also contribute to improvement of yield quality. Next, there is a speculation that biologically active sulfur and boron of Albit also participates in increase of yield quality since sulfur and chelate boron are also known to be a factors of gluten content increase. There are plenty of examples of successful Albit application to increase of gluten content. For example, collective farm Zarya and farmer Tutov A.V. of Rostov obiast were using Albit in 1999-2004.

Seeds of winter wheat were treated with Aibit + halved dose of chemical seed treatment agent. Albit was also applied together with fertilizing. The average yield in 1999-2002 was 45-49 centners/hectare, gluten content was 26-29%, gluten deformation index was 65-95. Neighbor farms which did not use Albit, harvested grain mostly of forage quality. In tests performed by Kursk Regional Plant Protection Station on winter wheat in 2002, gluten content in set with application of Albit was 30.4% that is 4.8% higher than in control (25.6%), tensility of gluten increased from 13 to 14 cm. Presowing seed treatment of spring wheat with Albit increased gluten content by 1.4% whereas foliar spraying increased it by 2,1%. In experiments of Kurgan Institute of Grain Growing, treatment of spring wheat with Albit increased gluten content from 23.4 to 26.3% in 1997, in 2001 gluten content increased by 5.1%. In Ryazan oblast (Experimental Farm Aleshkinskoe, 1998), Albit treatment increased content of protein in grain from 13.8 to 15.2%, gluten content - from 26,1 to 28.1%. Nowadays, some dealers of Albit make treaties with farms on terms of guaranteed increase of gluten content (to 28-30%) and yield on the assumption of their general scientific guidance.

Application of Albit on beer barley often raises a question about possible influence of this bioformulation on protein content in yield (analogously to wheat). Till now there were no cases of protein content excess over critical brewing threshold of 12%. Possibly, different reaction of barley and wheat to Albit treatment (regarding to protein content) depends on genetically determined features of these cultures.

Under Albit treatment the content of mycotoxins in yield is significantly decreased on such cultures: oats, wheat, barley (for details, see here).

Besides eared cereals, Albit also increases yield quality of the other agricultures:

- Marketability of potato tubers is increased by 9.6%.
- Decreasing the amount of rotten tubers and increasing fraction of large potato (>50 g) are observed. In some varieties of potatoes under Albit treatment the content of starch in tubers is increased by 12-15 rel. %.
- According to All-Russia Flax Institute, Albit treatment increases quality of flax fiber: percent-number is increased by 26.1-55%, number of long fiber is increased by 0.5-1.3%.
- Sugar content of sugar beet is increased by 0.3-2.5%. (According to data of trials performed by All-Russia Institute of Plant Protection, presowing seed treatment with Albit increased just yield, whereas foliar sprayings with Albit increased both yield and sugar content.)

Content of vitamins in vegetables (tomatoes, cucumbers, vegetable marrows, peppers, eggplants, lettuce, carrot, table beet) is increased averagely by 11-29 relative % under the influence of Albit; increase of ß-carotin and ascorbic acid (vitamin C). In field trial VNIISSOK on onion of open ground, Albit treatment increased the content of vitamin C in production by 13-41 %, in field trial Mendel University in Brno (Brno, Czech Republic) on lettuce – by 6-22 % (and total antioxidative capacity (TAC) of lettuce leaves – by 43-257 %). In leaves of lettuce, Albit increases ascorbic acid content simultaneously with decrease of nitrates content. As result, relation between ascorbic acid and nitrates becomes higher than medical safety threshold of 2:1, *i.e.* Albit makes plants grown in greenhouse conditions safe for human consumption. Analogous effect was detected on Chinese cabbage, tomatoes, cucumbers and onion (data of All-Russia Institute of Vegetable Selection and Seed Breeding, 2001-2004). Under the influence of Albit, nitrate content decreases by 16-26 relative % in majority of vegetables. Decrease of nitrate content in leaves of onion might reach of 74-77 relative %. Also, Albit provokes increase of dry weight of onion leaves (by 7.4-10.7%), that indicates increase of their nutritive value.

Trials performed by North-Caucasian Institute of Horticulture and Viticulture of Krasnodar kray demonstrated that Albit increases weight and plumpness grape bunch by 21-24%. Field trials of Crimea (National Institute of Grapevine and Wine «Magarach», Yalta) demonstrated that use of Albit improves of organoleptic properties and gains of variety aromas of berries of table varieties of grape vine (total organoleptic score is increased by 21,5 %). Biochemical base of this effect - under Albit treatment in berries of grape vine the content of main flavonoid compounds is increased (flavonols by 5,9 %, anthocyanins by 24,4 %, stilbenes by 32.8 %). The trace amount of pesticides in production is decreased by 51,5 %. Albit inhibits sugar and organic acids decay during storage: losses of flavouring and aroma are decreased by 75 %, decrease of berries weight is observed by 39,5 %. In apples, currant and gooseberries due to the treatment with Albit, content of vitamin C is increased by 5.8-13.8%; quality and marketability of apples and berries is also improved. Treatment with Albit decreases number of rotten and dried berries of currant and cherry and increases size of berries by 10-20%. In farm Keith Smith under Albit treatment increase of vitamin C content in blueberry is observed during industrial cultivation by 13,8 %. Weight of a particular strawberry is increased averagely by 1.3 g. In field trial ACPEL yield of commodity products of melon under Albit treatment increased by 6 % (at the same time, in variants with use the most of biological and chemical standards –decreased by 4-25 %). In farm Terkin Seraphimovsky region of Volgograd oblast under Albit treatment wholesale price of watermelons in 2010 increased by 2 rubles/kg, in 2011 by 1,25 rubles due to increase of fraction of commercial yield.

Treatment of flowers (roses, gladioluses, pinks, cyclamens,. Kaffir lily, amaryllis, violets, *etc.*) with Albit besides induction blossoming improves decorative appearance, increases of number of flowers per plant, increases production of top-quality flowers (by 10-15%), provides more bright and saturated coloration of flowers and increases resistance to diseases.

EFFECT OF BIOSTIMULANT ALBIT® ON MYCOTOXIN CONTENT IN GRAIN CROPS

The World has been facing an acute problem of mycotoxins in recent years, though its history goes back for more than one hundred years. Cases of mold toxins poisoning in humans and animals were known as far as in the Middle Ages. FAO estimates show that up to a quarter of the world's foods contain mycotoxins. One billion tons of agricultural products are under the potential risk of mycotoxin contamination. In Russia, the amount of grain contaminated by mycotoxins has increased tenfold in recent years. Up to 45.0-74.7% of grain samples in the Central, Central Chernozjom, Volga-Vjatka and Volga regions of Russia are contaminated by fungi of the genus *Fusarium*. The analyses of feed, vegetation, various agricultural products, food products show high (up to 80-100%) contamination by microscopic fungi, specifically by toxigenic fungi in 40-60% of cases, and in 21% of cases production of mycotoxins in hazardous concentrations was detected. More than 30% of feed and other agricultural products are contaminated by mycotoxins. Livestock sector suffers serious economic losses from reducing productivity and reproduction of farm animals due to disorders caused by mycotoxins. A significant contamination of grain by microscopic fungi of the genera *Fusarium* and *Alternaria* in most regions of Russia shapes the urgency of the mycotoxin issues for the grain-producing regions of the country, and stresses the need to control the quality of raw grain.

To date, there are more than 250 species of fungi that produce several hundred mycotoxins, of which trichothecenes, aflatoxins, patulin, ochratoxins, and zearalenone are the most common and dangerous to the health of humans and animals. The representatives of the trichothecene group include toxins such as T-2, diacetoxyskirpenol, HT-2, deoxynivalenol, nivalenol, roridine A, krotocin. They are produced by both obligately saprophytic fungi (*Stachybotrys alternans*) as well as plant pathogenic fungi (*Trichoderma roseum, Myrothecium verrucaria, Fusarium* spp.). Major producers of T-2 toxin were isolated from feed and food raw materials which caused foodborne toxicoses in agricultural animals and humans. These include *Fusarium poae, F. acumination, F. sporotrichioides*. Deoxynivalenol (DON, vomitoxin) is mainly produced by various strains of *F. graminearum, F. culmorum, F. nivale*. It should be emphasized that one and the same kind of producer fungus can synthesize several mycotoxins. Trichothecenes exhibit teratogenic, cytotoxic, immunodepressive, dermatotoxic properties, affect hematopoietic organs, central nervous system, cause leukopenia, hemorrhagic syndrome, trigger a number of food mycotoxicoses in humans and animals. The toxic properties are due to their involvement in the suppression of protein biosynthesis. Among the mycotoxins produced by *Fusarium* fungi, special consideration is given to deoxynivalenol (DON), zearalenone and fumonisin in the European Union. Ochratoxin A, B and C are produced by *Aspergillus ochraceus* and *Penicillium viridicatum* fungi. Ochratoxin A is the most toxic agent while it is most often present in foods. Mycotoxins of this group have nephrotoxic, teratogenic and immunosuppressive effects. They inhibit protein synthesis and interrupt glycogen metabolism. Ochratoxins trigger renal disease in pigs.

Pigs are very sensitive to mycotoxins produced by fusaria. Diseases usually appear in Winter and Spring and in most cases have an implicit zonal dependence. In case of an acute form of the disease, animals die within 15-20 hours after eating their food containing the toxin. Toxic effects of T-2 toxin appear differently in poultry, depending on the duration of its presence and concentration in the feed. For example, in case of layers, if T-2 toxin is present in the feed for 14-18 consecutive days, there is a decrease in egg production and egg weight. In addition to the mentioned symptoms, prolonged use of the mycotoxin-containing feed results in eggshell thinning, reduced hatchability, craw mucosa and gizzard stomach lesions. Importantly, a simultaneous presence of mycotoxins such as aflatoxin and T-2 toxin in the feed has the greatest immunosuppressive effect – a synergy is observed.

Mycotoxin content in food and feed varies widely and can reach hundreds of micrograms / kg, that's why their levels are subject to regulations in many countries. Hygienic requirements for the quality and safety of food raw materials and food products accepted in Russia (SanPin 2.3.2.1078-01) limit the level of T-2 toxin in grain to 0.1 mg/kg and provide for its absence in baby cereal-based food (< 0.005 mg/kg). Wheat grain may be used for food purposes if it contains deoxynivalenol (DON) of not more than 0.7 mg/kg. The maximum allowable concentration of zearalenone in grain, grain products, nuts, seeds, oil seeds, fats, oils, protein isolates is 1 mg/kg and it should be absent in baby food and dietary food. The content of ochratoxin A in bread grain, including wheat, rye, oats, barley, rice and grits, flakes, wheat flour, pasta should not exceed 0.005 mg/kg.

The protection of crops against mycotoxin contamination using conventional direct-action fungicides is not always productive, as the toxins are produced not only by phytopathogenic but also by saprophytic, endophytic, mold fungi. Moreover, our studies have shown that the application of chemical fungicides may even lead to increased levels of ochratoxin in yield of barley and HT-2 toxin in wheat. Perhaps, this is due to the suppression of non-toxigenic microbes.

The application of immunizing fungicides (including biostimulant Albit) seems to be promising in this respect. Immunizing effect of the biostimulant allows it to control the development of those plant pathogenic fungi which do not manifest themselves through symptoms of disease. In particular, this applies to toxicogenic fungi developing in ear of cereals. According to the previous studies, Albit has biological efficacy against *Fusarium* head blight in the range of 35-45%, against *Alternaria* blight up to 50%. Besides that, Albit does not have general biotoxic effect.

The purpose of our work was to evaluate the effect of Albit treatment on mycotoxin content in grain harvest. Mycotoxins in grain samples were identified by using conventional HPLC-based techniques in the state-certified laboratory TEST-Pushchino, Pushchino, Moscow oblast'. Samples were taken from field trials conducted in different regions of Russia.

Primary research results are shown in the Table. The application of Albit in recommended dosages has been found to reduce the content of ochratoxin and trichothecenes on the background of both pure check and chemical pesticides, in case of grain contamination above threshold levels.

Table: Effects of Albit on mycotoxin contamination of grain harvest in 2010-2011 field experiments

Crop	Test site, year	Treatment type	Micotoxins (mg/kg)		
			Ochratoxin A	T-2	HT-2
Winter wheat	OSC Chapaev breeding farm, Krasnodar kraj, 2010-2011	check	< 0.0001	< 0.005	< 0.005
		Albit spraying (40 ml/ha) in the tillering stage with herbicide	< 0.0001 (100*)	< 0.005 (100)	< 0.005 (100)

Winter wheat	Nastional Institute of Biological Plant Protection, Krasnodar, 2010-2011	check	< 0.0001	< 0.005	0.012
		Tebuconazole-based fungicide seed dressing followed by propiconazole + cyproconazole-based fungicide spraying	< 0.0001 (100)	< 0.005 (100)	0.022 (183.3)
		40 ml/ton Albit seed dressing + 40 ml/ha twofold spraying	< 0.0001 (100)	< 0.005 (100)	0.012 (100)
Spring barley	Orjol Nobel-Agro LLC, Kolpnjanskij district of Orjol oblast', 2011	check	< 0.0001	< 0.005	–
		2-fold carbendazim-based fungicide spraying	0.00231 (2310.0)	< 0.005 (100)	–
		2-fold Albit (40 ml/ha) + carbendazim-based fungicide spraying	< 0.0001 (100)	< 0.005 (100)	–
Winter wheat	L.I. Tumakov's Farm, Tokarevskij district of Tambov oblast', 2010-2011	check	0.003	< 0.005	< 0.005
		Albit seed dressing (30 ml/t)	< 0.0001 (3.3)	< 0.005 (100)	< 0.005 (100)
Oats	National Institute of Plant Protection, Ramon' district of Voronezh oblast', 2011	check	–	0.0199	0.0577
		Albit seed dressing (20 ml/ton) + spraying (20 ml/ha)	–	0.0144 (72.4)	0.0430 (74.5)

* in brackets – % of the check level. (–) samples were not available

The further more detailed studies were conducted in oats. Oats is consumed by animals and humans (instant cereals) in a practically unprocessed form, so mycotoxin content in this crop is crucial compared to other grains. Furthermore, our previous experience has shown that the concentration of toxins in the oats grain was higher than in other cereals.

Detailed experiment on different Albit dose applications to the Lion oat was conducted in 2012 in the experimental field of the National Plant Protection Institute of the Russian Academy of Agricultural Sciences in the Voronezh oblast'. Albit was used by the scheme proven for most cereals: pre-sowing seed treatment and further spraying with herbicide in the tillering stage. Pure herbicide treatment (a.i. tribenuron-methyl) served as a check. The Albit dosage is ranged from 10 to 100 ml / ton (ha). The most dangerous trichothecene mycotoxins T-2, HT-2, DON and ochratoxin A were selected for evaluation.

All the studied Albit dosages led to an increase in the yield of oats (by 8-12% at the check level of 3.55 t/ha). The development of diseases peculiar to oats in this area (leaf red-brown spot and Septoria leaf spot) was observed to a very small extent (incidence 1-2%). However, mycotoxin contamination of grain was sufficiently large (about 1 mg/kg). The analysis of the grain harvest has shown that Albit application has helped to reduce the content of almost all the studied mycotoxins in it. Mycotoxin level reduction compared with the check without

Albit treatment ranged from 25% to 71%. However, when using certain Albit dosages, the levels of mycotoxins in grain conversely exceeded the check levels.

Table: The effect of Albit on the yield and mycotoxin content in oats grain depending on the application rate in the 2012 experiment (% of check level)

Albit application rate, ml/ton(ha)	Yield	Mycotoxin content in grain, mg/kg				
		DON	T-2	HT-2	Ochratoxin A	Mycotoxins total
0	100.0	*0.91**	*0.012*	*0.016*	*0.0002*	*0.9382*
10	112.4	28.6	41.7	51.3	600.0	29.2
20	107.6	60.4	41.7	56.3	50.0	60.1
40	112.1	48.4	241.7	356.3	50.0	56.1
60	111.3	144.0	41.7	56.3	50.0	141.1
80	107.6	42.9	75.0	106.3	50.0	44.4
100	109.9	49.5	125.0	100.0	50.0	51.3

*absolute mycotoxin content levels (mg/kg of grain) are given for the check variants.

The lowest rates of Albit application has demonstrated the maximum efficiency against trichothecene mycotoxins and a sharp jump in the level of ochratoxin A. Alternatively, higher dosages of the biostimulant have shown increased efficiency against ochratoxin and several maximum levels of trichothecenes deoxynivalenol. Thus, lower dosages of Albit are more effective against those toxins produced by *Fusarium* spp. and are inactive against toxins produced by fungi of the genera *Aspergillus* and *Penicillium*. At the same time, Albit at high application rates, is less active against fusariotoxins and is more active against ochratoxin A. According to data aggregate, the Albit effect was the highest at the 20 ml / ton (ha) application rate since its use was not marked by an excess of any of the mycotoxins studied.

In assessing the Albit effect on the total content of mycotoxins in grain, the averaged picture shows the maximum of Albit anti-mycotoxin activity at low application rates and the minimum at 60 ml/ton (ha) dose.

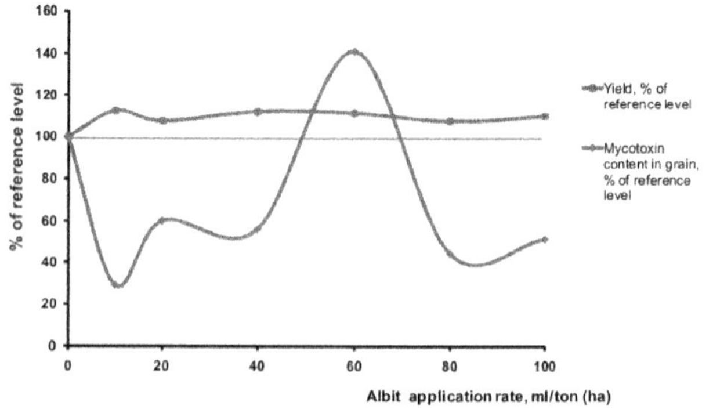

Fig. : The Albit application rate effect on the total content of mycotoxins (DON, T-2, HT-2, ochratoxin A) in oats grain in field experiment (2012)

The data obtained are consistent with the 2011 oats experiment and allow us to recommend the application of Albit in the dosage of 20 ml / ton of seeds + 20 ml/ha to reduce mycotoxin content in oats harvest. Mycotoxin content was reduced under Albit application by approximately 25% in the 2011 experiment with an overall higher mycotoxin content, and by 40-60% in the 2012 experiment with a lower level. The application of other dosages of Albit may, in contrast, increase contamination of grain by some mycotoxins.

Thus, the results of the studies conducted have shown a significant effects of Albit against mycotoxins in grain harvest. 20 ml/ton (ha) Albit application makes sense for oats and standard recommended drug doses of 30-40 ml / ton (ha) are reasonable for spiked cereals (wheat, barley).

FUNGAL INHIBITION

In order to inhibit *Fusarium oxysporum cubense* growth, several fungal growth inhibitors will be produced when fusaric acid is sensed. Those being 2,4-DAPG, chitinase, DMDS, DMTS and pyoverdine. 2,4-DAPG or 2,4-Diacetylphloroglucinol is an antibiotic against plant pathogens. Chitinase is a lytic enzyme that breaks down fungal cell walls. Dimethyldisulfide (DMDS) and dimethyltrisulfide (DMTS) are sulfur by-products produced by *Pseudomonas putida* and have been shown to stimulate plant growth and inhibit *F.oxyposrum* respectively. And lastly, pyoverdines are iron-chelating compounds that are produced when iron is limited in order to scavenge for iron to induce iron competition. With these fungal growth inhibitors our engineered *P. putida* will be able to better inhibit *F. oxysporum*.

2,4-Diacetylphloroglucinol(2,4-DAPG)

2,4-DAPG, full name 2,4-diacetylphloroglucinol, is an antibiotic that is widely used in the agricultural industry against pathogens. It is a broad spectrum antibiotic that has been shown to play a key role in the biological control of various plant pathogens including *F. oxysporum*. In addition to that, 2,4-DAPG has also shown to induce systematic resistance in plants. Since *P. putida* does not produce DAPG by itself, a gene cluster obtained from *Pseudomonas fluorescens* was introduced into *P. putida*. The *phl* gene cluster contains eight genes, from *phlA* to *phlH*. Gene cluster *phlABCDE* was used for this project as literature has been shown that *phlABCD* are the DAPG synthesis genes and *phlE* codes for an efflux pump. *phlABCD* has been expressed before in *P. putida* and has been shown to produce 2,4-DAPG.

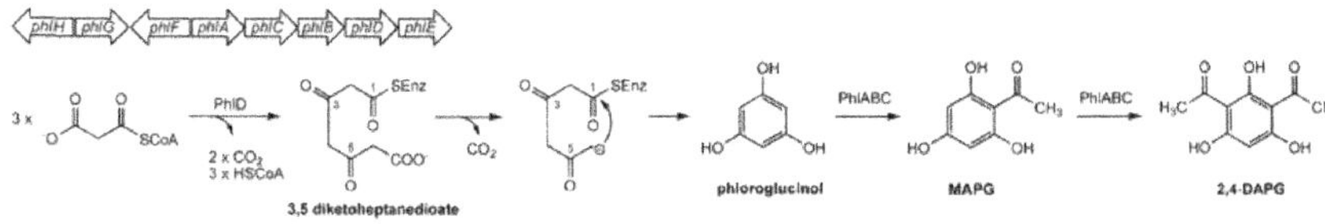

Fig. : 2,4-DAPG synthesis pathway

Dimethyldisulfide (DMDS) and Dimethyltrisulfide (DMTS)

When breaking down methionine for ammonium, methanethiol gets formed as a by-product. Then methanethiol gets oxidized into dimethyldisulphide (DMDS) and dimethyltrisulphide (DMTS). DMTS was shown to have an inhibitory effect on *F. oxysporum* with an inhibition at the start of 78, but then slowly declined to 26%. DMDS is used as plant growth promoter and at the same time also has been shown a slight inhibition to *F. oxysporum* of 10%. Both DMTS and DMDS are naturally produced in *P. putida* but a higher production is desired. One thing that causes low yield of DMDS is the low affinity of methionine-γ-lyase to methionine, which means less conversion of methanethiol, which then leads to low DMTS and DMDS production. So, we decided to overexpress an enzyme that has a higher affinity towards methionine and will therefore increase DMDS and DMTS production, this enyzme is a methionine-γ-lyase from*Brevibacterium linens*, which has been shown to increase DMDS production in *Lactocossus lactis*. This gene was codon-optimized and synthetsized by IDT for *P. putida*.

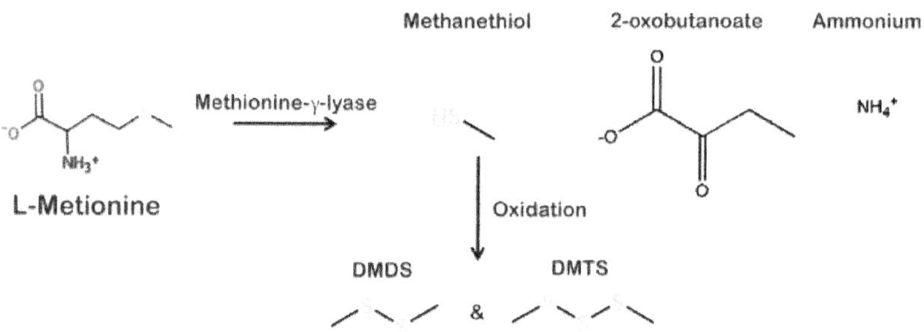

Fig. : Dimethyldisulfate (DMDS) and dimethyltrisulfate pathway.

Pyoverdine

Pyoverdines are siderophores produced by *P. putida*. Siderophores are green/yellow fluorescent compounds that have high affinity to iron(III), which by scavenging free iron can in the end lead to iron limitation. Due to iron starvation, the growth of pathogenic fungi and bacteria in the rhizosphere will be restricted. It was shown that there was a direct correlation of siderophore production and their inhibition to germination of chlamydospores of *F. oxysporum*. In addition to that, siderophores have also been shown to induce resistance in radish plants.

However the effects of siderophore decreases when the disease incidence increases above 74%. *P. putida* WCS358 is able to produce another sort of siderophore, pseudobactin 358 (PBS358), which has been shown to be involved in inhibition of *Fusarium*. Pyoverdine production is iron dependent as it is regulated by a ferric uptake regulator protein (Fur). The Fur protein inhibits expression of PfrI and PfrI is a transcription activator that is needed for activation of genes involved in pyoverdine synthesis. So an overexpression was done of PfrI, which we expected to result in pyoverdine production, even when in an iron abundant environment.

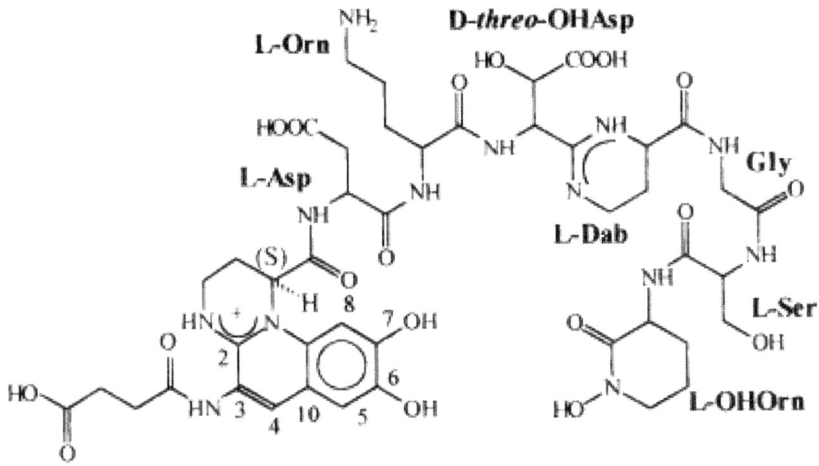

Fig. : Pyoverdine structure.

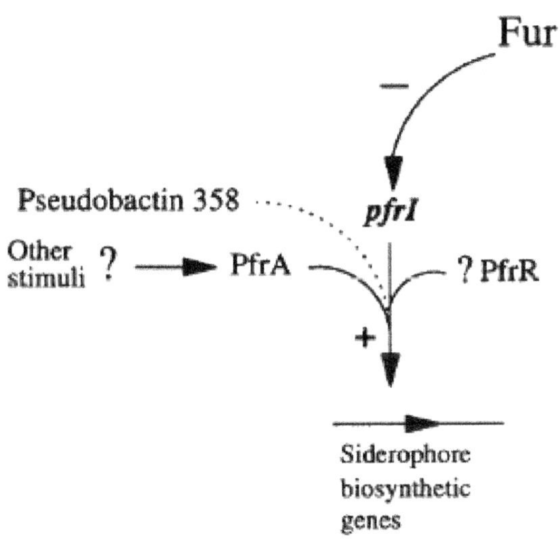

Fig. : Fur regulation of siderophore genes.

Chitinase

Chitinase is a hydrolytic enzyme that breaks down hydrolytic bonds in chitin and is produced in both bacteria and plants. Chitinase has been shown to be useful in biological control against fungi. In bacteria their function is to attack shellfish animals or fungi, and degrade their chitin cell walls. In plants they are known as pathogen related proteins that are involved in the induced systematic resistance of plants in order to defend themselves against pathogens. *P. putida*KT2440 has a lytic enzyme ***PP3066*** that is predicted to have chitinase activity, so this gene was overexpressed to increase chitinase activity.

2,4-Diacetylphloroglucinol(2,4-DAPG)

PhlABCDE was successfully cloned and put into plasmid pSEVA254, which has a kanamycin resistance, a RFS1010 ori and a *lacIq* with a P*trc* promoter. A SEVA plasmid was used instead of the iGEM plasmid because it has been shown to work in *P. putida,* it contains a LacI that controls an IPTG inducible Ptrc promoter and it has terminators behind the gene insert. Both *E. coli* and *P. putida* KT2440 were transformed with a pSEVA containing the *phlABCDE* gene cluster. This strain was further referred to as 2,4-DAPG *P. putida*. Succesfull transformants were verified via colony PCR. During colony PCR of *P. putida,* we were not able to get an expected 5.4 kbp band. However, when we used a primer pair that forms a 1 kbp product, we could see positive transformants. For detection of 2,4-DAPG we successfully set up a High-performance liquid chromatography (HPLC) method. Standards were detected and a standard curve was made. However, verifying production of 2,4-DAPG via HPLC turned out to be difficult due to a lot of background noise of the samples, even after an extraction step.

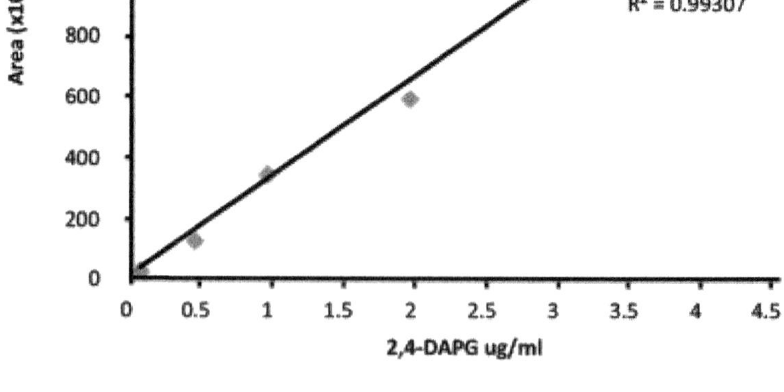

Fig. : 2,4-DAPG HPLC standard curve.

In order to test the inhibition of pure 2,4-DAPG against *F. oxysporum*, an assay was done on Komada agar plates. Different concentrations of pure 2,4-DAPG were plated and then inoculated with a 5mm *F. oxysporum* plug. Plates were then incubated at 25°C for several days.

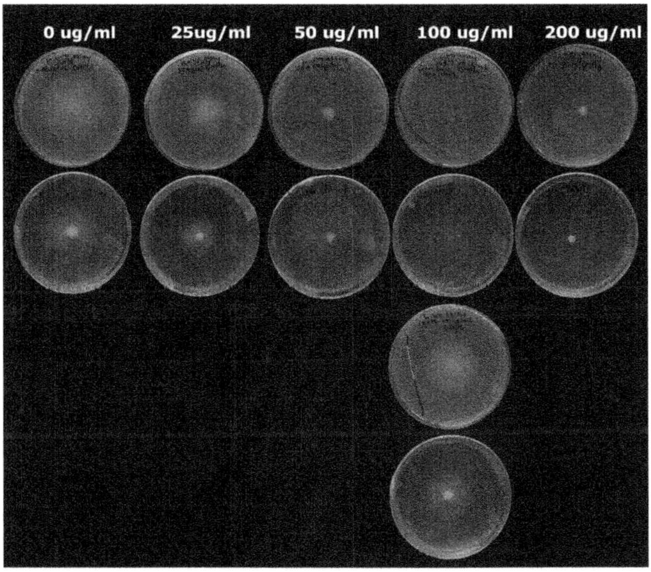

Fig. : Inhibition assay with pure 2,4-DAPG concentrations (0, 25, 50, 100 and 200ug/ml) inoculated with 5 mm *F. oxysporum*TR4 plug. t = 6 days, upper row = face down plates, lower row = face up plates.

From figure it can be seen that with an increase of 2,4-DAPG there is a decrease in *F. oxysporum* growth. However when it reached to 100 ug/ml results were not reproducible with one plate having more growth than the other. Afterwards, another inhibition experiment was set up to test 2,4-DAPG concentration from 100-400 ug/ml. However here spores (3.87E+07 spores/ml) were used as an inoculum, instead of a *F. oxysporum* plug, due to unavailability of a fresh *F. oxysporum* plug on agar plate. Plates were then incubated in 25°C for several days.

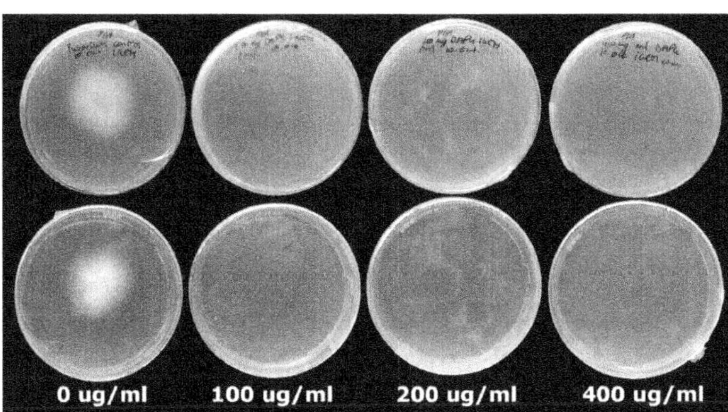

Fig. : Inhibition assay with pure 2,4-DAPG with *F. oxysporum*TR4 spores (3.87E+07 spores/ml) with concentration range (100,200 and 400 ug/ml), t = 5 days, upper row = face down plates, lower row = face up plates.

When comparing figures it can be seen that when inoculated with spores even in 5 days there is absolutely no growth of *F. oxysporum* at a concentration of 100 ug/ml 2,4-DAPG, whereas when for the *F. oxysporum* plug there is growth after 6 days with mixed results at the same concentration of 100ug/ml 2,4-DAPG. After testing inhibition with pure 2,4-DAPG, another inhibition experiments was done using our 2,4-DAPG *P. putida*, containing the *phl* gene cluster for 2,4-DAPG production. An overnight 2,4-DAPG *P. putida* liquid culture was spread on a LB agar plate containing 1mM IPTG for induction, grown overnight at 30°C and then inoculated with a 5 mm *F. oxysporum* plug.

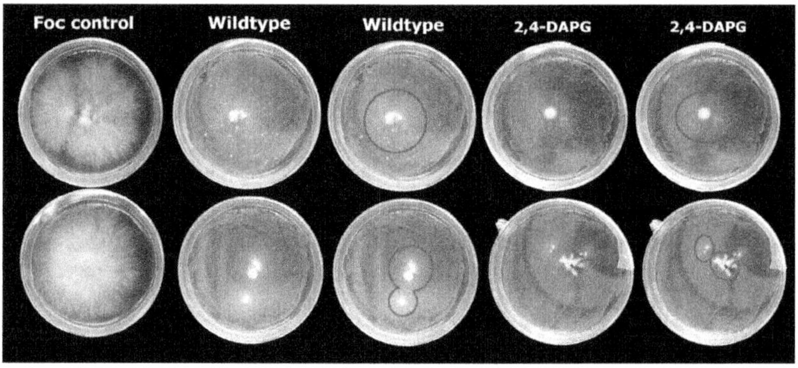

Fig. : Inhibition profiles of *F. oxysporum*TR4 by 2,4-DAPG *P. putida*. A 5mm *F. oxysporum* TR4 plug was used, tests done in duplo (upper & lower row), t = 7 days. Red circle indicates area occupated by *F. oxysporum* growth. Foc control = only *F. oxysporum* TR4, Wildtype = *P. putida* KT2440 and DAPG=2,4-DAPG *P. putida* containing *phlABCDE* gene cluster.

In figure, plate pictures were modified with a red circle to better visualize the fungal disc. It can be seen that the wild type *P. putida* KT2440 already inhibits *F. oxysporum* very well. Our 2,4-DAPG *P. putida* seems to show a slightly smaller *F. oxysporum* disc when comparing it with the wild type. It is very hard to see a big difference between the wild type and our 2,4-DAPG *P. putida*, as the wild type has been shown to already inhibit *F. oxysporum* greatly when comparing wild type and *foc* control. Please note that *phlABCDE* was not yet made into the BioBrick standard, due to too many illegal sites and insufficient time to remove them.

Dimethyldisulfide (DMDS) and Dimethyltrisulfide (DMTS)

Methionine-γ-lyase was successfully made into a BioBrick standard format. It was then put into plasmid pSEVA254 plasmid, which was transformed in both *E. coli* and *P. putida* KT2440. Succesfull transformants were verified via colony PCR for *E. coli* and *P. putida*. Transformants are further referred to as DMDS/DMTS *P. putida*. DMDS/DMTS *P. putida* was grown, induced and harvested after 3 hours. Cell free extracts were obtained via sonification and were then used for an assay. DMDS and DMTS compounds were supposed to be measured via Gas Chromatography (GC). But it was in the end not possible due to unexpected problems with the GC. However an inhibition assay was done co-inoculating our DMDS/DMTS *P. putida* with *F. oxysporum*.

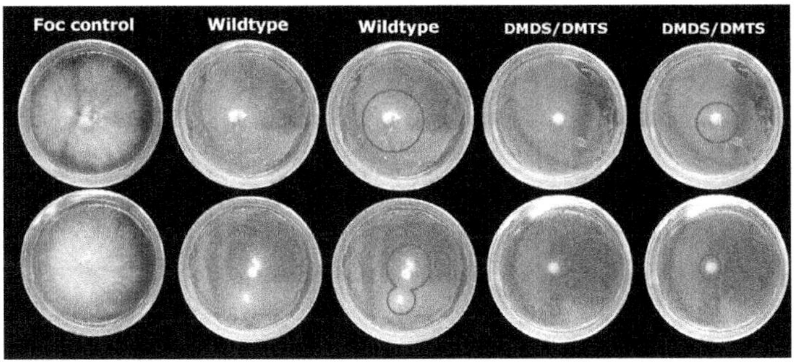

Fig. : Inhibition profiles of *F. oxysporum* TR4 by DMDS/DMTS *P. putida*. A 5mm *F. oxysporum* TR4 plug was used, tests done in duplo (upper & lower row), t = 7 days. Red circle indicates area occupated by *F. oxysporum* growth. Foc control = only *F. oxysporum*TR4, Wildtype = *P. putida* KT2440 and DMDS/DMTS = DMDS/DMTS *P. putida* overexpressing Methionine-gamma-lyase.

In figure it can be seen when comparing our DMDS/DMTS *P. putida* the wild type *P. putida* there is a much smaller fungal disc, probably cause by a higher inhibition of our DMDS/DMTS *P. putida*.

Pyoverdine

The *pfri* gene was successfully made into a BioBrick standard format , validated and characterized. *pfri*was cloned and put into plasmid pSEVA254. Then transformed in both *E. coli* and *P. putida* KT2440. Successful transformants were verified via colony PCR. And are further referred to as pyoverdine *P. putida*. Afterwards growth experiments were done in minimal M9 medium supplemented with iron (31μM) and the pyoverdine compound was measured using spectrophotomery. When grown overnight it was possible to see that pyoverdine *P. putida* were slightly greener when compared to the control. Control used here was a *P. putida* containing an empty pSEVA254 plasmid.

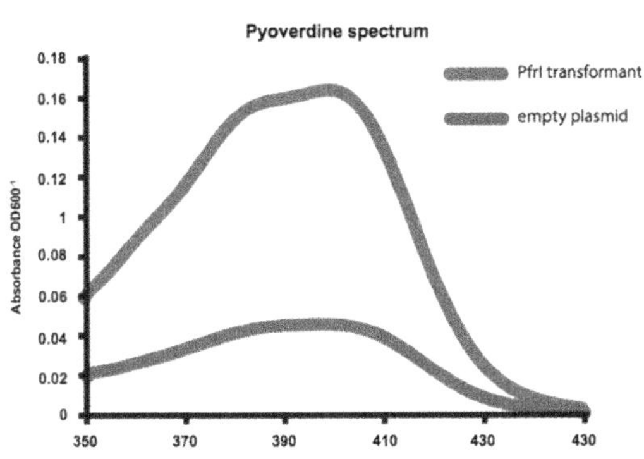

Fig. : Pyoverdine spectrum(350-400nm)OD corrected average from biological duplicate.

In figure it can be seen that the peak (400nm) of pyoverdine *P. putida* is higher than the peak from a *P. putida*containing an empty plasmid.

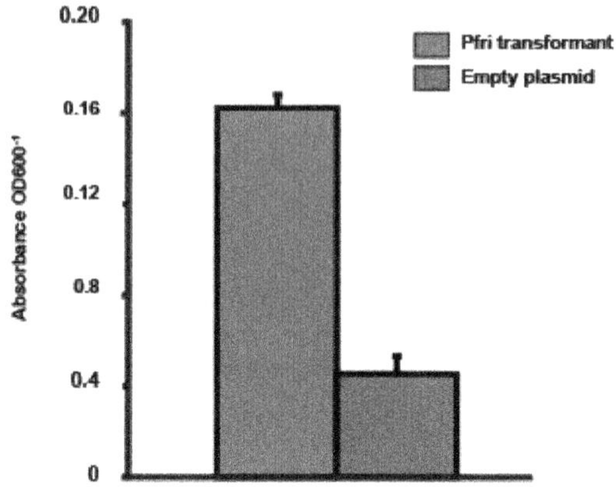

Fig. : Pyoverdine production in M9 medium supplemented with 31 μM iron with error bars,OD corrected. empty plasmid = *P. putida*containing an empty plasmid for control.

When looking only at the aborbance at 400 nm , it can be seen that there is a 4-fold increase of pyoverdine production when *pfri* is overexpressed in P. putida. In addition to this experiment, inhibition assay was also done with pyoverdine *P. putida* co-inoculated with a 5mm *F. oxysporum* plug.

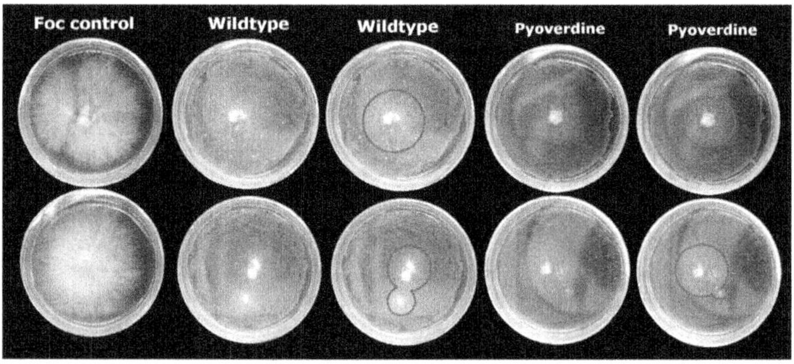

Fig. : Inhibition profiles of *F. oxysporum* TR4 by pyoverdine *P. putida*. A 5mm *F. oxysporum* TR4 plug was used, tests done in duplo (upper & lower row), t = 7 days. Red circle indicates area occupated by *F. oxysporum* growth. Foc control = only *F. oxysporum*TR4, Wild type = *P. putida* KT2440 and pyoverdine = pyoverdine *P. putida* with an overexpression of *pfri*.

In figure it can be seen that pyoverdine *P. putida* has a smaller *F. oxysporum* disk when compared to the wild type. Meaning it's inhibition is increased but only ever so slightly.

Chitinase

PP3066 was succesfully cloned into broad host range plasmid (pSEVA254) and transformed into *P. putida*. Successful tranformants were verified via colony

PCR. And are further referred to at chitinase *P. putida*. Chitinase *P. putida* was grown, induced and harvested 3 hours after induction. Cell free extracted were obtained via sonification of cells, sadly there was not enough time to do a chitinase assay. Chitinase *P. putida* was co-inoculated with Fusarium in an inhibition assay, results in figure.

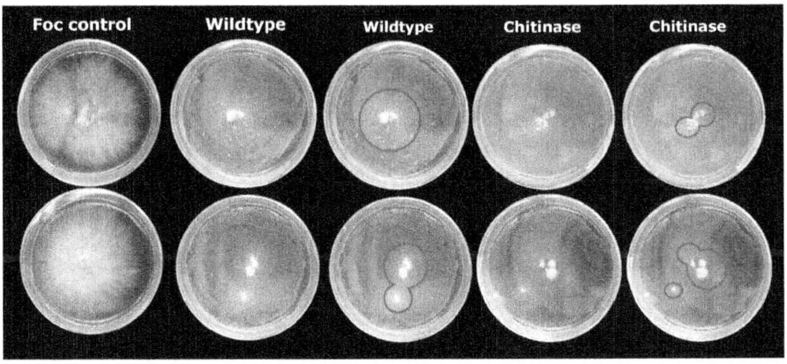

Fig.: Inhibition profiles of *F. oxysporum* TR4 by chitinase *P. putida*. A 5mm *F. oxysporum* TR4 plug was used, tests done in duplo (upper & lower row), t = 7 days. Red circle indicates area occupated by *F. oxysporum* growth. Foc control = *Fusarium oxysporum cubense* TR4, Wild type = *P. putida* KT2440 and chitinase = chitinase *P. putida* with an overexpression of *PP3066* predicted to have chitinase activity.

In figure, it can be seen that chitinase *P. putida* seem to have a smaller fungal disc as when compared to the wild type.

All Fungal Growth Inhibitors Together

In addition to testing just one tranformant with *F. oxysporum* TR4, two culture mixes were made.

- Chitinase, DMDS/DMTS and pyoverdine
- 2,4-DAPG, Chitinase, DMDS/DMTS and pyoverdine

These two culture mix were then spread on agar plates using overnight liquid culture, grown overnight in 30°C and co-inoculated with *F. oxysporum* the following day.

BIOLOGICAL FERTILISERS

Biofertilisers can be derived from microorganisms (and in some cases are actually live microbial inoculants), algae (seaweed) or higher plants. This workshop will cover much of these forms and discuss the potential modes of action and the best way to utilize them.

Humic and fulvic acid products are also increasingly available. These are also biologically derived albeit from eons ago when coal was formed. They have a variety of uses that range from significant cation exchange effects in the soil through to stimulatory and chelation effects to improve plant growth and nutrient uptake. These are not covered in the workshop but there does seem to be beneficial effects from their use. Humic acids may be useful if applied in bulk to improve nutrient holding capacity of soils or spayed on as a plant stimulant, they can also be mixed with liquid preparations of nitrogen fertiliser to perhaps provide a slower release of nitrogen. Fulvic acids are smaller in size and soluble, they have a significant effect on improving nutrient uptake and decomposition of pesticide residues and stimulating plant growth.

Some definitions of biological fertilizers include chemical fertilizers whose main action is to stimulate soil biological activity. These are not covered in this workshop but they include: Fine limeflour (*e.g.* 1 to 80 micron, applied @ 10 to 200 kg/ha) in suspension or by careful spreading (avoiding too much loss of the small particles) to stimulate beneficial soil microbial activity. And liquid fertilisers, which can be used strategically to apply some trace elements that are not taken up efficiently from some soils, and apply major elements at key limiting times for fast uptake and effect. In many cases the liquid fertiliser's effect can be largely due to a biological stimulant present in the product which along with some of the chemical nutrient components are able to improve plant growth and quality.

Mode of Action of Biofertilisers

Biofertilisers act to stimulate plant growth and the plant immune systems. They help overcome periods of stress for the plant and help the plant achieve its potential. They enable a more efficient utilisation of nutrients. It is important to stress that the amount of mineral nutrients in biofertilisers (especially liquid ones) may be to low to be the sole major element source for a long term sustainable nutrient budget.

Biofertilisers usually contain a range of growth promotion factors with some of the main ones being forms of cytokinin (a microbial and plant hormone responsible for promoting cell division and growth) including betaines. Plants produce their own cytokinin though this production may be limited at times of stress or simply below optimum, and plants can respond positively to the application of cytokinin alone. The theory with biofertilisers is to provide a range of growth promotion factors and compounds that also increase the uptake of nutrients *e.g.* through stimulation of fine feeder roots and general root production and/or an increase in cell permeability.

The claims of benefits usually include some of:

- increased plant growth
- higher Brix levels (more "energy" in the plant)
- better plant nutritional quality
- increased ability for nutrient uptake
- increased fine feeder root formation

- particular benefit for legumes (*e.g.* in pasture)
- increased nodulation in legumes
- increased chlorophyll content (for photosynthesis)
- lower free nitrogen content in plant tissue (higher quality)
- reduced pest and disease pressure.

Most of these aspects lend themselves to testing and in the case of commercial products, manufacturers may or may not be able to provide research data of their own or from independent testing. It is possible to set up simple trials to go some way towards establishing whether the product is worthwhile in a particular system.

Many growers will make their own products which saves money at the expense of taking time and energy. Such "home-made" products might utilize resources that are on-farm, nearby or relatively free.

Some of the techniques used by commercial companies might make a higher value product that home-made products (this could be tested) and the extra financial outlay may be worthwhile – a case by case pragmatic decision. This does not, however, sit well with the idealistic aim of reducing outside commercial inputs.

All of the above benefits have some basis in scientific literature but results from biofertilisers are characteristically variable perhaps due to product quality and probably also due to environmental fluctuations and the vagaries of not understanding the actual modes of action.

Biofertiliser Use

- *Time of Day:* Best applied on overcast or drizzly days or in the early morning or late afternoon to avoid the excessive death of any live microbes and to improve the amount of foliar uptake of the biofertiliser.
- *Timing with Crop:* Best applied at key growth stages such as germination, emergence, stem elongation flowering and fruit formation.
- *Time of Year:* Least likelihood of effectiveness at times when growth is limited by climatic conditions *e.g.* too cold or too dry. On the other hand some biofertilisers may be useful in reducing frost damage and/or helping to overcome the stresses (and low plant hormone production) of dry conditions.
- *Method of Application:* Spraying over plants is generally better than applying just to the soil as there can be more direct benefit to the crop plants. A finer spray is generally better.
- *Mixing:* Most liquid products should be diluted at least 1 to ten to help avoid damage to crop plant tissue. Live microbial products should not be mixed with other materials (*e.g.* copper, lime sulphur, high concen-

trations of nutrient etc) that will inactivate or kill the microbes (if an inoculant effect is desired).

SOURCES OF BIOFERTILISERS

Types of Microorganisms

Some products contain single species (often selected strains of microorganisms and others can contain a wide range to gain potentially wider and synergistic effects. Some products may simply contain extracts of microorganisms that include growth promotion factors.

Bacteria

Many different types of bacteria are used as either live inoculum or providers of biofertiliser extracts. Fluorescent pseudomonads have been extensively researched for their growth promotion effects and some other types are detailed below.

Cyanobacteria

These are primitive bacteria capable of photosynthesizing and producing oxygen. Their benefits in the soil system are many fold and include, provision of oxygen to aid plant root health and nutrient uptakes as well as aiding general soil biological activity through the provision of oxygen, chelating agents (for improving nutrient uptake), growth stimulating factors including hormones and simple proteins), production of sticky polysaccharides for improved soil structure and water holding capacity. Some forms are also capable of nitrogen fixation (free living in the soil or associated with plant roots and nodules. A few forms are capable of solubilising phosphorus from its mineral form making it available to microbes and plants.

Lactic Acid Bacteria

Probably best known for their pivotal fermentative role in yoghurt making, they are also an important part of the manufacture of quality silage. In the soil system they can be beneficial by aiding the fermentative decomposition of organic matter and producing growth promoting substances that benefit many microorganisms and plants. They also produce bacteriocins and other natural toxins that restrict the growth and activity of other soil microorganisms particularly harmful ones.

Actinomycetes

Actinomycetes are a range of hardy bacteria of which many are filamentous (grow in threads). Much of the microbial smell of "good" soil is due to actinomycetes.

Some species of actinomycetes have been used to provide extracts useful as biofertilisers with high levels of growth promoters. These are usually species that do not produce antibiotics.

Purified antibiotics not counted as part of the biological's spectrum because the action is mainly one of antibiosis rather than stimulation.

This can lead to antibiotic resistance and causes problems for some forms of beneficial bacteria. The actinomycetes can produce ionophore antibiotics which interestingly increase the uptake of nutrients including cations (to the detriment of susceptible organisms). In a biofertiliser, the presence of or stimulation of organisms that antagonize harmful microorganisms may be one of the modes of action. This differs from the use of a specific purified antibiotic in that there is arrange of modes of action, many of which are stimulatory rather than antagonistic and the lack of purification reduces the likelihood of resistance developing.

Fungi

Two of the main types of fungi used in biofertiliser production are mycorrhizal and yeast. Other types of fungi *e.g. Aspergillus oryzae* and *A. niger* may also be used, often for their extracts which contain growth promoters. Nematode predating fungi may are available in some countries for potentially reducing nematode problems.

Mycorrhizal Fungi

Almost all crop plants are capable of a special beneficial relationship with fungi partly inhabiting ("infecting") the plant roots and partly growing out into the soil. The mycorrhizal fungi increase nutrient uptake for the plant (especially phosphorus which is not picked up efficiently by many crop plant roots without mycorrhizal "infection") reduce the susceptibility of the plant to root disease.

There are some products available that provide a range of mycorrhizae suitable for most crop plants. In most cases the right organisms will already be present in the soil and may just need extra stimulation and the provision of the right soil conditions (including not too much soluble phosphorus fertiliser use) to achieve good levels of root "infection".

Yeasts

Most yeasts are very fast growing and many are useful for providing good levels of growth promoters for use in biofertilisers. Low alcohol strains of beer yeast (*Saccharomyces cerevisiae*) are sometimes used in the manufacture of biofertilisers and are often included in animal probiotics.

Trichoderma spp.

Selected strains of various *Trichoderma* spp. are very efficient at dominating soil and plant habitats and controlling plant disease. They have also been shown to stimulate plant growth including root production. Several are commercially available in New Zealand.

Algae (Seaweed)

Many people make their own seaweed brews from beach collected seaweed. In commercial preparations, there is often an emphasis put on live harvesting (*e.g.* using deep sea divers) to gather highly active material (higher levels of growth promoters).

Plants

One of the principles of selecting plants as biofertilisers is to try fast growing plants. Although seaweed is generally much faster growing, successful commercialized biofertilisers have been made from…

Giant Knotweed (Reynoutria sacchilinensis)

A relative of dock, commercial preparations are available in Europe and North America. Used to induce resistance to fungal diseases (including powdery mildew and botrytis). Found to increase plant growth and yield, provide more intense flower colour in flower crops and increase chlorophyll content (photosynthesis capacity) of crop leaves. Giant knotweed is from the far north of Japan naturalized as a large weed in areas of the South Island West Coast. A water extract can be made with either fresh leaves or dried crushed leaves.

Yucca (Yucca schidengo)

This Central American plant is commonly used to reduce odours from in dung and other waste material. It has also been suggested to aid soil biological activity through provision of a surfactant action protection of soil microorganisms from water stress. Sometimes used as an additive to other materials in a biofertiliser product. Grows naturally in Central America but with the growth of use of this plant in health and agricultural products, the plant has suffered over harvesting and is now largely grown in plantations. In New Zealand Yucca material can be purchased from importers.

River Hemp (Sesbania cannabina)

River hemp is an Australian member of the pea family and is used for biomass production and also due to high levels of growth promoter compounds is used as a biofertiliser.

Comfrey (Symphytum spp.)

Comfrey is commonly used to make comfrey liquid manure to provide small balanced levels of nutrients and levels of microorganisms. Leaves are left to ferment in water for a matter of weeks. Usually the range of microorganisms depends on those present on the leaves to start with and the conditions (rather anaerobic) in the container.

Willow (Salix spp.)

This fast growing tree genus is used to make willow water (leaves soaked in water) which can be used as a stimulant for plant cuttings.

Dock (Rumex spp.)

A water extract of dock leaves has been shown to be effective against powdery mildew and due to its relatedness with giant knotweed, may also be useful as a stimulant of plant growth.

Stinging Nettle

Water extract of stinging nettle is used to simulate strong plant growth sometimes with a belief that the silica contained in nettle is useful in strengthening the plant cells against disease attack.

Potential Contradictions in Microbial Inoculant Theory

Anaerobic is Bad?: Some of the writings on biofertilisers and microbial inoculants seem contradictory. Many describe anaerobic as bad and aerobic as good. This is a simplification which though largely true, neglects the benefits of many anaerobic organisms and also might overlook the fact that there will always be copious anaerobic pockets in the soil which are best being filled with beneficial anaerobic organisms rather than harmful or wasteful ones.

Microbial Inoculants Will Suffer in Soil Conditions: But there is a strong element of truth that conditions should be improved to aid beneficial organisms. Indeed if some of the action is actually a stimulatory one then it is achieving just that. Many of the natural processes of beneficial organisms will aid the provision of better conditions for beneficial organisms. On a large scale it can, for instance, be seen that if earthworm activity is increased by stimulatory means then the earthworm activity will improve soil conditions (including soil structure, soil aeration, soil pH, nutrient availability, microbial activity) yielding a longer lasting effect of the stimulation than may have been initially anticipated.

FUNGI USES IN PEST MANAGEMENT

Fungi are useful biological controls for pests in crops and urban environments and have been commercially available since 1981. They can be used asherbicides, insecticides and fungicides. Mycopesticides are interesting organic alternatives to chemical pesticides due to their enhanced safety and their high specificity for the non-targeted species such as humans, pets and desired insects.

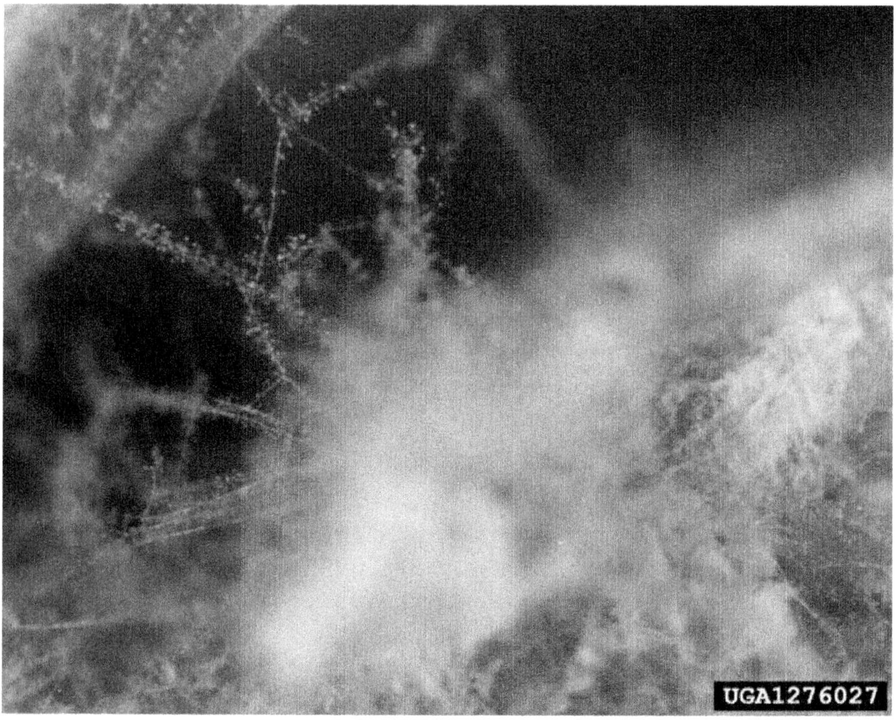

Fig. : *Lecanicillium lecanii* on hemlock woolly adelgid cadaver.

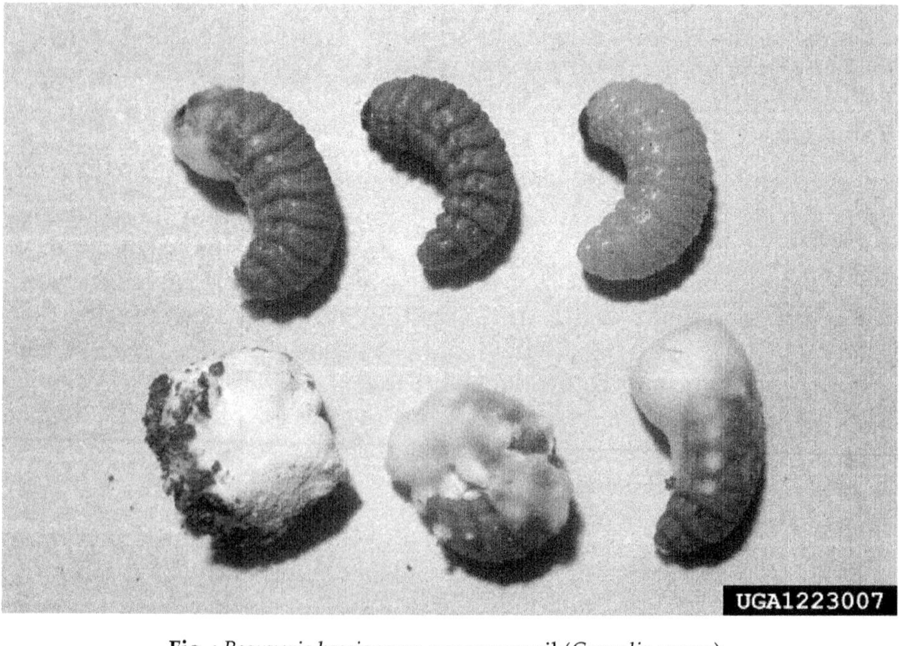

Fig. : *Beauveria bassiana* on pecan weevil (*Curculio caryae*).

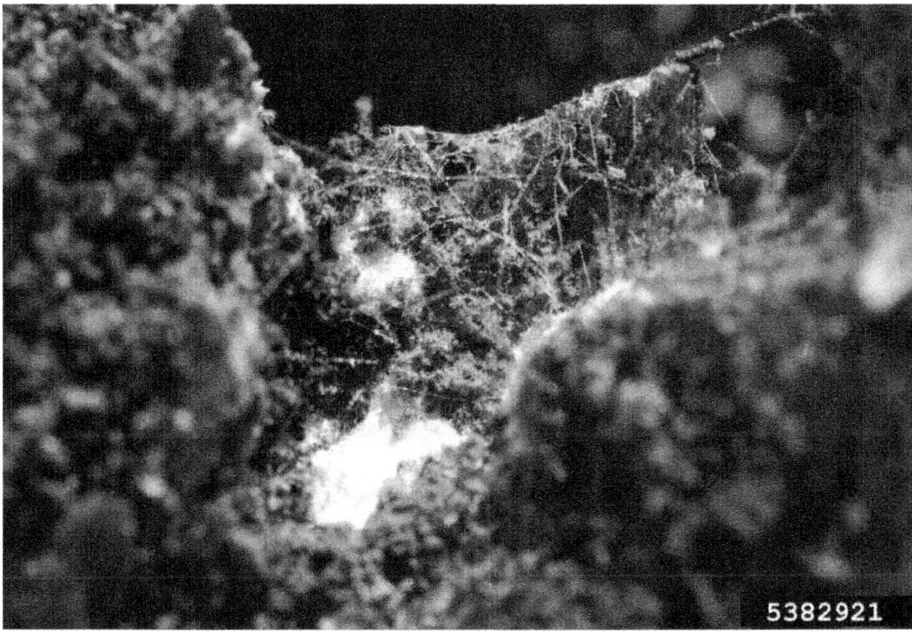

Fig. : *Trichoderma spp.* on Norway spruce (*Picea abies*).

They share advantages and disadvantages with other biopesticides. *Lecanicillium lecanii*, *Beauveria bassiana* and *Trichoderma* spp. are amongst the most commonly used species and have been widely studied in spite of the short history of this field of study. With the growing enthusiasm for organic commercial farming and justified environmental concerns in the population, mycopesticide have a definitely a bright future.

History

For over 100 years, scientists have been studying many different species of fungi with entomopathogenic characteristics but the use of these organisms in the field has been relatively elusive. Many of these fungal organisms are not only entomopathogens, but also aid in general plant growth. Since the late 1940's, chemical insecticides have been the main tool that growers have utilized, in order to control potentially damaging pests and diseases. Some of the earliest recorded studies of entomopathogenic fungi occurred in the early 1800 when the silkworm industry in France was devastated by *Beauveria bassiana*. As a consequence of this insight, further research began looking into the use of fungal insect pathogens to manage a wide range of pests.

Most recently, DNA sequence analysis has lead to a much better understanding between the differences in entomopathogenic fungi and other types of fungi. This new research has revealed that some of these entomopathogenic fungi also play the role of beneficial rhizosphere associates, plant growth promoters, as well as plant endophytes. These findings have now lead to further research to determine if some of these entomopathogenic fungi can be used for more than just one purpose.

The use of Entomopathogenic fungi in pest management covers a wide variety of Eumycota subdivisions. These subdivisions of Eumycota include;Mastigomycotina, Zygomycotina, Ascomycotina,and Deuteromycotina. In 1981, the first mycoinsectiside registered in the U.S. was *Hirsutella thompsonii* and was given the registered name; Mycar. This species has been noted to cause epizootics as early as 1920 in some species of spider mites.

With the rising cost of synthetic chemical pesticides and increasing cases of pesticide resistance, the search for natural biologically based forms of pest management has been a key area of research for the last decade. With the world population rising over 7 billion and the drastic effects of climate change, the need for alternative forms of pest control will become essential. Although insects are key in the performance of many ecosystems, they play a large role in the $35 billion worth of crop that is lost in the United States each year.

There is enormous potential for pest and vector control in pest management in the future due to an increasing amount of knowledge in the area of mycology.

Types of Mycopesticides

Mycoherbicide

Fungi can be used as an herbicide to exterminate specific weeds and other plants that are detrimental to the effective growth of crops. In order to use a fungus for this purpose, a mycoherbicide is created with the fungi as its active ingredient. The use of mycoherbicides focus on the theory of the disease triangle. The disease triangle breaks down the conditions that need to be apparent for infection to occur, those being the environment, pathogen and host. The following will fall under one or more of these categories. Mycoherbicides act by releasing phytopathogens to suppress weed growth. The phytopathogens release phytotoxins that can kill weeds in up to five weeks time. A common phytotoxin found in the fungi species *Penicillium* is vulculic acid. The fungi species most commonly used as herbicides in North America are *Collectrichum gloesporioides* (Collego®) and *Phytophthora palmivoraa* (De Vine®). A specific weed strain commonly known as arrowhead (*Sagitaria trifolia*) is the cause of the largest rice plantation problem in Iran. Arrowhead is notably resistant to chemical herbicides. Collego® is used in the USA as the solution to arrowhead. Other types of fungi that can be used as mycoherbicides are rusts and mildews. These fungi use spore distribution to cause infection in their host plants.

Mycoinsecticide

Certain species of fungi can act as parasites of insect. When a fungus is used as insecticides, it is called mycoinsecticide. In recent years, crop protection has been trending towards integrated pest management (IPM) using bacteria and fungi as insecticides. Approximately 750 species of fungi are pathogenic to insects and only 12 are being utilized as insecticides. Two prominent species of fungi used as insecticides are *Beauveria bassiana*, and *Metarhizium anisopliae*. Mycoinsecticides function by first being applied to the insects in spray form. The fungi then use their hyphae to burrow into the insects. The hyphae spread the insectotoxins throughout the insect to activate them, eventually leading to the insect's death.

Mycofungicide

There are many cases of mycoparasitism in nature. In fact, a large amount of dormant spores like sclerotia or chlamydospore are found in the wild already parasitized by another fungi. This is made possible by the ability of most fungi to release lytic enzymes like chitinase that allows them to break down chitinuous cell walls. The parasitism of a fungi on another one occurs generally through four main stages: chemotaxi, recognition, attachment and penetration. The type of relationship that will follow such an invasion will be either necrotrophic or biotrophic. The most commonly used species are in the *Trichoderma* genus.

Mycopesticides in Urban Environments

Entomopathogens are key components for integrated pest management solutions. There is a large market for urban pest control that has been dominated by the use of pesticides and insecticides. These chemical pesticides pose a danger not only to the person applying them, but other people, or pets, which may come into contact with the chemicals in an urban environment. Previously used in agricultural settings as deterrents, fungi such as *Beauveria bassiana*, *Metarhizium anisopliae* and *Paecilomyces fumosoroseus* may have possible pest control applications in urban settings as well.

Advantages and Disadvantages

The negative impacts of chemical pesticides is what has guided new attention towards increased emphasis on biological control agents with respect to using fungi through the use of naturally occurring pathogens of arthropods. Many of these naturally occurring pathogens produced by fungi are already being employed, mostly on a small scale controlling arthropod pests in greenhouse crops, orchards, ornamentals, turf and lawn grasses, stored products and forests, and also for moderation of vectors of animal and human diseases.

Although the utilization of the full potential of fungal biocontrol agents against arthropod pests has had limited commercial success, the major advantages of these biocontrol agents are:

- Effectiveness and potentially high specificity;
- Acceptable naturalness;

- Safety for humans and other non target organisms;
- Reduced chemical pesticide use and consequential reduction of residues in food and the environment;
- Protection of natural enemies of the pest;
- Protection of biodiversity in managed ecosystems.

Despite these promising advantages of using fungi as biological control agents in pest management, there are some disadvantages as well such as:

- Can be very costly to produce for commercial use and in quantity;
- They can have a short shelf life;
- The pest must be present before the pathogen can be usefully applied thus making preventative treatment difficult.

Furthermore, the disadvantages of the use of entomogenous fungi as biocontrol agents against arthropod pests has been lacking by the need for specific environmental conditions (humidity over 80% and above) during the prolonged period in which the fungi are required to have spores germinate and then penetrate the surface of the arthropods cuticle. This is different from bacterial pathogens in which they go through the host's digestive tracts. Researchers have been trying to overcome this problem through the development of oil based and other formulations of fungal spores for use in biological control. Thus we can see that a intricate set of interacting processes, both environmental and biotic, are necessary for or inhibitory to development of epizootics caused by entomopathogenic fungi. These include microbial antagonists; host behavior, physiological condition, pathogen vigor and age; presence of pesticides; and appropriate temperature, humidity, and inoculum thresholds.

Novel Uses of Mycopesticides

Throughout history, there have been many instances of harmful pesticides being released into the environment without fully understanding the implications of such actions. One example of this was the use of DDT, a chemical that has an extremely long half-life and a high degree of bioaccumulation directly linked to a high lipophilicity, which causes eggshell thinning in several birds, including birds of prey. With global environmental awareness increasing, there has been a greater demand for less harmful pesticides. Alternatives to pesticides, biopesticides, generally have a lower toxicity, increased safety, as well as a high efficacy in pest control.

Consequently, the demand and commercial application of microbial biopesticide has increased proportionally to 0.98% of the total worldwide pesticide market, growing at a rate of just over 13% per annum, representing an increase of 47% between 2004/2005 and 2007/2008. Fungal biopesticides represent a large proportion of the rapidly growing sector of pest management.

Several commercial applications of mycopestocides have been investigated, as described below.

Culicoides Nubeculosus Biting Midge and the Bluetongue Virus

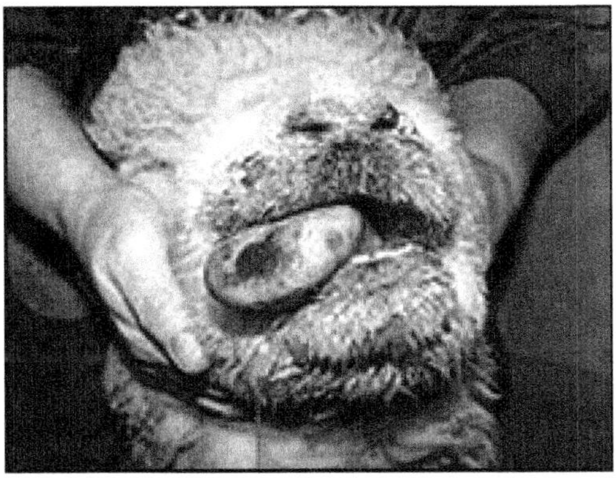

Fig. : Sheep infected with Bluetongue Virus.

The Bluetongue Virus (BTV), transmitted through populations of cattle by *Culicoides nubeculosus* causing Bluetongue disease, is a potentially death-inducing virus. When an animal is infected, it can develop a fever, high salivation, and cyanosis of the tongue, hence the name 'Bluetongue disease'. Losses have been estimated of causing upwards of $3 billion per year.

BTV awareness has increased in recent years due to expansion of its range into newly colonized areas, exacerbated by climate change. As a result of the expanding range of the virus, direct losses from sheep and cattle as well as indirect losses from quarantining have caused great economical losses. Although vaccination can eventually control an outbreak such as this, a vaccine for a recent outbreak only made it into the field 18 months after the initial epidemic. *Metarhizium anisopliae* strain V275 and several other fungi were tested to see their efficacy of control on *Culicoides nubeculosus*, the disseminator of BTV. *M. anisopliae* strain V275 ensured a 100% mortality rate within the colonies of the biting midge, with sporulation occurring on the midge cadavers. The high infectiousness and virulence of the fungi will limit the amount of blood meals taken, therefore reducing the probability of the disease vector being acquired and dispersed. *Metarhizium anisopliae* strain V275 is already commercially available in the USA, sold as F52 by Novozymes. The initial infrastructure requirement is low, so this method can be a cost-effective way of limiting arboviruses. The fungi poses no risk to other wildlife, including humans, birds, fish, and the like, so environmental damage is limited to the pest itself. Although the fungus shows great efficacy as a method of biting midge management, further large-scale trials are needed to better determine the possibility as well as method of using *Metarhizium anisopliae* strain V275 as a mycoinsecticide in a commercial environment.

Adult Housefly Management in Poultry Farms Using Beauveria Bassiana

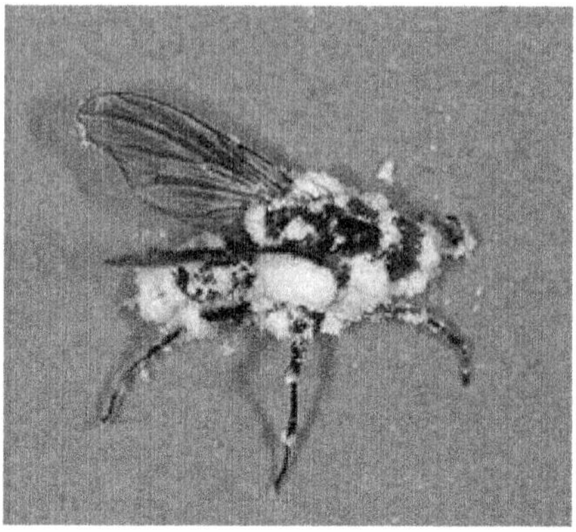

Fig. : Fly colonized by *Beauveria bassiana*

The common housefly is a constant pest to most people, having a short life cycle and being able to reproduce large amounts of offspring. Poultry manure, being slightly moist, is an optimal habitat for the development of large populations of the housefly. When flies occur in a large-scale, high rise, caged-layer poultry facilities, exponential population expansion can occur. The large numbers of flies are potentially released into the environment, causing distress for neighbours and local businesses. Flies also carry several pathogenic organisms affecting humans, poultry and several other species, and can act as a vector for their dispersal. The defecation and regurgitation that flies perform can also cause blotching on the structure and equipment in the poultry farm, on the light fixtures, and also on the eggs, therefore reducing the potential marketability of the final product.

As an attempt to reduce the amount of flies present in a commercial poultry facility, a commercially available product, balEnce, was used to determine the efficacy *Beauveria bassiana* as a mycopesticide. Since the fungus is non-pathogenic to humans, poultry, and many other organisms, *B. bassiana* was sprayed throughout a facility. This allowed efficacy of *B. bassiana* to be compared to another well-known insect neurotoxin, pyrethrin. Adult fly populations were lower in the *B. bassiana* facility *vs.* the pyrethrin facility, as well as reducing the number of fly larvae by over one-half when compared to the pyrethrin.

When the fungi is applied in the manure pits, it colonizes the pit completely and is applied when flies are breeding, killing any flies that come in contact with the manure, as well as any emerging from it. The use of *B. bassiana* also does not negatively impact the levels of beneficial insects in the population, many of which are natural enemies of the fly. With adulticides like pyrethrin, the chemical does not affect the larvae; therefore, repeated applications are required. Therefore, it can be determined that fewer applications of *B. bassiana* will lead to a better pest management system, reducing costs. *B. bassiana* answers the questions on the minds of many large-scale poultry producers: how to kill large amounts of household flies without pesticide residue to damaging my livestock.

SYSTEMIC FUNGAL INFECTION

Systemic fungal infection is becoming more and more common in modern hospitals. In this chapter we will concentrate on candidiasis and aspergillosis, and hardly mention other systemic fungal infections such as Histoplasmosis, Blastomycosis, Coccidioidomycosis and so on. (Cryptococcus neoformans deserves a whole page on its own)! Severe systemic fungal infection in hospitals is commonly seen in three major settings:

- Neutropaenic patients following chemotherapy, and other oncology patients with immune suppression;
- Persons immune compromised due to Acquired Immune Deficiency Syndrome caused by HIV infection;
- Patients in intensive care (ICU), who are not necessarily neutropaenic, but are compromised due to the presence of long-term intravascular lines or other breaches in their integument, severe systemic illness or burns, and prolonged broad-spectrum antibiotic therapy.

Other (quoted) predisposing factors are:

1. APACHE score > 10;
2. renal dysfunction;
3. haemodialysis;
4. surgery for acute pancreatitis, or even possibly splenectomy;
5. recurent GIT perforation;
6. Hickmann catheters.

Systemic fungal infections cause ~ 25% of infection-related deaths in leukaemics. Infections due to Candida species are the fourth most important cause of nosocomial bloodstream infection. In certain other circumstances, fungal infections are also a major problem. Serious fungal infections may cause 5-10% of deaths in those undergoing lung, pancreas or liver transplantation Acquired fungal sepsis occurs in up to 13% of very low birthweight infants. Small retrospective studies of neonates suggest an association between prolonged third-generation cephalosporin use and Candida infection.

There is a strong suggestion that invasive fungal infections have become more common in recent years, with a nearly 500% increase in the incidence of blood-stream infection with Candida spp. since the 1980s. Most systemic fungal infections are in fact due to Candida, but Aspergillus infections are also seen. Other causative organisms are less common, although in people with AIDS, a host of different fungi are important causes of disease and death - for example, Cryptococcus neoformans, as well as Histoplasma capsulatum, and strange organisms such as Penicillium marneffei, trichosporonosis, and fusariosis.

One should also not forget that Pneumocystis carinii, so common in AIDS that prophylaxis is not only warranted but mandatory, is now regarded by many as a fungus, although it is not susceptible to many conventional antifungals because it contains cholesterol and not ergosterol in its cell membrane!

Candida Albicans

C. albicans is an asexual, diploid, dimorphic fungus that is widespread on humans and in their environment. We still don't understand why this common commensal sometimes becomes pathogenic, although impaired host defence mechanisms seem crucial. A variety of fungal virulence factors is being actively explored by researchers, but fungal adherence may be most important. In the future it may even be possible to disrupt this adherence, but at present, we're stuck with antifungal agents.

Much of this page will touch on Candida infection, as this is by far the most important fungal infection in modern hospitals. Candidaemia is said to be the fourth leading cause of bloodstream infection in the USA. Colonization usually precedes candidaemia. Health care workers commonly transmit yeasts between patients by hand, as shown by molecular typing of strains. Candida spp. have been isolated from 15-54% of hands of health care workers in the ICU setting.

There is a large amount of information on the web about C. albicans. Some good sites are listed *below.*

A Note on Terminology

Fungal terminology is confusing to the uninitiated (me)! Here are a few terms to whet your appetite:

1. blastoconidium ('blastospore') - round, oval or elongated fungal cell, which reproduces by budding;
2. hypha - a chain of cells broadly attached to one another;
3. pseudohypha - a chain of elongated yeast cells, with marked constriction of the cells where they attach;
4. conidium - a non-motile, asexual 'spore';
5. chlamydospore - a resting conidium, formed as the enlargement of a hyphal cell; the term is technically incorrect, although commonly used (The 'correct' term is actually a chlamydoconidium)!
6. germ tube - a short hypha;
7. arthrospore - hypha-like chains, but "jointed" with cells not broadly joined at their bases.

How Do We Distinguish C. Albicans from Other Fungi?

This depends on:

- Morphology: When the organism is grown on cornmeal agar, *chlamydospores* are indicative of C. albicans, but take up to 4 days to develop. Pseudohyphae without chlamydospores indicate another Candida sp.; arthrospores suggests Trichosporon. C. glabrata does NOT form pseudohyphae on cornmeal agar.

C. albicans colonics at 30 ° on glucose peptone agar are white or cream, smooth and glistening or occasionally dull and rough, features which are not really helpful in distinguishing it from other spp. True hyphae, pseudohyphae and chlamydospores, with clustered blastospores along the lengths of hyphae, are seen on cornmeal agar.

C. tropicalis is similar (also with true hyphae, but no chlamydospores. There may also be internodal blastospores. Biochemistry is useful, as C. tropicalis can ferment sucrose. C. dubliniensis is morphologically and biochemically *very* similar to C. albicans.

C. krusei doesn't have true hyphae (although the pseudohyphae may look very hypha-like), with ellipsoidal blastospores.

C. lusitaniae has long pseudohyphae, few branches, prolific small oval blastospores, and characteristically assimilates rhamnose.

C. parapsilosis has pseudohyphae that often terminate in a swollen cell, with occasional roundish blastoconidia.

C. glabrata doesn't form pseudohyphae (as noted above), for which reason it was formerly known as Torulopsis. It has small round to oval blastoconidia. (C. famata and C. inconspicuosa are similar, as is Saccharomyces cerevisiae). }

- Biochemistry. Sugar fermentation can give a rapid presumptive diagnosis. This is *not* good enough on its own (at present) - morphology should always be examined too.

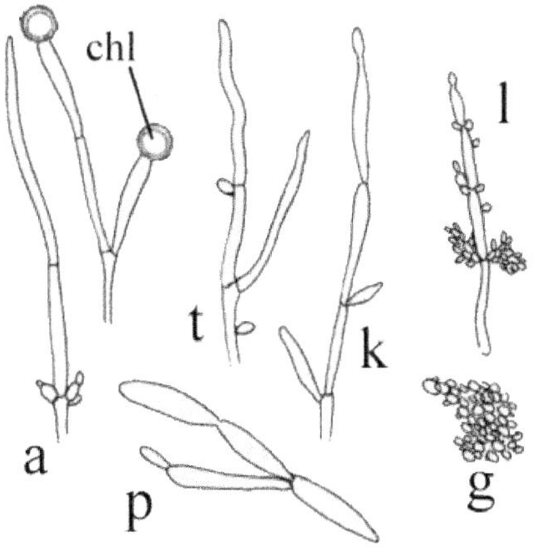

Candida morphology on cornmeal agar
a = albicans, t = tropicalis, k = krusei, l = lusitaniae, g = glabrata, p = parapsilosis; chl = chlamydospore

- "The serum germ tube test" is a quick presumptive test for C. albicans! A light inoculum of cultured yeast is incubated in horse serum for 2-3 hours at 37 °C. The test is positive if there are short hyphae *without a constriction where the hypha joins the parent cell.* False negatives occur.
- Molecular techniques will probably revolutionise the identification of Candida species.

Non-albicans Candidiasis

Not all fungal infections are due to C. albicans - more than a third are due to other organisms such as C. glabrata, C. krusei, C. parapsilosis, C. tropicalis, C. pseudotropicalis, and occasionally very resistant species like C. lusitaniae. In the European SENTRY study, only 53% of 170 episodes of candidaemia were attributable to C. albicans. Prominent organisms are:

- o C. glabrata (formerly Torulopsis glabrata). Resistance to fluconazole is common, apparently mediated by an "ATP binding cassette transporter gene CgCDR1". Over-the-counter azole antifungals (miconazole, clotrimazole) may promote resistance to fluconazole. The current drug of choice is probably still amphotericin B. Voriconazole appears to be much more effective than fluconazole against C. glabrata.
- o C. krusei is also intrinsically resistant to fluconazole. Here too, voriconazole may be of use.
- o Other fungi: C. parapsilosis infection is becoming more and more important in neonatal ICUs, and the immune compromised. C. tropicalis is *not* usually a commensal, (unlike C. albicans) *i.e.* isolation suggests *infection* , and the organism may be fluconazole resistant. C. lusitaniae is interesting (and worrying) as it may be resistant to amphotericin B - surprisingly, fluconazole may work!

Invasive Aspergillosis

Substantial outbreaks of this disease may occur in immune compromised patients following on inhalation of Aspergillus spores. See for example. In addition, the disorder may be more common than previously thought in patients with chronic obstructive pulmonary disease! In a Spanish study of AIDS patients, 1.12% had invasive pulmonary aspergillosis - antiretroviral treatment increased survival substantially.

Invasive aspergillosis (IA) is often fatal. One review reported an overall fatality rate of 58%, with rates of 87% in bone marrow transplant patients. Another review of 595 patients with probable or confirmed IA showed complete responses to amphotericin B in only one quarter of patients, with similar high death rates, especially in the sicker patients who received AB therapy alone, rather than itraconazole +- AB.

Diagnosis of Systemic Fungal Infection

The problem with diagnosis and treatment of systemic fungal infection can still often be summed up as "too little too late". Conventional diagnosis of these infections, based on blood cultures or culture of the offending organism from multiple sites, often delays therapy. Patients have frequently been on many different antibiotics for long periods of time in the vain hope that the systemic signs of infection or 'sepsis' will subside. When antifungal therapy is instituted, inappropriate dosing seems quite common.

There is controversy about when you should treat for apparent systemic fungal infections, especially Candida "infection". Most but perhaps not all authors would treat based on isolation of the organism from the bloodstream. There is not even agreement that intravascular lines should be replaced if fungaemia is detected, although most would also regard this as imperative. It has been said that density of infection (some talk about the "colonisation index"), number of positive cultures, isolation from non-contiguous sites, type of organism (C. tropicalis), and isolation from usually sterile fluids all predict the likelihood of severe systemic fungal infection.

Candiduria is also controversial, although most would agree that it should be treated once (a) confirmed and (b) risk has been stratified appropriately. This is unfortunately seldom done in clinical practice

Newer tests that have been advocated for early diagnosis of systemic fungal infection include:

- Sandwich ELISA for circulating galactomannan
- Polymerase chain reaction (PCR) shows promise in the diagnosis of Candida infections, even unusual species.

Diagnosis of Invasive Aspergillosis

Invasive aspergillosis may be difficult to diagnose, especially if one is not aware that it might be present. The presence on high-resolution CT scan of a "halo" or "crescent air" sign is thought to be diagnostic, but in one study the crescent sign was only seen in 10/21 patients; others allege that the halo sign is reliable with early CT, but disappears later on! Screening the blood for galactomannan may be very valuable with a sensitivity of up to 90% and specificity of 98%. In one prospective study, 14% of 215 patients on chemotherapy appear to have developed invasive pulmonary aspergillosis. Polymerase chain reaction may also be of value in early diagnosis.

Antifungal Therapy

We will first discuss the drugs used in treatment, and then treatment of specific fungal infections:

Amphotericin B

The mainstay of antifungal therapy for severe systemic mycoses is still Amphotericin B (often abbreviated to AB). This predominantly fungicidal agent is available in various forms:

1. Conventional AB = AB desoxycholate
2. Abelcet (AB lipid complex, ABLC).
3. ABCD = AB colloidal dispersion (Amphocil). This is formulated with sodium cholesteryl sulphate. It appears similar in effectiveness to conventional AB, but with perhaps a slightly better therapeutic index, although big studies are lacking.
4. AmBisome = liposomal AB, a small unilamellar liposome preparation (45 to 80 nm). This seems to be somewhat less toxic than conventional AB, but is extremely expensive. Efficacy is similar.

Because of the cost of the lipid formulations, and little evidence of increased efficacy (despite better tolerance), most would reserve these for cases where AB desoxycholate is causing severe toxicity. Some have advocated mixing AB in "Intralipid" but the mixture does not appear to be very stable.

Amphotericin B is useful in treatment of infection with Blastomyces, Coccidioides, Histoplasma, Paraoccidioides, Candida, Cryptococcus, but does have substantial associated toxicity. It is a "polyene", and works on fungi by binding to ergosterol in the fungal cell membrane, disrupting the membrane and killing the fungus. Note that AB appears to have a marked post-antifungal effect (PAFE) of up to ten or more hours, although the clinical relevance of this is controversial.. AB may also have immune stimulatory effects! The drug doesn't work reliably for Trichosporon beigelii, Pseudoallescheria boydii, certain Fusarium spp., and Candida lusitaniae.

Trials of therapy have often been poorly designed, but there is a fair amount of recent evidence that conventional Amphotericin B does work (although you would appear to have to treat 25 patients to save one life), as does liposomal AB. Meta-analyses were based on 8 trials showing that IV A/B lowers mortality (RR 0.72, CI 0.51 - 1.02), and three trials of AmBisome. Another analysis suggests that routine prophylaxis with antifungals does not influence mortality, a suggestion that few oncologists will happily swallow!

AB's oral bioavailability is under 5%. Systemic (intravenous) administration is associated with a long list of adverse effects, including marked fever (in about 50%), anaphylaxis (about 1 in 100), nausea, vomiting, and nephrotoxicity. Nephrotoxicity includes lowered GFR, increased loss of potassium and magnesium, and even distal renal tubular acidosis. Nephrotoxicity is said to be irreversible if therapy is prolonged. Other nephrotoxic agents may worsen toxicity. It is alleged that nephrotoxicity may be diminished by sodium-loading. Rarely, severe hypertension has been reported in association with AB therapy! The desoxycholate form has a long half life (24-48 hours), and a volume of distribution of about 4 L/kg.

Because it precipitates in normal saline, AB must be mixed in 5% dextrose water, and given via a central line to prevent phlebitis. It is not necessary to adjust dosage in liver failure, but many lower the dose in renal failure(?). The drug is not dialysable. CSF penetration is poor (2-4% of serum levels), but meningeal penetration may be better, and the drug enters the CSF of infants extremely well. We often tend to underdose our patients with (say) 0.5 mg/kg - one should probably aim for closer to 1 mg/kg or more, especially with more resistant fungi such as Aspergillus. Such doses are achieved with fewer side-effects with newer formulations such as liposomal amphotericin, but the cost is immense. Far higher doses have been used with lipid formulations of AB (even 7mg/kg).

The old tradition of giving a test dose of AB has been questioned, and some would ignore this approach. (Dose: 1 mg in 100 ml 5% D/W over 30 min). Conventionally the daily dose is given over about 6 hours, but some argue that it should be given over a longer period.

Fluconazole and Other Azoles

Azoles (such as fluconazole) *inhibit* fungal growth by preventing formation of ergosterol, vital for fungal cell membrane integrity. They do this by inhibiting fungal cytochrome P-450 (14-a sterol demethylase), so it is not surprising that these agents have powerful inhibitory effects on some human CYPs.. Fluconazole is usually regarded as a "first generation azole", and may well be supplanted in the next few years by newer azoles such as vorizonazole, which have a wider spectrum of activity. Azoles generally have "non-concentration dependent activity" against most fungi.

Fluconazole is mainly used for C. albicans infection (and some other susceptible Candida spp. but NOT C. krusei, and has inconstant activity against C. glabrata). C. albicans may acquire resistance, especially with chronic or recurrent treatment in AIDs patients. Fluconazole may be effective against Cryptococcus neoformans meningitis, and Coccidioidomycosis.

Fluconazole has good GIT absorption, and is renally excreted. CVVHD causes substantial removal, even more than normal kidneys, and far greater than with CVVH or intermittent dialysis. There is a substantial potential for drug interactions owing to *inhibition of CYP2C9*, and also 3A4. Volume of distribution may be substantially increased in critically ill patients. There is good cerebrospinal fluid penetration. Adult doses used for severe infections are often in the range of 200 mg to 400 mg/day, although 800 mg/day(+) has sometimes been used. Half life (at 200 mg/day) is about 27-37 hours, with a V_D of 0.7 to 1 L/kg.

Susceptibility testing may be of value in guiding dosing, with the AUC:MIC ratio perhaps a good predictor of response. Fluconazole has no post-antifungal effect, unlike AB. Both human serum and GM-CSF may enhance killing of fungi by fluconazole.

For a review of fluconazole and itraconazole in C. albicans infection see. In neonates, fluconazole has been given at 5 to 6 mg/kg, but long dosing intervals are required in very low birthweight infants.

Other Agents

itraconazole (capsules poorly bioavailable, elixir better, but far from tasty!; an intravenous form has recently become available. There is complex hepatic excretion. The agent may be an alternative to AB for Aspergillosis, and works for Sporotrichosis, but with treatment of severe disease, blood levels of itraconazole and hydroxyitraconazole are recommended! Pharmacodynamics appear very complex).

1. voriconazole {= UK 109,496; Pfizer}
2. posaconazole { = Sch 56592; Schering-Plough} - similar to itraconazole, with good Aspergillus fungicidal activity, a really broad-spectrum azole (which may even work against Trypanosoma cruzi)!
3. ravuconazole { = BMS-207147; Bristol-Myers Squibb}
4. echinocandins (inhibit fungal glucan assembly), preventing fungal cell wall synthesis; *e.g.* cilofungin {not currently being used}; V-echinocandin = LY303366 = VER-002 {Eli-Lilly}; MK-0991 = L-743,872 = caspofungin; Have concentration-dependent antifungal activity, with a substantial post-antifungal effect. They seem like a fair bet for the future, but need more research - for example, the combination of V-echinocandin and steroids is lethal in mice.
5. FK-463, which is echinocandin-like, appears promising.
6. Flucytosine. Not readily available in many countries, this is mainly used in combination with AB, as otherwise resistance rapidly develops. Gastrointestinal and haematologic toxicity are common.
7. in "early development":
    - sordarins;
    - chitin synthetase inhibitors (polyoxins, nikkomycin Z);
    - topoisomerase inhibitors;
    - lipopeptides ;
    - pradimycins (quinones that complex with cell wall mannoproteins of Candida, Cryptococcus and Aspergillus spp). They are toxic and poorly soluble.
    - Liposomal nystatin!

Voriconazole

Voriconazole is an azole antifungal. Its plasma protein binding is about 70%, it is mainly cleared by the liver (? CYP 3A4), and has good, rapid oral bioavailability. There is substantial first-pass metabolism, which is saturable. Good for Candida spp, Aspergillus. This includes fluconazole resistant organisms such as C. glabrata. Voriconazole is *fungicidal* for aspergillus at twice MIC. It is well absorbed orally (about 90%). Voriconazole is also effective against:

1. Histoplasma, Coccidioides, Blastomyces, Paracoccidioides, Cryptococcus, and dermatophytes;
2. Scedosporium apiospermum;
3. Pseudallescheria boydii;
4. Penicillium marneffei, Fusarium spp;
5. Acremonium strictum.

It is NOT useful against Sporothrix schenckii and zygomycetes.

Dosage should probably be ~200 to 400 mg/day. In one study, the loading dose was 6 mg/kg BD for 24 hours, followed by 3 mg/kg BD. Side effects include (reversible) mild-to-moderate visual disturbance in about 14% of people (enhanced brightness or blurred vision), and raised bilirubin/alkaline phosphatase in some. Rash occurs in about 4%, and photosensitivity and erythema have been reported.

Treatment of Systemic Candida Infection

The mainstay of therapy is still amphotericin B, but this *may* change with the advent of newer azoles such as voriconazole. C. lusitaniae, which is often resistant to AB, is fortunately still fairly uncommon. The major problem is the toxicity of AB, which is substantial. The expense of lipid formulations of AB practically precludes their use.

Some have advocated the use of fluconazole, often in high doses, for putative or confirmed Candida infection. The problem with this is that not only are C. krusei and C. glabrata often resistant to fluconazole, but with the advent of repeated or long-term fluconazole therapy/prophylaxis (in, for example, AIDS patients), even C. albicans may be resistant. Dosing of fluconazole is highly controversial.

Treatment of Invasive Aspergillosis

Amphotericin B is probably still the drug of choice for invasive aspergillosis, although voriconazole may be very effective, and posaconazole has shown promise in experimental models of the disease, as has ravuconazole. Some would even recommend surgical resection for selected cases of invasive pulmonary aspergillosis. A review of 2121 published cases suggests that amphotericin B combined with either rifampicin or flucytosine may be better than amphotericin B monotherapy. The 'evidence' is not cast in stone.

Treatment of Aspergillus infection in the immune-compromised is urgent, and should be based on clinical or radiological criteria, without waiting for microbiological confirmation.

Sequential Therapy

This has become quite popular - initial aggressive amphotericin B therapy may be followed by *e.g.* long-term fluconazole. This approach is best-documented in cryptococcal meningitis in AIDS patients, but a similar approach (AB followed by itraconazole) has also been used with Aspergillus infection.

Treatment Failures

These are still very common with all systemic mycoses. Often they are thought to reflect the severity of the underlying disease, but there is some concern that more often than not, we underdose our patients with antifungal therapy, use the wrong agents, or start too late. Secondary resistance may also arise. Treatment failure occurs in about twenty to thirty percent of patients with candidaemia, and the death rate in patients is high. In one study of 415 cases of candidaemia, overall mortality was 45%, being lower in children. In invasive aspergillosis, treatment failures are still more common (about 50%).

Secondary resistance is usually seen in more chronic situations, especially in AIDS patients where there is recurrent or prolonged therapy with fluconazole.

Assessment of resistance to antifungal therapy is difficult. The National Committee for Clinical Laboratory Standards (NCCLS) has defined guidelines for testing of azole antifungals, but standards for amphotericin B are not clear.

For fungal resistance see also. Primary resistance to fluconazole is common in C. glabrata, and often emerges (or may be primarily present) in C. krusei. C. lusitaniae is primarily resistant to amphotericin, but fortunately, such infections are uncommon.

PLANT ROOTS AND SOIL BIOTA

What is the Rhizosphere?

The rhizosphere is the region of soil that surrounds, and is influenced by, plant roots. A major influence in the rhizosphere is the release of organic molecules into the soil from roots. Although the exudates and mucilaginous materials released by roots diffuse through the soil, they always occur in their highest concentrations adjacent to the root surface. Because soil organisms are attracted to these exudates as a source of food, the abundance of soil organisms also increases close to roots. This is a dynamic relationship. The rhizospheres of old and young roots provide very different habitats for soil organisms. As roots age, they release different types and quantities of carbon substrates, which in turn affect soil microorganisms.

The rhizosphere also differs from the remainder of the soil in other ways. For example, the amount of inorganic nutrients can be either higher or lower in the vicinity of roots compared with that in the soil some distance away from roots. The concentration or depletion of nutrients around roots depends on physiological processes associated with root function, including nutrient uptake by plants. In addition, roots release hydrogen or bicarbonate ions into the soil, causing the pH in the rhizosphere to decrease or increase. Roots also consume or release oxygen. These contrasting effects, in addition to the considerable exudation of carbon compounds from roots into the soil and the compression of soil as roots grow in length and breadth creates a dynamic environment for soil organisms.

Some nutrients move towards roots in soil water by the process of mass flow. Other nutrients are less mobile because they are adsorbed onto soil particles. The availability of adsorbed nutrients such as phosphorus and copper to plants depends on the extent to which roots explore the soil and intercept these ions. This in turn affects the extent of the rhizosphere and the spatial distribution and activity of soil microorganisms.

Water use by plants adds to the dynamic nature of the rhizosphere. Daily cycles of water uptake and loss by roots cause roots to expand and shrink. This adds to physical changes induced by seasonal cycles of wetting and drying or freezing and thawing. These daily and seasonal cycles alter the soil habitat, which influences the growth and activity of soil organisms.

As a root develops, the number of organisms near the root surface increases rapidly. For many plants there is not a clear distinction between the root surface and the soil. Roots are dynamic, growing and casting off cells and navigating soil cavities and rough surfaces. As a consequence, the root surface can be highly irregular.

What is the Origin of Root Exudates?

A variety of organic compounds released from roots provide a source of energy and carbon for soil organisms. In a study of exudation from wheat roots, the plant was labelled with 14C. This technique allows the carbon released from the plant to be followed precisely. The greater the amount of 14C released into the rhizosphere of the wheat plant, the greater the quantity found in the microbial biomass. The quantity of the compounds released from roots is not the same along the length of a root. Older roots can be more 'leaky' than younger roots, although in this experiment there was no difference, at least for one line of wheat. The release of relatively large quantities of organic carbon into soil under normal conditions stimulates microbial activity and provides a source of energy that can accelerate the decomposition of organic matter.

How Do Roots Affect Soil Organisms?

Roots provide an important habitat for bacteria, fungi and very small soil animals. The number of organisms can be about 500 times more in the rhizosphere than in the bulk soil.

The surface of a root is not completely covered with bacteria and fungi, except in special circumstances, such as where roots are colonised by an ectomycorrhizal fungus. Usually less than 10% of the root surface is colonised. Bacteria accumulate in crevices along the edges of outer root cells or where the root surface is damaged.

Some organisms form loose associations with roots and simply depend on the plant releasing molecules that they can use. Other organisms have more specific interactions with roots. For example, root nodule bacteria can stimulate the root to form additional cell tissue that proliferates to form a nodule.

Root nodule bacteria live and multiply inside these root cells. When roots begin to disintegrate with age or through disease, there are more chances for opportunistic bacteria and fungi to enter through crevices between the surface cell layers or dead root cells.

Roots indirectly affect soil animals because they respond to changes in the abundance of bacteria and fungi associated with roots. The types of organisms present in the rhizosphere and in the bulk soil change over time. These changes are associated with the intensity of grazing by soil animals and the abundance of other bacteria and fungi. Furthermore, there is a wide range in numbers of different types of animals. Overall, there is great diversity in type and number of organisms in the vicinity of roots.

How are Roots Affected By Soil Organisms?

Microorganisms can indirectly increase or decrease root growth. For example, organisms influence the movement of carbon from shoots to roots. This can alter root growth as well as stimulate the release of carbon into the rhizosphere. Some organisms increase root growth by producing compounds that function as root growth hormones. Organisms may decrease root growth by creating a significant drain on carbon reserves in the plant. This reduces the quantity of carbon in the root that would otherwise be used to form more roots.

The effects of soil organisms on roots can be either indirect or direct. For example, earthworms indirectly promote root growth because their burrows provide channels for roots to grow more deeply into the soil profile. In contrast, some rhizosphere bacteria and fungi impede root growth. Roots may grow more extensively if such microorganisms are absent than if they are present. Ectomycorrhizal fungi also reduce root growth by colonising the surface of roots and stunting their growth. In addition to the thickening of the root, the roots often branch in response to hormones produced by the fungus.

Arbuscular mycorrhizal fungi can reduce root growth. In a study of maize, microorganisms decreased both root hair length and root hair number. However, the extent of the decrease depended on whether arbuscular mycorrhizal fungi were present or not. In the presence of these fungi, both root growth and root hair number were considerably impeded in comparison to when these fungi were absent.

Generalisations about the effects of soil organisms on root growth are difficult to make because different species of plant and different types and combinations of soil organisms interact in different ways. Bacteria in the genus *Azospirillum* are interesting examples of microorganisms that have the potential to influence root growth. Species of *Azospirillum* commonly live in the rhizosphere of grasses. Some bacteria in the genus *Azospirillum* increase root growth by contributing nitrogen to the plant and others stimulate root growth by hormone production. Strains of species of *Azospirillum* vary in their capacities to fix atmospheric nitrogen or produce hormones. Nitrogen fixing strains of*Azospirillum* are not expected to increase root growth if nitrogen levels in the soil are already adequate for the particular plant species; if root growth is affected, other mechanisms must be involved.

How do Roots Function in the Presence of Soil Organisms?

Changes in root structure alter the capacity of root systems to absorb nutrients and water. Principally, this is related to the way that roots are distributed in soil. However, close associations between roots and microorganisms enhance nutrient uptake by plants if organisms have access to nutrients to which the plant does not. An example of this is the association between roots and nitrogen-fixing bacteria. Another example is the association between various types of mycorrhizal fungi and roots that allow root systems access to nutrients well beyond the rhizosphere.

Root pathogens commonly interfere with nutrient and water uptake by damaging the root surface and root cortex or by blocking the xylem vessels in the vascular system inside the root. Damage to the cortex or the vascular system will reduce nutrient and water transport between roots and shoots. Soil animals such as nematodes may also damage roots. Parasitic nematodes induce gall formation on some plants and animals such as springtails graze on small roots and damage their surface. Disturbances of this kind create opportunities for a variety of fungi and bacteria to enter roots that would not otherwise be able to do so.

MYCORRHIZAL ASSOCIATIONS

What is a Mycorrhizal Association?

The majority of plant species form mycorrhizal associations. Mycorrhiza means 'fungus root' and refers to several specialised associations between fungi and roots. There are four major types of mycorrhizas: ectomycorrhizas, arbuscular mycorrhizas (also called vesicular arbuscular mycorrhizas), ericoid mycorrhizas and orchid mycorrhizas. These associations are generally beneficial to both the plant and the fungus, with nutrient exchange between the partners. However, under some conditions, at least for arbuscular mycorrhizal, there is no beneficial effect and plant growth can even be reduced.

Orchid mycorrhizal associations are very different from the other three types. The minute orchid seed has no reserves of carbon and is totally dependent on the fungus during the early stages of its growth.

Almost all plant species form one of the four types of mycorrhizal associations. Each plant species usually forms only one type of mycorrhizal association. In the less common situation in which more than one type of mycorrhiza is formed on the same plant, the different mycorrhizas usually occur on different roots or are present at different times during the life cycle of the plant.

One type of mycorrhiza is usually dominant in each ecosystem. For example, plants that form ericoid mycorrhizas often dominate heathland ecosystems. Trees that form ectomycorrhizal associations commonly dominate forests in Europe and North America. Tropical forests may contain more equal proportions of plant species that form arbuscular and ectomycorrhizal associations. The type of mycorrhiza present generally depends on the diversity of plant species that occur there.

The eucalypt trees of Australian forests predominantly form ectomycorrhizal or arbuscular mycorrhizal associations. For example, in a jarrah forest in south-west Australia, seventy two percent of the 93 species of plant studied formed either or both of these two mycorrhizas. The 25% found with no mycorrhizal association were mostly members of the family Proteaceae, which does not form mycorrhizas but they can form specialised roots 'proteoid' roots which facilitate nutrient uptake.

Most annual agricultural plants form arbuscular mycorrhizas. Ectomycorrhizal associations are less common in disturbed ecosystems and are more frequently observed on perennial plants. Horticultural species form a variety of mycorrhizal associations, depending on whether they are ornamentals (all types of mycorrhizas are represented in this group) or vegetable and orchard plant species (most of these form arbuscular mycorrhizas).

What are Arbuscular Mycorrhizas?

Arbuscular mycorrhizas are the most common type of mycorrhiza and they are formed by a fungal group that occurs in most soils. The distinctive fungi that form arbuscular mycorrhizas have co-evolved with plants over millions of years. However, worldwide, there are relatively few species of arbuscular mycorrhizal fungi in comparison with other groups of soil organisms. Fewer than 200 species have been described, although many others occur which have not yet been formally named and described. These 200 species all occur within a relatively small number of genera including: *Glomus, Acaulospora, Archaeospora, Gigaspora and Scutellospora.* The genus of each fungus can be identified from the characteristics of their spores and molecular analysis is now also used to distinguish among species and genera.

The most characteristic feature of arbuscular mycorrhizal fungi is the arbuscule. Arbuscules occur within cells of the plant and are key sites for nutrient exchange between the two partners of the association. The distinctive vesicles formed inside roots by some, but not all of these fungi contain high concentrations of lipids and appear to be storage structures.

Biology of Arbuscular Mycorrhizal Fungi

The roots of most plants are colonised by several species of arbuscular mycorrhizal fungi at the same time. There is little evidence of specificity between particular plants and these fungi, although some plants may be colonised to a greater extent by some of them. The relative abundance of different arbuscular mycorrhizal fungi within roots depends on soil conditions, root development and characteristics of the fungi such as the rapidity of spore germination and growth of hyphae.

Arbuscular mycorrhizal fungi are dependent on the plants they colonise for their survival. The fungi are "biotrophic" which means that they are unable to complete their live cycle without a plant. Because of this unusual characteristic of arbuscular mycorrhizal fungi, it is difficult to grow large quantities of arbuscular mycorrhizal fungi on artificial media in the laboratory without a plant.

When newly formed hyphae of AM fungi intercept a root, the fungus enters with no apparent alteration to the root surface. Species, and even strains, of the same fungus colonise roots to different extents, but generally the more hyphae there are in the soil the greater the extent of mycorrhizal root formed.

Arbuscular mycorrhizal fungi can survive for long periods in the soil as spores or as pieces of hyphae, even when the soil is dry or frozen. Most species produce spores that have relatively thick walls that are resistant to desiccation and are not easily eaten by organisms in the soil or colonised by pathogens. They have been shown to contain bacteria that appear to be natural inhabitants. Spores of arbuscular mycorrhizal fungi germinate when soil conditions become suitable, providing they are not in a dormant phase. Pieces of hyphae can survive in dead root pieces or directly in the soil.

Effects of Arbuscular Mycorrhizal Fungi on Plants

An important impact of AM fungi on plants is the alleviation of phosphate deficiency. Arbuscular mycorrhizal fungi form a network of hyphae in soil surrounding roots, and in doing so, effectively extend the root system. This can increase uptake of phosphorus from soil into roots. For example, some plants that are well colonised by arbuscular mycorrhizal fungi can grow just as well with only half the amount of phosphorus available in soil as that required by plants that do not have these fungi within their roots.

The extent to which a plant benefits from its arbuscular mycorrhizal fungi depends on the availability of nutrients and water. In terms of phosphorus nutrition, only plants growing where there is an inadequate supply of phosphorus will receive any benefit from having arbuscular mycorrhizas (this also requires that water and other nutrients are not limiting to growth). In soils where there is very little available phosphorus or where adequate phosphorus is available, there is generally no direct nutritional benefit of arbuscular mycorrhizas. Plant root architecture also influences the ability of a plant to supply itself with phosphorus, and therefore its dependence on an association with arbuscular mycorrhizal fungi. Plants such as grasses have fibrous root systems and are capable of thoroughly exploring soil for phosphorus. As a result, they do not benefit nutritionally from mycorrhizas to the same extent as a plant such as clover which has a coarser root system.

This effect of arbuscular mycorrhizal fungi on plants through increased phosphorus uptake has usually been studied only for young plants or plants grown in controlled conditions. The contribution of arbuscular mycorrhizal fungi to the growth of older plants, especially those growing in field soils, is not well understood.

Obviously the relationship between the way the symbiosis functions and the nutritional status of the plant is complex and the plant does not always benefit. For example, under some conditions, arbuscular mycorrhizal fungi reduce plant growth, possibly because they take a large amount of carbon from the plant. Alternatively, *Scutellospora calospora* can increase plant growth and phosphate uptake into plants when the level of phosphorus in soil is extremely deficient. But, it can also decrease plant growth when the quantity of phosphorus in the soil is almost sufficient for the plant to reach its maximum potential growth. At higher levels of available soil phosphorus, the fungus appears to remove excessive carbon from the plant to an extent that is detrimental to the growth of the plant.

The abundance of arbuscular mycorrhizal fungi within plant roots changes the supply of phosphorus. Very low levels of phosphorus can be insufficient for the growth of the plant as well as the fungus, keeping fungal numbers low. As more phosphorus becomes available, both the plant and the fungus increase in growth. There can be an increase in the length of root colonised by arbuscular mycorrhizal fungi and a corresponding decrease in the proportion of the root colonised. This occurs if the fungus does not grow fast enough to keep up with the growth of the roots. In soil where phosphorus is present at an adequate level for maximum plant growth, the quantity of arbuscular mycorrhizal fungi in roots decreases substantially. This may either be due to a detrimental effect of excessive amounts of phosphorus on the fungi in the soil or to inadequate carbon supply in the roots for fungal growth.

Clearly, it is not always practical to classify a fungus as being either effective or ineffective at enhancing plant growth. The effectiveness of fungi can be altered by changing the amount of fungus in the soil, by changing the amount of phosphorus in the soil or by altering soil conditions that affect the growth of the fungus. In addition, strains of the same fungus may respond differently, so the diversity of fungi present is also significant.

Phosphorus is not the only nutrient taken up by the hyphae of arbuscular mycorrhizal fungi. For example, the supply of Cu (copper) and Zn (zinc) to some plants is improved in the presence of arbuscular mycorrhizal fungi. However, arbuscular mycorrhizal fungi do not increase the supply of nitrogen to plants. The reason that AM fungi enhance the uptake of some nutrients and not others is related to the way that nutrients interact with soil particles. Nutrients (such as nitrate) that move with water towards plant roots in most soils are not likely to become more available to a plant if it has arbuscular mycorrhizas. In contrast, P, Cu and Zn become adsorbed to soil particles and diffuse through a soil rather than move with water towards roots. Roots need to explore the soil to intercept these nutrients that move slowly by diffusion. By acting as an extension of the root system, the hyphal network of arbuscular mycorrhizal fungi increases the likelihood that roots have access to nutrients that diffuse slowly in soil.

Even if no nutrient benefit is gained from the presence of arbuscular mycorrhizal fungi, there can be a benefit to the physical state of soil if hyphae help to stabilise soil aggregates. Hyphae of arbuscular mycorrhizal fungi may also protect roots from disease.

There is some evidence that arbuscular mycorrhizal fungi assist plants in water uptake, especially under dry conditions. Demonstration and interpretation of such benefits is confounded because the fungi also take up phosphorus. Enhanced growth may be due to increased phosphorus uptake as well as to water uptake if the soil is phosphate deficient. Therefore, increased water uptake may be an additional benefit in some circumstances, but the effects related to water and phosphorus are difficult to separate.

Within soil, arbuscular mycorrhizal fungi form extensive networks of hyphae that connect the roots of many plant species into an integrated system. This provides a pathway for nutrient exchange between different plant species. The extent to which this occurs, and its importance for plant growth needs further investigation. One particularly interesting area is the role of these networks in aiding seedling growth. Seedlings emerging from a soil containing a network of arbuscular mycorrhizal hyphae may gain access to nutrients from existing plants and grow better than might be expected if they are able to connect into this network. In addition, arbuscular mycorrhizal fungi may increase the chance of survival of seedlings in ecosystems where plants compete for nutrients and light. This is especially important for slower growing plants. Therefore, successful establishment of arbuscular mycorrhizas may help maintain or increase plant species diversity in a natural ecosystem.

What are Ectomycorrhizas?

Ectomycorrhizal fungi are the dominant mycorrhiza in many natural ecosystems, even though they form associations with a smaller number of plant species than do AM fungi. Plants that form ectomycorrhizas include well-known trees and many smaller perennial plants.

The main features of an ectomycorrhizal association are the hyphal mantle on the root surface, the Hartig net formed between the cells of the root closest to the root surface and the extension of hyphae into the soil. The hyphae of ectomycorrhizal fungi do not enter root cells in the same way as AM fungi. Rather, the Hartig net is formed when the ectomycorrhizal fungi enter roots between epidermal cells and spread out to form a network of hyphae through the root. Ectomycorrhizal fungi are usually clearly visible on the surface of roots. They promote root branching and restrict root extension which results in a great diversity in architecture of ectomycorrhizas. Roots become covered with hyphae to varying degrees.

Ectomycorrhizas are formed by many different species of fungi, in contrast to the relatively small number of species that form arbuscular mycorrhizas. Most belong to families of the basidiomycetes and a number to families of the ascomycetes. Some are clearly visible in forests and other reasonably undisturbed ecosystems because they produce mushrooms and similar fungal spore-containing structures above the soil surface. Others form their reproductive structures below the soil surface. Small mammals dig up and eat these structures. Truffles are a well-known example of fungus-produced underground fruiting structures.

Some ectomycorrhizal fungi form associations with many different plant species , while others form associations with fewer plants. For example, most species of *Amanita* have an intermediate or broad host range, whereas most of species of *Suillus* have a narrow host range. In contrast, 25% of species of *Russula* have a narrow host range and 30% have a broad host range.

Compared to AM fungi, many species of ectomycorrhizal fungi can grow in artificial, laboratory media, making them easier to study, although they often grow very slowly. Others are either difficult to grow or have not yet been grown at all under laboratory conditions.

Biology of Ectomycorrhizal Fungi

It is not easy to define a simple life cycle for ectomycorrhizal fungi because many genera of fungi are involved. Ectomycorrhizal fungi are commonly associated with perennial host plants that are relatively long-lived. Therefore, ectomycorrhizal fungi within the roots of living plants have the potential to survive for long periods.

Ecological successions of ectomycorrhizal fungi occur on the same root systems leading to a gradual change in dominance of fungi. Some plants form more than one type of ectomycorrhizal association, either on different roots of the same plant or at different times. In this study, fruiting structures of the fungus*Laccaria* sp. generally occurred further from the tree base than those of *Lactarius*. However, the presence of these structures is not a direct measure of the extent of mycorrhizal fungus present in soil or in roots. Recent molecular studies have highlighted the lack of a quantitative relationship between fungi in roots and the number of fruiting structures formed above the ground.

New roots usually become colonised by the hyphae of ectomycorrhizal fungi attached to existing mycorrhizal roots on the same or on a nearby plant. The spores that are formed by these fungi do not appear to be very important for initiation of new mycorrhizas.

Various degrees of specificity exist between ectomycorrhizal fungi and host plants. A process of molecular recognition takes place prior to colonisation of the root by the hyphae. Colonisation of the root by the fungus proceeds only if the fungus and root are compatible. Recognition of compatibility between a root and a fungus involves molecular signalling using genetic information in both the fungus and plant.

Effects of Ectomycorrhizal Fungi on Plants

It is more difficult to estimate the contribution of ectomycorrhizal fungi to plant growth than for AM fungi, except at early stages in plant development. Studies on seedlings clearly illustrate the contribution of ectomycorrhizal fungi in the early stages of tree development, but extrapolation of these benefits to older trees is not reliable.

Ectomycorrhizal associations can improve the phosphate status of plants in soils that are phosphate deficient. Under some circumstances, ectomycorrhizal fungi make nitrogen more available to plants and increase plant resistance to drought and disease. These benefits can lead to a more robust plant that has a greater chance of survival.

There is substantial evidence for benefits of ectomycorrhizal fungi in nitrogen nutrition of plants where nitrogen is deficient. Some ectomycorrhizal fungi can use simple or more complex forms of organic nitrogen. For example, all three fungi in this study were effective at using glutamine as a nitrogen source, but none could use histidine, which is a very complex form of nitrogen. In contrast, *Elaphomyces* was able to use nitrogen from arginine, but the other two fungi were not. These differences are likely to be important in soils with low levels of inorganic nitrogen and high levels of organic matter. The nitrogen that is taken up by the ectomycorrhizal fungi is passed onto their plant hosts, improving what would otherwise be relatively poor growing conditions for the plant.

Ectomycorrhizal fungi also appear to directly help plants avoid disease and survive periods of drought. Where the hyphal mantle covers the root, it provides protection from pathogenic fungi, which can not penetrate the hyphal layer. In addition to this, due to the presence of the ectomycorrhizal fungi changes occur in root physiology that minimises the susceptibility of the root to infection by pathogenic fungi or bacteria. The hyphal networks around mycorrhizal roots enables water to be taken up from parts of the soil that the roots cannot reach. Protection from both disease and drought are contributions that ectomycorrhizal fungi may make to various degrees at different stages in the growth cycle of the host plant. The contributions depend on the type of fungus and the soil conditions.

What are Ericoid Mycorrhizas?

Ericoid mycorrhizas are the specialised mycorrhizal associations formed with members of the plant families Ericaceae. These mycorrhizas have a different morphology from the mycorrhizas already mentioned and play an important role for their host plants. Ericoid mycorrhizal fungi colonise the individual cells of very fine hair-like roots directly from the soil, forming coils of hyphae within the root cells. The hyphal connections to the soil coalesce around the roots as a fine network. The hair roots colonised by ericoid mycorrhizal fungi are short lived. Therefore, ericoid fungi need to continuously re-establish the mycorrhizal associations as the roots grow.

There appears to be a high degree of specificity between the fungi that form ericoid mycorrhizas and their host plants. Details of the nature of the recognition process between the fungus and plant are well understood.

Some ericoid mycorrhizal fungi grow in pure culture, and analysis of these cultures shows that ericoid fungi are very different from each other. This was demonstrated by comparing the electrophoretic pattern of enzymes that decompose pectin after they were separated in a gel. The migration of enzymes in the gel indicated differences in the molecular weight of pectic enzymes from some fungi. This is indicative of differences in genetic characteristics among the fungi.

There is evidence that ericoid mycorrhizal fungi have certain characteristics similar to wood degrading fungi that may allow them to degrade complex components of organic matter in the soil. In comparison to a selection of ectomycorrhizal fungi, two ericoid mycorrhizal fungi had greater potential to degrade lignin and soluble phenolic molecules. This conclusion was based on the response of the fungi in four biochemical tests that measured characteristics of fungi important in degradation of lignin and other complex plant molecules.

Effects of Ericoid Mycorrhizal Fungi on Host Plants

Ericoid plants occur in nutrient poor soils and the mycorrhizal associations that they form are likely to be important in providing them with nutrients from the organic matter that would otherwise remain unavailable to the plant. In particular, nitrogen appears to be released from organic matter by the activity of ericoid mycorrhizal fungi. There is also a possibility that organic phosphorus is mineralised from organic matter by these fungi.

Although ericoid mycorrhizal fungi have characteristics that are very different from those of both AM and ectomycorrhizal fungi, all three types of mycorrhizas coexist in many natural ecosystems. This illustrates how plant-microbial associations respond to the nutrient resources in a soil.

How do Orchids Depend on their Fungal Association?

The morphology and physiology of orchid mycorrhizas are very different from those of the other three types of mycorrhizas described above. Orchid seeds are highly dependent on fungi. During the early stages of seedling growth, the fungi supply carbon to the growing orchid seedling. This is the reverse of what occurs in other mycorrhizas. Some orchid mycorrhizal fungi have been isolated and grown in culture for many years and many belong to the genus *Rhizoctonia* or to related genera.

NITROGEN-FIXING ASSOCIATIONS

What is the Range of Associations between Nitrogen-Fixing Microorganisms and Plants?

Nitrogen gas (N_2) constitutes nearly 80% of the atmosphere, but in this form nitrogen is not available to plants. Several groups of soil bacteria convert nitrogen gas into ammonium (NH_4^+), which is a form of nitrogen that can be used directly by plants. This transformation is catalysed by the bacterial enzyme nitrogenase. Very few bacteria have the capacity to activate the nitrogenase enzyme system, and those that do occur as free living bacteria in soil, as loose associations with root surfaces or within highly specialised, symbiotic associations with plants.

Bacteria in the genus *Azospirillum* live in a loose association with root surfaces of grasses and other plants. This contrasts with the highly specific associations that form between cyanobacteria and fungi in lichens. Highly specific nitrogen-fixing associations are also formed between some tree species and bacteria (actinomycetes) in the genus *Frankia*. The more primitive gymnosperms such as cycads also form characteristic associations with cyanobacteria. The most widely studied of all nitrogen fixing associations are formed between legumes and bacteria from the family Rhizobiaceae.

The biological transformation of N_2 gas to ammonium is equivalent to the industrial transformation of N_2 that requires excessive application of heat and pressure. This highlights the uniqueness of biological nitrogen fixation and its practical benefit.

What is 'Associative' Nitrogen Fixation?

Nitrogen-fixing bacteria that live in loose associations with roots are called 'associative' nitrogen fixers. Associative nitrogen fixing bacteria grow on the surface of roots and may also colonise the outer layers of a root by entering between epidermal cells.

A common genus of associative nitrogen fixing bacteria is *Azospirillum*. Species of*Azospirillum* are not all equally effective at converting atmospheric nitrogen to ammonium. There is conflicting evidence about the extent to which nitrogen fixed by these bacteria is available to plants. Even for the species of *Azospirillum* that are very effective at fixing atmospheric nitrogen, the nitrogen fixed is not immediately available to the plant. It is not until the bacteria die and are mineralised that the nitrogen becomes available to the plants.

Nitrogen fixed by *Azospirillum* is not immediately available to plants because of the complex processes of ammonium transfer through the bacterial cell wall. Ammonium excess to the requirements of the bacterium can be transported to the outside surface of the cell wall. However, another process can transfer the ammonium back into the cell, slowing the transfer of ammonium to the plant. Some mutant strains of *Azospirillum* are effective at permanently transferring ammonium from the bacteria.

The movement of molecules into and out of bacteria and fungi are complex processes. In the case of *Azospirillum*, an understanding of these processes shows why nitrogen is not immediately available to plants from bacteria that fix nitrogen in the root zone, at least while the bacteria are alive.

The benefits of associative nitrogen fixers have been estimated in agricultural soils. The quantity of nitrogen fixed varies greatly. In very nitrogen deficient soils, the quantities of nitrogen fixed are not usually sufficient to overcome nitrogen deficiency in agricultural plants.

How do Associations Between Legumes and Nitrogen Fixing Bacteria Function?

Legumes are important plants in both natural as well as managed ecosystems. The majority of legumes form symbiotic associations with nitrogen-fixing bacteria. As a consequence of these associations, distinctive structures (called nodules) form on the roots of legumes. The bacteria multiply within the nodule cells.

Many kilograms of nitrogen per hectare can be converted every year by agriculture plants including crop (*e.g.* soybean) and pasture (*e.g.* clover) plants because of their associations with nitrogen-fixing bacteria. These associations are of great importance as a substitute for nitrogen fertiliser. Nitrogen fixing symbioses are also common in natural ecosystems.

Bacteria that form associations with legumes live in the soil as part of the microbial community. However, they do not fix nitrogen directly in the soil. Nitrogen fixation only occurs when bacteria form a nodule on a legume root. When an appropriate legume is present, only one in a million bacteria are necessary to form a nodule. In forming a nodule, the bacterium undergoes a highly specific interaction with the plant, involving several stages and complex signals between the bacterium and plant.

Associations between particular bacteria and their legume host plants are usually highly specific. For example, the bacteria that fix nitrogen in association with species of clover cannot fix nitrogen in association with species of acacia. Nevertheless, the same bacteria can form associations with different clover species and occasionally different bacteria will infect the same plant. However, nodules generally contain only one strain of bacterium that multiplies to fill the central core of nodule cells.

Although there is considerable specificity among root nodule bacteria and legumes, it is interesting that the same bacteria form nitrogen-fixing associations with lupin (*Lupinus* spp.), a crop legume and serradella (*Ornithopus* spp.), a pasture legume. Because the architecture of both the roots and shoots of these two plants is very different, the bacteria which colonises both are evidently highly adaptable.

The shape and size of nodules varies considerably, with each determined by the plant, not the bacteria. In some plants, nodules can continue to grow throughout the life of the plant, as for example with clover. Where this occurs on perennial species, layers of new nodule tissue are added annually. In contrast, nodules formed on other legumes have a limit to their growth and are termed determinant nodules. Soybean is an example of a legume that forms determinate nodules.

The bacteria that form symbiotic associations with legumes differ in their ability to fix nitrogen. The effectiveness of these bacteria can also be influenced by soil conditions. Therefore, when planting agricultural legumes it is necessary to select the most effective bacteria that will do best in the environment. This is also true when planting tree species for revegetation and forestry projects.

The bacteria that fix the most nitrogen when in association with a legume are not necessarily those that form the most nodules. Conversely, it is not uncommon to find nodules formed on agricultural plants by bacteria that are ineffective at fixing nitrogen. In soils where there are different strains of bacteria that can nodulate the same legume, many factors influence the success or failure of nodulation. Therefore, the full potential for nitrogen fixation is not always achieved, even if highly effective strains of bacteria occur in the soil.

Diversity Among Root Nodule Bacteria

There is considerable diversity among root nodule bacteria. Bacteria that form nodules on legumes differ in the rate at which they multiply on artificial laboratory media. Acacias form nodules with groups of bacteria with very different characteristics and this can affect the detection of bacteria on growth media if the rapidly growing bacteria overgrow those that grow more slowly.

There is a difference in the DNA structure among root nodule bacteria. In some, but not all, DNA is present as circular segments, separate from the main chromosome. These segments, called plasmids, contain the genetic information required to initiate symbiotic nitrogen fixation. Plasmids can replicate separately from the chromosome and transfer from one bacterial cell to another.

Genetic Basis for Nodulation and Nitrogen Fixation

The location of genes on the DNA that code for different stages in the nodulation process illustrates complex interactions that take place between legumes and root nodule bacteria during nodulation. Separate regions of bacterial DNA are responsible for recognition of the legume root surface, initiation of root hair curling, initiation of the infection thread formation, and root cell multiplication to form the nodule.

Bacteria present in the rhizosphere stimulate the plant to produce molecules called flavanoids which are released in root exudates. In turn, these molecules can, to varying degrees activate the nodulation genes in the bacteria. For successful nodulation to occur, the flavanoid molecules need to be present in sufficient quantities to induce a response in the bacteria. Some plants are better able to do this for some bacteria than others. In this example, the maximum activity of flavanoids produced by *Medicago murex* occurred at pH 6.0 whereas flavanoids produced by *Medicago littoralis* were most active at a higher pH. This response partly explains why nodulation changes with soil pH for some legumes. Even though the bacteria may grow well at a certain soil pH, the lack of production of exudates that will initiate nodule formation may hinder their ability to form a nodule.

Legumes in natural ecosystems grow in soil containing indigenous, diverse communities of root nodule bacteria. Similarly, highly managed agricultural systems may contain many different strains of bacteria that have been introduced with agricultural legumes. High diversities of root nodule bacteria can result from a lengthy history of growing different legume species on the land, multiple introductions of commercial inoculant root nodule bacteria, and ongoing genetic diversification of bacteria in soil.

An analysis of the range of strains of root nodule bacteria that occur in Oregon pastures illustrates the level of genetic diversity that is possible. In this study, the bacteria were characterised using serological methods with four distinct serogroups identified and one most abundant in the surface 10 cm of the soil profile. Subsequent investigations using DNA analysis confirm the diversity of strains of rhizobia in soils.

When bacteria selected for a particular legume are introduced into a soil, they have to survive as part of the community of soil bacteria. Otherwise, the only way they will be maintained in the soil is by regular re-introductions. The newly introduced bacteria also need to compete with existing strains of similar bacteria for nodulation sites on the legume root. For example, the strain of bacterium that was one of the original inoculants for soybean in the USA was subsequently found to be less effective at fixing nitrogen than some strains introduced later. The original bacterium was very competitive at forming nodules on soybean in comparison to the more recently identified effective strains.

It is difficult for a newer bacterium to out compete older bacteria that have already colonised the soil and are highly competitive at forming nodules. This is clearly a major problem for the introduction of strains that are more effective at fixing nitrogen. A similar obstacle occurs when two clover species are grown in the same pasture. Nitrogen fixation occurs most effectively for each with different strains of *Rhizobium leguminosarum* biovar *trifolii*. However, it is not possible to ensure that each clover species is nodulated by the most appropriate strain of bacterium. The result could be that neither clover species fixes nitrogen according to its potential.

Bacteria that form legume symbioses can survive in a soil in the absence of suitable plants. Once the roots of a suitable plant reappear, the bacteria can re-initiate the symbiotic association. Some bacteria can survive without a plant for longer than others can. Even if nodule formation does not occur as soon as the plant reappears, the plant can stimulate the growth of the bacteria and make it more likely that nodulation takes place when other soil conditions are suitable.

Which Non-Legume Plant Species form Nitrogen-Fixing Associations?

Frankia is a genus of bacteria (actinomycetes) that has the capacity to fix nitrogen gas in association with several tree species. Examples of such trees include the genera *Casuarina* and *Allocasuarina* in the family Casuarinaceae and *Alnus* in the family Betulaceae.

The associations formed by species of *Frankia* are generally host-plant specific, but the same bacteria can nodulate closely related plant species. *Frankia* associations with plants range from those that are highly effective to those that are relatively ineffective at fixing nitrogen.

Nitrogen deficiency in *Casuarina* can be remedied by inoculation with strains of*Frankia* that are effective at fixing nitrogen. Two of the five sources of *Frankia* studied were effective at increasing growth of C. *equisitifolia*. Another was highly effective with C. *cunninghamiana*.

The association between *Frankia* and roots of their host plant is highly specific. The infection process is initiated through molecular signalling between the bacteria and root as for legume symbioses. When there is a compatible combination of bacteria and plant, molecular communication leads to the formation of a nodule on the plant root that develops a colony of bacteria within it. The bacteria stimulate the plant to form nodule tissue. Nodules can live for a long time, with a new layer of actively growing root cells and bacteria added each year. The bacteria fix nitrogen gas that diffuses through the walls of the nodule cells; the fixed nitrogen is transferred into the root cells and from there into the rest of the plant through the vascular system.

A variety of other associations also occur between nitrogen-fixing organisms and plants. For example, *Macrozamia riedlei*, a member of the family Cycadaceae, forms an association with nitrogen fixing cyanobacteria. This plant forms highly branched roots that contain the cyanobacteria.

SOIL-BORNE PLANT PATHOGENS

Soils contain diverse communities of microscopic organisms that are capable of damaging plants. A detrimental interaction between a soil organism and a plant is often highly specific. For example, a fungus that causes root-rot of wheat may have no effect on the roots of another plant growing in the same soil. Highly specialised interactions between soil organisms and plants can kill seedlings and even adult trees. Many organisms target younger plants but others appear as problems at later stages in the life of the plant. Other pathogens are able to cause disease in many different plant species.

The soil organisms that have the potential to be plant pathogens include fungi, bacteria, viruses, nematodes and protozoa. Some pathogens of the above ground parts of plants (leaves, stems) survive in the soil at various stages in their life cycles. Therefore, a soil phase of a plant pathogen may be important, even if the organism does not infect roots.

In spite of the potential for severe damage to be inflicted on plants by soil organisms, most plants do not display serious symptoms of disease. Disease usually occurs when conditions are particularly unfavourable, or when a soil organism is accidentally introduced where a highly susceptible plant species occurs. Intensive production in agriculture, horticulture or forestry increases the opportunities for diseases to develop compared with undisturbed natural ecosystems. Planting of similar plant species together in monoculture increases the probability of disease outbreak. In contrast, the damage caused by the fungus *Phytophthora cinnamomi* in many different plant species in diverse natural ecosystems in Australia demonstrates the damage that can be caused by a pathogen that infects the roots of many unrelated plants.

Some plant pathogens depend on their host plant for survival and are unable to complete their life cycle without infecting their host plant. Biotrophic organisms of this type are often difficult to grow in laboratory media.

Disease-causing microorganisms and soil animals are a natural component of the soil community. The organisms are normally present in relatively low numbers. An outbreak of disease commonly follows either an increase in the abundance of the pathogen or a change in the susceptibility of the host to the pathogen. Important questions are:

- Why do outbreaks of disease occur?
- What are the soil conditions that allow disease to develop or that suppress disease?
- What are the soil conditions that suppress disease?
- How do management practices influence the soil as a habitat for plant pathogens?
- How do plant pathogens survive in the absence of their specific host plant?

What is Plant Disease?

Plant disease is the result of a complex interaction between the host plant, the pathogen and the environmental conditions including those in the soil. Soil conditions that influence the development of disease are pH, moisture, oxygen, and nutrients. In addition, other soil organisms interact with the pathogen. The level of this interaction is also influenced by soil conditions, which affects the growth and activity of the non-pathogenic organisms. Consequently, whenever the soil environment changes, there is potential for either a positive or negative influence on soil-borne plant pathogens.

Plant disease reduces plant survival and productivity, but the symptoms of plant disease are often difficult to recognise. One reason for this is that they are not necessarily the same in plants of different ages. The amount of a potential pathogen in a soil and the environmental conditions can also alter the symptoms displayed by a plant. Furthermore, symptoms may be related to infection by more than one organism. Interactions between organisms can produce symptoms that are not the same as those that occur when each type of organism is present alone.

Similar symptoms can be caused by different pathogens and furthermore, the symptoms of some pathogens resemble those of various nutrient deficiencies. Therefore, it is not always easy to identify a pathogen responsible for a disease simply from the symptoms on the plant. A great deal of experience and knowledge of the range of symptoms produced by different pathogens and knowledge of symptoms of nutrient deficiency is essential for an accurate identification of most pathogens.

Examples of Major Soil-Borne Plant Diseases:

(i) Disease caused by Phytophthora cinnamomi: Phytophthora cinnamomi causes a serious disease that threatens forests and other ecosystems especially in

south-eastern and south-western Australia. Hundreds of different plant species are killed by this introduced pathogen. Vehicles are often responsible for the widespread distribution of the pathogen by disturbing and transporting infected soil. There is no simple solution to *Phytophthora*disease in the forest. Quarantine methods have been introduced to limit the spread of the fungus and cleaning of vehicles is mandatory.

P. cinnamomi and related fungi also cause disease of horticultural plants such as azalea, pineapple and avocado. In horticultural systems, organic mulches have been used to stimulate the community of soil organisms and reduce the negative impact of the pathogens.

(ii) *Take-all Disease of Wheat:* Take-all disease is caused by the fungus *Gaeumannomyces graminis* var. *tritici*. This pathogen infects the vascular tissue of wheat roots and restricts the transport of water and nutrients within the plant. Severely infected plants have stunted root systems. In addition to root rot, a severe symptom is 'white heads' which occurs if plants survive seedling damage and grow to maturity. Such plants form seed heads with poor grain development that are characteristically white. The fungus survives in the soil on decaying plant material and relies on this material as a carbon source to sustain it until it is able to infect new roots in the following wheat crop or alternate hosts.

The take-all fungus is a major root disease world-wide and is estimated to cause millions of dollars in lost wheat yield each year. The main method for control is by crop rotation with non-host plants and removal of weeds that may act as alternate hosts. There has been little success in breeding for resistance in wheat to the fungus *Gaeumannomyces graminis* var. *tritici*. Under some conditions, soils become naturally suppressive to the fungus. This has been observed in long-term monocultures of wheat. Reduced disease is likely to be due to changes in activity of other soil microorganisms.

(iii) *Crown Gall:* Crown gall occurs on many genera of plants and is characterised by the formation of root tumours caused by the bacterium *Agrobacterium tumefaciens*. The bacteria infect the root and induce plant cells to divide; a tumour-like swelling is formed that contains infected cells around its outer surface. Bacterial DNA is transferred to the host plant.

(iv) *Root Knot Nematode Disease:* Root knot nematodes cause disease on hundreds of plant species, especially horticultural species, in warmer climatic zones. Species of nematodes in the genus *Meloidogone* induce the formation of numerous galls throughout the root system. The damaged roots also have malfunctioning root tips which reduce root growth, resulting in considerable yield losses.

How is the Source of Plant Disease Identified?

The fundamental approach to identifying the organism or organisms responsible for a disease is to follow the protocol known as Koch's Postulates :

Step 1. Visual inspection of the plant to identify signs of the presence of a pathogen and symptoms on the host (observed without the aid of a microscope.

Step 2. Observation of the diseased portion of the plant using a microscope.

Step 3. Isolation and purification of the pathogen (growth of the organism on artificial media).

Step 4. Inoculation of plants of the same type with organisms isolated from the diseased plant to observe symptoms of disease, for comparison with the original observations in Step 1.

The purpose of using Koch's Postulates is to identify the organism responsible for disease in a rigorous manner. This is the first step in working out how to overcome the problem. Not all organisms can be isolated from plants and grown in culture and for many diseases it is difficult to identify the responsible organism.

Once a potential pathogen has been identified, a variety of procedures can be used to determine the conditions under which the disease is likely to be most severe. An example of this approach was used to identify the causal organism of brown spot and root rot of lupin in Western Australia as described below.

Case Study: Pleiochaeta Root Rot of Lupin

The pathogen responsible for a serious root disease of lupin in Western Australia was not immediately identified when it was first detected. In fact, several species of fungi were commonly reported from diseased lupin roots, including species of *Fusarium*, *Pythium*, *Rhizoctonia* and *Pleiochaeta* further hindering the control of the disease.

The protocol used by Sweetingham to identify the causal pathogen was to

- *Observe and describe* the disease symptoms and occurrence in the field.
- *Identify* the fungi isolated from diseased roots. More than 10 species of fungi were isolated, but none were clearly identified as the possible pathogen.
- *Perform* pathogenicity tests to identify the degree of disease caused by each fungus when it was present by itself on lupin roots. Some fungi were more pathogenic than others, but none of the fungi, when present alone, was distinctly more pathogenic than any other.
- *Test* the effect of selected fungicides on the presence of different fungi and on the extent of disease using field soil. This technique was most successful in identifying the likely cause of the root rot.

The use of fungicides that targeted the major fungi present in soil where disease was observed demonstrated that *Pleiochaeta* was the most likely pathogen. The fungicides Ridomil, Benlate, Rovral and a mixture of all three were used. Ridomil eliminated *Pythium* without causing a major decline in the disease rating of roots. Benlate had little effect on *Pythium*, but reduced *Pleiochaeta* and *Fusarium* in the roots leading to a reduced disease rating. Rovral eliminated *Pleiochaeta* but left *Furarium* and *Pythium* and these roots had a much reduced disease rating. This set of fungicide treatments was successful in discriminating among the fungi that were present in the diseased roots.

This case study illustrates the complexity involved in identifying the causal agent of root disease, especially when more than one fungus may be involved. It cannot be presumed that the presence of an organism implicates it as a causal agent of disease. Non-pathogenic fungi often enter damaged roots and therefore a number of organisms may be present in diseased roots without being responsible for the disease.

How are Disease-causing Organisms Dispersed?

Disease occurs under the following conditions:

(i) if there are enough infective hyphae, spores or other fungal structures such as sclerotia of a pathogen present in the soil,

(ii) if there is a suitable environment for growth of the organisms, and

(iii) if the plant is susceptible to infection.

If the organism is present in low numbers, it may have little impact on the plant. If soil conditions are suitable for the multiplication of the pathogen, there will be a greater chance of disease if other factors are also appropriate for disease development. In general, the more organisms present, the more severe the disease , but plant and soil conditions are also important.

The hyphae of some fungal pathogens can grow long distances through the soil. For example, hyphae of the fungus *Armillaria* extend many metres horizontally in soil and simultaneously infect the roots of several plants. Some fungi, for example species of *Phytophthora,* form narrow and short hyphae that do not spread the fungus through the soil. Often these fungi are disseminated in water or in soil during erosion and by other mechanisms including transport of soil on vehicles or footwear.

Spores of fungi that are pathogens of leaves and stems may have a soil phase during which spores remain dormant. Once the dormancy period ceases, spores survive in the soil until conditions are appropriate for their germination and infection of a new plant. Growth of some pathogens is stimulated by the presence of roots which release exudates or emit other signals.

Spores of some soil-borne fungi are dispersed by wind. Dispersal of bacterial plant pathogens is common when soil is transported from one place to another. Some disease-causing organisms are dispersed on plant debris. *Pleiochaeta setosa* spores are dispersed in soil via dead organic material from the previous crop.

How Do Pathogens Enter Roots?

Pathogenic invasion of plant roots involves several steps. The first of these is recognition between the pathogen and the plant, which involves complex molecular mechanisms. Recognition occurs in the diffuse interface between the root and the soil created by root exudates and sloughed off root cells. This interface between soil and root may be a very suitable habitat for growth of the pathogen before it enters the root.

This initial interaction can also be the stimulus for the fungus to penetrate the plant cell wall. Some organisms produce enzymes that dissolve cell wall layers, allowing the hyphae to enter. Pathogens can also enter roots at points damaged by soil animals or other disturbances. A plant may react to the presence of a pathogen by producing chemical or physical barriers that prevent entry or confine potential pathogens to a small part of the root.

Once inside the root, pathogens occupy specific cell types. Some fungi enter the cells that conduct nutrients, carbon and water throughout the plant. If these cells are damaged, important processes that control the life of the plant are disrupted. Other fungi predominantly colonise the cortical root cells that are important in nutrient uptake from the soil.

How Do Roots React to Pathogens?

Plants differ in their susceptibility to plant pathogens. Even cultivars of one plant species can be more or less resistant to a pathogen. Similarly, species or strains of pathogens differ in their capacity to infect a plant. Differences in the degree of disease can therefore be related to:

(i) the susceptibility of the plant;
(ii) the virulence of the pathogen;
(iii) the stage of development of the plant;
(iv) the abundance of the pathogen; and
(v) the environmental conditions.

How Does Plant Nutrition Affect Disease?

The nutritional status of a plant influences many of its physiological processes and this can alter the response that plants have to the presence of pathogens. Changes in the concentration of nutrients such as phosphorus in a plant can alter root and shoot growth as well as the movement of carbohydrate and other molecules within the plant. These changes have the potential to influence the development of a disease.

Fertilisers can also alter plant susceptibility to disease. In this study, in the absence of manganese fertiliser, barley grew less well if nematodes were present in the soil than when manganese was added at sufficient levels to meet the requirements for growth of the plant This occurred even when the numbers of nematodes in both treatments were similar.

What is 'Biological Control' of Plant Disease?

The control of pathogens and prevention of plant disease is a natural soil biological process. Indeed, in most situations, plant disease is not strongly evident even when potentially pathogenic fungi are present in a soil. This suppressive characteristic of soil is common in natural ecosystems, but becomes less apparent with high levels of disturbance and the introduction of plant monocultures. Disturbance creates soil conditions and a high density of susceptible roots that encourages the multiplication of pathogens. Once potentially damaging organisms become present in high numbers in a soil, they may be difficult to eradicate. Management practices are required that create conditions in the soil that are not favourable to pathogens so that their growth is limited and therefore, disease it restricted.

There are many demonstrations of the potential for soil organisms to reduce disease. As demonstrated in this study, when apple trees were replanted in a soil previously prone to apple replant disease, increased tree growth occurred when various strains of bacteria were added to the soil, indicating a beneficial effect. Similar effects were observed at two sites, but because different quantities of the bacteria were added in the inoculum it is difficult to compare the effectiveness of each organism.

In some environments, take-all disease of wheat has declined naturally. These areas of natural suppression of disease have been investigated extensively to determine how the pathogen is suppressed. Specific bacteria that are capable of limiting the growth of the fungus, including *Pseudomonas* species, have been isolated from disease suppressive soils. But it is still unclear whether these bacteria are responsible for the decline in take-all disease.

Handbook of

Control of Fungal Plant Pathogens in Agricultural Biotechnology

Handbook of

Control of Fungal Plant Pathogens in Agricultural Biotechnology

Contributors :
Hye Lin Park,
Youngchul Yoo, *et al.*

Volume 2

KOROS PRESS

KOROS PRESS LIMITED
London, UK

Handbook of Control of Fungal Plant Pathogens in Agricultural Biotechnology
Contributors : Hye Lin Park *and* Youngchul Yoo, *et al.*

Published by Koros Press Limited

www.korospress.com

United Kingdom

Printed in 2017 for Sale in the Indian Subcontinent

Handbook of Control of Fungal Plant Pathogens in Agricultural Biotechnology

ISBN: 978-1-78569-231-4

British Library Cataloguing in Publication Data
A CIP record for this book is available from the British Library

Exclusively distributed by CBS Publishers & Distributors Pvt. Ltd.

Sales & Distribution Rights only for India, Pakistan, Bangladesh, Sri Lanka, Nepal and Bhutan.This book is not to be sold outside these territories.

PREFACE

Fungi are a kingdom of Eukaryotic Organisms. The fungi are heterotrophic organisms characterised by a chintinous cell wall, and in majority of species, filamentous growth as multicellular hyphae forming a mycelium; while some fungal species grow as single cell. Sexual and asexual reproduction is via pores, often produced on specialised structures or in fruiting bodies. For example, yeasts, moulds, and mushrooms are fungi. Fungi are found all over the world, and grow in a wide range of habitats, including desert. Most fungi grow in terrestrial environments, but several species occur only in aquatic habitats. Fungi along with bacteria are the primary decomposers of organic matter in almost all terrestrial ecosystems, worldwide.

Aquatic fungi contribute to the energy flow and productivity of an aquatic and semi aquatic ecosystems by their active role in the utilization and biodeterioration of organic materials. These fungi also possess the ability to parasitize aquatic plants and animals including fishes under certain conditions. Physiology, environmental needs and activity of these fungi vary with their morphological characters. The taxonomy of these fungi can not remain static. The morphological variability of these species ultimately be linked to the genetic control of their characters. A considerable amount of taxonomic studies has been carried out by a number of mycologists all over the world.

This book blends information on classical fundamental aspects with recent development in fungal systematics. The textbook of fungi presents information on the morphology, life cycle and their economic uses in human life. Special attempt has been made on the biological activities of the microbial products. They produce several types of drugs including antibiotics, drugs that reduce high blood pressure. They also produce many of the enzymes used in the food, chemicals and pharmaceutical industries.

This page left intentionally blank.

CONTENTS

This page left intentionally blank.

List of Contributors

Hye Lin Park

Graduate School of Biotechnology, Kyung Hee University, Yongin 446-70, Korea;

E-Mails: hlpark@khu.ac.kr (H.L.P.)

Youngchul Yoo

Department of Plant Molecular Systems Biotechnology & Crop Biotech Institute, Kyung Hee University, Yongin 446-70, Korea; E-Mail: yooyc@khu.ac.kr

Tae-Ryong Hahn

Graduate School of Biotechnology, Kyung Hee University, Yongin 446-70, Korea;

E-Mails: trhahn@khu.ac.kr (T.-R.H.)

Seong Hee Bhoo

Graduate School of Biotechnology, Kyung Hee University, Yongin 446-70, Korea;

E-Mails: shbhoo@khu.ac.kr (S.H.B.)

Sang-Won Lee

Department of Plant Molecular Systems Biotechnology & Crop Biotech Institute, Kyung Hee University, Yongin 446-70, Korea

Man-Ho Cho

Graduate School of Biotechnology, Kyung Hee University, Yongin 446-70, Korea

This page left intentionally blank.

Chapter 8

BIOFUNGICIDES

WHAT ARE BIOFUNGICIDES?

Biofungicdes are formulations of living organisms that are used to control the activity of plant pathogenic fungi and bacteria. The concept of biofungicides is based upon observations of natural processes where beneficial microorganisms, usually isolated from soil, hinder the activity of plant pathogens. Biocontrol microorganisms are free-living fungi, bacteria, or actinomycetes that are active in root, soil, and foliar environments. These microorganisms produce a wide range of antibiotic substances, parasitize other fungi, compete with other fungi, and induce localized or systemic resistance to a variety of plant pathogens. The use of composts and suppressive growing medium, which both contain living microorganisms, to ameliorate disease is another example of this disease management option.

How Biofungicides Work:

- **Rhizosphere Competence -** The most successful of the strains of biocontrol microorganisms exhibit rhizosphere competence, the ability to colonize and grow in association with plant roots. They can colonize entire root surfaces for several months. Here they effectively compete with plant pathogens for nutrients, infection sites, and space. Competition for glucose in the soil is involved in disease suppression. Biofungicide species also metabolize seed and root exudates that normally stimulate pathogen germination or zoospore attraction.
- **Parastitism -** Parasitism, the ability of species to attack and consume plant pathogens, has been well studied. Mycoparasitism of biocontrol microorganisms includes directed growth, contact and binding, coiling of hyphae around the host fungus, penetration and degradation. Production of cell wall degrading enzymes is almost always part of the process.

- **Antibiosis -** Antibiosis occurs when one microorganism produces molecules that directly affect other organisms negatively by toxicity or growth inhibition. These compounds are called antibiotics and are commonly produced by a wide range of soil dwelling microorganisms in the course of their growth. A familiar antibiotic, streptomycin, is produced by *Streptomyces* species which are classified as actinomycetes or filamentous bacteria.
- **Inducing Metabolic Changes -** An important mechanism of biocontrol microorganisms is the ability to induce metabolic changes in plants that increase their resistance to a wide range of plant pathogenic microorganisms (fungi and bacteria). Systemic Acquired Resistance (SAR) improves the plant response to pathogen attack by priming the metabolism of plant defense compounds. This capacity to induce resistance to a wide range of diseases in a variety of plants appears to be widespread. Enhanced resistance is systemic because disease resistance occurs at sites distance from the location of the biocontrol microorganisms.
- **Plant Growth Promotion -** A final way in which these organisms act is through plant growth promotion. Beneficial root-colonizing microorganisms promote plant growth and productivity. Many resistance-inducing fungi and bacteria promote both root and shoot growth in the absence of disease causing fungi. In the greenhouse industry, biofungicides are applied preventively to growth media or as a seed treatment for disease control and can be as effective as chemical fungicides. Biofungicides are safer for growers, more persistent, and less expensive. When applied as seed treatment, biofungicides increase root development in a number of plants and improve drought resistance. Improvements in plant growth result from effects on soil microflora and direct effects on the plant. Biofungicides can also improve nutrient uptake (copper, phosphorous, iron, and manganese).

Biofungicides Used in Greenhouse Production

Plant Shield, Root Shield, PreStop, SoilGard, and T-22 PlanterBox are commercial formulations of *Trichoderma* species and the closely related *Gliocladium* species. Species of *Trichoderma* are the best studied of all biocontrol organisms. These species are also abundant in many types of compost and are the most frequently isolated soil fungi from all temperate and tropical soils. The production of antibiotics by these species is well documented. In fact, *Trichoderma* species have been found to be highly resistant to a variety of toxins and antibiotics produced by soil microorganisms, antimicrobial compounds produced by plants, and even chemical fungicides. This makes these fungi active colonizers of toxic environments and a strong competitor.

Companion and Cease are bacteria in the species *Bacillus*. Many biofungicides that consist of bacteria are labeled mainly for soilborne diseases, though there are some that are also labeled for foliar diseases like leaf spots, powdery mildew, and downy mildew. See table. Other bacteria are termed Plant Growth Promoting Rhizobacteria (PGPR) and while not strictly biofungicides, their growth promoting effects can contribute to disease tolerance. Actinovate, Actino-Iron, and MycoStop are composed of species of *Streptomyces*, an actinomycetes or filamentous bacteria. They are labeled for both root and foliar diseases. *Agrobacterium radiobacter* 84 (Galltrol A) is registered for control of Crown Gall (*Agrobacterium tumefaciens*) and there are also a few *Pseudomonas* species present in some biofungicides.

BIO-FUNGICIDES

The biological control process sparks many questions. First, what is biological control of plant disease? It is the involvement of the use of beneficial microorganisms, such as specialized fungi and bacteria, to attack and control plant pathogens and the diseases they cause. So what are these "specialized fungi and bacteria" that can attack and control plant pathogens? They are microorganisms that are part of the normal microbiological environment of most "healthy" soils.

They are not genetically engineered. In their native habitat, these beneficial microorganisms compete with other microorganisms for space and food. In some cases, they are parasitic on other microorganisms and/or they produce toxic substances that kill other soil-inhabiting microorganisms such as Pythium sp., Phytophthora sp., Rhizoctonia sp. and other plant pathogens. Scientists are well aware of these beneficial microorganisms and have studied them for many years. They have shown that these microorganisms play a vital role in the makeup of the soil environment and are part of the normal checks and balances that make up a "healthy" soil.

Many beneficial fungi and bacteria have been isolated from the soil and tested in private and university-based laboratories as to their ability to control plant pathogens. Recently, some of the more promising of these beneficial fungi and bacteria have been further developed and marketed to ornamental plant growers as an alternative to traditional chemical-based fungicides. Below is a better understanding of the current crop of beneficial organisms (biofungicides) -- how they work and, most importantly, their advantages and disadvantages when compared to traditional chemical fungicides.

How They Work

There are four different mechanisms by which beneficial or biocontrol agents interact with other microorganisms. Most biocontrol agents apply only one of these four mechanisms; however, some may employ more than one.

Direct Competition

In this case, the biocontrol agent out-competes the target organisms for nutrients and space. This is typically a fungus or bacteria that grows very fast and overwhelms the target organism with sheer numbers. The target organism is suppressed due to lack of food and space. The target organism may not die out completely, but its population becomes so low it is no longer a legitimate threat to the host plant. In order for this type of biocontrol agent to be most effective, the environmental conditions must favor the growth and reproduction of thc biocontrol agent.

Antibiosis

With antibiosis, the biocontrol agent produces a chemical compound such as an antibiotic or some type of toxin that kills or has some sort of detrimental effect on the target organism. Many microorganisms produce antibiotics and toxins. Some of the more common antibiotics humans use to warrant-off infections came originally from common soil-inhabiting fungi and bacteria. In some cases, antibiosis can be accompanied by other detrimental mechanisms. Antibiosis is one of the most effective methods of controlling microorganisms.

Predation or Parasitism

This is the mechanism that most of us envision when we think of biocontrol agents. In this case, the biocontrol agent attacks and feeds directly on the target organism, or the agent produces some sort of toxin that kills the target organism and then feeds on the dead target. Like direct competition, the environment must favor growth and development of the predator or parasite since populations need to be high enough to overwhelm the target organism.

Induced Resistance of the Host Plant

Scientists have known for decades that once a plant is infected with a pathogenic microorganism, infection triggers some sort of biochemical reaction in the infected host plant that helps keep it from being infected with further pathogens (super infection). The infected plant becomes more "resistant" to other infections. Plants do not have immune systems to protect them from infection as we do; however, they do have physiological and biochemical systems that help inhibit infection and spread of pathogens within tissues of the affected plant. Some biocontrol agents are known to trigger these mechanisms, and in the case of induced resistance, host plants are purposely inoculated with this agent in an effort to trigger the resistant response. The microorganism that triggers the response is usually not a severe pathogen of the host. If it were, it would defeat the whole purpose. Induced resistance is not highly understood and is currently a very exciting area of research throughout the scientific community.

Advantages and Disadvantages

Even though it appears as if these biocontrol agents are the cure-all, there are distinct advantages and disadvantages to using them, when compared to traditional chemical controls.

Advantages.

1. If used properly, they help reduce the use of chemical-based fungicides. This is good for the environment and is one of the most important reasons to consider their use.
2. They help reduce the risk of developing pathogen resistance to traditional chemicals. Due to the overuse of certain chemical fungicides, some common plant pathogens such as Pythium sp. and Botrytis sp. have become resistant to these fungicides. This is less likely to happen with biocontrol agents because the beneficial organism co-evolves along with the target organism and adapts to the changes. Something a chemical cannot do.
3. In most cases, they are safer to use. Most biocontrol agents have very low or no toxicity to humans and other mammals. This is a tremendous benefit in this day and age.
4. They tend to be more stable than chemical pesticides if stored properly. These are living organisms and must be stored as such. If they spoil, they are no longer affective.
5. In most cases, they have lower re-entry interval (R.E.I.) times. This is a significant factor especially when it is necessary to enter the production facility immediately following application.
6. In most cases, they are less phytotoxic. Because they are "natural" they are less likely to cause toxic effects on the host plant, especially if mistakes are made and rates are miscalculated.

Disadvantages.

1. Biocontrol agents tend to be more difficult to implement when compared to chemicals. Since most of these products have to be implemented prior to the onset of disease, greater preparation by the user is necessary. Biologicals work best in greenhouses that routinely scout for diseases and insects and detect problems early.
2. In most cases, they have a narrower target range. Most are not broad-spectrum products. Identification of the correct target organism is imperative.
3. They may not work as quickly as chemicals. Since their populations need to take time to build up they can take more time to be effective. That is why it is necessary to apply them prior to the onset of severe disease outbreak.

4. These products do not eradicate the pathogen or rescue the host from infection. They have to be administered prior to the onset of disease, in most cases at preplant.
5. They may have a shorter shelf life if not stored properly. Remember, these are living organisms that don't take well to extreme temperatures.
6. In most cases, biocontrol products are more expensive to use. This includes both time and money. They may be a bit more expensive to purchase initially, and they take more time to initiate, if used properly.
7. They may not be compatible with the use of other chemical fungicides and bactericides. The product label should be checked to see with what chemicals the product is compatible. Many of these beneficials are fungi, and some of the more common greenhouse fungicides have the potential to kill these beneficial microorganisms.

The Products

Currently there are close to 40 commercial products that are marketed as biological controls worldwide. Not all of these are available in the United States. For greenhouse floriculture and perennial production, there are about a half dozen products that are currently popular. Of these, PlantShield appears to be the most widely used. Plantshield is the T-22 strain of the soil inhabiting fungus Trichoderma harzianum (TH). TH's mode of action against the target organism is multifaceted. It uses both antibiosis and predation against many common soil-inhabiting fungi that cause root and crown rots such as Pythium, Rhizoctonia, Fusarium and Sclerotinia. It appears to be one of the most popular biofungicides in the greenhouse industry and can be an asset to a disease management program if used properly.

Keys to Success

In order for any of these biological control agents to work for you, two simple rules must be followed. First, all of these products must be used in conjunction with standard disease cultural controls. Cultural controls include: growing plants in a well-drained media; not over watering; keeping the greenhouse relative humidity below 85 percent; practicing strict sanitation; and making sure that the nutrient and pH conditions of the host plant are within the ideal range for proper growth and development. This will help assure that the environment is favorable for the growth and development of the beneficial organism.

Second, all of these biocontrol products must be applied at preplant or prior to the onset of disease. In most cases, they will not rescue plants that are already infected. If you abide by these two critical conditions, the likelihood of you having success with a biocontrol agent is good. If you don't, they won't work. Manufactures who have traditionally been the source of chemical fungicides will be producing and marketing biofungicides. Growers need to be aware of what products are available, the way they work and their limitations. It will be a while before we see a biofungicide that controls Pythium sp. as good as Subdue.

BIO-FUNGICIDES: ORNAMENTAL PLANTS

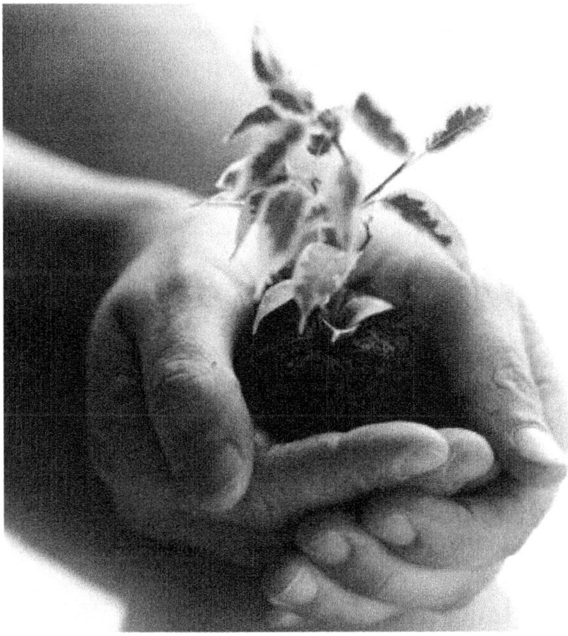

Bio-fungicides are an alternative to chemical fungicides to keep plants healthy.

Bio-fungicides are soil fungi or bacteria that prevent or reduce plant diseases. These fungi and bacteria normally inhabit healthy soil and play a vital role in the natural soil environment. After isolation of some of these beneficial fungi and bacteria, the microorganisms were developed for use by vegetable and ornamental plant growers. Bio-fungicides are an alternative to chemical fungicides. To qualify as bio-fungicides, just as with chemical fungicides, studies need to show safety and effectiveness, and bio-fungicides need to be registered with the U.S. Environmental Protection Agency. Beneficial fungi or bacteria use several mechanisms to prevent plant diseases and increase growth, often using more than one mechanism. Consider the advantages and disadvantages of bio-fungicides before using them.

Competition

One way beneficial fungi fight plant diseases is by out-competing the pathogenic or disease-causing organisms for nutrients and living space. The beneficial microorganism must grow quickly to out-compete and reduce the numbers of the pathogenic organism. The beneficial fungi may take up space on the roots or leaves of the ornamental plant, preventing pathogenic organisms from attaching. Bio-fungicides containing non-pathogenic Fusarium species or Trichoderma harzianum -- T-22 strain -- work by competition. For best results, apply competitive bio-fungicides liberally.

Antibiosis

In antibiosis, the beneficial microorganism uses antibodies or toxins to kill or injure a pathogenic organism. Such beneficial fungi or bacteria do not need to be present in high numbers for the bio-fungicide treatment to work. Bio-fungicides that work by antibiosis are highly effective in protecting ornamental plants. An example of a beneficial fungi that works by antibiosis is Gliocladium virens -- GL-2 strain. Like all bio-fungicides, antibiosis bio-fungicides work best if used before plant disease starts.

Parasitism

During parasitism or predation, the beneficial microorganism directly attacks or feeds on the disease-causing organism. This is a relatively rare mechanism used by bio-fungicides. For best results, the beneficial organism has to be present in the roots or leaves of the ornamental plant before the disease-causing organism infects the plant. Bio-fungicides containing Trichoderma harzianum -- T-22 strain -- and Amelomyces quisqualis -- AQ-10 isolate -- use this mechanism.

Induced Plant Resistance

Research shows that some plants, when triggered by a beneficial microorganism, produce salicylic acid -- a derivative of aspirin -- which diffuses to different parts of the plant and causes the plant to turn on its natural defenses, notes University of Connecticut Integrated Pest Management. Studies show Trichoderma and Bacillus species induce plant resistance, but research as to the exact mechanism is still ongoing.

Nutrient Availability

The microorganisms in bio-fungicides can contribute to the health and growth of ornamental plants because the microbes help the plants to take up more nutrients from the soil. Ornamental plants treated with some bio-fungicides grow larger than untreated plants, even when no disease-causing organism is around. Trichoderma harzianum -- T-22 strain -- and Gliocladium virens -- GL-2 strain -- are examples of fungi in bio-fungicides that enhance nutrient uptake.

Advantages and Disadvantages

Bio-fungicides are good tools for organic gardeners and growers because they reduce the need for chemical fungicides. For the most part, bio-fungicides are safe to use and have fewer toxic effects on ornamental plants than chemical fungicides. Users of bio-fungicides also don't have to worry about pests developing resistance to chemical fungicides. One disadvantage of bio-fungicides is they have to be used before ornamentals become diseased. It also takes longer for bio-fungicides to become effective. Bio-fungicides are usually more expensive than chemical fungicides.

BIOLOGICAL FUNGICIDES: DO THEY WORK AND ARE THEY SAFE?

The term biofungicide can have several different meanings, but it is most frequently used to refer to fungicides that contain a microorganism (usually a bacterium or fungus) as the active ingredient.

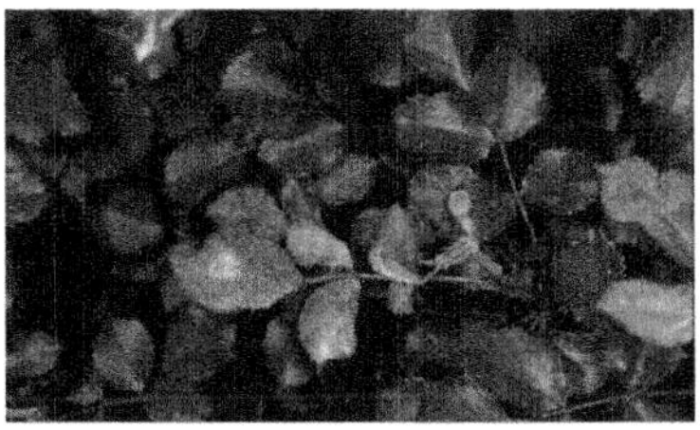

Fig. : Powdery mildew in roses can be managed with foliar applications of the biofungicide Bacillus subtilis.

These microbially-based biofungicides are the focus of this chapter. Biofungicides can control many different kinds of fungi and water molds, although each separate active ingredient controls only certain pathogens. Some also control bacterial diseases. Virtually all of the organisms used in biofungicides on the market today occur naturally in soil or on plant surfaces, and most are approved for use in organic production.

Advances in fermentation technology have allowed mass production of highly specialized microbes that previously could only be grown in small batches on highly specific substrates, such as on roots infected with pathogens. Consumer demand for organically certifiable pesticides and increased regulatory pressure on older synthetic pesticides, especially in Europe, has

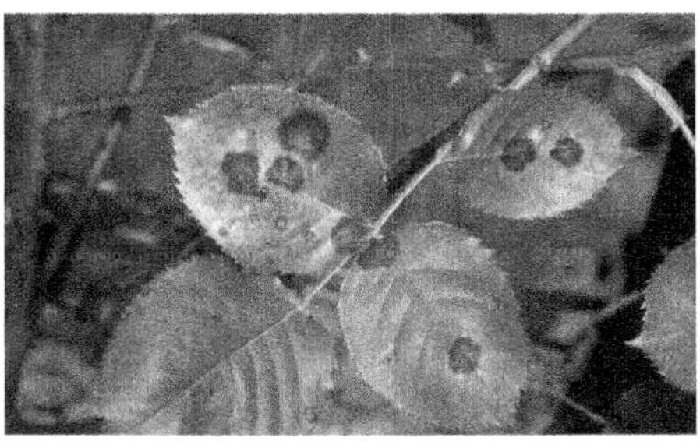

Fig. : Black spot on rose can also be managed with foliar applications of Bacillus subtilis.

fostered increased commercial interest in the production of living organisms that can suppress or kill pathogens. For these reasons and also because as natural products biofungicides generally have few negative impacts on health and the environment, the number available will likely continue to increase.

Since microbial biofungicides contain living organisms, their modes of action differ from those of synthetic fungicides. Some of these mechanisms include:

- *Competition*: The biocontrol agent is more effective than the pathogen at gathering critical nutrients or space and, therefore, must be in place before disease onset.
- *Antibiosis*: The biocontrol agent produces a chemical compound of some type (antibiotic or toxin) that acts against the pathogen.
- *Predation or parasitism*: The biocontrol agent directly attacks the pathogen.
- *Induction of host plant resistance*: The biocontrol agent triggers a defensive response in the host plant that limits the ability of the pathogen to invade the plant.

Most biofungicides use one or more of the above mechanisms to target only one or a few specific pests. As such, applicators should both read the label and diagnose the problem carefully to insure that the product will be effective. Biofungicides work best when applied preventively. Application after a plant is already infected has little chance of significantly altering the course of the disease for that plant, although it may decrease the ability of the pathogen to move from that plant to other plants, especially if the pathogen has to move through the soil to do so. Thus, an application of biofungicide is not likely to cure an infected plant; but it may protect other nearby plants in the field.

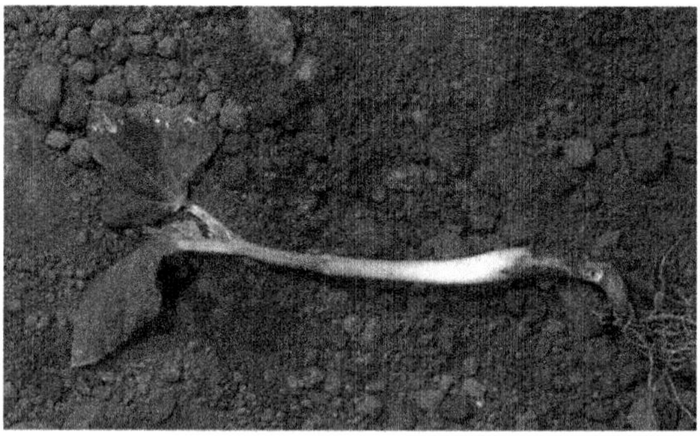

Fig. : Biofungicides containing Streptomyces or Trichoderma can be used to prevent infection of plants with damping off or root or seed rot pathogens such as Pythium.

Although independent testing by university researchers and others has verified some manufacturer claims made for these products, efficacy data for many other products against pathogens on ornamentals is unavailable. For instance, as of early 2014, independent testing has not demonstrated adequate field efficacy of any biofungicide for landscape or agricultural use against Armillaria root rot (otherwise known as oak root rot). However, a number of biologicals have been found effective for control of *Pythium, Phytophthora, Verticillium,* and other pathogens on a variety of plant hosts. With these types of products, eradication of the pathogen is not the goal and is probably never achieved. Instead, biofungicides rely on a core tenet of the IPM philosophy: keeping pest levels below damaging thresholds and using biofungicides (when necessary) in combination with cultural practices that promote healthy plant growth.

Biofungicides cannot take the place of proper cultural care. They are a valuable tool for keeping a strong plant healthy, but they cannot forestall the inevitable. If your client's Japanese maples are routinely drowned, allowed to wilt, and then drowned again, adding a biofungicide will not prevent them from contracting *Phytophthora* if it is present in the soil.

If biofungicides are a useful and environmentally friendly tool in the landscape, why aren't they more widely used? One reason is that these fungicides rely on living organisms for efficacy, so they must be stored appropriately in order to retain their fungicidal properties over time. However, a more commonly cited reason is that the personal protective equipment needed to apply them is more involved than for some other compounds. Routine exposure to the proteins found in the spray mists of some biofungicides can result in the development of allergic reactions. To keep commercial applicators safe, they must wear NIOSH approved respirators when mixing, loading, or applying biofungicides in agricultural or landscape settings. This may not be immediately obvious when reading the labels, as a quick scan often only shows the following required personal protective equipment (PPE):

- Long sleeved shirt and long pants
- Shoes plus socks
- Waterproof gloves

The respirator requirement is only evident when reading the text following the list. Biofungicides are safe to use as long as mixer/loaders and applicators have and use a respirator as part of their PPE. However, a NIOSH approved respirator requires proper training and fitting in order to be effective.

Table: Sample Biofungicides available for use in California landscapes.

Active Ingredient	Sample Trade Name	Sample Diseases on Label	Types of Plants on Labels	Application Notes
Bacillus subtilis	Serenade, Rhapsody	Powdery mildew, bacterial spot and many foliar fungal and bacterial pathogens	Annual, perennial and woody plants	Foliar application
Streptomyces lydicus	Actinovate Actono-Iron	Turf diseases including Pythium, Rhizoctonia, Fusarium, powdery mildew, grey mold	Turf primarily, although used on vegetables in commercial agriculture.	Soil drench or foliar treatment
Trichoderma harzianum	Rootshield Home and Garden Fungicide	Many soil pathogens including Pythium, Rhizoctonia and Fusarium	Many woody or herbaceous plants	Application on seeds or roots prior to planting. Mostly used in nurseries.
Pseudomonas fluorescens A506	BlightBan A506	Fireblight	Apple, pear	Foliar. Used primarily in commercial agriculture. A bactericide.

When used properly and with forethought, biofungicides can be an important part of an IPM program to prevent or mitigate problems with plant pathogens in the landscape. However, nothing can ultimately take the place of proper plant selection and care.

EVALUATION OF BIOFUNGICIDES FOR CONTROL OF SCLEROTINIA DISEASE IN CARROTS

Sclerotinia rot of carrots is a major problem for Manitoba's carrot industry. Prior to the 2010 crop year, no chemical fungicides were registered for use in-field in Canada. Two new biofungicides, Contans™ and Serenade ASO™ have recently been registered for use against Sclerotinia in carrots. Contans™ is a naturally occurring strain of*Coniothyrium* fungi (*Coniothyrium minitans* strain CON/M/91-08) and is registered for soil application. This fungus parasitizes the dormant form of sclerotinia, the sclerotia and reduces the amount of sclerotia germinating and producing mycelia infection or ascospores. Serenade ASO™ (a bacterium *Bacillus subtilis* QST 713 strain) is registered for foliar application on carrots to control Sclerotinia. In order to evaluate the efficacy of these products in commercial scale carrot fields in Manitoba, Manitoba Agriculture, Food & Rural Initiatives (MAFRI) coordinated a project with a carrot producer and United Agri Products (UAP).

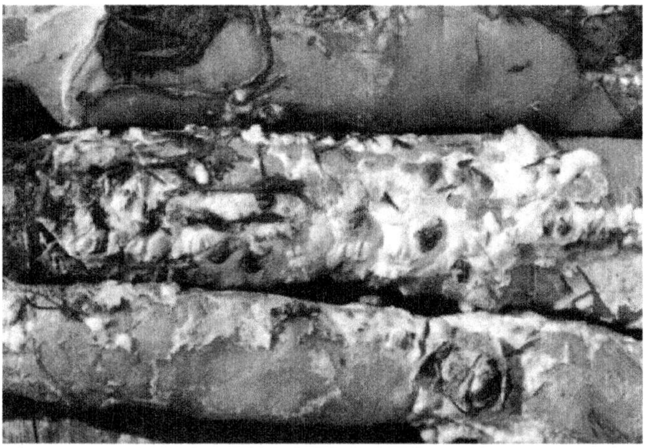

Fig. : Sclerotinia on Carrots in Storage.

Description of Field Trials

Two fields with a history of sclerotinia disease were selected for evaluation. One field was planted to carrots in 2010 and the second field is planted to wheat in 2010 and will be planted to carrots in 2011.

The field planted to carrots in 2010 had four treatments applied in 3 replicates on the field. Soil samples were taken prior to the application of the treatments to quantify the background levels for sclerotinia in each field. The treatments applied to the carrot field in 2010 are;

- Contans WG™
- Serenade ASO™
- Contans WG™ with Serenade ASO™
- Untreated control (no spray)

The field planted to wheat field this year had three treatments applied to it in 2010. The treatments applied in 2010 were;

- Untreated control
- Contans WG™ (0.4 kg/ha)
- Contans WG™ (0.8 kg/ha)
- There will be a fourth treatment applied in 2011 of Contans WG™ and will be incorporated with tillage.
- All four of these treatments will be split in half in 2011. One half will have no further treatments applied and the other will have Serenade ASO™ applied.

The Contans WG™ treatments on both the carrot and wheat fields were applied on May 28, 2010. There were rain events after the Contans WG™ was applied on each field which incorporated the product into the soil (1 inch plus was recorded on May 28 and another inch plus on May 29). Aside from the biofungicide applications each of the fields are managed by the cooperating producer as per their standard management practices.

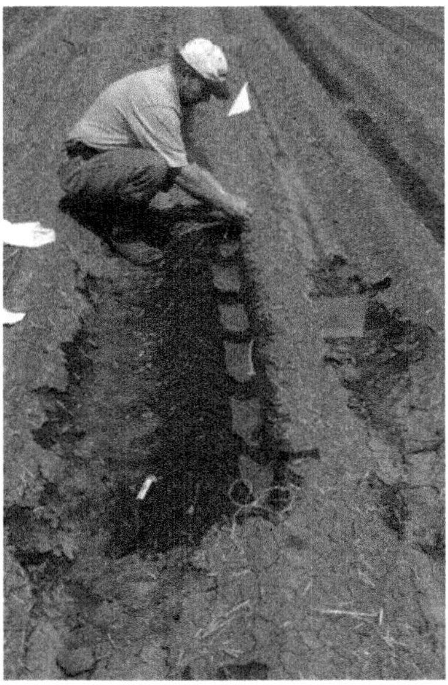

Fig. : Mesh Bags Containing 50 Sclerotia Each Buried in Carrot Beds.

Nylon mesh bags containing 50 sclerotia each were prepared in the lab. On June 4, 2010 six bags were buried 1½ to 2 inches below the top of the soil in the carrot bed and ½ to 2 inches below the top of the soil the wheat field. In the fall of 2010 four bags will be removed from each treatment before freeze up (approximately late August, mid September, mid October and early November) and in the early spring of 2011 the remaining bags will be removed (approximately mid March and mid April). Each time that a bag is removed, the sclerotia will be evaluated in the lab to determine the efficacy of the treatments in reducing the germination of sclerotia.

Data loggers were installed in each of the fields on June 4, 2010. The loggers collect data on the following parameters;

- Soil temperature
- Soil moisture
- Air temperature
- Relative humidity

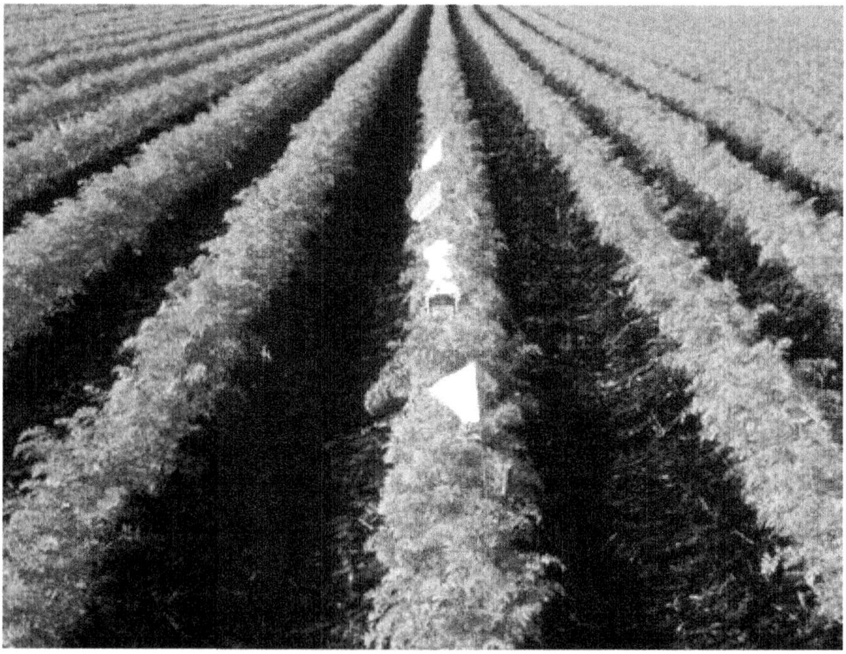

Fig. : Data Logger in Carrot Field.

Samples of carrots will be hand harvested from each plot prior to the producer commercially harvesting the field. The samples will be evaluated for yield. The percentage of Sclerotinia incidence and severity will be calculated for each plot. Soil samples will be taken from each treatment in the fall (immediately after the carrots are commercially harvested) and assayed to determine the number of sclerotia present.

Over the winter of 2010/11 the data collected will be statistically analyzed. A report on the 2010 portion of the evaluation will be prepared and made available to Manitoba carrot producers, UAP and any others interested in the project. In 2011 the trial will be completed, the data statistically analyzed and a final report will be prepared and presented to Manitoba carrot producers.

FUNGICIDE

Fungicides are biocidal chemical compounds or biological organisms used to kill or inhibit fungi or fungal spores. Fungi can cause serious damage in agriculture, resulting in critical losses of yield, quality, and profit. Fungicides are used both in agriculture and to fight fungal infections in animals. Chemicals used to control oomycetes, which are not fungi, are also referred to as fungicides, as oomycetes use the same mechanisms as fungi to infect plants.

Fungicides can either be contact, translaminar or systemic. Contact fungicides are not taken up into the plant tissue and protect only the plant where the spray is deposited. Translaminar fungicides redistribute the fungicide from the upper, sprayed leaf surface to the lower, unsprayed surface. Systemic fungicides are taken up and redistributed through the xylem vessels. Few fungicides move to all parts of a plant. Some are locally systemic, and some move upwardly.

Most fungicides that can be bought retail are sold in a liquid form. A very common active ingredient is sulfur, present at 0.08% in weaker concentrates, and as high as 0.5% for more potent fungicides. Fungicides in powdered form are usually around 90% sulfur and are very toxic. Other active ingredients in fungicides include neem oil, rosemary oil,jojoba oil, the bacterium *Bacillus subtilis*, and the beneficial fungus *Ulocladium oudemansii*.

Fungicide residues have been found on food for human consumption, mostly from post-harvest treatments. Some fungicides are dangerous to human health, such asvinclozolin, which has now been removed from use. Ziram is also a fungicide that is thought to be toxic to humans if exposed to chronically. A number of fungicides are also used in human health care.

Natural Fungicides

Plants and other organisms have chemical defenses that give them an advantage against microorganisms such as fungi. Some of these compounds can be used as fungicides:

- Tea tree oil
- Cinnamaldehyde
- Citronella oil
- Jojoba oil
- Nimbin

- Oregano oil
- Rosemary oil
- Monocerin
- Milk

Whole live or dead organisms that are efficient at killing or inhibiting fungi can sometimes be used as fungicides:

- *Bacillus subtilis*
- *Ulocladium oudemansii*
- Kelp (powdered dried kelp is fed to cattle to help prevent fungal infection)
- *Ampelomyces quisqualis*

Resistance

Pathogens respond to the use of fungicides by evolving resistance. In the field several mechanisms of resistance have been identified. The evolution of fungicide resistance can be gradual or sudden. In qualitative or discrete resistance, a mutation (normally to a single gene) produces a race of a fungus with a high degree of resistance. Such resistant varieties also tend to show stability, persisting after the fungicide has been removed from the market. For example, sugar beet leaf blotch remains resistant to azoles years after they were no longer used for control of the disease. This is because such mutations often have a high selection pressure when the fungicide is used, but there is low selection pressure to remove them in the absence of the fungicide.

In instances where resistance occurs more gradually, a shift in sensitivity in the pathogen to the fungicide can be seen. Such resistance is polygenic – an accumulation of many mutations in different genes, each having a small additive effect. This type of resistance is known as quantitative or continuous resistance. In this kind of resistance, the pathogen population will revert to a sensitive state if the fungicide is no longer applied.

Little is known about how variations in fungicide treatment affect the selection pressure to evolve resistance to that fungicide. Evidence shows that the doses that provide the most control of the disease also provide the largest selection pressure to acquire resistance, and that lower doses decrease the selection pressure.

In some cases when a pathogen evolves resistance to one fungicide, it automatically obtains resistance to others – a phenomenon known as cross resistance. These additional fungicides are normally of the same chemical family or have the same mode of action, or can be detoxified by the same mechanism. Sometimes negative cross resistance occurs, where resistance to one chemical class of fungicides leads to an increase in sensitivity to a different chemical class of fungicides. This has been seen with carbendazimand diethofencarb.

There are also recorded incidences of the evolution of multiple drug resistance by pathogens – resistance to two chemically different fungicides by separate mutation events. For example, *Botrytis cinerea* is resistant to both azoles and dicarboximide fungicides.

There are several routes by which pathogens can evolve fungicide resistance. The most common mechanism appears to be alteration of the target site, in particular as a defence against single site of action fungicides. For example, Black Sigatoka, an economically important pathogen of banana, is resistant to the QoI fungicides, due to a singlenucleotide change resulting in the replacement of one amino acid (glycine) by another (alanine) in the target protein of the QoI fungicides, cytochrome b. It is presumed that this disrupts the binding of the fungicide to the protein, rendering the fungicide ineffective. Upregulation of target genes can also render the fungicide ineffective. This is seen in DMI-resistant strains of *Venturia inaequalis*.

Resistance to fungicides can also be developed by efficient efflux of the fungicide out of the cell. *Septoria tritici* has developed multiple drug resistance using this mechanism. The pathogen had 5 ABC-type transporters with overlapping substrate specificities that together work to pump toxic chemicals out of the cell.

In addition to the mechanisms outlined above, fungi may also develop metabolic pathways that circumvent the target protein, or acquire enzymes that enable metabolism of the fungicide to a harmless substance.

Fungicide Resistance Management

The fungicide resistance action committee (FRAC) has several recommended practices to try to avoid the development of fungicide resistance, especially in at-risk fungicides including *Strobilurins* such as azoxystrobin.

Products should not be used in isolation, but rather as mixture, or alternate sprays, with another fungicide with a different mechanism of action. The likelihood of the pathogen's developing resistance is greatly decreased by the fact that any resistant isolates to one fungicide will be killed by the other; in other words, two mutations would be required rather than just one. The effectiveness of this technique can be demonstrated by Metalaxyl, a phenylamide fungicide. When used as the sole product in Ireland to control potato blight (*Phytophthora infestans*), resistance developed within one growing season. However, in countries like the UK where it was marketed only as a mixture, resistance problems developed more slowly.

Fungicides should be applied only when absolutely necessary, especially if they are in an at-risk group. Lowering the amount of fungicide in the environment lowers the selection pressure for resistance to develop.

Manufacturers' doses should always be followed. These doses are normally designed to give the right balance between controlling the disease and limiting the risk of resistance development. Higher doses increase the selection pressure for single-site mutations that confer resistance, as all strains but those that carry the mutation will be eliminated, and thus the resistant strain will propagate. Lower doses greatly increase the risk of polygenic resistance, as strains that are slightly less sensitive to the fungicide may survive.

It is also recommended that where possible fungicides are used only in a protective manner, rather than to try to cure already-infected crops. Far fewer fungicides have curative/eradicative ability than protectant. Thus, fungicide preparations advertised as having curative action may have only one active chemical; a single fungicide acting in isolation increases the risk of fungicide resistance.

It is better to use an integrative pest management approach to disease control rather than relying on fungicides alone. This involves the use of resistant varieties and hygienic practices, such as the removal of potato discard piles and stubble on which the pathogen can overwinter, greatly reducing the titre of the pathogen and thus the risk of fungicide resistance development.

TYPES OF FUNGICIDE: USING FUNGICIDES IN YOUR GARDEN

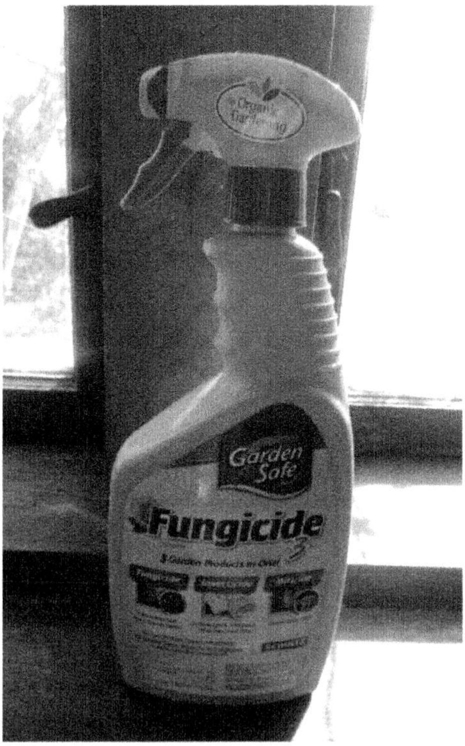

When and how to use fungicide on your plants can be tricky without the right knowledge. Getting professional help beforehand can help determine if using fungicides in your garden is even necessary and, if so, what types of fungicide are available.

When to Use Fungicide

When using fungicides in your garden, it is important to first determine if your plant actually needs a fungicide. Many symptoms can be from other causes, so using garden fungicides may do more harm than good.

The first thing that should be done is to contact an expert, either at a local nursery or *agricultural extension office*. They can help you determine what is wrong with your plants and can also recommend the proper types of fungicide to use.

Keep in mind that garden fungicides are used to prevent the problems from starting or spreading. They can't cure the problems. Once a determination is made that a fungicide is needed, when to use fungicide on your plants depends on the type of fungus.

Types of Fungicide

There are different types of fungicide to deal with different forms of fungus. There are mold fungicides and lawn fungicide. One can't be substituted for the other. There are both chemical and natural fungicides, and even *homemade fungicides for gardens*.

That being said, not all fungicides work the same, as they require different distribution methods. Some are dust powders, some are liquid, some are wettable powders (active only after wet), and flowables. No matter which type you use, caution is recommended. It is best to wear protective gear to limit your own exposure to the chemicals, if going this route.

How to Use Fungicide

All garden fungicides come with specific directions. Using too much is just as harmful as not using enough. Some people prefer to avoid the chemicals completely and opt for natural fungicides. Even when using a natural fungicide, you will still need to follow the directions carefully.

The correct amount, distribution method and time of year are important considerations in how to use fungicide properly. Certain plants require certain types of fungicide.

Now that you know more about using fungicides in your garden, you'll be better able to combat any fungal issues that may crop up with ease.

FUNGICIDE INJECTION TO CONTROL DUTCH ELM DISEASE

In some situations, injecting trees with fungicides is an effective treatment for the management of Dutch elm disease (DED). Several injection products are on the market, and various means of application are recommended. Each product and method has pros and cons. The "best" product depends on the individual tree— its current condition, the objectives of the treatment, and the resources available. The purpose of this chapter is to bring the options and documentation together into one package, so that you can make an informed decision on what you will recommend for injection.

Basic Principles of why and when Injection Works

To understand why and when injection works, you need to understand how the DED fungus gets into and kills elms. The fungus infects the vascular tissue of elms, causing the vessels in the active, outer rings of xylem to become clogged. The fungus gets into an uninfected elm in one of two ways: either through roots grafted to diseased elms, or by elm bark beetles feeding in the branches or upper crown of the tree. When an elm becomes infected through root grafts, the fungus can spread very rapidly and extensively throughout the tree's vascular transport system. Injection of currently available fungicides is not effective in protecting trees from root graft infection, or in therapeutically treating trees that have become infected through root grafts. Injection can be effective in preventing or treating infection caused by bark beetle inoculation.

When a bark beetle that is contaminated with DED fungus spores feeds on a healthy elm, several factors determine whether the tree will become infected by the fungus, or if infected, die. These factors include the inoculum load and point of introduction (this can vary by beetle species), the aggressiveness of the pathogen (at least three species of Ophiostoma cause DED, and they differ in aggressiveness), the physiology of the tree (vitality, vessel structure, *etc.*), the suitability of the environment for fungal growth (temperature, moisture, chemistry, *etc.*), and the ability of the tree to compartmentalize the infection (may differ by elm species or cultivar, health of the tree, *etc.*). Injection of fungicide into trees can be effective by either making the infection court unsuitable, or by stopping fungal growth within the tree. The former is the basis of preventive fungicide injection; the latter is the basis of therapeutic injection.

Kondo (1978a), Campana (1977) and Stipes (1988) addressed some of the many factors that limit the effectiveness of fungicide injection. To be effective in preventing infection, a fungicide must inhibit or kill the DED fungus, and it must be present in adequate concentration at all potential points of infection. Even when injected at fairly high dosage, the quantity of chemical present at the points of potential introduction of the DED fungus is quite low . The chemical, dosage and means of application are critical to success.

For therapeutic treatment, the fungicide must be applied before the fungus has caused extensive damage to the vascular system of the tree, so early detection and timely treatment are critical to success. Sherald and Gregory (1980) and Lanier (1988), among many others, specifically addressed therapeutic injection.

Chemicals for DED control have been researched since the 1940's. There are extensive bodies of literature on the subject, including major portions of symposia proceedings.

A Bit More About Methods of Injection

There are two common ways of injecting the available fungicides into the vascular system of elms. Microinjection is forceful injection of a low volume of concentrated chemical into holes drilled into the stem or base of the tree. Macroinjection is the injection (under pressure) or infusion (without pressure) of large volumes of dilute chemical solutions into holes drilled in the stem or base of the tree .

Kondo (1978b) refined the macroinjection system for injection into the excavated root flare of trees. He found that if the root flare was excavated the circumference of the stem was greater and more injection holes could be well-spaced around the stem, resulting in better chemical distribution in the crown of the tree. He also observed that the wood tissue in this stem-root transitional area seemed functionally different from stem tissue, and drill wounds in this area closed more rapidly with less wetwood problems than wounds higher on the stem.

The holes are drilled with a 7/32 to ¼ inch drill bit, and plastic tees are inserted into the holes. The tees are all connected by tubing to a pressurized container (10-20 p.s.i.) of the fungicide solution. The amount of chemical and volume of solution to inject are based upon the diameter of the tree. In small diameter trees, there is proportionately less vascular tissue in the tree per unit of diameter, so care must be taken not to overdose small trees. This method has been demonstrated to thoroughly distribute chemical in the crown .

However, there are some drawbacks to macroinjection: injection wounds, if repeatedly inflicted, may eventually result in significant discoloration and decay . The chemicals may also damage the cambium around the injection site. The chemicals may also cause foliar phytotoxity, especially on smaller diameter trees.

Microinjection for DED treatment is accomplished with pre-packaged canisters of chemical. Generally, the tips of the injection canisters are placed into holes drilled into the trunk or root flare of the tree and then are pressurized by the squeezing of a built-in plunger. The products are self-contained and require no extra water, which provides an advantage of convenience. Our literature search did not reveal any documentation that microinjection provides adequate distribution and effective concentrations of the chemical to consistently prevent or arrest DED infections. Microinjection has the same disadvantages as macroinjection in regards to phytotoxicity and injection wounds.

There is an art and a science to properly injecting chemicals. The procedure should be done by a certified arborist or skilled tree care specialist who has been specifically trained in the procedure.

Currently Available Chemicals

Six chemicals (in various formulations) are currently registered in the USA for injection to manage DED: three benzimidazole compounds (carbendazim phosphate, thiabendazole hypophosphite, and Debacarb), two triazole compounds (propiconazole and tebuconazole), and a patented formulation of copper sulphate pentahydrate. Another benzimidizole product, with carbendazim hydrochloride, is currently pending registration.

As a group, benzimidazoles are systemic and are active at low concentrations. Benzimidazole compounds affect fungal growth by interfering with mitotic cell division. In low concentrations, benzimidazoles are fungistatic (prevent growth, but don't outright kill) toward the DED fungus. At higher concentrations, they are fungicidal (kill the fungus) . Their acid salts are water soluble, and thus can be adapted for use in injection systems. The benzimidazole salts vary in their effectiveness against the DED fungus and in their chemical behavior in the tree. The effectiveness of these compounds depends on how well they distribute in the crown, the rate at which the chemical is applied (and thus concentration in the plant tissues), and how well they persist in the tree. Four different formulations of benzimidazoles are (or soon will be) available for DED management.

Carbendazim is a breakdown product of benomyl, which is insoluble in water. Acid salts of carbendazim are water soluble, and many have been tested for usefulness in DED injection; two have been available as commercial products. Carbendazim hydrochloride (originally available as Lignasan®) and carbendazim phosphate (originally available as Lignasan BLP®) were both shown to be effective against the DED fungus without high phytotoxicity to elm plant tissue. Carbendazim phosphate readily moves throughout the tree in the transpiration stream, but unfortunately does not move into new wood as it is produced, even when applied at high dosage rates, so preventive treatments using this chemical must be applied annually. Clifford *et al.* (1977) found little difference in persistence between carbendazim phosphate and carbendazim hydrochloride; both compounds declined rapidly between 2-3 months after injection.

Carbendazim phosphate is no longer available as Lignasan BLP®, but can be acquired as Elm Fungicide®. Proper root-flare injection at rates ranging from 0.98 to 3.2 g of carbendazim phosphate per cm of tree DBH results in levels of carbendazim in the twigs during the season of injection at levels inhibitory to the DED fungus. The current label rate for Elm Fungicide® is 0.16 to 0.35 g carbendazim phosphate per cm tree DBH, which is far below the documented effective rate.

The original carbendazim hydrochloride formulation (Lignasan®) has not been commercially available in the USA for many years, but a new product called Eertavas® is pending registration. This product contains 4.7% carbendazim hydrochloride, which is a much higher concentration of carbendazim than the currently available carbendazim phosphate product. At the time of this writing, we have no information on the label dosage rate for Eertavas®.

Thiabendazole hypophosphite (available as Arbotect 20S®) has also been shown to be effective against the DED fungus. When injected at 5.6 g active ingredient per cm tree DBH, thiabendazole continues to appear in new wood in concentrations high enough to be detected by bioassay through three growing seasons . Protection of mature trees from artificial inoculation with the DED pathogen has been shown to last for two seasons in northern climates . It is worth noting that in southern climates, with longer growing seasons, the effective period of the chemical may be shorter (Bruce Fraedrich, Personal Communication). Many arborists have successfully used the highest label dosage of Arbortect 20S® on a 2-3 year rotation to protect high value elms from DED for 15 years. Thiabendazole also has demonstrated effectiveness for therapeutic injection. Unfortunately, however, the injection solution can be very damaging to the cambium as well as to parenchyma cells in a column of wood surrounding the injection site.

Debacarb is another benzimidazole fungicide available for injection. It is available in combination with carbendazim in microinjection canisters, as Fungisol®. The label does not indicate whether the carbendazim is formulated as carbendazim hydrochloride, carbendazim phosphate or a different salt. Lanier (1987) reported that Fungisol® did not significantly prevent infection of artificially inoculated branches, but that there did appear to be some effect on symptom progression within the tree. Lanier (1987) also tested this product as a therapeutic treatment and was not able to demonstrate effectiveness. Our literature search revealed no published documentation of the distribution or concentration of the active ingredients in elms trees treated with Fungisol®.

The current label rate for this fungicide is approximately 0.016 g active ingredient per cm tree DBH. A study on carbendazim salts in other formulations determined that the minimum application level at which carbendazim phosphate could be detected in elm shoots by bioassay is 0.98 g active ingredient per cm tree DBH . Elliston and Walton (1979) found that carbendazim applied at low levels (0.16 g active ingredient per cm tree DBH) resulted in low or undetectable recovery of the chemical in twigs. The documented effective dosage rate for carbendazim is 60 times higher than the current label rate for Fungisol®, which brings the efficacy of this product into question.

The triazole fungicide propiconazole is effective in management of oak wilt disease , and it is also labeled for management of DED. It is a highly systemic sterol inhibitor that prevents fungal growth by interfering with cell wall formation. The commercially available formulation of propiconazole (Alamo®) is microencapsulated to make it soluble in any clean water near neutral pH. Stipes has demonstrated propiconazole to be effective in preventing DED infection following challenge inoculations by the DED fungus. The rates of propiconazole used in Stipes' studies number 6, 9, 10, and 11 varied from 1.1 to 3.6 grams of active ingredient per cm tree DBH. The current highest label rate for Alamo® is equivalent to 1.1 gram active ingredient per cm of tree DBH.

Propiconazole at the highest label rate may provide protection for multiple seasons, as two mature elms which were challenge inoculated multiple times at multiple points were protected against DED infection for an equivalent length of time as trees treated with thiabendazole hypophosphite . However, according to recent findings by Stipes, residual activity of propiconazole is considerably shorter than that of thiabendazole hypophosphite. Therapeutic treatment of 24 mature symptomatic American elms in 1996 and 1997 with the highest label rate of Alamo® resulted in 79% survival by the end of the 1998 growing season, so propiconazole has demonstrated therapeutic value. Foliar toxicity of propiconazole is low, even at rates of up to 3.8 grams per cm tree DBH, though severe phytotoxicity may occur with high dosage rates on small diameter trees when treated early in the growing season. Propiconazole does not require high dilution rates with water so treatment is considerably faster than with thiabendazole hypophosphite, and there is less tissue injury at the injection site.

Another triazole fungicide, tebuconazole, has very recently been registered for use against DED. Mauget's Tebuject® is a microinjection product containing this fungicide. There are no published research data on the use of tebuconazole to manage DED.

The activity of copper sulphate pentahydrate is based on the fungicidal effect of metallic copper. Knutson (1991) reported a 22% higher level of copper in the leaves of a Phyton 27® treated elm compared to leaves of untreated elms at 15 months after treatment. The distributors of Phyton 27® claim protective effect against DED for at least 36 months, but there are no published data on this treatment. Leaf abscission is common following Phyton 27® injection, but is usually followed by refoliation. Lanier (1987) reported severe vascular tissue discoloration and damage at Phyton 27® injection sites and only seemingly marginal fungicidal effect within the tree.

An injected non-fungicide protective treatment has recently been developed. It is a suspension of live spores of the fungus *Verticillium dahliae*, which is injected into the tree with a specially developed "gouge pistol". Treatment reportedly protects against new infections by the DED fungus during the year of treatment by inducing resistance in the tree. This product, marketed as Dutch Trig®, is currently available in U.S. under test exemption and is being tested primarily by Bartlett Tree Research Laboratories. The product was developed by ARCADIS Heidemij in the Netherlands and is distributed in the U.S. by Innovative Tree Services L. L. C., Tampa, FL.

There are pros and cons to this new "Dutch Trig®" treatment. The injection holes are small and seal quickly. The application procedure is relatively rapid; A 75 cm DBH tree can be treated in approximately 10 minutes. Single treatment cost may be considerably less than fungicide (price in US is still to be finalized), but the treatment must be performed every year. Early results of tests on American elm in the U.S. have shown a protective, but not therapeutic effect. Another potential drawback is that the fungus *Verticillium dahliae* is a plant pathogen, and the isolate used in this product is of European origin. This isolate cannot be recovered from trees one season after inoculation. APHIS did consider the risk of introducing an exotic strain of a plant pathogen prior to allowing experimental use of this product in the U.S. We also do not yet know the cumulative effect of annual treatments on the health of the tree: treatment causes discoloration of the annual ring, so apparently the physiology of the host tree is affected. Table. A summary comparison of chemicals available for injection to manage DED.

Table: A summary comparison of chemicals available for injection to manage DED.

Common name	Chemical name	Commercial products (Trade name, source and % active ingredient in formulation)	Current Label Dosage Rate for commercial product	Comments
Carbendazim phosphate (MBC PO_4)	Methyl-2-benzimidazole carbamate phosphate	Elm Fungicide® (formerly Lignasan BLP) Elm Research Institute, Harrisville, NH (0.7% Carbendazim phosphate)	Approximately 126 ml per inch of diameter Note: 126 ml/ inch DBH=0.35 g active ingr. per cm DBH	Benzimidazole fungicide. Low phytotoxicity. Does not persist in newly formed wood; treatments must be applied annually. Effectiveness against DED fungus documented in scientific literature, but at a rate higher than current label.
Carbendazim Hydrochloride (MBC HCl)	Methyl-2-benzimidazole carbamate hydrochloride	Eertavas® St. James Tree Service Winnipeg, Manitoba, Canada (4.7% carbendazim hydrochloride)	Not yet known.	Benzimidazole fungicide. Low phytotoxity. Low persistence.
Thiabendazole hypophosphite (TBZ H_2PO_2)	2-(4-thiazolyl) benzimidazole hypophosphite	Arbotect 20-S® Novartis Corp., Greensboro, NC. (20% Thiabendazole)	"3 yr rate" is 12 oz. (72 ml) per inch of diameter Note: 12 oz/ inch DBH=8.1 g active ingr. per cm DBH	Benzimidazole fungicide. May cause phytotoxicity at injection site and in crown. Effective for up to 3 seasons (in North) at highest label rate. Preventive and therapeutic effectiveness against DED fungus documented in scientific literature.

Debacarb (DEBC)	2-(2-Ethoxyethoxy) ethyl 2-benzimidizole carbamate	Fungisol®, Abasol®, and Imisol® micro-injectors J.J. Mauget Company, Burbank, CA (1.7% Debacarb and 0.3% Carbendazim)	2.0 ml per inch of diameter Note: 2.0 ml/ inch DBH=0.016 g active ingr. per cm DBH	Both active ingredients are benzimidazole fungicides. Very limited published data indicates some effectiveness for preventive of progression of symptoms. Label dosage rates are far below rates proven effective for similar chemicals.
Propiconazole	1-[[2-(2,4-dichlorophenyl)-4-propyl-1, 3-dioxolan-2-yl]methyl]-1H-1,2,4-triazole	Alamo® Novartis Corp., Greensboro, NC (14.3% encapsulated Propiconazole)	Microinjection: One 10 ml unit per inch DBH. Macroinjection: 6 to 10 ml per inch DBH (preventive); 20 ml per inch DBH (preventive and therapeutic) Note: 20 ml/ Inch DBH=1.13 g active ingr. per cm DBH	Triazole fungicide. Available in both microinjectors and as liquid for macroinjection. Low phytotoxity. May be effective for more than one season at highest label rate, though persistence of chemical in elm tissue is low. Preventive and therapeutic effectiveness against the DED fungus has been documented in scientific literature.
Tebuconazole	alpha-[2-(4-chlorophenyl)-ethyl]-alpha-(1,1-dimethylethyl)-1H-1,2,4-triazole-1-ethanol alpha-[2-(4-chlorophenyl)-ethyl]-alpha-(1,1-dimethylethyl)-1H-1,2,4-triazole-1-ethanol	Tebuject®microinjectors J.J. Mauget Company, Burbank, CA (% active ingredient unknown)	Unknown	Brand new product. Triazole fungicide. No information available on dosage or effectiveness at time of press.
Copper sulphate pentahydrate	Copper sulphate pentahydrate	Phyton 27® by Source Technology Biologicals, Inc. of Edina, MN (21.36% copper sulphate pentahydrate)	Approx 4.0 ml per inch of diameter (actual rate is based on a table of diameter classes) Note: 4.0 ml/ inch DBH=0.34 g active ingr. per cm DBH	Fungicidal activity based on metallic copper. Effectiveness of this chemical to prevent or arrest DED is not documented in independent scientific literature. Manufacturer will provide data upon request.

Fitting it all Together into a Management Strategy

Injection is only one part of an overall management strategy for Dutch elm disease, but it does provide some options for protecting or saving high value individual trees. The recent USDA Forest Service publication "How to Identify and Manage Dutch elm Disease" provides information on how various management activities can be used to interrupt the DED disease cycle. A copy of this publication is enclosed with this issue of PDQ.

With the exception of DDT sprays to prevent smaller European elm bark beetle feeding, all of the available evidence indicates that every historically successful Dutch elm disease management program has depended almost exclusively on sanitation to reduce bark beetle populations and the available inoculum of the DED fungus. Even with injection of selected trees, no program will succeed without the sanitation necessary to minimize disease pressure. Conversely, individual owners of high value elms may not be able to rely on sanitation alone for protection if their city does not enforce mandatory removal of diseased landscape or wild elms. Subsequently, where stringent sanitation practices cannot be followed, fungicide treatments play an important role.

Therapeutic treatment is only an option for early stages of infection, but it is a potentially powerful tool when added to successful sanitation programs that pivot around thorough inspections and prompt removals. It is not always effective, but a success rate as low as 50% may be more than enough to justify the cost of the effort. The cost of tree removal is high, and the value of large stately elms is even greater.

There are risks to tree health in injecting trees. A long-term preventive injection program may cause significant stem damage to a valuable elm. Consider whether early detection (and thus opportunity for therapeutic treatment) is likely for a high value elm. Consider whether bark beetle and DED fungal populations are high in the surrounding area. As with any resource management decision, it is important to weigh the risks against the benefits.

FUNGICIDES MADE SIMPLE

Disease is a major source of plant damage in roses with the most common rose diseases being caused by fungus. While some rarely-contracted rose diseases are caused by viruses or bacteria, more often than not, it is fungi that wreak havoc in our rose gardens. Nationwide, blackspot fungus (*Diplocarpon rosae* Wolf) is the most common rose disease. Powdery mildew (*Alphitomorpha pannosa*) runs a close second in its commonality, but unlike blackspot outbreaks tend to be seasonal. Occasionally, outbreaks of downy mildew (*Peronospora sparsa* Berkeley) and botrytis blight (*Botritis cinerea*) may be found, but these diseases appear much less frequently than either blackspot or powdery mildew.

Damage due to fungal attack can range from one extreme to another on the seriousness scale – from repeated loss of leaves (weakening the plant) to less damaging symptoms of minimal lesions (spots) on foliage and canes. Roses with high levels of infection produce less new growth and fewer blooms because their leaves (which are needed for photosynthesis) are affected by disease and fall off. Because of the important role that foliage plays in the overall health of the rose bush, repeated defoliation ultimately impacts the health and longevity of the plant.

Roses differ in their susceptibility to fungal attack with modern hybrid teas being the most susceptible. If left untreated, infected susceptible roses can lose a large percentage of their leaves.

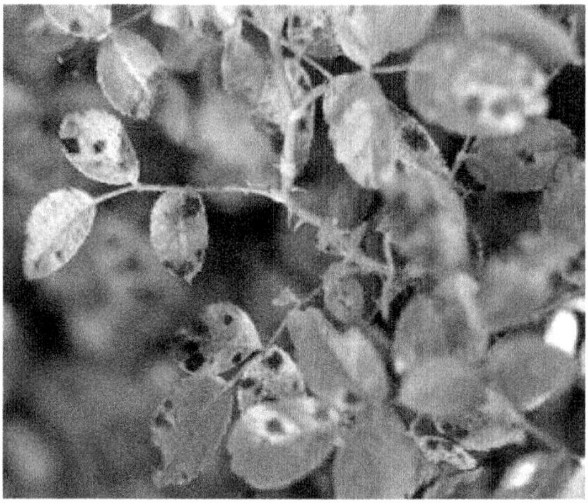

Fig. : Blackspot fungus on rose plant.

Fifty years ago, Dr. Griffith Buck, identified the correlation of leaves to plant hardiness finding that roses with the ability to hold onto their leaves even when infected by fungus were hardier, more disease resistant landscape plants. This correlation has been reinforced through the Earth-Kind® Rose Research Program being conducted through a partnership between the Texas AgriLife Extension Service and the Houston Rose Society.

Managing Fungal Diseases

Fungal diseases are best managed through a multi-disciplinary approach that includes:

(1) plant selection,

(2) planting times,

(3) level of fertility,

(4) sanitation and

(5) applications of fungicides.

Some roses exhibit more tolerance than others to attack by common diseases. The susceptibility of a rose variety to disease will dictate the management practices that must be employed to maintain the health of the garden site.

There are many roses on the market that have exhibited tolerance to disease, however, there is an even larger population of roses that are highly susceptible to fungal diseases. The gardener's job is to balance plant selections so that management of fungal diseases in their roses can be attained within the maintenance parameters the gardener has available. Planting young new roses at times when environmental conditions are ripe for disease development without protecting those plants against attack tends to invite the disease process. Some hypothesize that excessive fertilization in early spring and late fall (when weather conditions support disease formation) can initiate the process. There is some logic to this theory as blackspot and powdery mildew tend to attack new plant growth first. Heavy fertilization encourages the development of new growth that, if left unprotected, is fair game for fungal pathogens seeking a host. However, cultural practices alone will not eliminate diseases from the garden.

"Fungicides" are a specific type of pesticide that controls disease by inhibiting or killing the disease-causing fungus. They work by :

(a) attacking and damaging cell membranes,

(b) interfering with energy production, or

(c) interfering with the life cycle of the fungus.

Some of the newer disease treatments on the market do not directly affect the fungus itself, but instead boost the plant's natural defense system causing the plant to produce thicker cell walls and anti-fungal proteins so that the plant is better able to defend itself when disease pressures are high. Examples of this would be products like Messenger and similar "biorational" treatments.

While most fungicides are only capable of protecting uninfected growth from disease, only a handful are effective against pathogens once infection occurs. For this reason, to be effective *most* fungicides should be applied:

(a) when environmental conditions are right for disease development,

(b) before disease occurs, or

(c) at the first appearance of symptoms.

Types of Fungicides

Fungicides fall into two categories – "curatives" (products that attack fungi that have already infected the plant) or "preventatives" (products that inoculate plant tissue so that the plant is better able to fight off disease. Fungicides with curative properties would include products sold under the trade names, Mancozeb, Manzate and Fore and products containing Maneb. Preventative fungicides would include products such as Rose Pride (formerly known as Funginex), Rose Defense, Daconil and products containing Neem oil.

Some homeowners start applying fungicides to their roses only when the plants lose all their leaves and then stop spraying once new growth emerges. Other gardeners use fungicides for the wrong purpose – for example using a preventative on a bush exhibiting symptoms of heavy blackspot infection. These treatment approaches are wholly ineffective in controlling fungal diseases.

Once defoliation occurs, preventative fungicides become ineffective. In this instance, applications of a curative fungicide, such as Mancozeb, applied every five days for three weeks (or as directed on the manufacturer's label) is critical to establish disease control. I have found that it takes three weeks of regular applications of a curative fungicide in addition to good garden housekeeping to reestablish disease control. The need to spray should be dictated by four factors:

(1) the health of the rose,
(2) the environmental conditions, including the level of disease pressure that may be present,
(3) the disease susceptibility of the rose, and
(4) the desired level of plant perfection.

Disease Control With Fungicides

Fungicides come in powder, granular and liquid form. Most are mixed with water and applied by spraying the mixture onto the plant. A few fungicides are applied to the soil where they break down into compounds taken up by the plant making the plant toxic to the disease. For spray applications, coverage of all parts of the plant (upper and lower leaf surfaces and from the top to the bottom of the plant) is critical because very few fungicides have the ability to be absorbed into plant tissue and move through the plant. One exception would be Aliette, which is applied as a spray treatment and then translocates through the plant's system from the leaves to the roots and back to the foliage on a "seek and destroy" mission.

On susceptible roses, effective control of fungal diseases may necessitate multiple applications of fungicides – sometimes as frequently as once weekly from the first flush of new growth in the spring until the first hard frost in the fall. These repeated applications are necessary to protect emerging new growth and to replace fungicide product lost by decomposition, degradation by sunlight and/or removal by wind and water. Unfortunately, repetitive applications of the same fungicide can lead to the fungus developing resistance to the treatment – making treatment applications ineffective.

Keeping susceptible roses healthy requires some application of fungicides, especially when conditions support disease development. Blackspot spores germinate and infiltrate cell tissue when temperatures are between 65 – 85° F and when moisture is present on the foliage for seven hours or more. Powdery mildew tends to appear when humidity is high but conditions are dry and cool (warm days / cool nights).

Fungicide labels contain the words "contact" and "systemic", which is a general description of the method by which the product kills fungal pathogens. "Contact" fungicides are applied to and remain on the plant surface and do not penetrate into plant tissues. Contact fungicides *must* come in contact with the disease-causing pathogen to be effective. Examples of contact fungicides are Mancozeb, Fore, Manzate and Daconil.

Often these types of fungicides leave a spray residue on leaves and stems and only the parts of the plant with spray residue are protected from infection. Contact fungicides are very sensitive to the environment and are usually effective for 7 to 14 days, however product life depends the amount of rain/irrigation the bush receives and the amount of UV rays the chemical is exposed to. One to 2 inches of rain will reduce the residue of contact fungicides by half, but rainfall of 2 inches or more will eliminate the product from the plant. The product label will provide direction on application frequency.

Systemic fungicides, sometimes called "penetrants", are absorbed into the plant and have the ability to move from the application site (similar to how blood moves through our bodies). The distance that systemics are able to move within the plant is dependent on the nature and type of fungicide used. Some systemic fungicides stay in the leaf tissue. Others have the ability to travel from the leaves to the root system, but not back up through the plant structure. Only a few systemic fungicides, like Aliette, can travel up and down freely within the plant. Some of the common systemic fungicides used in rose gardening are Aliette, Fertilome Liquid Systemic Fungicide, Monterey Fungi-Fighter, Rose Pride (Funginex) and Bonide Systemic Fungicide.

Mode of Action – How They Work

How a fungicide works is called its "mode of action." Fungicides are manufactured in such a way that their modes of action attack either "single" or "multiple" sites within a fungus. Single-site fungicides target one critical component (usually an enzyme or protein) needed by the fungus for survival. The slightest mutation of the fungus impedes the effectiveness of the fungicide treatment by camouflaging the product's target. Fungicides with single-site modes of action are Rose Pride (Funginex) and Green Light Systemic Fungicide.

Genetic mutation of fungi is more common than we may think. There are 54 known variations of blackspot fungus in North America and these variations or "races" of the disease are frequently geographically specific. Roses resistant to one race of blackspot fungus may prove susceptible to other races of the disease.

Fungicides designed to attack multiple sites are usually affective against different types of disease components within a fungus. These types of fungicides treat a variety of disease components so that in the event that there is a mutation of one particular enzyme/protein in the pathogen's make-up, there are usually other non-mutated disease components for the fungicide to attack. Contact fungicides typically affect multiple sites in fungi. Examples of multi-site fungicides are Mancozeb, Manzate, Aliette, Fertilome Liquid Systemic Fungicide, Ortho Garden Disease Control and Daconil.

To Spray or Not to Spray

In North Texas it is not uncommon for hybrid tea roses to require 15 to 20 applications of fungicides per year to maintain plant health. Along the Gulf

Coast, environmental conditions can dictate even more applications and for a longer period of time. Sometimes even hardy roses may require a few spray applications when environmental conditions support disease development and disease pressures around the garden are high.

A good disease management approach includes cultural and environmental controls, in conjunction with the right type of fungicide, including:

- Plants must get six to eight hours of full direct sunlight;
- Having adequate spacing between plants (a minimum of 2 feet between mature plants);
- Avoiding wetting the leaves during irrigation and not watering plants at night;
- Removing leaf litter in and around the bushes;
- Moderate fertilization; and
- Rotation of fungicide products that have different "modes of action" to delay the development of the resistance to chemical treatments.

A gardener need not hold a Ph.D. in chemistry, toxicology or plant pathology to effectively treat fungal diseases. With all of the rose care products on the market today it may seem a daunting process to select the right product for the disease process in your own garden.

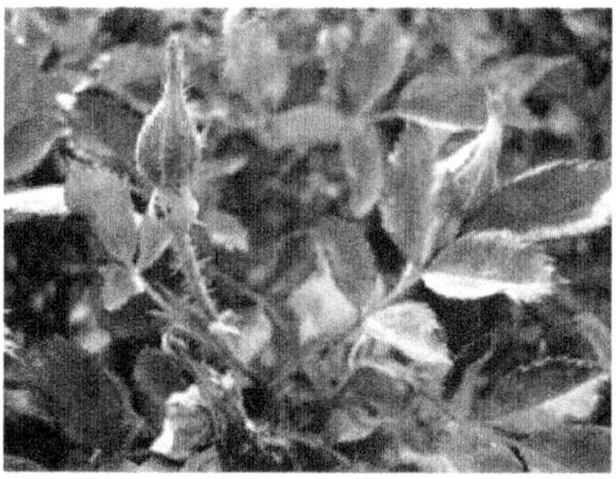

Fig. : Powdery mildew on rose bud and leaves.

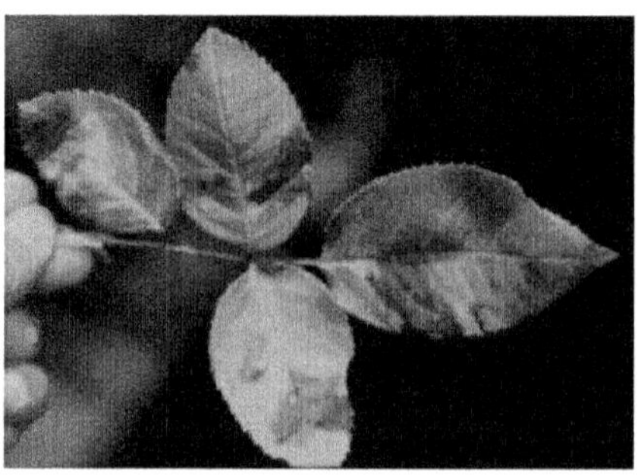

Fig. : Downy mildew on rose leaf – note rectangular shape stopping at vein margins.

FUNGICIDES USED IN BANANA PLANTATIONS

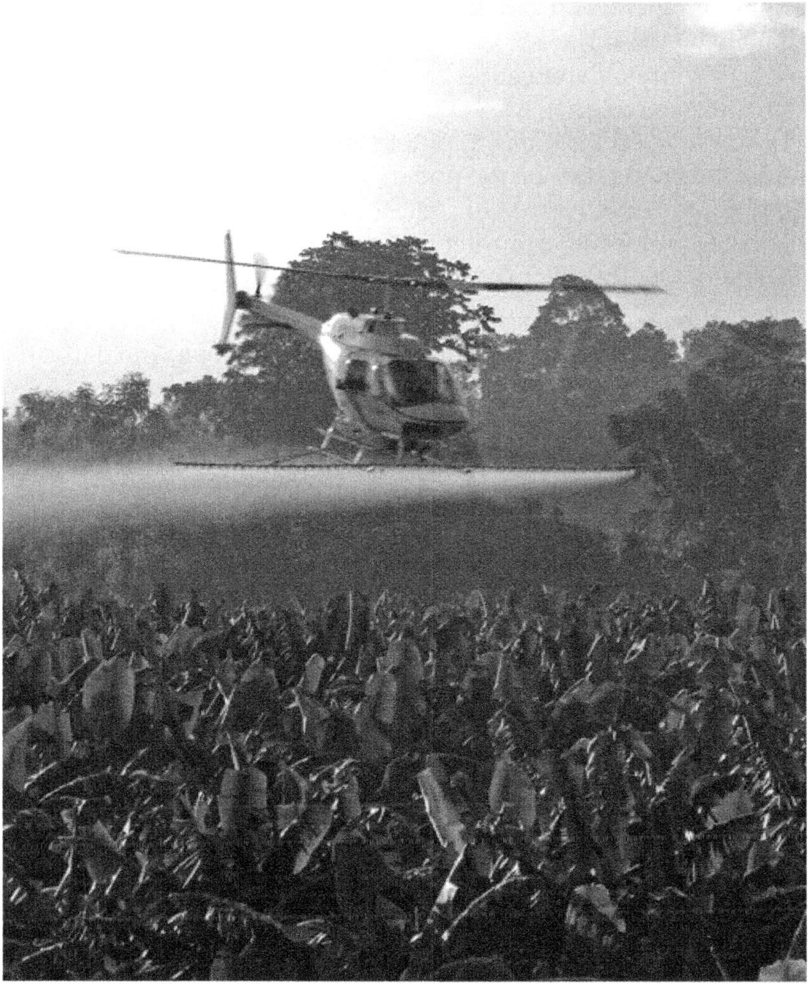

A fungicide is a specific type of pesticide used in controlling fungal diseases by inhibiting or killing the causal agent. Fungicides are routinely used in commercial plantations to control black leaf streak (BLSD) and Sigatoka leaf spot, as well as post-harvest diseases. However, not all diseases caused by fungi can be controlled by fungicides. Fusarium wilt, for example, cannot be controlled by fungicides, one of the reasons that led to the demise of Gros Michel as the export banana whenrace 1 strains of the Fusarium fungus entered Latin America.

Fungicides are used as a formulated product consisting of an active ingredient plus inert ingredients that improve the performance of the product. Against leaf diseases, fungicides are mixed with water, oil or an oil-water emulsion and then applied using equipment ranging from small hand-held and back-pack sprayers to large spray units carried by tractors or aircraft.

Contact Fungicides

Contact fungicides (also called protectants) are used against foliar diseases. They remain on the surface of plants and act by preventing spore germination. However, they are only effective on the leaves present at the moment of application and can be easily washed away by rain. Their main drawback is the need for repeated applications to protect the new leaves. In the case of BLSD, since most infections start on the underside of the leaf during the unfurling of the cigar leaf, the effectiveness of contact fungicides depends on the quality of the coverage during the unfurling and on the interval between applications. On the other hand, they do not induce the selection of resistant fungal strains.

There are only a few types of contact fungicides. The main chemical classes of contact fungicides used in banana plantations are benzene derivatives and carbamates.

Benzene Derivatives

Chlorotalonil

Chlorotalonil is a broad spectrum contact fungicide used in banana plantations to control BLSD. It can only be mixed with water as it becomes toxic to the plant when mixed with oil. It's unlikely to present acute hazard in normal use. According to the banana working group of the Fungicide Resistance Action Committee (FRAC), Chlorotalonil can be used solo or in mixtures with partners at manufacturers' recommended effective rates. There are no limitations or restrictions concerning the number of applications, the timing or the sequence as long as it is within the limits of the manufacturers' labels.

Carbamates

Mancozeb

Mancozeb is a fungicide in a subclass of carbamate pesticides called dithiocarbamates. As a cholinesterase inhibitor, it can have effects on the nervous system. It's unlikely to present acute hazard in normal use. According to FRAC banana working group, Mancozed can be used solo or in mixtures with partners at manufacturers' recommended effective rates. There are no limitations or restrictions concerning the number of applications, the timing or the sequence as long as it is within the limits of the manufacturers' labels.

Systemic Fungicides

Systemics (also called penetrants) are absorbed into plants. Some only penetrate the leaf, while others can be distributed to the entire banana plant. They also have a more or less pronounced curative effect. Against BLSD, the curative effect is more pronounced on stages 1 and 2. They don't have any effect on the necrotic stages (5 and 6), although sporulation may be temporarily decreased.

Since the mode of action of these systemic fungicides is very specific, small genetic changes in the fungus can overcome their effectiveness. Disease management strategies that rely heavily upon these fungicides often lead to pathogen populations becoming less sensitive or resistant to them. In their biennial review of fungicide sensitivity and use, the FRAC banana working group focuses exclusively on the control BLSD and makes recommendations on eight main classes of fungicides used against the fungal disease.

Benzimidazoles

Benzimidazoles (BCMs) are broad-spectrum fungicides that interfere with cellular division. The following fungicides are used in banana cropping to control BLSD: benomyl, carbendazim, thiophanate, thiophanate-methyl.

BMCs were introduced the late 1960s. At the time, they represented a ground-breaking class of fungicides with high systemic and curative activity that allowed long intervals between applications. However, within a few years loss of disease control was beginning to be recorded in many crops. *Mycosphaerella fijiensis* began to show resistance to this class of fungicides within 2 to 3 years of their first use.

The FRAC banana working group recommends that they should be used only in mixtures and only in full alternation with other, non-cross resistant modes of action. No consecutive BCM-applications (blocks) should be applied. To reduce selection pressure, the total number of applications a year should not exceed 3 and should not represent more than 33% of total number of sprays. They are preferably used at lower disease pressure and the sprays must be separated by at least 3 months of a BCM-free period.

Quinone Outside Inhibitors

Quinone outside inhibitors (QoIs) in the Strobulirin chemical class of fungicides that inhibit mitochondrial respiration by binding to the cytochrome B enzyme complex. The following QoI fungicides are used in banana cropping to control BLSD: azoxystrobin, pyraclostrobin, trifloxystrobin.

Resistance to QoIs in *Mycosphaerella fijiensis* and some other fungi is conferred by a point mutation in the mitochondrial cytochrome B gene. Resistance to QoIs has been observed in Ecuador, Colombia, Guatemala, Costa Rica and the Philippines. QoIs should be used only in mixtures and only in full alternation. To reduce selection pressure, the total number of applications a year should not exceed 3 and should not represent more than 33% of total number of sprays. They are preferably used at lower disease pressure and the sprays must be separated by at least 3 months of a QoI-free period.

Demethylation Inhibitors

Demethylation inhibitors (DMIs) are inhibitors of ergosterols synthesis. The following DMI fungicides are used in banana cropping to control BLSD: bitertanol, difenoconazole, epoxiconazole, fenbuconazole, myclobutanil, propiconazole, tebuconazole, tetraconazole, triadimenol. Shifting of DMI sensitivity over the baseline sensitivity has been observed up to 2009. Sensitivity has since stabilized at a higher level compared to pre 2009 in Belize, Ecuador, Colombia, Guatemala, Costa Rica, Honduras, Panama and the Philippines. Mechanism of resistance to DMIs has been elucidated to be based on mutations in the cyp 51 gene of M. fijiensis. Additional reduction of sensitivity may come from over-expression of the cyp 51 gene. All the active ingredients belonging to the DMI class of fungicides are considered to be a single product group, amongst which there is in general some degree of cross resistance.

Mixtures of two or more DMIs can be applied to provide good biological efficacy; however, they do not provide an anti resistance strategy and must be treated as a solo DMI for resistance management. The FRAC banana working group recommends that they should be used only in mixtures and only in full alternation with other, non-cross resistant modes of action. To reduce selection pressure, the total number of applications a year should not exceed 8 and should not represent more than 50% of total number of sprays. Applications should preferably start at the onset of the annual disease progress curve.

Amines

Amines are another group of ergosterols synthesis. The following amine fungicides are used in banana cropping to control BLSD: spiroxamine, fenpropimorph, fenpropidin and tridemorph. The sensitivity to amines is at high levels and did not change significantly between 2010 and 2014.The FRAC banana working group says that Amines can be used solo or in mixtures, but recommends applications in mixtures. A maximum of 2 consecutive sprays (block) containing amines can be used. Full alternation is preferred. Amines should be used at a maximum of 15 applications, but not more than 50% of the total number of sprays.

Anilinopyrimidines

Pyrimethanil is the only active ingredient from the group of anilinopyrimidines, which is currently used in banana cropping to control BLSD. The FRAC banana working group recommends that Pyrimethanil should be used only in mixtures and only in full alternation with other, non-cross-resistant modes of action. Consecutive or so-called "block" applications are not recommended. To reduce selection pressure, the total number of applications a year should not exceed 8 and should not represent more than 50% of total number of sprays.

Succinate Dehydrogenase Inhibitors (SDHIs)

The following SDHI fungicides are used in banana cropping to control BLSD: boscalid, fluopyram, fluxapyroxad and isopyrazam. The FRAC banana working group recommends that SDHIs should be used only in mixtures and only in full alternation with other, non-cross resistant modes of action. No consecutive SDHI-applications (blocks) should be applied. To reduce selection pressure, the total number of applications a year should not exceed 3 and should not represent more than 33% of total number of sprays. Applications should preferably start at the onset of the annual disease progress curve and be applied at times of lower disease pressure. Applications should be separated by at least 3 months of a SDHI-free period.

Where an SDHI is used as a soil drench for nematode control then, as a precautionary measure, it should be counted as one of the permitted SDHI applications. Another mode of action should be used for the first foliar fungicide application providing satisfactory disease control against *Mycosphaerella fijiensis* within the first 7 days after the soil drench application. Foliar applications with alternating mode of actions should be continued for the remaining growing period.

Guanidines

To control BLSD, the FRAC banana working group recommends that Guadinines should be used only in mixtures and only in full alternation with other, non-cross-resistant modes of action. To reduce selection pressure, the total number of applications a year should not exceed 6 and should not represent more than 33% of total number of sprays. Applications have to be separated by at least 6 weeks of a Guanidine-free period.

N-Phenylcarbamates

To control BLSD, the FRAC banana working group recommends that N-Phenylcarbamates should be used only in mixtures and only in full alternation with other, non-cross-resistant modes of action. To reduce selection pressure, the total number of applications a year should not exceed 3 and should not represent more than 33% of total number of sprays. Applications should preferably start at the onset of the annual disease progress curve and be applied at times of lower disease pressure. Applications have to be separated by at least 3 months of a N-Phenylcarbamates-free period.

OCCURRENCE OF STROBILURIN RESISTANCE AND IMPACT ON MANAGING POWDERY MILDEW OF CUCURBITS

Resistance to strobilurin fungicides in the cucurbit powdery mildew fungus was detected in the USA in 2002. This was not surprising because strobilurin resistance had already developed elsewhere in the world. The cucurbit powdery mildew fungus has demonstrated a high potential for developing resistance. Resistance has developed to each chemical class active for powdery mildew that is at risk for resistance somewhere in the world following repeated use. Strobilurins are in fungicide group 11, quinone outside inhibitors (QoI).

Strobilurin resistance was documented in four fungicide efficacy experiments conducted in research fields in GA, NC, VA, and NY. In these fields, strobilurins, when used alone on a 7-day schedule (use pattern not labeled), did not effectively control cucurbit powdery mildew compared to previous years. Strobilurin efficacy was seen to decline dramatically after the second application in NY. Efficacy also was reduced in commercial fields in KY and research fields in AZ, CA, KY, IL, and MI in 2002 where strobilurins were used predominantly or exclusively. One resistant isolate was obtained from both the AZ and CA fields; however, this is not considered conclusive because of the small sample size.

Resistance can be detected in research fields more easily than commercial fields, especially when control is not greatly affected. Reduced control is often the first indication that resistance has developed. Detecting reduced control is easier in research fields because there are also plants treated with other fungicides not affected by resistance and non-treated plants. Comparing all the treatments and considering how these treatments have performed in previous experiments reveals whether efficacy is indeed reduced. Poor application timing is another possible reason for reduced control; however, this would affect efficacy of other treatments as well. Resistance also can be more difficult to detect in commercial fields because other fungicides used with strobilurins in a program designed for managing resistance might provide enough control of powdery mildew to mask the presence of strobilurin resistant strains, especially if they are at a low frequency.

The type of resistance detected to strobilurins was qualitative, which means individuals of the cucurbit powdery mildew pathogen were either highly sensitive to strobilurins or highly resistant. With qualitative resistance, control cannot be regained by applying the fungicide more frequently and/or at a higher rate or by switching to a more active fungicide in the same chemical class, in contrast with quantitative resistance. All strobilurins will be affected by resistance.

When strobilurins were first introduced, several people involved with their development felt they had a moderate to low resistance risk. These are site-specific compounds, which has often indicated a high resistance risk; however, they have a new mode of action (inhibition of mitochondrial respiration) that was thought to be difficult for fungi to overcome. Resistance developed much more quickly than expected, but resistance did develop first in a pathogen with a past history of developing resistance to site-specific compounds.

Resistant strains of the cucurbit powdery mildew fungus were found in 1999 after just two years of commercial use in four countries. Unexplainably, resistance was not detected in North America, despite the size of the pathogen population, until the fifth year of commercial use. Just as unexpectedly, resistance to strobilurins developed first in the USA in the fungus causing gummy stem blight and black rot in cucurbits. This was not expected based on past history with benomyl resistance in these pathogens. In the USA, control failure associated with resistance to benomyl was observed in the cucurbit powdery mildew fungus just 1 year after its registration for use on cucurbits, but it was not observed in the gummy stem blight fungus until 23 years after its registration.

Current general recommendations for managing fungicide resistance include using a diversity of fungicides having different modes of action, within an integrated disease management program that includes non-chemical practices, such as use of resistant cultivars. Fungicides at risk for resistance should be applied as infrequently as possible. Using a resistance management program will also minimize yield loss when resistance occurs. It is extremely important to evaluate disease control when using any fungicide that is at risk for resistance development. Contact extension staff or the fungicide manufacturer promptly, while the crop is still there, if control is inadequate and there are no other feasible explanations, such as poor application timing.

The fungicide program that has been recommended by extension specialists in the USA for managing powdery mildew and resistance is a strobilurin fungicide (azoxystrobin formulated as Quadris® or trifloxystrobin formulated as Flint®) applied in alternation with the DMI fungicide myclobutanil (formulated as Nova® or Rally®) tank-mixed with a protectant fungicide. This program uses two strategies for managing resistance:

1. alternation among systemic fungicides in at least two chemical groups and
2. inclusion of protectant fungicides which are not at risk for resistance development because they have multi-site mode of action. Myclobutanil is a triazole fungicide in the DMI activity group, which is fungicide group
3. Growers need to know what chemical group fungicides are in to avoid alternating among products in the same group.

Although isolates were not tested from commercial production fields, it is prudent for growers to consider improving their resistance management program. Strobilurin resistance likely occurred in commercial fields as well as research fields in 2002. Strobilurin resistance appears to be widespread in the USA. It was confirmed in GA, VA, NC, and NY. It is likely in AZ and CA. And efficacy of strobilurins was reduced in some mid-western states. The cucurbit powdery mildew fungus produces spores wind-dispersed over large areas. Inoculum for powdery mildew developing on cucurbit crops is thought to be wind-dispersed northwards through the eastern and mid-western USA each year. Occurrence of resistance in commercial fields will reduce the utility of strobilurins, including those not yet registered, and eliminate an important tool for managing DMI resistance. Strobilurins and DMIs are the only systemic fungicides registered for cucurbit powdery mildew in the USA.

One suggested change to improve resistance management is to apply a contact fungicide with strobilurins as well as with DMIs. Based on research with Microthiol Disperss® and JMS Stylet-oil®, sulfur and mineral oil are recommended for resistance management because they are more effective than chlorothalonil and other contact fungicides for powdery mildew on the lower leaf surface. Quadris applied in alternation with Nova and Microthiol Disperss was more effective than Quadris alternated with Nova and Bravo in an experiment conducted in Riverhead, NY, in 2002. However, chlorothalonil is recommended when other diseases are a concern. Sulfur is very inexpensive, but can be phytotoxic to melon. Oil is not compatible with chlorothalonil, captan, sulfur, Kelthane, and possibly foliar nutrients; these materials should not be applied within 2 weeks of an oil application.

Strobilurins should be used as little as possible for managing powdery mildew to delay build up of resistant isolates. This will be challenging because these fungicides are effective against several different plant pathogenic fungi that are often present at the same time as powdery mildew. This group is unique in that these fungicides are the first synthetic, site-specific compounds to provide significant control of plant diseases caused by pathogens from all three major groups of fungi. Consequently they have quickly become a very valuable tool for managing diseases. Early in disease development is the best time to apply strobilurins for powdery mildew because they have outstanding ability to inhibit spore germination, however, other diseases controlled by strobilurins begin to develop later in the growing season than powdery mildew. No more than half of the applications in a season should include strobilurins. A limit of one-third is recommended in other areas of the world.

Strobilurins could be used more wisely if the proportion of the powdery mildew pathogen population that is resistant to strobilurins was known before the first application of these fungicides and the impact was known of resistance management programs on the pathogen population. This information will be obtained in NY in 2003 through work supported by the Friends of Long Island Horticulture Grant Program and the New York State Integrated Pest Management Program.

Managing resistance to DMIs is also extremely important. Resistance to this group of fungicides is quantitative, in contrast with the qualitative resistance detected to strobilurins in the USA. With quantitative resistance the pathogen exhibits a continuous range in sensitivity including highly sensitive individuals, moderately insensitive individuals controlled with high but not low rates of the fungicide, and resistant individuals that are not controllable with the fungicide.

Presently in the USA the degree of resistance to DMIs is such that Bayleton, the first DMI fungicide registered, is no longer effective while new DMIs, Nova and Procure, are still highly effective, especially when used at high rates. These fungicides reportedly are no longer adequately effective in other areas of the world, which indicates the cucurbit powdery mildew pathogen has the genetic potential to develop a higher degree of resistance to the DMIs in the USA. To manage DMI resistance, apply these fungicides at the manufacturer's higher label rates and shorter application intervals.

Managing DMI resistance may be challenging. The strobilurin-resistant isolates examined in 2002 also exhibited reduced sensitivity to DMIs. Most individuals in the powdery mildew fungal population in 2002 that were insensitive to one of these chemical groups must have been insensitive to the other, consequently, applying either a strobilurin or a DMI fungicide shifted the population towards insensitivity to both.

It is important to recognize that the systemic fungicides registered recently for cucurbit powdery mildew are not new chemistry. Cabrio is a strobilurin while Procure is a DMI fungicide. Some feel Procure may be sufficiently different from Nova that cross resistance between these may not always occur.

The specific directions on fungicide labels must be adhered to -- they supersede these recommendations, if there is a conflict. Any reference to commercial products, trade or brand names is for information only; no endorsement is intended.

Chapter 9

MOLECULAR BIOLOGY OF BIOCONTROL TRICHODERMA

TRICHODERMA

Trichoderma is a genus of fungi that is present in all soils, where they are the most prevalent culturable fungi. Many species in this genus can be characterized as opportunistic avirulent plant symbionts. This refers to the ability of several *Trichoderma* species to form mutualistic endophytic relationships with several plant species. The genomes of several *Trichoderma* species have been sequenced and are publicly available from the JGI.

Taxonomy

The genus was described by Christiaan Hendrik Persoon in 1794, but the taxonomy has remained difficult to resolve. For a long time it was considered to consist of only one species, *Trichoderma viride*, named for producing green mold.

Subdivision

The genus was divided into five sections in 1991 by Bissett, partly based on the aggregate species described by Rifai:

- *Pachybasium* (20 species)
- *Longibrachiatum* (10 species)
- *Trichoderma*
- *Saturnisporum* (2 species)
- *Hypocreanum*

With the advent of molecular markers from 1995 onwards, Bissett's scheme was largely confirmed but *Saturnisporum* was merged with *Longibrachiatum*. While *Longibrachiatum* and *Hypocreanum* appeared monophyletic, *Pachybasium* was determined to be paraphyletic, many of its species clustering with *Trichoderma*. Druzhina and Kubicek (2005) confirmed the genus as circumscribed was holomorphic. They identified 88 species which they demonstrated could be assigned to two major clades. Consequently, the formal description of sections has been largely replaced by informal descriptions of clades, such as the Aureoviride clade or the Gelatinosum clade.

Species

The belief that *Trichoderma* was monotypic persisted until the work of Rifai in 1969, who recognised nine species. Currently there are 89 accepted species in the *Trichoderma*genus. *Hypocrea* are teleomorphs of *Trichoderma* which themselves have *Hypocrea* as anamorphs.

Characteristics

Cultures are typically fast growing at 25–30 °C, but some species of *Trichoderma* will grow at 45 °C. Colonies are transparent at first on media such as cornmeal dextrose agar (CMD) or white on richer media such as potato dextrose agar (PDA). Mycelium are not typically obvious on CMD, conidia typically form within one week in compact or loose tufts in shades of green or yellow or less frequently white. A yellow pigment may be secreted into the agar, especially on PDA. Some species produce a characteristic sweet or 'coconut' odor.

Conidiophores are highly branched and thus difficult to define or measure, loosely or compactly tufted, often formed in distinct concentric rings or borne along the scant aerial hyphae. Main branches of the conidiophores produce lateral side branches that may be paired or not, the longest branches distant from the tip and often phialides arising directly from the main axis near the tip.

The branches may rebranch, with the secondary branches often paired and longest secondary branches being closest to the main axis. All primary and secondary branches arise at or near 90° with respect to the main axis. The typical *Trichoderma* conidiophore, with paired branches assumes a pyramidal aspect. Typically the conidiophore terminates in one or a few phialides. In some species (*e.g. T. polysporum*) the main branches are terminated by long, simple or branched, hooked, straight or sinuous, septate, thin-walled, sterile or terminally fertile elongations. The main axis may be the same width as the base of the phialide or it may be much wider.

Phialides are typically enlarged in the middle but may be cylindrical or nearly subglobose. Phialides may be held in whorls, at an angle of 90° with respect to other members of the whorl, or they may be variously penicillate (gliocladium-like). Phialides may be densely clustered on wide main axis (*e.g. T. polysporum, T. hamatum*) or they may be solitary (*e.g. T. longibrachiatum*).

Conidia typically appear dry but in some species they may be held in drops of clear green or yellow liquid (*e.g. T. virens, T. flavofuscum*). Conidia of most species are ellipsoidal, 3–5 x 2–4 μm (L/W = > 1.3); globose conidia (L/W < 1.3) are rare. Conidia are typically smooth but tuberculate to finely warted conidia are known in a few species.

Synanamorphs are formed by some species that also have typical *Trichoderma* pustules. Synanamorphs are recognized by their solitary conidiophores that are verticillately branched and that bear conidia in a drop of clear green liquid at the tip of each phialide.

Chlamydospores may be produced by all species, but not all species produce chlamydospores on CMD at 20 °C within 10 days. Chlamydospores are typically unicellular subglobose and terminate short hyphae; they may also be formed within hyphal cells. Chlamydospores of some species are multicellular (*e.g. T. stromaticum*).

Trichoderma genomes appear to be in the 30–40 Mb range, with approximately 12,000 genes being identifiable.

Teleomorph

Teleomorphs of *Trichoderma* are species of the ascomycete genus *Hypocrea*. These are characterized by the formation of fleshy, stromata in shades of light or dark brown, yellow or orange. Typically the stroma is discoidal to pulvinate and limited in extent but stromata of some species are effused, sometimes covering extensive areas. Stromata of some species (Podostroma) are clavate or turbinate. Perithecia are completely immersed. Ascospores are bicellular but disarticulate at the septum early in development into 16 part-ascospores so that the ascus appears to contain 16 ascospores. Ascospores are hyaline or green and typically spinulose. More than 200 species of *Hypocrea* have been described but few have been grown in pure culture and even fewer have been described in modern terms.

Occurrence

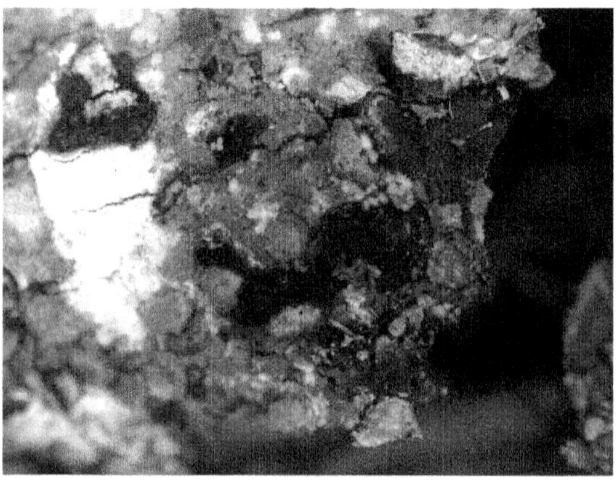

Fig. : *Trichoderma* colony in nature.

Trichoderma species are frequently isolated from forest or agricultural soils at all latitudes. *Hypocrea* species are most frequently found on bark or on decorticated wood but many species grow on bracket fungi (*e.g. H. pulvinata*), Exidia (*H. sulphurea*) or bird's nest fungi (*H. latizonata*) or agarics (*H. avellanea*).

Biocontrol Agent

Several strains of *Trichoderma* have been developed as biocontrol agents against fungal diseases of plants. The various mechanisms include antibiosis, parasitism, inducing host-plant resistance, and competition. Most biocontrol agents are from the species *T. harzianum,T. viride* and *T. hamatum*. The biocontrol agent generally grows in its natural habitat on the root surface, and so affects root disease in particular, but can also be effective against foliar diseases.

Causal agent of Disease

T. aggressivum (formerly T. harzianum biotype 4) is the causal agent of green mold, a disease of cultivated button mushrooms.. *Trichoderma viride* is the causal agent of green mold rot of onion (evidence?). But, Dieback of *Pinus nigra* seedlings caused by a strain of *Trichoderma viride* is known.

Toxic House Mold

The common house mold, *Trichoderma longibrachiatum*, produces small toxic peptides containing amino acids not found in common proteins, like alpha-aminoisobutyric acid, called trilongins (up to 10% w/w). Their toxicity is due to absorption into cells and production of nano-channels that obstruct vital ion channels that ferry potassium and sodium ions across the cell membrane. This affects in the cells action potential profile, as seen in cardiomyocytes, pneumocytes and neurons leading to conduction defects. Trilongins are highly resistant to heat and antimicrobials making primary prevention the only management option.

Medical Uses

Cyclosporine A (CsA), a calcineurin inhibitor produced by the fungi *Tolypocladium inflatum* and *Cylindrocarpon lucidum*, is an immunosuppressant prescribed in organ transplants to prevent rejection.

Industrial Use

Trichoderma, being a saprophyte adapted to thrive in diverse situations, produces a wide array of enzymes. By selecting strains that produce a particular kind of enzyme, and culturing these in suspension, industrial quantities of enzyme can be produced.

- *Trichoderma reesei* is used to produce cellulase and hemicellulase

- *Trichoderma longibrachiatum* is used to produce xylanase
- *Trichoderma harzianum* is used to produce chitinase.

PLANT-TRICHODERMA INTERACTION

The genus ***Trichoderma*** comprises a large number of rhizocompetent filamentous fungal strains found in a large variety of ecosystems. Some rhizosphere-competent strains of ***Trichoderma*** can colonize entire root surfaces with morphological features reminiscent of those seen during mycoparasitism and can be defined as opportunistic plant symbionts.

Like beneficial plant rhizobacteria, *Trichoderma* can induce priming for enhanced defense in plants. In primed plants, defense responses are not activated directly by the beneficial microorganism but are accelerated upon pathogen or insect attack, resulting in enhanced resistance against the attacker.

We are focusing on the study of different aspects of the interaction. Among the questions we are seeking to answer are:

1. How does Trichoderma mask its presence from the plant innate immune system during colonization of the root?
2. How does Trichoderma reprogram plant gene expression during colonization? (This happens both locally in the root and also systemically in the leaves, far from the site of colonization by the entophyte).
3. What are the metabolomic and transcriptomic changes that occur during induced systemic resistance (priming)?

To answer these questions we are using metabolomic and transcriptomic approaches in combination with transgenic approaches of knockout and overexpressing genes, both in plants and in the entophyte.

TRICHODERMA: A BIO-CONTROL AGENT FOR MANAGEMENT SOIL BORN DISEASES

Trichoderma is a very effective biological mean for plant disease management especially the soil born. It is a free-living fungus which is common in soil and root ecosystems. It is highly interactive in root, soil and foliar environments. It reduces growth, survival or infections caused by pathogens by different mechanisms like competition, antibiosis, mycoparasitism, hyphal interactions, and enzyme secretion.

General Characteristics

Colonies, at first transparent on media such as cornmeal dextrose agar (CMD) or white on richer media such as potato dextrose agar (PDA). Mycelium typically not obvious on CMD, conidia typically forming within one week in compact or loose tufts in shades of green or yellow or less frequently white. Yellow pigment may be secreted into the agar, especially on PDA. A characteristic sweet or 'coconut' odor is produced by some species.

Conidiophores are highly branched and thus difficult to define or measure, loosely or compactly tufted, often formed in distinct concentric rings or borne along the scant aerial hyphae. Main branches of the conidiophores produce lateral side branches that may be paired or not, the longest branches distant from the tip and often phialides arising directly from the main axis near the tip. The branches may rebranch, with the secondary branches often paired and longest secondary branches being closest to the main axis. All primary and secondary branches arise at or near 90° with respect to the main axis. The typical *Trichoderma* conidiophores with paired branches assumes a pyramidal aspect.

Phialides are typically enlarged in the middle but may be cylindrical or nearly subglobose. Phialides may be held in whorls, at an angle of 90° with respect to other members of the whorl, or they may be variously penicillate (gliocladium-like). Phialides may be densely clustered on wide main axis (*e.g. T. polysporum, T. hamatum*) or they may be solitary (*e.g. T. longibrachiatum*).

Conidia typically appear dry but in some species they may be held in drops of clear green or yellow liquid (*e.g. T. virens, T. flavofuscum*). Conidia of most species are ellipsoidal, 3-5 x 2-4 μm. Conidia are typically smooth but tuberculate to finely warted conidia are known in a few species.

Synanamorphs are formed by some species that also have typical *Trichoderma* pustules. Synanamorphs are recognized by their solitary conidiophores that are verticillately branched and that bear conidia in a drop of clear green liquid at the tip of each phialide.

Chlamydospores may be produced by all species, but not all species produce chlamydospores on CMD at 20° C within 10 days. Chlamydospores are typically unicellular subglobose and terminate short hyphae; they may also be formed within hyphal cells. Chlamydospores of some species are multicellular (*e.g. T. stromaticum*).

Teleomorphs of *Trichoderma* are species of the ascomycete genus *Hypocrea* Fr. These are characterized by the formation of fleshy, stromata in shades of light or dark brown, yellow or orange. Typically the stroma is discoidal to pulvinate and limited in extent but stromata of some species are effused, sometimes covering extensive areas. Stromata of some species (*Podostroma*) are clavate or turbinate. Perithecia are completely immersed. Ascospores are bicellular but disarticulate at the septum early in development into 16 part-ascospores so that the ascus appears to contain 16 ascospores. Ascospores are hyaline or green and typically spinulose. More than 200 species of*Hypocrea* have been described but only few have been grown in pure culture and fewer have been redescribed in modern terms.

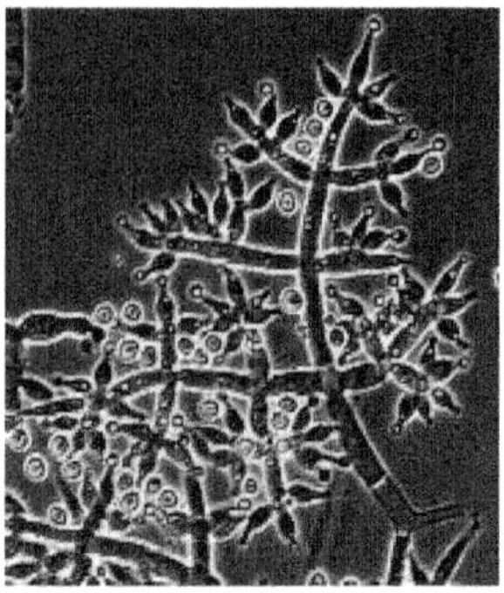

Fig. : *Trichoderma harzianum.*

Benefits of Trichoderma

1. **Disease Control:** Trichoderma is a potent biocontrol agent and used extensively for soil born diseases. It has been used successfully against pathogenic fungi belonging to various genera, *viz.* Fusarium, Phytopthara, Scelerotia *etc.*
2. **Plant Growth Promoter:** Trichoderma strains solubilize phosphates and micronutrients. The application of Trichoderma strains with plants increases the number of deep roots, thereby increasing the plant's ability to resist drought.
3. **Biochemical Elicitors of Disease:** Trichoderma strains are known to induce resistance in plants. Three classes of compounds that are produced by Trichoderma and induce resistance in plants are now known. These compounds induce ethylene production, hypersensitive responses and other defense related reactions in plant cultivars.
4. **Transgenic Plants:** Introduction of endochitinase gene from Trichoderma into plants such as tobacco and potato plants has increased their resistance to fungal growth. Selected transgenic lines are highly tolerant to foliar pathogens such as Alternaria alternata, A. solani, and Botrytis cirerea as well as to the soil-borne pathogen, Rhizectonia spp.
5. **Bioremediation:** Trichoderma strains play an important role in the bioremediation of soil that are contaminated with pesticides and herbicides. They have the ability to degrade a wide range of insecticides: organochlorines, organophosphates and carbonates.

Biocontrol Mechanisms of Trichoderma

The Trichoderma may suppress the growth of the pathogen population in the rhizosphere through competition and thus reduce disease development. It produces antibiotics and toxins such as trichothecin and a sesquiterpine, Trichodermin, which have a direct effect on other organisms. The antagonist

(*Trichoderma*) hyphae either grow along the host hyphae or coil around it and secrete different lytic enzymes such as chitinase, glucanase and pectinase that are involved in the process of mycoparasitism. Examples of such interactions are *T. harzianum* acting against *Fusarium oxyporum, F. roseum, F. solani, Phytophthara colocaciae* and *Sclerotium rolfsii*. In addition, Trichoderma Enhances yield along with quality of produce. Boost germination rate. Increase in shoot & Root length Solubilizing various insoluble forms of Phosphates Augment Nitrogen fixing. Promote healthy growth in early stages of crop. Increase Dry matter Production substantially. Provide natural long term immunity to crops and soil.

Method of Application

1. **Seed treatment:** Mix 6 - 10 g of Trichoderma powder per Kg of seed before sowing.
2. **Nursery treatment:** Apply 10 - 25 g of Trichoderma powder per 100 m2 of nursery bed. Application of neem cake and FYM before treatment increases the efficacy.
3. **Cutting and seedling root dip:** Mix 10g of Trichoderma powder along with 100g of well rotten FYM per liter of water and dip the cuttings and seedlings for 10 minutes before planting.
4. **Soil treatment:** Apply 5 Kg of Trichoderma powder per hector after turning of sun hemp or dhainch into the soil for green manuring. Or Mix 1kg of Trichoderma formulation in 100 kg of farmyard manure and cover it for 7 days with polythene. Sprinkle the heap with water intermittently. Turn the mixture in every 3-4 days interval and then broadcast in the field.
5. **Plant Treatment:** Drench the soil near stem region with 10g Trichoderma powder mixed in a liter of water

Trichoderma Formulations

Important commercial formulations are available in the name of Sanjibani, Guard, Niprot and Bioderma. These formulations contain 3 x 10^6 cfu per 1 g of carrier material. Talc is used as carrier for making powder formulation.

Uses

Used in Damping off caused by *Pythium* sp. *Phytophthora* sp., Root rot caused by *Pellicularis filamentosa*, Seedling blight caused by *Pythium*, Collar rot caused by *Pellicularia rolfsii*, Dry rot caused by*Macrophomina phaseoli*, Charcoal rot caused by *Macrophomina phaseoli*, Loose smut caused by Ustilago segetum, Karnal bunt diseases, Black scurf caused by *Rhizoctonia solani*, Foot rots of Pepper and betel vine and Capsule rot of several crops. Effective against silver leaf on plum, peach & nectarine, Dutch elm disease on elm's honey fungus (*Armillaria mellea*) on a range of tree species, Botrytis caused by *Botrytis cinerea*,

Effective against rots on a wide range of crops, caused by *fusarium, Rhizoctonia,* and *pythium,* and sclerotium forming pathogens such as *Sclerotinia* & *Sclerotium*

Recommended For

Trichoderma is most useful for all types of Plants and Vegetables such as cauliflower, cotton, tobacco, soybean, sugarcane, sugarbeet, eggplant, red gram, Bengal gram, banana, tomato, chillies, potato, citrus, onion, groundnut, peas, sunflower, brinjal, coffee, tea, ginger, turmeric, pepper, betel vine, cardamom *etc.*

Precautions

- Don't use chemical fungicide after application of Trichoderma for 4-5 days.
- Don't use *trichoderma* in dry soil. Moisture is a essential factor for its growth and survivability.
- Don't put the treated seeds in direct sun rays.
- Don't keep the treated FYM for longer duration.

Compatibility

Trichoderma is compatible with Organic manure Trichoderma is compatible with biofertilizers like *Rhizobium, Azospirillum, Bacillus Subtilis* and *Phosphobacteria. Trichoderma* can be applied to seeds treated with metalaxyl or thiram but not mercurials. It can be mixed with chemical fungicides as tank mix.

TRICHODERMA LONGIBRACHIATUM

Trichoderma longibrachiatum is a fungus in the genus *Trichoderma.* In addition to being a distinct species, *T. longibrachiatum* also typifies one of several clades within *Trichoderma* which comprises 21 different species. *Trichoderma longibrachiatum* is a soil fungus which is found all over the world but mainly in warmer climates. Many species from this clade have been adopted in various industries because of their ability to secrete large amounts of protein and metabolites.

Taxonomy and Nomenclature

Trichoderma is a diverse genus with other 135 species in Europe alone. This species was first characterized by Mien Rifai in 1969. It is an exclusively anamorphic species complex allied with the sexual species, *Hypocrea schweinitzii.* Evolutionarily *T. longibrachiatum* is the youngest clade of *Trichoderma.*

Growth and Morphology

Trichoderma longibrachiatum is a fast-growing fungus. It typically produces off-white colonies that change to greyish green with age. This species is able to grow over a wide range of temperature; however the optimal temperature for growth is ≥ 35 °C. *Trichoderma longibrachiatum* is a clonal species that reproduces through 1-celled, smooth-walled conidia.

Metabolism

Trichoderma longibrachiatum occurs commonly on decaying plant material where its ecological role ranges from that of a strict saprotroph to a parasite of other saprotrophic fungi. *Trichoderma longibrachiatum* uses cellulases to digest cellulose from decaying plant biomass, and chitinases to digest the chitinous walls of other fungi. It is also able to digest proteins with the aid of aspartic proteases, serine proteases, and metalloproteases. *Trichoderma longibrachiatum* produces many secondary metabolites including:peptaibols, polyketides, pyrones, terpenes and diketopiperazine-like compounds.

Distribution and Habitat

Trichoderma longibrachiatum is a soil fungus often found on dead wood, other fungi, building material and sometimes animals.

Toxicity

Trichoderma longibrachiatum is not thought to pose risk to human health, although it has been isolated as an indoor contaminant with high allergenic potential. This species has also been implicated in the colonization of immunocompromised people and has been found in the blood cultures derived from a neutropenic patient with lymphoma, bone marrow transplant patients, and patients with severe chronic kidney disease.

Industrial Use

Trichoderma species are useful in industry because of their high capacity to secrete large amounts of protein and metabolites. It has been suggested that *Trichoderma longibrachiatum* could be used as a biocontrol agent for its parasitic and lethal effects on the cysts of the nematode *Heterodera avenae*. Because *T. longibrachiatum* is a mycoparasite, it has also been investigated for use in combating fungal diseases on agricultural crops. Its enzymatic capacity could potentially be useful in bioremediation, for use in remediation of polycyclic aromatic hydrocarbons (PAHs) and heavy metals.

Other industrial uses include using the various cellulases for staining fabrics in the textile industry, increasing digestibility of poultry feed, and potentially in the generation of biofuels. *Trichoderma longibrachiatum* has also been reported in promoting plant growth by increasing nutrient uptake, inhibiting the growth of plant parasites, increasing carbohydrate metabolism, and phytohormone synthesis.

Chapter 10

BIOLOGICAL CONTROL OF FUNGAL DISEASES

FUNGI

Some insect species, including many pests, are particularly susceptible to infection by naturally occurring, insect-pathogenic fungi. These fungi are very specific to insects, often to particular species, and do not infect animals or plants. Fungal growth is favored by moist conditions but fungi also have resistant stages that maintain infection potential under dry conditions. Fungi have considerable epizootic potential and can spread quickly through an insect population and cause its collapse. Because fungi penetrate the insect body, they can infect sucking insects such as aphids and whiteflies that are not susceptible to bacteria and viruses.

Several fungal species have potential as microbial insecticides and, in some countries, are commercially available in formulations that can be applied using conventional spray equipment.

Habitat (Crops)

Most crops including soybeans, greenhouse crops, vegetables, cotton, citrus, and ornamentals; also interior plantscapes and forests. An aquatic fungus infects mosquito larvae of some genera.

Pests Attacked

Most insect pests are susceptible to fungal pathogens. Some fungi, such as the *Entomophthora* and related species, are fairly specific with regard to the groups of insects affected; others, such as *Beauveria*, have a wider host range.

Mode of Action

Fungi invade insects by penetrating their cuticle or "skin." Once inside the insect, the fungus rapidly multiplies throughout the body. Death is caused by tissue destruction and, occasionally, by toxins produced by the fungus. The fungus frequently emerges from the insect's body to produce spores that, when spread by wind, rain, or contact with other insects, can spread infection.

Symptoms

Infected insects stop feeding and become lethargic. They may die relatively rapidly, sometimes in an upright position still attached to a leaf or stem, perhaps in an elevated location or concentrated near crop borders. The dead insect's body may be firm and "cheese-like" or an empty shell, often but not always with cream, green, red, or brown fungal growth, either enveloping the body or emerging from joints and body segments. Infected aphids can be swollen and discolored; infected root maggot flies may be clustered on shoot tips, tall grasses, or other prominent features.

Relative Effectiveness

Insect-pathogenic fungi usually need moisture to enable infection, and natural epizootics are most common during wet or humid conditions. The effectiveness of these fungi against pest insects depends on having the correct fungal species and strain with the susceptible insect life stage, at the appropriate humidity, soil texture (to reach ground-dwelling pest species), and temperature. The fungal spores, which can be carried by wind or water, must contact the pest insect to cause infection. Naturally occurring fungal epizootics may decimate aphid, root maggot fly, caterpillar, leafhopper, and thrips populations. They can be an important natural control of aphids in potatoes and other crops.

Many insect-pathogenic fungi occur in the soil. There is evidence that application of some soil insecticides, fungicides, and herbicides can inhibit or kill these fungi. For example, even quite low concentrations of some herbicides can severely limit the germination and growth of *Beauveria bassiana* fungal spores in soil samples.

Commercially Availability

Metarhizium anisopliae is registered in the U.S. for control of household cockroaches. *Beauveria bassiana* Strain GHA (trade names Mycotrol GH-OF and Mycotrol GH-ES) is registered to control grasshoppers, locusts, and Mormon crickets on rangeland, improved pastures, alfalfa, corn cotton potatoes, rapeseed, safflower, small grain crops, soybeans, sugarbeets, and sunflowers. *Paecilomyces fumosoroseus* Apopka Strain 97 has been approved for use on ornamentals, non-food crops in greenhouses, and interiorscapes to manage whiteflies, aphids, thrips, and spider mites and is sold by Thermo Trilogy Corp. In addition, there are fungal antagonists available to combat various fungal pests.

Some Common Insect-Pathogenic Fungi

Entomophthora muscae infects flies. Susceptible pest species include the adults of the onion maggot, cabbage maggot, and seedcorn maggot. The fungus multiplies within the body of the adult fly which becomes enlarged; yellowish bands of fungal spores stripe the abdomen.

Related species include *Pandora neoaphidis*, a common, naturally occurring pathogen of aphids that can be particularly effective during moist periods. *Zoophthora radicans* is another common fungus, with a broad host range. Strains have been isolated from several caterpillars, including diamondback moth, leafhoppers, aphids, and some weevils. Other related species infect thrips.

Beauveria bassiana is an insect-pathogenic fungus found naturally on some plants and in the soil. Epizootics are favored by warm, humid weather. It is known as the white muscardine fungus because infected insect larvae eventually turn white or gray. *Beauveria* is used as a fungal microbial insecticide in some countries. It has an extensive host list that includes such important pests as whiteflies, aphids, grasshoppers, termites, Colorado potato beetle, Mexican bean beetle, Japanese beetle, boll weevil, cereal leaf beetle, bark beetles, lygus bugs, chinch bug, fire ants, European corn borer, codling moth, and Douglas fir tussock moth. It has even been found infecting the lungs of wild rodents, and the nasal passages of humans. Unfortunately, natural enemies, such as lady beetles, are susceptible too. One possible application method that may avoid harming beneficial insects is the use of fungus-contaminated insect baits that are attractive to pest species only. There are many different strains of the fungus that exhibit considerable variation in virulence, pathogenicity and host range. It occurs in the soil as a saprophyte.

Metarhizium spp. is being tested as a natural enemy of corn rootworm, white grubs (scarabs), and some root weevils. It has a very broad host range and is extensively used in Brazil against spittle bugs in sugar cane and alfalfa.

Other species of insect-pathogenic fungi have been tested as microbial insecticides for the control of pests. *Verticillium lecanii* is used in Europe against greenhouse whitefly and thrips and aphids, especially in greenhouse crops. *Neozygites floridana* has been successfully tested against spider mites, and Nomuraea rileyi has activity against green cloverworm, cabbage looper, imported cabbageworm, armyworms, corn earworm, and tobacco budworm. *Hirsutella thompsonii* infects mites and was previously registered for use in the United States.

FUNGI FOR THE BIOLOGICAL CONTROL OF INSECT PESTS

Fungi are a diverse group of organisms with close ties to agriculture. Some fungi create devastating diseases in crops, while others are crops themselves (mushrooms). Other fungi are used successfully to protect crops from a variety of pests. Among the most prominent of these are the fungi that are used against insects and other related pests.

Most fungi used for the control of insect pests belong to the group hyphomycetes. Some species have been developed as commercial products because of their ability to be mass produced. Most fungi in this group are usually found in the soil and can cause natural outbreaks on their own when environmental conditions are favorable. They can infect a wide range of insect hosts. Specific fungal strains in commercial products target thrips, whiteflies, aphids, caterpillars, weevils, grasshoppers, ants, Colorado potato beetle, and mealybugs.

Currently (2008) allowable products containing the hyphomycete fungus, *Beauveria bassiana,* that are commercially available include Mycotrol O (Emerald BioAgriculture), Naturalis Home and Garden (H&G) and Naturalis L (Troy BioSciences, Inc.). Before applying any pest control product, make sure to include what you might want to use and how you intend to use it in your organic system plan and get your certifier's approval. (Caution: the use of an unapproved material can result in the loss of certification. Always check with your certifier before purchasing or using a new product or material to ensure that it is permitted for use in your organic farming system.

There is another commonly encountered group of fungi called the entomophthorales. Fungi in this group can cause natural outbreaks in the populations of their insect hosts, but they are difficult to mass produce and as yet are not in commercial production. They tend to be much more host specific; one well known species only infects aphids. Despite the difficulties in producing them commercially, they can still have a large impact on the pest populations they infect. There are currently no commercially-based products available for organic vegetable production.

Life Cycle of Fungi Infecting Insects

Fungi that infect insects are found in the environment as spores. Insects can become infected when they come into contact with spores on the surface of plants, in the soil, in the air as windborne particles, or on the bodies of already dead insects. Spores attach to the surface of the insect and infect by penetrating through the insect cuticle, often at joints or creases where the insect's protective covering is thinner. Once inside, the fungus grows throughout the body of the insect. Many fungi also produce toxins in the host that increase the speed of kill or prevent competition from other microbes. Usually after the insect has died, the fungus grows out through the outer covering (exoskeleton) of the insect, usually at thinner areas like joints or creases, and begins to produce spores. The spores of commercially developed fungi in the group hyphomycetes are spread passively by the action of the wind, rain, or contact with other hosts or animals in the environment. The spores of fungi that create natural outbreaks, in the group entomophthorales, are often actively ejected from the dead insect. Since many species in this group of fungi infect insects which cluster together, like aphids, this tactic can drastically increase the spread of the infection. Insects killed by fungi often have a "fuzzy" appearance, caused by the growth of the fungus out of the exoskeleton to produce spores. Most commercial strains of fungi produce spores that are either white or green in color, although the color of the fungi can change over time as the fungus grows and ages.

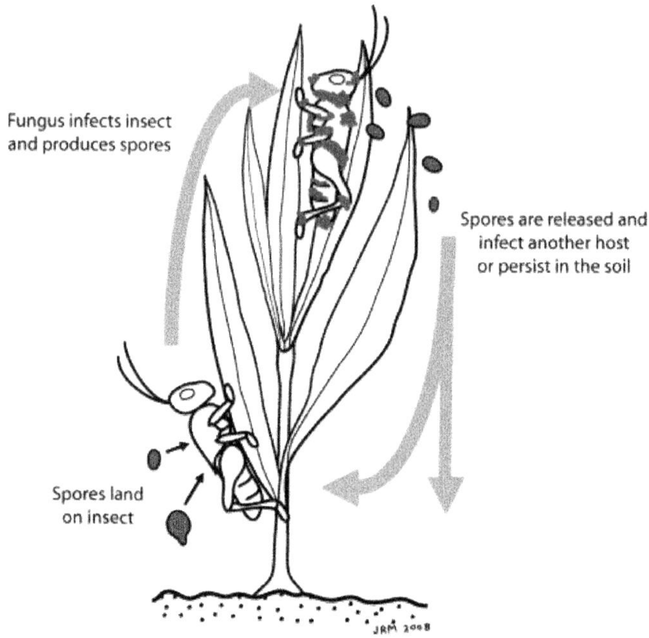

Fig. : Generalized life cycle of fungi infecting insects (spores not to scale).

Spores that do not encounter a host either die or persist in sheltered areas of crop plants or in the soil. Although some species of fungi can produce spores which can persist for years in the soil, most spores are only viable for a growing season or a year at most.

Advantages and Disadvantages of Fungi for Controlling Insect Pests

Advantages

Fungi make good biological control agents for a variety of reasons. They generally do not affect people or other mammals, making them extremely safe to use. It is relatively easy to mass produce spores of insect-parasitic fungi in the hyphomycete group, so they are comparably priced with other biological control agents, such as bacteria. Most commercial fungal products are formulated as spores, which are easily adapted to existing application technology, such as spray rigs. The relatively broad host range of many fungi means one can often achieve control of multiple pests with the same product. Finally, successful infections can spread to other hosts and lead to high rates of persistence within a growing season, even if between season persistence tends to be low for most types of fungi.

Disadvantages

High concentrations of spores are often needed to get adequate control of pests in a crop, which can cut down on the cost effectiveness of fungal products. The kill time is relatively long (~1 week for most fungi), although strains used for commercial products are chosen to kill as fast as possible. Their broad host range can sometimes be a problem, especially if beneficial insects (*i.e.* predators, parasitoids, and pollinators) are present in a crop; non-target mortality in these populations of beneficial insects can negatively impact the success of the overall biological control program. Environmental factors can also play an important role in the success of fungi. Moist conditions or high relative humidity in the canopy of the crop are often necessary for control to be effective. Prolonged exposure to sunlight can also inactivate spores, reducing persistence in the crop. Owing to these environmental limitations, natural outbreaks of fungi tend to be sporadic and very patchy in the environment, which can limit their effectiveness in controlling pests.

Suggestions for Application

As with all biological control agents, fungi work best as one component of a comprehensive integrated pest management program. There are several tactics growers can us to increase their effectiveness, though, especially for applied commercial products:

- Scout consistently and often. Apply only when the target pest is seen, not as a preventive application as residues are not long-lasting. The best time to apply fungi is before pest populations reach their peak, so early application can increase their effectiveness. Also, scouting can help determine the population levels of beneficial insects and pollinators so the timing of fungal applications will not impact them as strongly. Finally, scouting can help discover natural outbreaks of fungi (*e.g.* aphid fungi) in time to influence control decisions.
- Time applications of fungi to coincide with host life stages that are more likely to come in contact with the spores. In general, *B. bassiana* products are more effective against earlier compared with later stages of insects. For example, applying a fungal product for grasshoppers will be most effective when there are active nymphs present that have not grown into winged adults.
- Do not apply fungal products during droughts or dry spells since the environmental conditions will decrease their effectiveness.
- Be aware of fungicide applications in the area. Even if fungicides are not directly applied to the crop, drift from nearby fields could impact the success of a fungal biological control agent.
- Apply fungal inoculum carefully to get effective coverage. Cover all plants thoroughly. Also try to reduce of spill-over into refuge areas where natural enemies may be present.

- Do not apply fungal products during the heat of the day since this will diminish the potency of the spores. There have been some reports of phytotoxicity to young vegetable transplants with products formulated as an emulsifiable suspension. Also, do not apply on rainy days, when spores will be washed off of plant surfaces and may not come into contact with the target pest.
- Use cropping practices which encourage a diverse understory and soil surface, such as cover cropping or conservation tillage. These practices will help maintain fungi in the field and could increase persistence within and between seasons.

BIOCONTROL USING FUNGUS

Natural methods of controlling fungi and insects are being tested in India and Sri Lanka with increasing success. Worldwide, there are 380 fungal pathogens which attack tea. In Sri Lanka, scientists are focussing on antagonistic fungi and bacteria that parasitize pathogenic fungi and destroy the pathogen in one of several ways. According to Abhaya Balasuriya, a scientist at the Tea Research Institute (TRI) in Talawakelle, Sri Lanka, the first step in biocontrol using fungi involves searching for strains with potential for antagonism from areas where diseases are prevalent. "Isolates showing some promise are made into pure cultures before they are used for *in vitro* tests," he says. He notes that *Trichoderma harzianum* is a widely studied antagonist because it has the ability to out-compete pathogenic fungi, and to parasitize them by producing toxic compounds or enzymes that destroy pathogenic cells.

In the lab, Balasuriya and his colleagues have found some fungal biocontrol agents that perform comparably with chemical fungicides: *T. harzianum* was able to suppress the vegetative growth of black root disease by 85 per cent, compared to fungicides Pyraclostrobin and Propiconazole, both of which achieved 100 per cent control. When tested against red root disease and horse hair blight, *Trichoderma harzianum Rifai* showed better control than the systemic fungicide, Bitertanol. And in a field trial of horse hair blight control, Balasuriya found that a treatment containing *T. harzianum* was capable of bringing comparable control (about 50 per cent) to that of hydrated liming or the fungicide Propiconazole.

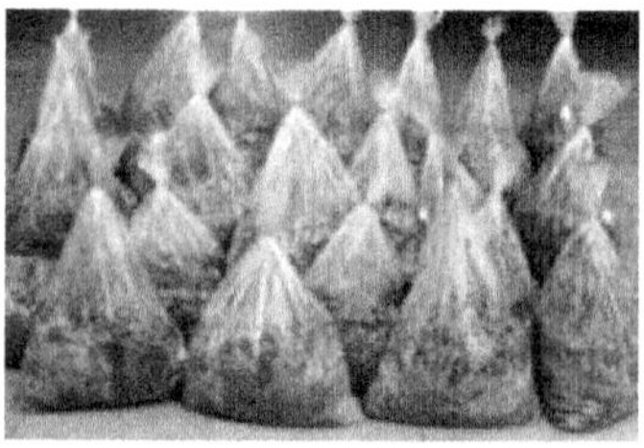

Fig. : Mass multiplication of Trichoderma harzianum on decomposed coffee waste.

Bacterial biocontrol of fungal pathogens is also proving effective and may be more successful than fungi in some cases because bacteria grow rapidly, are easier to handle and are able to survive under adverse conditions. At the Tea Research Foundation in Valparai, India, scientists studying grey blight have isolated two indigenous bacterial strains that produce lytic enzymes which can break down the cell wall of the fungal pathogen, effectively suppressing spore germination.

Insect Biocontrol

There are some 150 insect pests of tea, the most harmful being spider mites, live-wood termites, parasitic nematodes and shot-hole borer beetles. At the TRI, a broad research approach has been used to locate the 'weak' points in the life cycle of pests and their ability to infest the crop. Among techniques used are host-plant resistance breeding, methods of horticultural control, and biologically active insecticides such as hormone-mimics and chitin-synthesis inhibitors. Scientists are also using computer models to predict insect and mite outbreaks, which helps farmers to limit the use of chemical pesticides to these events.

While most biocontrol agents are native to the country of origin, Balasuriya notes that in Sri Lanka "there is a classic example of a parasitoid wasp (*Macrosentus homonae*), imported from Java in 1936 for the control of *Homona coffearia* (a leaf-eating caterpillar also known as the tea tortrix), which is still working." The TRI states that, "This success is regarded in scientific literature as the most outstanding example of classical biological control ever achieved in the world on a perennial crop."

Natural Pesticides

Pesticides derived from natural substances are less harsh and can at least partly replace synthetic pesticides. TRI scientists have discovered non-chemical alternatives to methyl bromide, an ozone-depleting substance being phased out across the world, which was used to eradicate nematode species that attack tea plants. Substituting nematode-infested soil with agricultural wastes, which have pesticidal properties (such as refuse tea, coir dust and paddy husk), reduces numbers of nematodes, as does drying nursery soil in the sun and planting nematicidal plants like wild sunflower. Meanwhile, in India, the Hindustan Lever company, using a bio-pesticide mixture made from sulphur and an extract of the montanova plant, has reduced total pesticide use on their plantations by 16 per cent from 2001 to 2003.

FUNGAL DISEASES OF CROPS

There are many different fungi that are capable of causing disease on plants. Our research aims to understand how this happens and how it could be combated. Most of our efforts are focussed on two plant:pathogen associations:

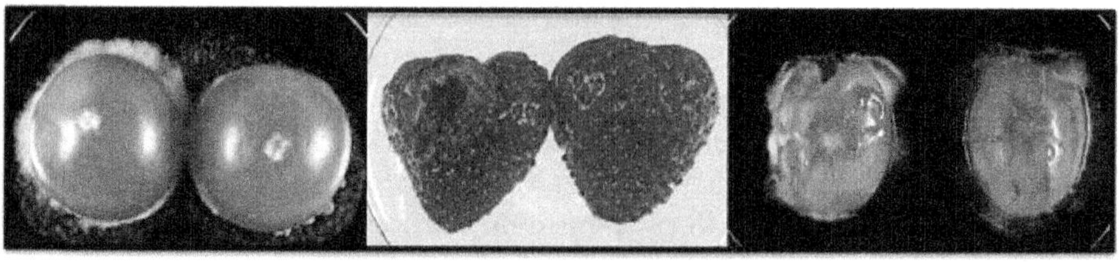

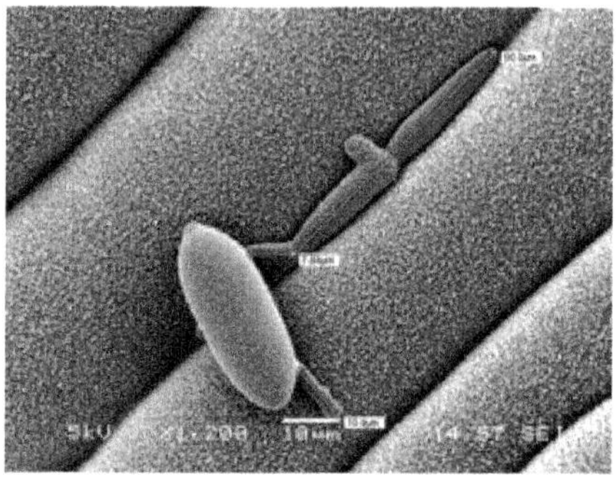

Mycosphaerella graminicola, also known as *Septoria tritici* causes leaf blotch on wheat, and is one of the most serious diseases of this crop. It is mainly controlled by the use of fungicides, but resistance can occur to these meaning that there is an ongoing need to develop new or improved active ingredients to control this disease. Recent research efforts have been centred on investigations into the secondary metabolism ability of this fungus, looking for new compound made by this species by chemical extraction and analysis from various conditions and also where and when the gene pathways are activated using promoter:GFP reporter fusions. Projects are also under way with Rothamsted Research investigating secreted proteins from this fungus.

Botrytis cinerea is the fungus responsible for gray mould on many fruit and vegetable crops. It has a wide host range and has world-wide distribution. We are interested in how this fungus performs post-transcriptional gene silencing and whether this is a defence against Mycoviruses - fungal viruses. These viruses usually have an RNA genome and this can be recognised and degraded in other eukaryotes, but the ability to do this has not been investigated in *B. cinerea*. If viruses can be identified which can avoid of overcome such defence processes, they may have the potential to be effective biological control agents for this disease. Some aspects of this research are in collaboration with Prof. Pearson's group in Auckland, New Zealand.

Other Interests in this Area Include

Disease Control

Past projects have included validation of fungicide targets using gene disruption to knock out the targeted gene and to determine the impact on virulence before committing to screening for fungicides that inhibit the same process. In addition, fungicide mode of action has been investigated using microscopy, transcriptional profiling and also analysis of resistant mutants derived from the field or laboratory.

Fungal Disease of Fungi

Agaricus bisporus, a basidiomycete fungus, commonly known as the white button mushroom has approximately 1.5 million tonnes produced for human consumption world-wide annually. Pathogenic threats to *A. bisporus* include bacteria, viruses, mites, insects and fungi, the most significant to the industry being*Lecanicillium (Verticillium) fungicola*.

L. fungicola infection is manifested in three types of symptoms, designated spotty cap, stipe blowout and dry bubble, all of which can drastically reduce yields and value of the crop. Despite strict hygiene, cultural practices and manipulation of environmental conditions, *L. fungicola* continues to be a major problem. Recent investigations have revealed a developing resistance to the single effective fungicide within *L. fungicola*. The industry is therefore in need of effective control systems due to the lack of availability of effective fungicides, which is combined with little or no intrinsic resistance within *A.bisporus*.

Major breakthroughs in *Lecanicillium (Verticillium) fungicola* research by our group.

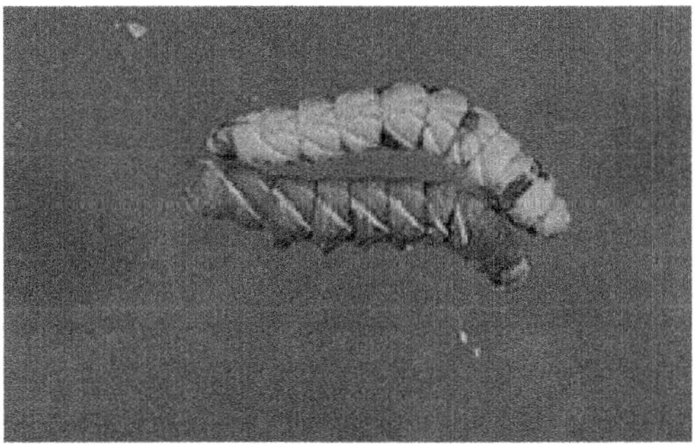

Fungal Disease of Insects

Several species of fungus can cause disease on insects and these may have the potential for use in biological control of such pests. Research has focussed on*Metarhizium*, a fungus with a broad host range that uses both enzymes and toxins to help it invade the insects and overcome any immune response.

EFFECT OF SOIL SOLARIZATION ON FUNGI AND BACTERIA

Disinfestation of soil of soilborne pathogens and pests has been mainly based on methods of soil fumigation and soil steaming , although soil flooding for various periods also has been effective. However, the report of Katan *et al.*

in 1976 on the use of solar heating for disinfesting soil of pathogens and pests, now known as soil solarization, attracted the attention of many agriculturalists and is now widely used. In contrast to soil steaming or chemical fumigation of soil, which are broad spectrum in their killing action of soilborne organisms , soil solarization is non-chemical, is selective in its action, and improves the filth and nutrient status of soil. Temperatures associated with soil steaming usually exceed the maximum temperatures (60° to 61.5°C) that can be tolerated by eucaryotic organisms and some non-spore-forming bacteria, although for some bacteria, life-limiting temperatures are over 100°C..

The application of chemical fumigants to soil for plant disease control has a long history dating back to 1869, and for soil steaming, back to 1893. However, the near-sterilization of agricultural soils by soil steaming or fumigation is now viewed with concern since these procedures destroy many of the beneficial and essential biotic and abiotic elements of soil, including biological control of plant pathogens, which are necessary for plant health. In this sense, soil solarization is an important advance because it targets mesophylic organisms, which include most plant pathogens and pests, without destroying the beneficial mycorrhizal fungi and the growth-promoting *Bacillus* spp. ; it contributes significantly to the principles of sustainable agriculture. Moreover, the increasing constraints of public opinion and regulations regarding chemical pollution of the environment by pesticides and the danger associated with their use, have focused attention on alternative methods for pest control, such as soil solarization and biological controls.

Technology of Soil Solarization

Soil solarization is a mulching process and is an extension of current mulching practices of using plastic films to warm soils to increase seed germination and seedling growth during cool weather or in the early spring months. The distinction, however, between soil solarization and the use of plastic films for warming soils, is that soil solarization reaches killing temperatures for many soilborne organisms, including pathogens, seeds, and weed seedlings. Soil solarization has been especially useful in regions with high solar radiation and where daily temperatures reach 32°C or higher, it is non-chemical, easily applied and is of no danger to workers; it encompasses the objectives of integrated pest management (IPM) and reduces the use of toxic chemicals for plant disease and pest control.

Soil Preparation

The soil in field or glasshouse areas to be solarized should be prepared to provide a smooth surface so that the plastic film will lie flat. The presence of large clods of soil or plant debris that raise the plastic film from the soil will cause air pockets which insulate the soil and reduce the rise in temperature needed for effective solarization. Also, laying the plastic film across deep furrows will reduce the effectiveness of soil solarization, especially in the furrows.

Soil can be prepared for solarization with or without raised planting beds. For raised beds, the plastic film is laid over the length of the bed and the edges are buried in the furrows. For crops grown without beds or on-the-flat, the edges of the plastic are anchored or buried in the soil about 20 cm deep. For large scale use of solarization on-the-flat, the plastic sheets can be glued together or folded together and buried in soil in such a way as to connect the plastic sheets in a continuous layer of film.

Soil Moisture

Soil solarization is a hydrothermal process and its success depends on moisture for maximum heat transfer to soilborne organisms. It has been shown that the temperature maxima of soils increase with increasing moisture content. Soil should be at least 70 percent of field capacity in the upper zones and moist to depths of at least 60 cm. If the field or soil area is level, irrigation under the plastic film can be accomplished by running water down shallow furrows made by tractor wheels. However, pre-irrigation of soil may be more practical, especially for uneven areas or fields; the plastic film should be laid within a few days or as soon as the soil will support machinery without compaction. When soil is bedded-up before solarization, the plastic film is laid over the bed and anchored on each side in the furrows. Irrigation water is then run in the furrows and the moisture will be absorbed into the raised beds.

Postplant solarization in newly planted or established orchards or vineyards has shown that soil under the canopy of perennial or woody plants can be effectively solarized without heat injury to the plants. Plastic polyethylene (PE) film up to 4 metres square is cut to the centre in order to place the film around the base of the tree. Plastic tape is then used to seal the PE around the trunk of the trees or vines and also to seal the cut edges together. The outside edges of the PE are anchored in the soil. In postplant solarization it is convenient to irrigate the soil before applying the PE film. After an initial pre-irrigation or irrigation under the PE film, no further irrigation is necessary since the soil will remain moist for several months. In semi-arid regions, postplant solarization of soil around fruit trees has greatly reduced the amount of water needed for irrigation.

Plastic Films

Polyethylene, polyvinylchloride, and ethlyene vinyl acetate are plastic films most frequently used for soil solarization. Among these, transparent polyethylene (PE) is ideal for soil solarization because it is transparent to solar radiation (280 to 2 500 nm) except for two narrow bands in the far infra-red; it is much less transparent to terrestrial radiation (5 000 to 35 000 nary), reducing the escape of heat from the soil. Additionally, PE films are flexible and have high tensile strength and resist puncturing and tearing.

Without ultra-violet stablizers in the plastic, PE films may deteriorate within four to six weeks, although fresh film (1 or 2 mil in thickness) has lasted as long as nine weeks under field conditions without major deterioration. In several studies, the thinner transparent PE films (I to 1.5 mil) have been more effective in solar heating of soil than thicker films (2 to 6 mil) or black PE film. Using a double layer of plastic creates an inner layer of air which acts as an insulator and the soil heats several degrees higher than soil being solarized with a single layer of plastic film. A similar principle is evident when glasshouse soils are solarized.

Soil Temperature

The heating of soil during solarization is dependent on several factors, such as soil moisture content, the duration and intensity of sunlight, air temperature, the thickness and transparency of plastic to solar radiation and to terrestrial radiation. The colour of soil is also important; dark soils reach higher temperatures during solar radiation than lighter coloured soils due to their greater absorption of solar radiation. Of major importance for effective soil solarization is the preparation of the soil surface so that the plastic film lies flat against the soil and is not raised up by clods and plant debris, following the line of deep furrows. Depending on soil depth, maximum temperatures of solarized soil in field areas are commonly between 42° to 55°C at the 5 cm depth and range from 32° to 36°C at the 45 cm depth. Solarized soil under glasshouse conditions or under double layers of transparent PE film, reaches higher temperatures than soil under a single layer of film in the field.

Effects of Soil Solarization on Soilborne Organisms

The decline in the viability of soilborne micro-organisms and weed seeds during solarization depends on both the soil temperature and exposure time, which are inversely related. For example, the ED90 of *Verticillium dahliae* at 37°C in soil with moisture at field capacity was about 14 hours whereas at 50°C it was approximately 9 minutes. Similar time/temperature relationships hold for most plant pathogenic micro-organisms which are grouped as mesophiles. However, some plant pathogenic fungi, such as *Macrophomina phaseolina* and *Pythium aphanidermatum* are less heat sensitive than most other soilborne plant pathogens; their control requires optimal conditions of temperature and moisture during solarization.

Sensitivity to High Soil Temperatures

Various lines of evidence have been reported regarding the ability of micro-organisms to withstand high temperatures ; however, the older concept that a high turnover of macromolecules is the basis for thermos/ability is no longer accepted. All macromolecules of thermophilic organisms that survive at temperatures up to 60 C appear to be stable at these high temperatures.

Direct Effects of Soil Solarization

The inability of organisms to tolerate high temperatures is related to an upper limit in the degree of fluidity of membranes, beyond which breakdown of membrane function may be associated with membrane instability. Additional causes for the thermal decline of microorganisms at high temperatures involve the sustained inactivation of respiratory enzymes. These are direct affects of high soil temperatures and account for a major share of the reduction in populations of soilborne micro-organisms and weed seeds.

Indirect Effects of Soil Solarization

Some effects of soil solarization are indirect. For example, cells of plant pathogens weakened by heat stress are more vulnerable by several orders of magnitude to soil fumigants, to antagonistic micro-organisms which are more able to tolerate high soil temperatures, and to changes in the gas environment which may develop during soil solarization.

During solarization of soil, changes occur in the structure or filth of soil, in soluble mineral substances available for plant and microbial growth, and in the populations of soilborne micro-organisms. These changes effect the inoculum density of plant pathogens, and also their aggressiveness and survival. Changes in the populations of other soilborne micro-oganisms occur during and after solarization which may influence the disease suppressiveness of soil and also the increased plant growth response associated with solarized soils.

Effects of Solarization on Soilborne Bacteria

During solarization of soil, populations of oxidaze negative fluorescent pseudomonads and gram positive bacteria, including *Bacillus* species, may be reduced by 78 to 86 percent compared with non-solarized soil ; whereas, populations of Actinomycetes may be reduced from 45 to 58 percent in solarized soil. Surprisingly, after solarization, *Pseudomonas* species quickly recolonize the soil and their populations reach high levels. Of great significance is the change in populations of *Bacillus* species during solarization; the percentage of colonies in solarized soil which exhibited antibiosis to *Geotrichum candidum* increased nearly 20-fold when compared with non-solarized soil. These bacteria are among those which are rhizosphere competent and are believed to contribute to the increased growth response of plants grown in solarized soil. Although initial populations of *Bacillus* species are greatly reduced, they are spore formers and are a major component of the soil microflora.

In contrast to the studies in California and Israel , studies in Western Australia showed that solarization increased the total numbers of bacteria and actinomycetes in soil. However, as in the California study where there was an increase in the proportion of antagonistic gram positive bacteria in solarized soil, the Western Australia study showed that the proportion of bacteria (actinomycetes) antagonistic to *Fusarium oxysporum, F. solani*, and *Rhizoctonia solani* was increased compared with non-solarized soil. In other studies , Actinomyces scabies was controlled by soil solarization.

Only limited studies have been made on the survival of plant pathogenic bacteria in solarized soil; however, *Agrobacterium* species are highly sensitive to solarization and populations are reduced up to 72 percent. Similarly, species of *Rhizobium* are heat sensitive and their populations are greatly reduced during soil solarization. Since none of the plant pathogenic bacteria form spores, their decline in solarized soil is expected to parallel that of green fluorescent pseudomonads.

However, pseudomonads are among the bacteria that quickly recolonize solarized soil and reach high populations ; some are rhizosphere competent and can provide a degree of protection against fungal root pathogens and stimulate plant growth. Like the pseudomonads, some strains of *Bacillus* species are rhizosphere competent and either through aggressive growth or the production of antibiotics, they appear to be major contributors to the disease suppressiveness of soils following solarization. *Bacillus* species are the predominant Gram-positive bacteria that survive soil solarization. Recolonization of solarized soils includes saprophytic bacteria which have less stringent nutritional requirements than plant pathogens.

Effects of Solarization on Soilborne Fungi

Plant pathogenic fungi are among the most sensitive soilborne organisms to soil solarization, especially species that are unable to grow at temperatures higher than 30° to 33°C; they are categorized as mesophiles. Immediately after soil solarization, the population densities of "total" fungi were reduced by 85 to 90 percent in different experimental plots. However, population densities of thermotolerant and thermophilic fungi remained relatively high following solarization, and increased to levels higher than present in non-solarized soil. The fungi most frequently isolated were thermotolerant *Aspergillus* and *Penicillium* species.

In regard to the effect of soil solarization on beneficial soilborne microorganisms, the mycorrhizal fungus *Glomus fasiculatus* survives solar heating of soil to the extent that there does not appear to be a reduction in its colonization of host roots in solarized soil.

Many plant pathogenic fungi, with the exception of *Macrophomina phasseolina, Pythium aphanidermatum*, and species of *Aspergillus, Penicillum* and *Glomus*. are differentially sensitive to moist heat and have been controlled by soil solarization. Depending on numerous variables in climate, soil conditions, populations and distribution of fungal propagules in soil, and technological differences in the use of soil solarization, different workers have been more or less successful in controlling plant pathogens and pests. A partial list of the fungal pathogens controlled by soil solarization is given in Table 1.

Weakening of Fungal Propagules and Biological Control

Soilborne propagules of fungi that are subjected to sublethal heat effects during solarization appear to have an increased sensitivity to antagonistic fungi and to bacteria which are less affected by soil solarization; the weakened propagules are also more sensitive to soil fumigants and other pesticides. Sublethal temperatures also may cause delays in germination of propagules and reduced virulence in host plants, that vary with temperature and the duration of exposure. Working with *Pythium ultimum, Rhizoctonia solanai, V. dahliae,* and *Thielaviopsis basicola,* Pullman *et al.* found that these effects of sublethal temperatures were most pronounced when the fungi were exposed to temperatures of 37° to 39°C. The longer a propagule was exposed to sublethal heating, the longer was the time required for germination. They suggested that this relationship indicates that heat damage accumulates gradually to a point beyond which the propagule cannot recover.

During sublethal heating, all living cells produce heat shock proteins. Heat shock proteins are associated with the acquistion of thermotolerance or thermos/ability; however, fungi have a transient heat shock response that is shortlived, even if they are maintained at high temperature. The overall effect of heat shock proteins on the survival of fungi during soil solarization is unknown.

Other effects of sublethal heating are well documented, especially in the case of *Sclerotium rolfsii* where the rind of sclerotia becomes cracked resulting in increased leakage of various substances. Weakened sclerotia are intensely colonized by *Trichoderma harzianum* and other micro-organisms. Another example where soil solarization may affect the germinability and aggressiveness of fungal propagules concerns *Rosellinia necratrix*. Postplant solarization of an apple orchard for the control of this pathogen provided evidence that the fungal propagules became highly vulnerable to colonization by *Trichoderma* species.

Suppressiveness of Soil after Solarization

The benefits of soil solarization involving pathogen and pest control and an increased growth response of crop plants commonly last for about two growing seasons. Among micro-organisms which survive the solarization process and contribute to the suppressiveness of soil, the predominant bacteria are *Bacillus* species whereas *Trichoderma* species and *Talaromyces flavus* are representative of the main fungal antagonists which inhibit the development of pathogenic fungi. Long-term effects of soil solarization have been observed for control of Verticillium and Fusarium wilts of cotton in Israel and in California , and for corky root of tomatoes and pink root and white rot of onion.

The development of disease suppressiveness in solarized soil reflects the re-colonization of the soil by aggressive mycoparasites, which are also rhizosphere competent. Microbial antagonists, such as Bacillus species, are also associated with the increased growth response of plants growing in solarized soil.

Enhancing the Effectiveness of Soil Solarization

Using soil solarization in combination with soil fumigants or with certain crop residues greatly shortens the lime needed to control pathogens and pests and extends the use of solarization to regions which are considered marginal because of air temperatures, length of day, and intensity of sunlight. For example, the combined action of metham-sodium (methylisothioc-yanate) and soil solarization was much more effective than either soil treatment used singly for control of *V. dahliae* and *Fusarium oxysporum* f. sp. *vasinfectum.* Whereas metham-sodium reduced the viability of propagules of *V. dahliae*to 30 percent of the control in one week and soil solarization alone reduced the count to 18 percent after six weeks, the combination of the two treatments begun in July, reduced the number of detectable propagules to 0 within one week. Additionally, when the comparisons were begun in August (lower air temperatures and shorter days), solarization or metham-sodium alone had little or no effect on the viability of propagules of *V. dahliae* after six weeks: however, solarization combined with metham-sodium reduced the viability of the propagules to 26 percent of the control after two weeks and to I percent after 6 weeks.

In other studies, amending field soils with dry cabbage (*Brassica oleracea*) leaves before soil solarization, resulted in a significantly larger decrease in both cabbage yellows and delectable propagules of *F. oxysporum* f. sp. *conglutinans* when compared with either solarization or cabbage leaves alone.

BIOLOGICAL CONTROL OF LATE LEAF SPOT OF GROUNDNUT

Groundnut (*Arachis hypogea L.*) is very important crop in more than 100 countries in the world. Its yields in developing countries are very low ranging from 0.3 to 0.9 t ha -1 compared to very high yields of 2.8 t ha -1 in USA. The occurrence of diseases and pests are the main causes for poor yields in groundnut. The crop is prone to many fungal diseases where ever it is grown. The most important fungal foliar diseases with attributed crop losses of more than 50% are late leaf spot (LLS) caused by Phaeoisariopsis personata and rust caused by Puccinia arachidis.

Chemical control of these diseases is being widely practiced to achieve higher yields all over the world. But the undesirable effects of chemical fungicides on ecosystem, development of fungicide tolerant strains, increasing cost of fungicides necessitated the search for the non-chemical/biological methods to control of these fungal diseases.

Biological Control Offers Promising Alternative for Chemical Fungicides/Pesticides

Several fungi and bacteria were reported to have broad spectrum antifungal activity and offered promising disease control in several crops under greenhouse and field environments. A few of these biocontrol agents (bacteria and fungi) are commercially exploited in several parts of the world. Thus, biological control appears to be a promising, eco-friendly and cost effective alternative disease management option for control of fungal diseases in crop plants.

Search for Rhizobacteria as Potent Antagonists to Late Leaf Spot Pathogen, Phaeoisariopsis Personata

One hundred and fifty bacterial strains were isolated from the rhizosphere of groundnut plants from different areas of Andhra Pradesh and Karnataka? These strains were numbered as GRS 1 to GRS 150 and were tested for in vitro antagonism against P. personata. Of the 150 bacterial strains tested, only 10 strains inhibited >90% conidial germination of P. personata.

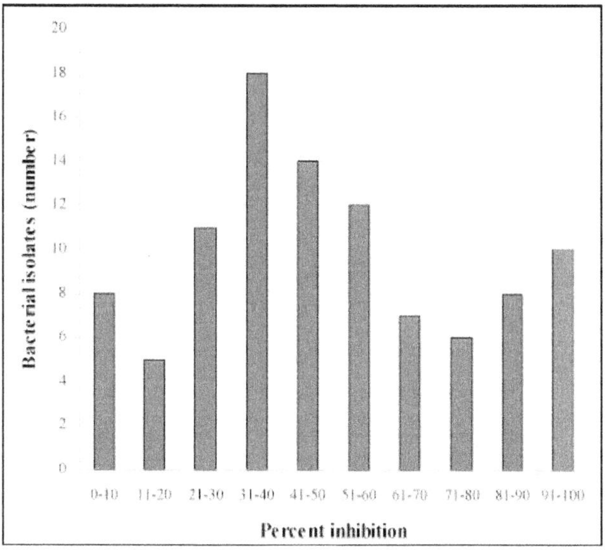

Fig. : Inhibition of P. personata conidial germination by rhizobacterial strains.

The length of the germ tube was also greatly reduced by these bacterial strains.

Antagonistic Rhizobacteria Offers Promising Control to Late Leaf Spot Under Greenhouse Environment

The ten bacterial strains which inhibited > 90% conidial germination of P. personata were tested for control of late leaf spot on susceptible cultivar TMV

2 under greenhouse environment. When the plants reached 35-days-old, the bacterial strains were applied as a foliar spray at three different intervals *i.e.* at 24h before, at 0 h (same time) and at 24h after the pathogen inoculation. The concentration of bacterial strains was log 9 cfu/ml and pathogen was 20000 conidia/ml. The severity of late leaf spot was measured as percent leaf area damaged, percentage of defoliation and disease score on 1-9 rating scale up to 30 days after pathogen inoculation. Percent leaf area damaged, percent defoliation and disease score were significantly less in the pots sprayed with all the ten bacterial strains than control. Two bacterial strains GRS 94 and GRS 100 were found effective even when applied 24 h before the pathogen inoculation. These two cultures were submitted to Indian Type Culture Collection, IMTECH and Chandigarh for strain identification.

Table: Late leaf spot score at 30 days after inoculation in simultaneous treatment of bacterial strains.

Bacterial strain	Leaf area damage (%)	Defoliation (%)	Disease score
GRS 2	7.4	10.4	3.2
GRS 16	11.7	10.4	3.0
GRS 18	13.8	20.8	3.0
GRS 20	5.8	18.8	3.3
GRS 33	7.0	14.6	3.3
GRS 49	2.8	8.3	2.8
GRS 94	3.4	4.2	3.0
GRS 100	3.3	2.1	3.0
GRS 134	6.1	14.6	3.5
GRS 149	5.9	16.7	4.3
Control	23.8	60.4	7.7
LSD (1%)	13.1	26.82	2.17

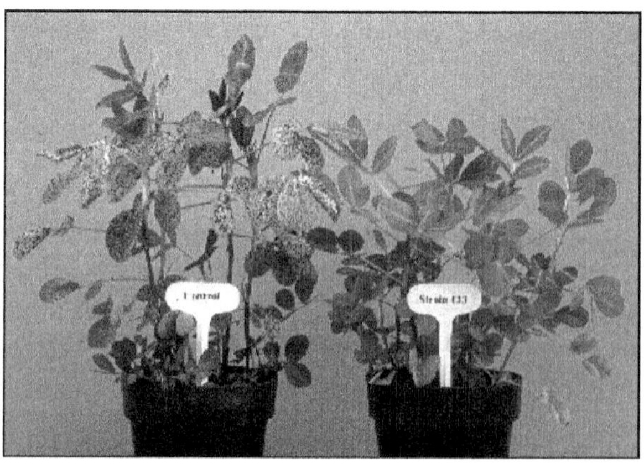

Fig. : Reduction in the severity of Late leaf spot by the foliar application of bacterial strain

Scope for Further Research

Determining the ability of the promising rhizobacterial strains to establish on the phylloplane of groundnut using genetically marked antibiotic resistant mutants will help to design an appropriate spray schedule of these bacterial strains. Development of formulations for long-term storage of these bacteria and suitable method for field application is essential. Also, determining the fungicide tolerance of the bacterial strains makes it feasible to integrate them with fungicides for effective control of late leaf spot.

NON-CHEMICAL CONTROL OF PLANT DISEASES IN THE HOME GARDEN

Plants are considered diseased when they do not grow and develop normally. The cause of such a condition may result from infection by disease-causing organisms or from environmental factors. Organisms known to cause disease are bacteria, fungi, nematodes, viruses, phytoplasmas and parasitic seed plants. Examples of environmental disorders are air pollution, poor soil, excessive heat, low or excessive nutrition, and drought.

When organisms cause disease, it is usually due to a parasitic relationship with a host plant. The parasite, or causal organism, is called a pathogen. The interaction between host and pathogen results in disease. This discussion is limited to diseases where a pathogen is involved. Most plant problems arising from environmental conditions are covered under such topics as fertilization, water management and other cultural procedures.

Prevention is the best approach to plant disease control when using either chemical or non-chemical (organic) methods. Prevention may involve suppressing the disease agent or avoiding the disease. Utilizing as many disease-preventative practices as possible will ensure the best possible control. Many of the following practices aid in controlling many different disease problems that arise. Examples regarding ways they have been used are given where clarification is needed. "Root Knot Nematode Control" is covered below.

Suppressing the Disease Agent

Rotation

Vegetables in the same family group are more likely to be susceptible to the same soil-borne diseases. Cantaloupes and watermelons, for example, have common diseases. If they follow each other in a rotation, a disease organism may be limited on the first crop but sustain enough carry-over in the soil to cause severe loss on the following crop. Vegetables from different family groups should be rotated because they are usually not susceptible to the same disease organisms. The groups listed below should be rotated so a selection from one group is not planted in the same location more than once every 2 to 3 years. Limited garden space may prevent an ideal rotation system.

Table: Crop Grouping for rotation to control soil-borne diseases

Group A	Group B	Group C	Group D	Group E	Group F
Cantaloupe, Cucumber, Honeydew Melon, Pumpkin, Squash, Watermelon.	Brussels Sprouts, Cabbage, Cauliflower, Collards, Lettuce, Mustard, Radish, Rutabaga, Spinach, Swiss Chard, Turnip.	Eggplant, Irish Potato, Okra, Pepper, Tomato.	Beet, Carrot, Garlic, Onion, Shallot, Sweet Potato.	Sweet Corn	Bean, Cowpea, Pea.

Organic Matter

Organic matter increases the number and kind of microorganisms in the soil. Many of these microorganisms compete with disease agents for nourishment. In most cases, the best organic matter is obtained by turning under a green cover crop, such as a small grain (wheat, oats, barley, cereal rye) or a legume.

Resistant Varieties

Agricultural scientists have made great strides in developing disease-resistant varieties. Resistance, however, is a relative term and does not indicate immunity. For example, a tomato plant resistant to Fusarium wilt may develop wilt if stressed but not become diseased to the degree that a susceptible variety might. Resistance could be expressed as slower disease development which allows resistant plants to produce an acceptable yield before or without losing vigor to disease attack.

Sanitation

Any crop residue destruction practice that reduces the disease agent's ability to reproduce or overseason could be included under sanitation. Examples are raking and burning diseased leaves and disposing of infected fruit. Root knot nematode-infested plants should be dug and as much of the root system as possible removed from the garden site. Root knot nematodes are harder to kill when protected by root tissue. Diseased leaves, fruits, and other plant parts should not be piled near the garden. Fungi, for instance, often produce thousands of spores in the reproductive process, and these spores can be wind-blown great distances. Diseased plant tissue should be buried, burned, or disposed of in some other way that prevents the dispersal of the disease agent. In most cases, it is best not to put diseased plant parts in a compost pile. This is especially true for nematode-infected plants. Although many disease organisms are destroyed by heat during composting, 100 percent kill is not normally obtained. Contaminated compost will only serve to spread disease organisms as the compost is being used. Compost for the garden could be made with grass clippings or shrubs and tree leaves, as the disease organisms on grass and leaves are not likely to affect garden plants.

Heat Treatment

Heating small amounts of soil is practical for potted plants and for growing seedling transplants. The soil should be moist but not saturated. Heat the soil at 180°F for approximately 3 hours in a standard oven. An average-sized Irish potato placed in the middle of the soil can act as an indicator. When it is cooked, the soil should be sterile. Microwave ovens have also been used for sterilizing small quantities. Using high power, heat the soil for 3 minutes. Do not overheat using either system. Excessive heating will release certain elements in the soil to toxic levels. A new heating approach called solar pasteurization or solarization was developed in Israel. It simply consists of covering well tilled, moist soil with clear plastic, sealing the edges, and leaving it for several weeks. Black plastic should not be used because soil heating will not be as great as heating under clear plastic. The best results are with fungi, principally those causing root rots. Certain weeds and nematodes are also controlled but not on a consistent basis. Solar pasteurization will work best during the hottest months. August should be ideal for home garden sites. For best results leave soil covered 30 days.

Fallowing

Leaving land idle and clean through the growing season will reduce disease agents in the soil. Fallowing is especially helpful if done in the summer months when soil temperatures are high. Frequent plowing will keep the soil dry and free of plant growth and expose soil-borne disease organisms, such as nematodes, to killing heat and excessive drying. Other benefits of fallowing are weed and insect control. This practice is most efficient if rainfall is low and temperatures are high.

Weed Control

Weeds harbor insects and serve as hosts for many virus diseases. For most viruses to survive they must remain in a living organism whether it be a host plant or insect. Destroying weeds in and around the garden may eliminate potential overwintering host plants.

Avoiding Disease

Change Planting Date: Some diseases can be controlled by changing the planting date. Spring-planted squash usually escapes mosaic virus, whereas fall-planted squash does not. Warm-season vegetables, like peas and okra, should be planted after soils warm sufficiently to avoid seedling disease.

Obtain Disease-Free Planting Stock

It is advisable to check transplants such as cabbage, tomatoes, peppers, and others for root knot nematode galls. Certified Irish seed potatoes are less likely to harbor a disease such as black leg. Some disease agents are seed transmitted; thus, only the best seed should be planted.

Cultural practices:

- **Select the best site:** Choose a site with deep, well drained soil.
- **Plant on raised beds:** Raised beds improve drainage. They also warm up faster than level soil which aids in faster emergence. Slow emergence increases chances of seedling disease.
- **Use proper plant spacing:** Crowded plants reduce air circulation which enhances leaf spot diseases. Septoria leaf spot on tomato is an example. However, in Far West Texas, with low humidity and few leaf spot diseases, crowding tomato plants may reduce losses to curly top virus by shading the plants and making them less attractive to the insect vector.
- **Overplant:** Over planting where possible will ensure enough production if some plants or fruit are lost to disease.
- **Proper fertilization:** Plants receiving all needed nutrients may resist some of the less vigorous disease- causing organisms.
- **Do not overwater:** Excessively wet conditions increase soil-borne fungal diseases like root rots and wilts.
- **Avoid overhead irrigation:** Leaf spot diseases develop rapidly when leaves are moist.
- **Do not prune roots:** Roots cut while cultivating reduce plant vigor and leave openings for root rot and wilt fungi to enter.

Root Knot Nematode Control

No doubt many gardens have become infested by planting contaminated transplants or by bringing in topsoil harboring root knot nematodes. Taking steps to prevent this problem is just as important as implementing the steps to control this once it has become a problem. Where soils are already infested and a garden is first put into production, nematode levels are generally low. After several years of gardening and growing susceptible plants, the nematode population increases to the point that damage becomes unacceptable.

Control practices that can be used are summer fallowing, rotation, adding organic matter, planting trap crops, removing diseased plants, and using resistant varieties. All these control measures are designed to reduce the nematode population in the soil. The benefits of summer fallowing have already been discussed. Removing all potential hosts and keeping the soil dry will reduce the nematode population, as the nematode requires a moist environment for survival. Additional years of fallowing will further reduce the nematode population.

Rotation with non-hosts or poor hosts of the root knot nematode is also another means of reducing the population. Sweet corn is a poor host and is good to use in a rotation, especially in an area where root knot has done severe damage. Onions, garlic, asparagus, and shallots are also poor hosts. Cool-season crops such as cabbage, Irish potatoes, greens (turnips), radishes, and broccoli are less likely to suffer yield loss from root knot nematodes. Even though these are susceptible plants, they grow best in the cooler time of the year which is not favorable for root knot nematode development.

High soil organic matter alone will not ensure root knot nematode control. The higher the organic matter, however, the better the chance that antagonistic organisms will develop. There are soil fungi that trap nematodes and utilize them as a food source. Some organic matter tends to work better than others. Turning under a green manure crop such as small grains or legumes several weeks before planting has shown to be the best. Additional nitrogen may be necessary for adequate crop production because decay organisms in the soil will utilize available nitrogen as they break down the green manure crop.

Sometimes people resort to using home remedies to control nematodes such as planting marigolds or mixing sugar or lye into the soil. Of these three, only French marigolds are effective in controlling nematode populations, and their effectiveness has limitations and is often questionable.

Some people think that marigolds secrete a toxic substance into the soil that kills nematodes and that planting a few marigolds around annual plants in infested soil will prevent infection. This is not true. Marigolds primarily act as a trap crop. Nematodes are able to enter their roots but are unable to complete their life cycle. The trapped nematodes die without reproducing.

The type of marigold is also important. French marigolds, Tagetes patula, are more effective in controlling root knot nematodes than the African marigold, Tagetes erecta, which is also referred to as the American, Big, or Aztec marigold. To be effective, marigolds must be planted as a solid crop and grown for 90 to 120 days to effectively reduce the nematode population sufficiently to grow annual plants without treatment. Marigolds should be planted in rows no further than seven inches between each plant, so that the roots penetrate the entire soil mass to trap as many nematodes as possible. If marigolds are planted close together, they form a dense canopy which helps retard the development of weeds and grasses. Many weeds and grasses serve as hosts for root knot nematodes. If the weeds are not controlled, the marigolds may be unable to suppress the nematode population.

Using marigolds in a manner other than that recommended can invite disaster. For example, planting a few marigolds here and there among tomatoes will encourage spider mites. The spider mite is one of the most difficult garden pests to control and can become nearly as serious a problem as the nematodes.

All nematode-infected plants should be removed from the garden as soon after production ceases as possible. Removing the root systems will eliminate many of the nematodes. To ensure that as much of the root system as possible is removed, use a shovel rather than just pulling the planting up by the stem.

Root knot resistant vegetable varieties are not plentiful. Fortunately, progress has been made in the development of root knot resistant tomatoes. The best resistance is found in the hybrid varieties that have been developed in recent years. Root knot resistant varieties are noted in seed catalogs by "N" following the variety name. Preceding the "N" are often "V" and "F." "VFN" stands for verticillium wilt resistance, fusarium wilt resistance, and nematode resistance. Hybrid tomato seed is more expensive than open-pollinated seed, but the benefits in disease resistance alone are enough to make it worth buying.

Diseases caused by root knot nematodes and other organisms can be prevented with non-chemical methods that either avoid the disease or suppress the disease agent. However, maximum control can only be assured by utilizing as many of these disease preventive practices as possible.

VEGETABLE DISEASES AND THEIR CONTROL

Damping-Off (Seedling Disease)

Seeds of many vegetables are susceptible to damping-off fungi when planted in infested soils. The seeds may fail to germinate, or seedlings may be attacked before or after emergence and turn brown, shrink, and finally die. Control measures include these: plant fungicide-treated seeds in well-drained areas; do not apply excessive amounts of nitrate forms of nitrogen fertilizers; and rotate the location of the vegetables.

Root Rot of Beans and Southern Peas

Root rot is severe on green beans, lima beans, and southern peas. The disease first appears as reddish or reddish-brown areas on stems and roots. As the disease advances, discolored areas spread until the entire root and lower stem are affected. Above-ground symptoms include stunting, yellowing, drooping of leaves, failure to produce normal pods, and death.

These control practices reduce losses from root rot:

- Use high-quality seeds treated with a fungicide like Arasan.
- In-furrow fungicides (Terraclor) help control root rot. Apply one- fourth of the material in the open furrow and the remainder in the covering soil during planting.

- During cultivation, do not throw soil against plant stems.
- Plant in a 4- or 5-year rotation with other vegetables.
- Plant in well-prepared soils with a pH of about 6.5, fertilized according to a soil test and treated for nematodes if recommended. Plant seeds 1 inch deep only during favorable weather, in warm soils, and on top of a bed to avoid "drowning."

Early Blight of Tomatoes

Early blight is a major disease of tomatoes in Mississippi. Symptoms first appear on lower, older leaves as circular, dark brown to black spots that often contain rings, giving a "target board" effect. As the disease progresses, leaves turn yellow, wither, and drop off. Frequently, only the upper half of the plant has green leaves, and in severe cases, the plant becomes completely defoliated.

Early blight also occurs on plant stems and sometimes on fruit. On seedlings, the disease may girdle the stem and give the appearance of damping-off.

Reduce losses to early blight by providing good ventilation in plant beds and watering when leaves have time to dry. Seed treatment with Thiram aids in controlling the damping-off stage. Do not set tomato plants where early blight occurred the year before, and remove and destroy all diseased plant debris in the garden after harvest.

Applications of chlorothalonil or mancozeb effectively control this disease. There is no waiting period after application until harvest for Bordeaux mixture or chlorothalonil, but there is a 5-day waiting period for mancozeb.

Begin applying when plants are 8 to 10 inches tall, and continue at 7-day intervals through the growing season. Applications of these fungicides also control some of the other leaf, stem, and fruit diseases of tomatoes.

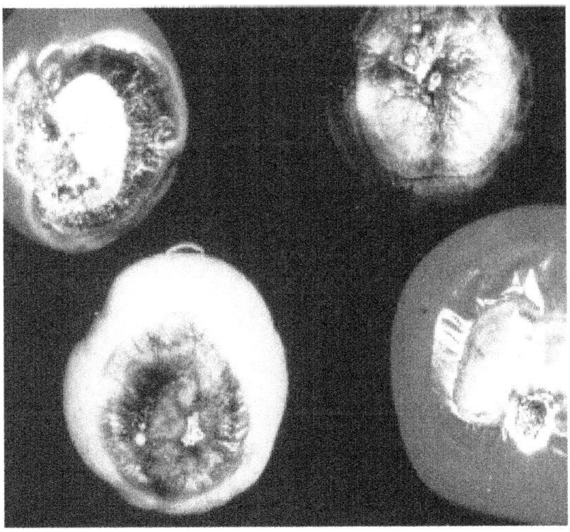

Blossom-End Rot of Tomatoes

Blossom-end rot occurs on the tomato fruit. It may also be a problem on peppers, squash, and watermelons. It is more common on fruit that is one-third to one-half grown and occurs on the blossom end of the fruit. It begins as a small, water-soaked spot that develops into a dark brown, leathery spot that may involve half the fruit. The surface of the spot shrinks and becomes flat or sunken.

Blossom-end rot is caused by a lack of calcium in the developing fruit. The uptake of calcium from the soil by the tomato plant can be reduced by fluctuations in soil moisture—either excessively wet soil or excessively dry soil. The disease commonly occurs when plants are grown rapidly and luxuriantly early in the season and are then subjected to prolonged dry weather. Because of drying conditions, the disease may be more serious on the windward side of a garden and on staked tomatoes than on unstaked or bushy plants.

Prevent blossom-end rot by maintaining a soil pH around 6.5, irrigating and mulching to maintain uniform soil moisture, and avoiding heavy applications of nitrogen. Control blossom-end rot by spraying with 4 tablespoons of 96 percent calcium chloride per gallon of water at 7- to 10-day intervals for 4 applications. Begin spraying with first appearance of symptoms. Overdosing plants with calcium chloride may result in leaf burn. Calcium chloride is suggested only for tomatoes.

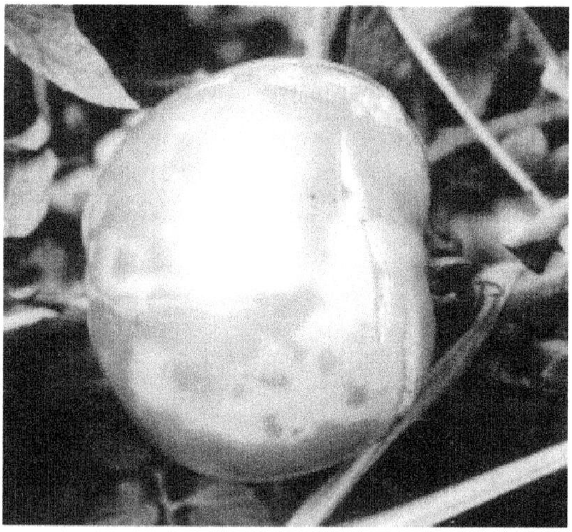

Spotted Wilt of Tomatoes and Peppers

This viral disease is transmitted by several species of thrips and may kill plants or drastically reduce fruit-set. Fruits from diseased plants are generally small and distorted. Tomatoes develop irregular yellowish blotches.

Initial symptoms appear as thickening of veins on younger foliage. Younger foliage generally exhibits a pronounced downward curling. Internodes become shortened, and immature fruit does not ripen. Dark purple streaks can occur on leaves, stems, and fruits. Other symptoms are blighting and blackening of young shoots. On individual leaflets, small, dark, circular dead spots may appear. Badly spotted leaves may turn dark and wither.

Some varieties are now being released with resistance. Check with your seed source. It is not clear how effective or long-lived these resistant cultivars will be.

Try these control practices:

- Remove and destroy diseased plants.
- Keep weed populations down in and around gardens to reduce movment of virus-carrying thrips from weeds to garden plants.
- Suppress thrips by applying approved insecticides (Malathion 50 or Diazinon 25 EC at 2 teaspoons per gallon).
- Further suppress thrips with shiny mulch materials around tomatoes and other susceptible vegetables. Apparently, light reflection from the mulch surface repels thrips and the chances of virus transmission. Conventional black plastic may be sprayed or hand painted with aluminum-colored paint. Oil-based paints adhere to plastic surfaces and may be easily applied. This technique gives best results when mulch is laid down at the time of planting and used in combination with other recommended control procedures.

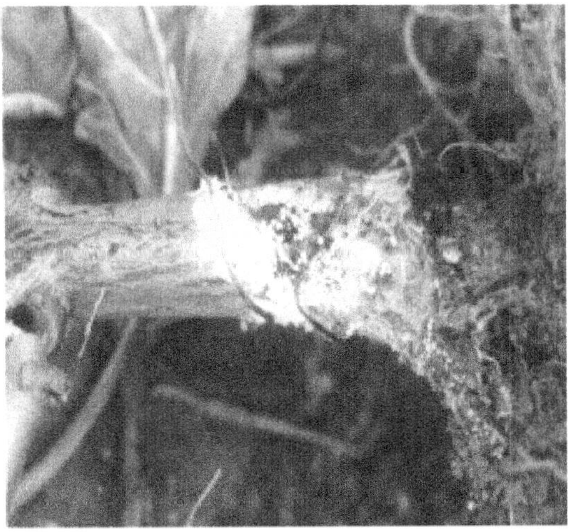

Southern Blight

Southern blight affects most garden vegetables. The fungus that causes southern blight attacks plant parts (roots, stems, leaves, or fruit) that are in contact with or just under the soil surface.

The first visible symptoms are usually an advancing yellowing and wilting of the foliage, beginning with the lower leaves. During warm, moist weather, a white fungus growth may appear on the lower stem near the soil surface and on organic debris in the soil. Later, light tan to dark brown mustard seed-like bodies called sclerotia develop in the mold. As the disease advances, several plants next to one another in the row die.

Southern blight is difficult to control, but you can reduce losses with these practices:

- Plow 6 inches deep in the fall to bury organic debris and the scle- rotia.
- Avoid throwing soil on the plants when cultivating.
- Where a few scattered plants are affected, remove them from the garden along with the soil 6 inches deep and 6 inches from the stem.
- Control foliar diseases, since dead leaves on the ground may trigger infection. Also control weeds early in the season for the same reason.
- Treat problem areas with PCNB (Terraclor) as a dust, granule, or wettable powder. Rates and methods of application are stated on the label.
- Wrap transplant stems with a 4- by 4-inch strip of aluminum foil and plant so that 2 inches of wrapped stems are below and 2 inches are above the soil.

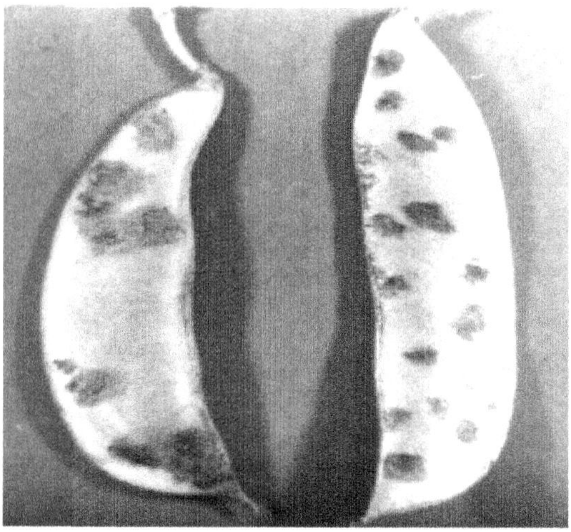

Stem Anthracnose of Lima Beans

Stem anthracnose is the most common disease of lima beans. The first stages of infection appear on pods as small, brick-red blotches. These blotches may spread over the entire surface of the pods. Later, the diseased areas become brownish to grayish and may have many tiny black specks which are fruiting bodies of the fungus. Occasionally, diseased pods fall from the plant.

A brick-red streaking may occur along the veins on the under side of leaves and on young stems.

Reddish spots occur on the lower leaf surface and enlarge and become noticeable on the upper leaf surface. Occasionally, leaves are killed and fall from the plant. Severely diseased plants are yellow and stunted.

Reduce damage from stem an- thracnose by following these practices:

- Because stem anthracnose can be carried over on seed to the next season, use only certified disease-free, western-grown seed.
- Never plant lima beans in the same location more than once in 3 years.
- Avoid fall planting of lima beans in an area of the garden where stem anthracnose was a problem the previous spring.
- Apply Bordeaux mixture or another copper-based fungicide on a 7-day schedule, beginning at full bloom.

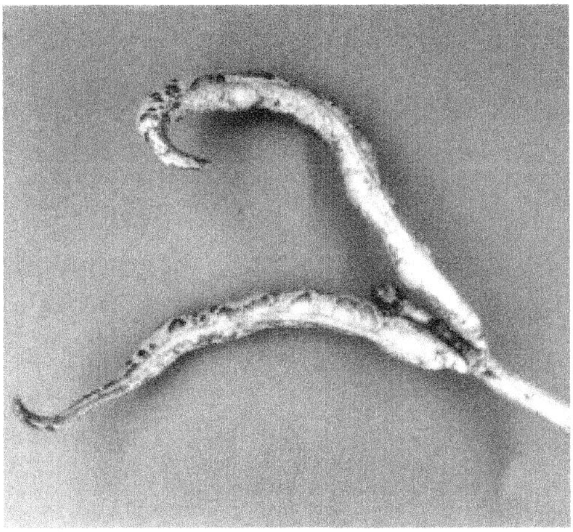

Mosaic

This virus disease commonly infects beans, sweet corn, squash, melons, cucumbers, peas, peppers, and tomatoes. Symptoms include the following:

- Misshapen leaves with light and dark green areas.
- Fruit with green specks, yellow and green mottling, or bumps.
- Distorted fruit.
- Overall stunted plants. Control of virus diseases is difficult. Reduce chances of mosaic in these ways:
    - o Plant resistant varieties when available.
    - o Control insects.
    - o Eliminate weeds.
    - o Remove diseased plants as they appear.
    - o Purchase certified transplants or buy western-grown seed.
    - o Do not use tobacco products when handling plants.

Phenoxy herbicide damage (such as 2,4-D) resembles symptoms of mosaic disease. Leaves and stems are typically twisted, deformed, curled, leathery, and excessively long and narrow. Apply herbicides carefully and correctly in and around the garden.

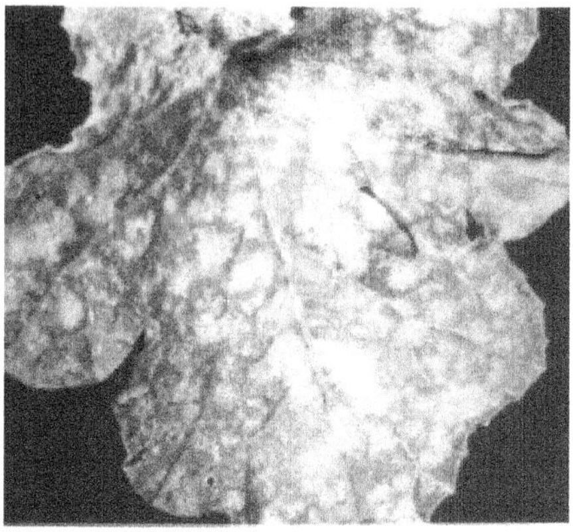

Powdery Mildew

Powdery mildew is caused by a fungus that commonly occurs as a white, powdery growth on leaves of cucumbers, squash, melons, beans, and English peas. Benomyl and chlorothalonil effectively control powdery mildew on vine crops, and sulfur provides control on beans and peas.

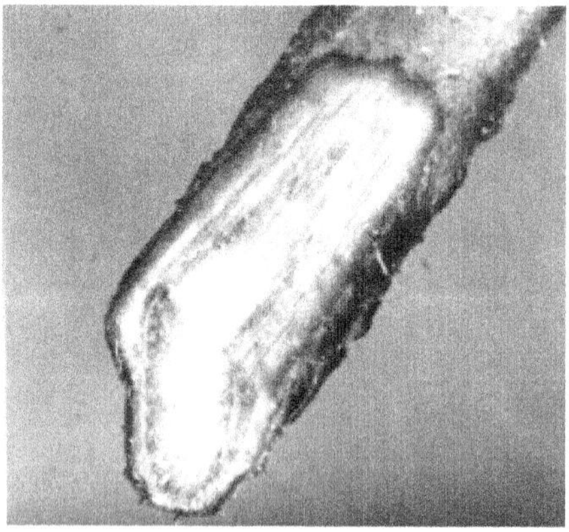

Fusarium Wilt

This fungal disease often infects watermelons, cabbage, tomatoes, sweet potatoes, beans, and peas. Lower leaves often turn yellow on one side of the plant only. Two brownish streaks that originate from the roots are exposed when the stems are split lengthwise. Infected plants are usually stunted and wilted. The best way to control fusarium wilt is using resistant varieties.

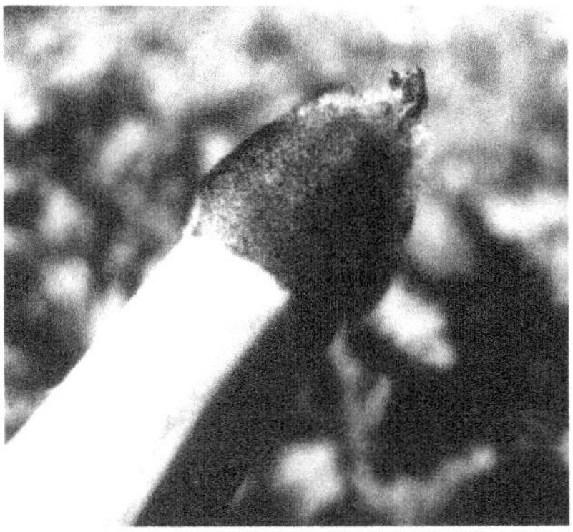

Fruit Rot

Bacteria and fungi often infect fruit, resulting in soft, slimy fruit with an offensive odor. You can reduce the occurrence of fruit rot by staking, mulching, avoiding mechanical injury to fruits, controlling insects, following a regular fungicide program, and removing mature fruit from the garden.

Rust

This fungus disease occurs commonly on beans and sweet corn as reddish-brown spots on leaves that rub off when touched. Apply fungicides like chlorothalonil or sprayable sulfur at the first sign of disease and at weekly intervals thereafter until the disease is under control.

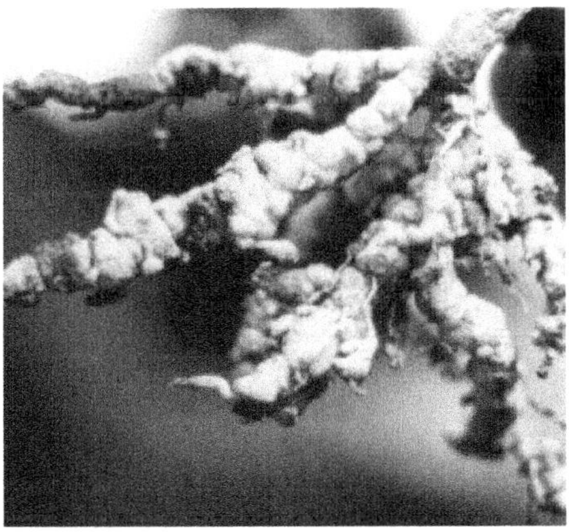

Nematode Diseases

Nematodes are slender, tiny, worm-like animals that feed on plant roots, stems, and leaves. Nematodes cannot ordinarily be seen with the naked eye and go unnoticed until plants become unthrifty and stunted. They seldom kill plants; however, they can reduce quality and yields of many vegetables, such as beans, beets, carrots, cucumbers, lima beans, okra, peas, squash, tomatoes, and watermelons.

Nematode injury to roots reduces uptake of water and nutrients from the soil. Typical above-ground symptoms are general stunting, yellowing, loss of vigor, and overall decline. The most common underground symptom in gardens is root-knot galling.

Nematodes other than the root-knot nematode also can cause severe plant damage. Some of the less obvious symptoms are stubby roots, tiny lesions, excessively branched roots, or complete loss of secondary roots.

The best time to determine if you have a nematode problem is during the fall when nematodes are most active. To detect root-knot nematodes in the growing season, observe the galled roots.

Once you know nematodes are present, you can use certain cultural practices to help reduce nematode populations. These include setting nematode-free transplants, rotating crops, fallowing, practicing good sanitation, controlling weeds, and planting resistant varieties. Vegetable varieties having resistance or tolerance to root-knot nematodes are marked in the list of recommended varieties. In addition, you can plant marigolds in gardens to help reduce nematode populations.

As an alternative to chemical treatment, solarization can reduce parasitic nematode populations. Solarization is the use of heat from the sun for killing nematodes in the soil prior to planting. To use solarization, place clear plastic (1 to 1.5 ml thick) on moist, tilled soil, and seal the edges with soil, bricks, or other materials. Apply the plastic in May or June. Leave it in place for at least 8 weeks. Remove the plastic in August in time to establish a fall garden, if desired. If not, remove it before cold weather begins.

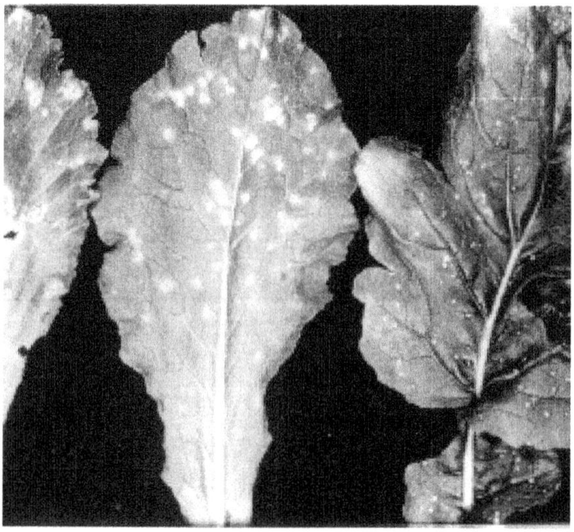

Leaf Spots

Leaf spots, caused by fungi or bacteria, commonly occur on many vegetables. They appear on leaves and sometimes stems as distinct, dark-colored or tan spots one-sixteenth to 1 inch in diameter. The regular application of a fungicide generally provides acceptable control of fungal leaf spots. Applying copper fungicides helps control bacterial as well as fungal leaf spots.

Bacterial Wilt of Cucumbers

This destructive disease is caused by a bacterium that overwinters in the bodies of adult striped and spotted cucumber beetles. As these beetles feed on young plants in the spring, bacteria are introduced into the vascular system. Here they are able to multiply rapidly and produce a sticky material that stops movement of moisture through the plant. As a result, leaves on an infected runner wilt rapidly, and within a short time all runners become permanently wilted. Plants can die within a week or two after initial symptoms appear. Yellowing is not normally associated with this disease.

A symptom of bacterial wilt is a thick, white, sticky substance that oozes from the cut stem of a wilted vine. If you press your finger tip against the cut surface several minutes after cutting and then slowly remove it, the bacterial ooze frequently remains attached and strings out in thin threads.

Since bacterial wilt resistant cucumber varieties are not commonly available, the best control is to keep cucumber beetle populations in check. A rigid spray schedule with recommended insecticides (refer to the Insect Control section) should reduce the incidence of bacterial wilt.

Black Rot of Cabbage

This disease attacks cabbage and other crucifer crops like collards, mustard, cauliflower, Brussels sprouts, kohlrabi, rutabagas, kale, and rape. Black rot may affect plants at any stage of growth but usually is most prominent close to maturity. On older plants, yellow wedge-shaped areas appear at leaf margins and expand toward the center. Blackened veins are apparent in affected areas. Vascular tissue within the stem also may become discolored.

Black rot causes head dwarfing, and soft rot frequently develops on affected heads. Practices important for controlling black rot include these:

- Use disease-free seeds that have been hot-water treated. This is the most effective treatment for ridding seeds of the causal bacteria.
- Purchase transplants that have been certified as disease-free.
- Rotate in the field so that at least 2 years, and preferably 3, elapse between cruciferous crops.

Yeast Spot of Lima Beans

Gray-brown sunken lesions on young or nearly mature seeds is a good indication of this disease. Yeast spot is more a problem in seasons when southern green stink bug populations are high because the yeast fungus enters seeds through pod punctures this insect makes. The spots, or lesions, develop within 2 to 3 days of inoculation. Bright, sunny days allow the stink bug to move from one plant to another, spreading the disease.

Chapter 11

CONTROL OF POSTHARVEST DISEASES OF FRUITS USING MICROBES

Postharvest diseases may start before or after harvesting. Plants or fruits infected in the field may not develop symptoms until stored. Once in storage, infections continue to develop on the fruits and vegetables. Wounds, cuts, or bruises caused during harvesting are common entry points for bacteria and fungi. Penetration can also occur during storage through natural openings, such as lenticels, or directly through the cuticle and epidermis.

Spread of the infection usually requires the presence of warm temperatures and high moisture, although some storage rots can continue to develop at low temperatures. *Penicillium,* a fungus which causes blue and green mold on fruits, can continue to grow slowly even at temperatures near freezing. Once produce is infected, the disease can spread to healthy produce by direct contact. Insects can also be involved in spreading disease. Fruit flies, attracted to fruits and vegetables infected with "sour rot" carry spores of the fungus *Geotrichum* from infected produce to healthy produce.

ORIGINS OF POSTHARVEST DISEASES

Symptoms of postharvest diseases develop during storage, but infection of fruit by decay-causing pathogens may occur prior to harvest, at harvest, or during the postharvest handling process and storage. Based on the nature of infection, postharvest diseases of pome fruits can be divided into two categories: those that originate from latent fungal infection of fruit in the orchard and those that originate from infection of wounds, such as stem punctures and bruises, by fungi at harvest or during the postharvest handling and packing process. The former category is exemplified by Sphaeropsis rot caused by *Sphaeropsis pyriputrescens,* Phacidiopycnis rot caused by *Potebniamyces pyri,* and Bull's eye rot caused by *Neofabraea*spp. The latter category is exemplified by blue mold caused by *Penicillium* spp., mainly *P. expansum*, and Mucor rot caused by *Mucor piriformis*. Understanding when infection occurs is an essential step for developing and implementing control measures to reduce storage losses due to decay.

Infection Prior to Harvest

- Infection of fruit by *Sphaeropsis pyriputrescens* (Sphaeropsis rot), *Potebniamyces pyri*(Phacidiopycnis rot) and *Neofabraea* spp. (Bull's eye rot) in the orchard. These three pathogens are commonly associated with dead bark and twigs and cankers on trees. Stem-end rot and calyx-end rot are common symptoms of Sphaeropsis rot on apples and Phacidiopycnis rot on d'Anjou pears. Infection of the stem and calyx of the fruit occurs before harvest in the orchard, and symptoms develop after a period of time during storage. Previous reports indicated that the Bull's eye rot fungus may infect fruit anytime between petal fall and harvest, but symptoms appear only after a few months of storage, and that fruit susceptibility to infection by the fungus increases as the growing season progresses.
- Infection of fruit by *Phytophthora cactorum*. *P. cactorum* is a soilborne and waterborne pathogen. Fruit are infected by pathogen-contaminated irrigation water. Fruit infected early by *Phytophthora* spp. may develop symptoms before harvest. Fruit infected by *P. cactorum*near harvest may develop symptoms in storage.
- Blossom infection early in the season. During the bloom period, the gray mold fungus*Botrytis cinerea* may colonize floral parts and cause latent infections. The fungus resumes growth when fruit become senescent after a period of time in storage and develop calyx-end rot. This has been reported on pears, but calyx-end gray mold is not commonly seen on fruit grown in the semi-arid climate of eastern Washington.

Infection at Harvest

Some decay-causing organisms are essentially wound pathogens. Damage on the fruit skin such as stem punctures, limb rubs and bruises may be created at harvest or during the fruit handling process. Decay-causing pathogens, particularly those airborne pathogens such as gray mold fungus, *Botrytis cinerea*, and blue mold fungus, *Penicillium expansum*, may contaminate wounds and later cause decay symptoms in storage.

Infection After Harvest

- Infection due to contamination of wounds by decay-causing pathogens during the drenching and packing processes. During the drenching process, spores of decay-causing pathogens such as *Penicillium* spp., *Mucor* spp., and perhaps *Botrytis cinerea*, accumulate in the circulated drench solution. Skin wounds or natural openings on the fruit could be inoculated with decay-causing pathogens during drenching and packing processes and later develop decay symptoms.

- Fruit-to-fruit spread during storage. One type of fruit infection during storage is fruit-to-fruit spread of decay in fruit storage containers (bins or boxes). Nesting of gray mold due to fruit-to-fruit spread is particularly common after an extended period of storage. *Botrytis cinerea* produces mycelia on decayed fruit, infects the surrounding sound fruit by mycelium through fruit-to-fruit contact, and consequently forms clusters of decayed fruit. Phacidiopycnis rot also has the ability to spread among fruit during storage. Thus, for fruit destined for a long-term storage, it is very important to control decay resulting from this type of infection because one decayed fruit can potentially destroy a considerable amount of healthy fruit.

HOW TO DIAGNOSE POSTHARVEST DISEASES

Postharvest diseases of apples and pears are caused by a variety of fungi. Some diseases can be easily recognized and separated from others based upon symptoms and signs of pathogens on decayed fruit. However, it can often be difficult to diagnose postharvest diseases based on symptoms alone because different diseases may have very similar symptoms. For example, symptoms of Phacidiopycnis rot in the early stage on d'Anjou pears are very similar to those caused by gray mold. On the other hand, the same disease may also have very different symptoms in the early stage than in the advanced stage. For example, Phacidiopycnis rot appears watersoaked in the early stage of its symptom development but turns black in the advanced stage. Some decay-causing pathogens do not sporulate under cold-storage conditions, increasing the difficulty to confirm diagnosis. The following three-step approach is a practical way for fruit quality control personnel in commercial packinghouses to separate common postharvest diseases, without isolation of causal agents.

The three-step approach involves "Look," "Feel," and "Sniff" described as follows:

Look: Look at the appearance of decayed fruit, including the following aspects:

1. Color of the decayed area
2. Sign of the pathogen: sporulation or fruiting bodies
3. Location of infection sites
4. Presence of wound
5. Size and shape of the decayed area

Feel: Feel the texture of the decayed area. Postharvest diseases can be divided into two groups:

1. soft or watery decay: blue mold and Mucor rot.

2. spongy or firm decay: gray mold, Sphaeropsis rot, Phacidiopycnis rot, Bull's eye rot.

Lesion of blue mold or Mucor rot has a sharp margin between the decayed tissue and healthy tissue. Decayed tissue of blue mold and Mucor rot is completely separable from the healthy tissue, whereas decayed tissue of spongy-type diseases is not separable from the healthy tissue.

Sniff: Decayed fruit of some postharvest diseases produce distinct odors. For example, blue mold produces an earthy, musty odor, whereas Sphaeropsis rot produces a distinct "bandage-like" odor. These characteristics can be used as diagnostic indications of these diseases.

Isolation of Causal Agents

Because of similarities in symptoms among postharvest diseases, isolation of causal agents from decayed tissue is often necessary to confirm diagnosis, particularly if no fruiting bodies of the pathogen are present on the decayed fruit for further examination.

GRAY MOLD

Gray mold is a common postharvest disease on apples and pears wherever these fruits are grown worldwide. This disease can cause significant losses on both apples and pears during storage. Losses as high as 20-60% due to gray mold are not uncommon after an extended period of storage, particularly on fruit that were not treated with fungicides prior to storage, because gray mold has the ability to spread from decayed fruit to surrounding healthy fruit through fruit-to-fruit contact during storage.

Symptoms

Gray mold originates primarily from infection of wounds such as punctures and bruises that are created at harvest and during the postharvest handling process. Gala apple fruit at harvest tend to split at the stem bowl area, which provides an avenue for *B. cinerea* to infect the fruit. Stem-end gray mold is common on d'Anjou pears and also occurs on apples. Calyx-end gray mold has been observed on pears in the Pacific Northwest but generally is not very common.

The decayed area appears light brown to dark brown and color is similar across the decayed area. The decayed area is spongy, and diseased tissue is not separable from the healthy tissue, which is different from blue mold (a soft decay). Under high relative humidity conditions, fluffy white to gray mycelium and grayish spore masses may appear on the decayed area. The internal decayed flesh appears light brown to brown at the margin. Generally, gray mold does not have a distinct odor, but in advanced stages decayed apples may have a "cedar-like" smell. In advanced stages, the entire decayed fruit may appear "baked" and eventually may turn softer than in the early stage.

In the early stage of symptom development, gray mold and Phacidiopycnis rot on d'Anjou pears are very similar, but the fruit flesh at the margin of Phacidiopycnis rot appears translucent and water-soaked, whereas internal decayed flesh of gray mold usually appears brown.

Causal Organism

The causal agent of gray mold is *Botrytis cinerea* Pers., teleomorph *Botryotinia fuckeliana* (de Bary) Whetzel. Isolates recovered from decayed apple and pear fruit vary greatly in sporulation on potato dextrose agar. Some isolates sporulate abundantly, whereas others produce only sclerotia or sclerotia with scarce conidia. Mycelium of *B. cinerea* grows at temperatures as low as -2°C. Conidia can germinate at 0°C.

Occurrence

Botrytis cinerea colonizes organic matter in the orchard. Conidia of the fungus are dispersed mainly by air currents and water splash. Gray mold of apples and pears primarily originates from infection of wounds that are created at harvest and during the handling processes. *B. cinerea* can infect stems of d'Anjou pear fruit and causes stem-end gray mold. d'Anjou pears have thick stems. At harvest, pear fruit are detached from the trees and fresh wounds at the stem are exposed to potential contamination by *B. cinerea*. *B. cinerea* is able to grow at pear storage temperatures of -0.5 to 0°C, move through the stem of the fruit to reach the fruit flesh after a period of time in cold storage, and then cause decay. The fungus can also invade stems of Delicious apples and cause stem-end rot. On pears, it has been shown that incidence of gray mold is correlated with spore levels on the fruit surface of d'Anjou pear fruit. *B. cinerea* has been reported to invade floral parts of pear fruit during bloom, causing calyx-end rot during storage, but calyx-end gray mold is not very common on apples and pears grown in the semi-arid climate of eastern Washington State.

Secondary infection of fruit through fruit-to-fruit contact during storage is commonly seen after a long period of storage and can cause significant losses. Mycelial growth and secondary infection through fruit-to-fruit spread is enhanced by high relative humidity. Fruit infected by gray mold due to secondary infection in storage bins may not have visible symptoms or lesions are very small at the time of packing, and thus infected fruit may be packed but symptoms develop on packed fruit during storage or transit.

Control

Orchard sanitation to remove decayed fruit and organic debris on the orchard floor helps reduce inoculum levels of *B. cinerea* in the orchard. Good harvest management to minimize punctures and bruises on the fruit skin helps avoid decay from wound infections. Gala apple fruit tend to split at harvest.

Harvesting Galas at the right maturity to minimize splits at the stem bowl would help avoid gray mold from infection at that site.

Preharvest fungicides such as thiram and ziram applied near harvest provide some control of gray mold. Research is in progress to evaluate more effective fungicides as preharvest treatments for control of postharvest gray mold. The vast majority of *B. cinerea* isolates from apple-related sources in Washington State are sensitive to thiabendazole. A postharvest drench treatment with Mertect (thiabendazole) applied prior to storage is effective to control gray mold, particularly for those that originate from infection of wounds. In 2004, two new postharvest fungicides, Penbotec (pyrimethanil) and Scholar (fludioxonil), were registered as postharvest treatments for control of postharvest diseases of pome fruits in the U.S. Both fungicides are labeled for use as either drench treatments or online sprays. It has been shown that both fungicides applied as prestorage treatments are effective to control gray mold from wound infections.

Photo Plate: Gray Mold

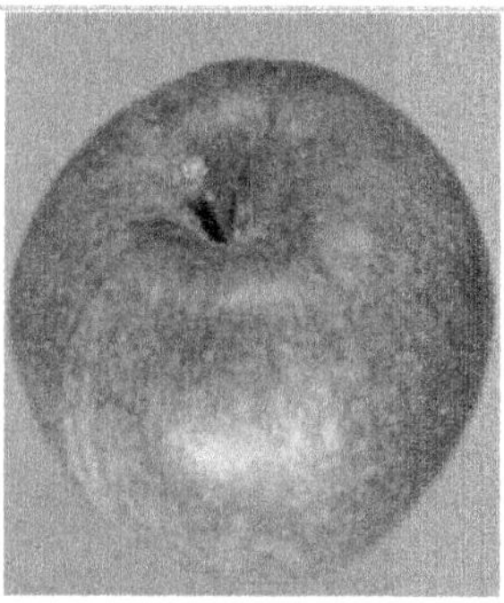

A: Gray mold originating from infection at the stem-bowl crack of a Gala fruit; decayed area brown, spongy to firm

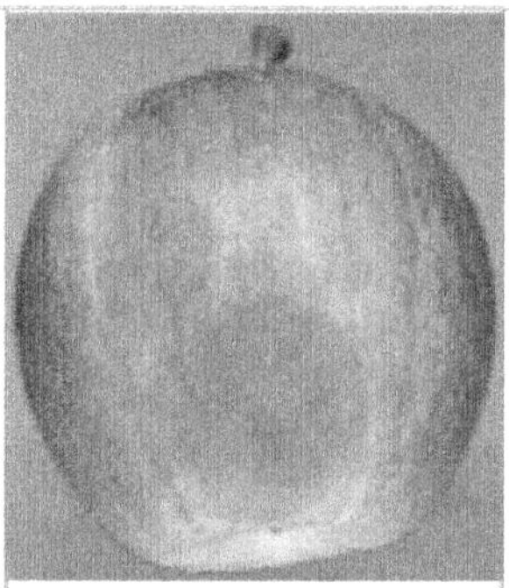

B: Early stage of gray mold originating from fruit-to-fruit spread

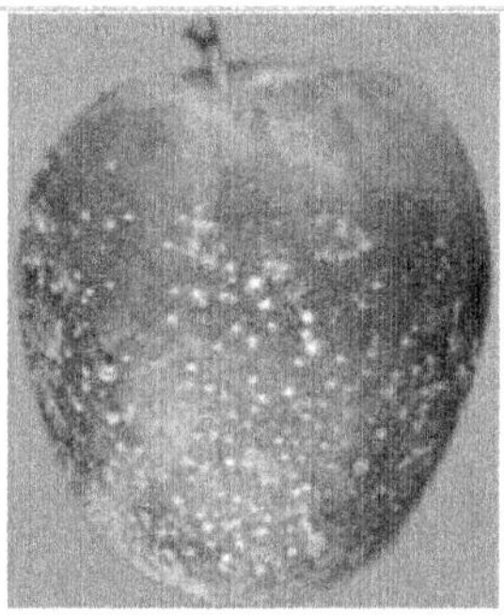

C: Advanced stage of gray mold; white to gray mycelium covering the decayed area under high humidity

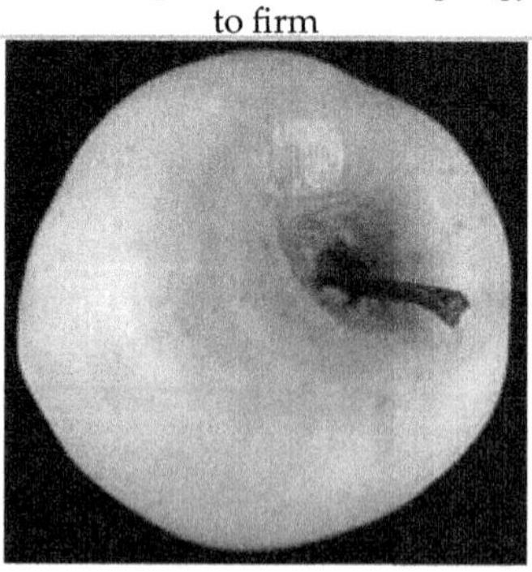

D: Gray mold originating from infection at stem or stem bowl; gray spore masses at the stem bowl area

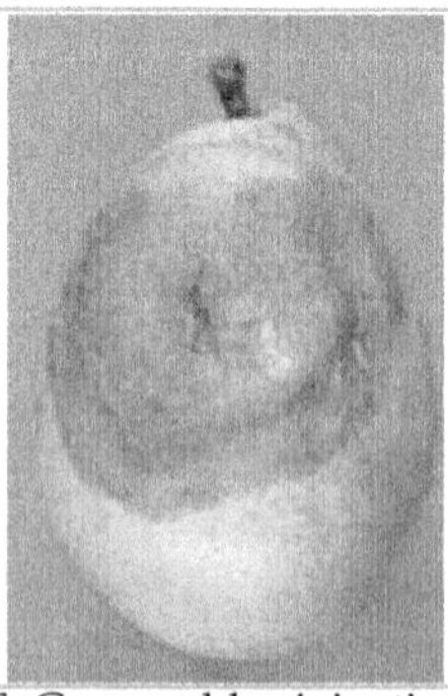

E: Gray mold originating from wound infection on a d'Anjou pear fruit

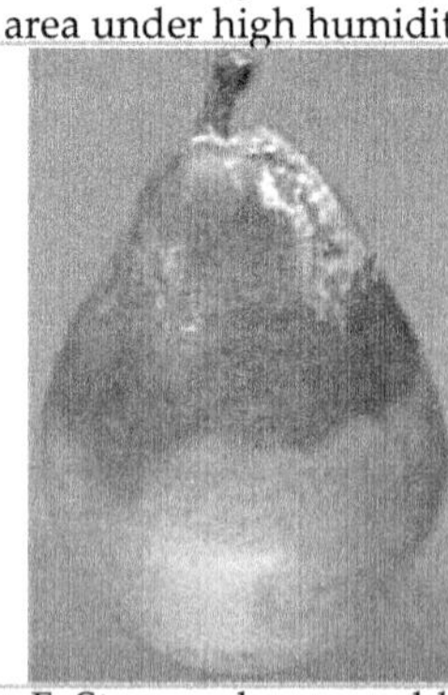

F: Stem-end gray mold commonly seen on d'Anjou pears; white to gray mycelium

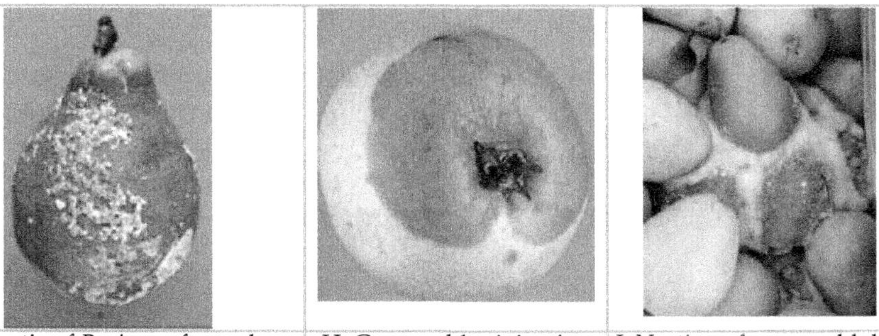

G: Sclerotia of *B. cinerea* formed on the surface of a decayed d'Anjou fruit at advanced stage

H: Gray mold originating from calyx infection of a d'Anjou fruit

I: Nesting of gray mold due to fruit-to fruit spread during storage; white to gray mycelium

Fig. : Symptoms and signs of gray mold *(Botrytis cinerea)* on apples and pears.

Disease Comparison

Table: Comparison between gray mold and blue mold.

Characteristics	Gray mold	Blue mold
Texture	spongy or firm; decayed tissue not separable from the healthy tissue	soft, watery; lesion with a sharp margin; decayed tissue completely separable from the healthy tissue, leaving it like a "bowl"
Color of decayed area	light brown to dark brown	light tan to dark brown
Signs of pathogen	fluffy white to gray mycelia; sporulation under high humidity; gray to brown spore masses; black sclerotia may form	white mycelia and blue or blue-green spore masses; sporulation often starts at the infection sites (wounds)
Color of internal flesh	light brown to brown	brown
Odor	generally not detectable	earthy, musty

Table: Comparison between gray mold and Phacidiopycnis rot on pears.

Characteristics	Gray mold	Phacidiopycnis rot
Texture	spongy or firm; decayed tissue not separable from the healthy tissue	spongy; decayed tissue not separable from the healthy tissue
Color of decayed area	light brown to dark brown; color similar across the decayed area	initially watersoaked, then light brown to brown, later black; color varies with age
Signs of pathogen	fluffy white to gray mycelia; sporulation under high humidity; sclerotia may form	white mycelia under high humidity; black pycnidia form on decayed fruit at advanced stages but often are immature under commercial cold-storage conditions
Color of internal flesh	light brown to brown at the margin	translucent, clear at the margin
Odor	generally not detectable	mild, distinct

BLUE MOLD

Blue mold is a common postharvest disease on apples and pears worldwide. This disease is an economic concern not only to the fresh-fruit industry but also to the fruit-processing industry because some strains of *Penicillium expansum* produce the mycotoxin patulin, which can rise to unacceptable levels and thus affect the quality of apple juice.

Symptoms

Blue mold originates primarily from infection of wounds such as punctures, bruises and limb rubs on the fruit. Blue mold can also originate from infection at the stem of fruit. Stem-end blue mold is commonly seen on d'Anjou pears, but it also occurs on apples such as Red Delicious. Calyx-end blue mold occurs on Red Delicious apples but is usually associated with fruit that are drenched prior to storage.

The decayed area appears light tan to dark brown. The decayed tissue is soft and watery and the lesion has a very sharp margin between diseased and healthy tissues. Decayed tissue can be readily separated from the healthy tissue, leaving it like a "bowl". Blue or blue-green spore masses may appear on the decayed area, starting at the infection site. Decayed fruit has an earthy, musty odor. The presence of blue-green spore masses at the decayed area and associated musty odor are the positive diagnostic indication of blue mold. Without the presence of spore masses of blue mold, blue mold can be misdiagnosed as Mucor rot, but a sweet odor is commonly associated with Mucor rot.

Causal Organism

Penicillium expansum Link is the primary cause of blue mold of apples and pears. Several other *Penicillium* species, including *P. solitum, P. commune, P. verrucosum, P. chrysogenum* and *P. regulosum,* have also been reported to cause decay on apples and pears.*P. expansum* can grow at temperatures as low as -3°C and conidia can germinate at 0°C.

Occurrence

In the orchard, *Penicillium* spp. survives in organic debris on the orchard floor, in the soil, and perhaps on dead bark on the trees. Conidia are also present in the air and on the surface of fruit. In the packinghouse facility, DPA- or fungicide-drench solutions, flume water and dump-tank water are common sources of *Penicillium* spores for fruit infection during the handling and packing processes. Spores of *P. expansum* are also commonly present in the air and on the walls of storage rooms.

P. expansum is essentially a wound pathogen. Wounds on the fruit skin such as punctures and bruises that are created at harvest or during the postharvest handling process are the major avenue of invasion by the fungus. Fruit with wounds can be inoculated with spores of thiabendazole-resistant isolates of *P. expansum* during postharvest drenching with diphenylamine and Mertect (thiabendazole, TBZ). Fruit may also be inoculated with*Penicillium* during the packing process. *P. expansum* can also cause decay through infection at lenticels, but this type of infection usually occurs on over-mature fruit or when lenticels have been injured. More than 50% of the *P. expansum* isolates recovered from decayed fruit collected from packinghouses in Washington State are resistant to thiabendazole, whereas only approximately 3% of the *P. expansum* isolates from apple orchards are resistant to TBZ. A prestorage application of TBZ is likely the major source of TBZ-resistant strains.

Control

Orchard sanitation to remove decayed fruit and organic debris on the orchard floor helps reduce inoculum levels of *Penicillium* spp. in the orchard. Good harvest and handling management to minimize punctures and bruises on the fruit helps prevent the fruit from infection at wounds by *P. expansum* and other *Penicillium* species.

Thiabendazole is commonly used as either a prestorage drench treatment or a line spray to control gray mold and blue mold. TBZ is effective to control gray mold but is not effective to control TBZ-resistant *Penicillium*. Two new postharvest fungicides, fludioxonil (Scholar) and pyrimethanil (Penbotec), can be used as drenches, dips or line sprays and have been reported to be effective to control blue mold originating from wound infections.

Biocontrol agent BioSave 110 (*Pseudomonas syringae*) applied on the packing line helps control blue mold from infection of wounds.

Sanitizing dump-tank and flume water is an essential practice to reduce infection of fruit by*Penicillium* spp. during the packing process. Fruit bins and storage rooms can harbor TBZ-resistant isolates. Bin and storage room sanitation may be beneficial in reducing TBZ-resistant populations in the packing facility.

Photo Plate: Blue Mold

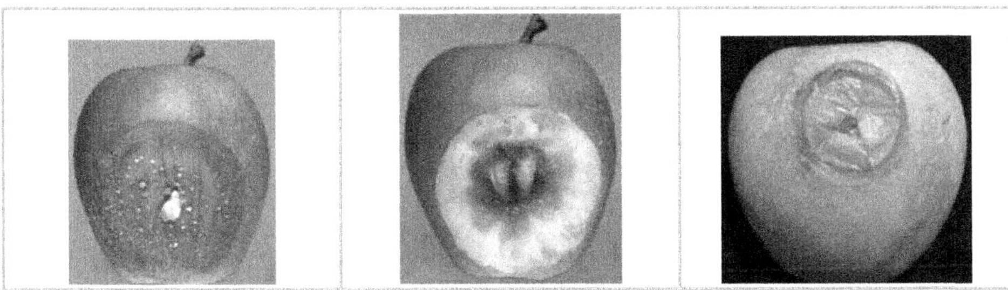

A: Decayed area brown, soft and watery, with sharp margin; mostly originating from infection of wounds; blue-green spore masses

B: Decayed tissue completely separable from the healthy tissue, leaving it like a "bowl"

C: Blue mold originating from infection of wound on a Granny Smith fruit; spore masses formed at the infection site

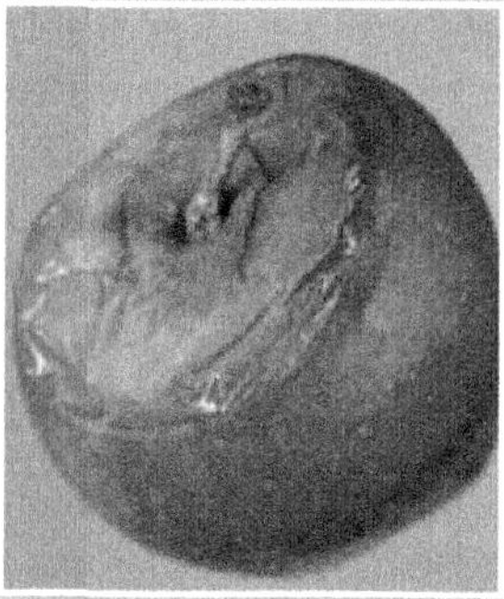

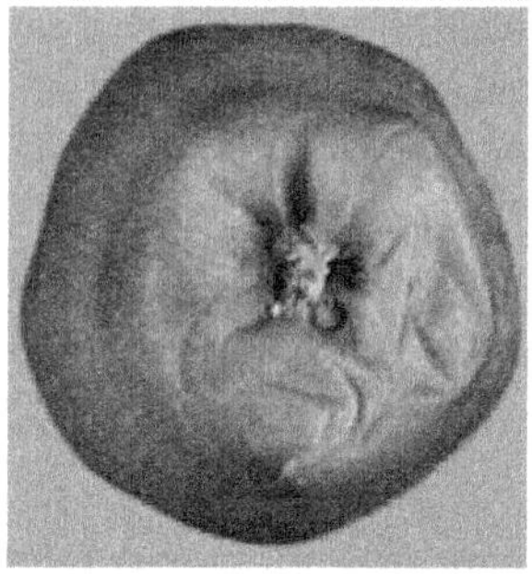

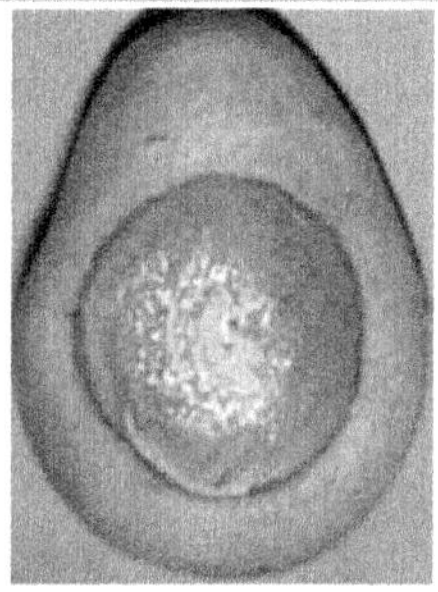

D: Blue mold originating from infection at the stem or stem-bowl area of a Red Delicious fruit

E: Calyx-end blue mold on a Red Delicious fruit; usually associated with drenched fruit

F: White mycelium and blue spore masses at the decayed area of a d'Anjou fruit

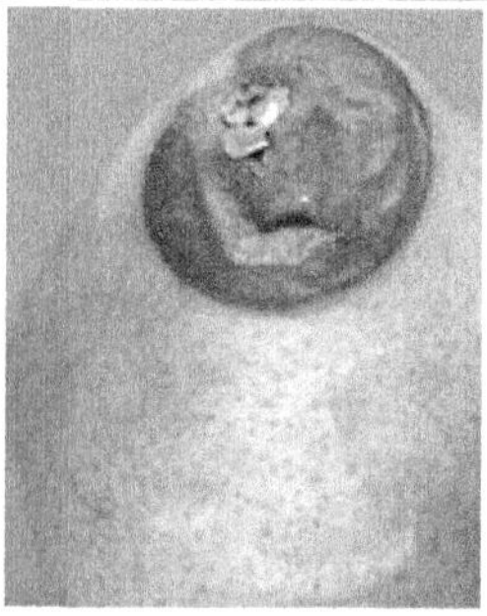

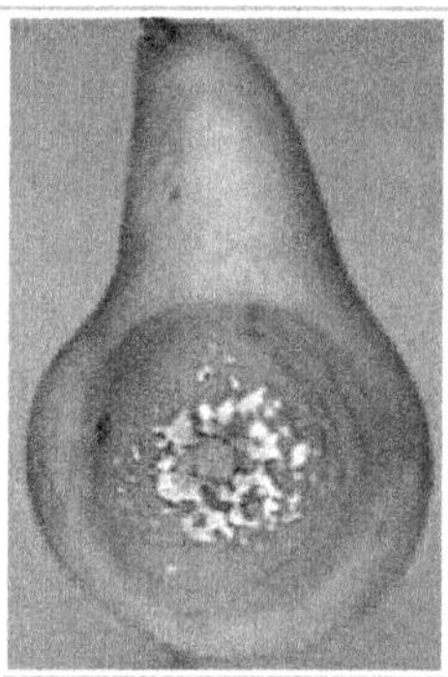

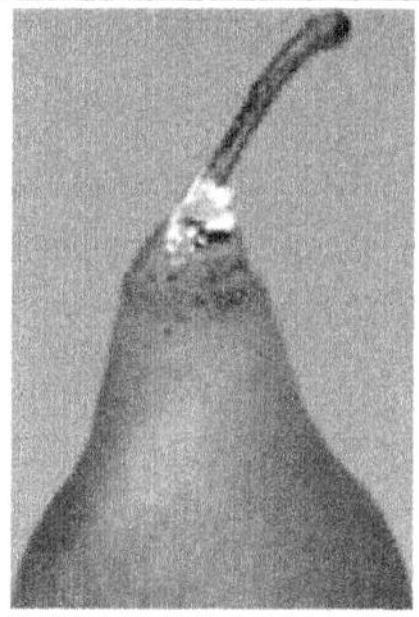

G: Stem-end blue mold commonly seen on d'Anjou pears, particularly after an extended period of storage

H: Blue mold on a Bosc fruit; watery decay lesion with white mycelium and blue spore masses

I: Stem-end blue mold on a Bosc pear fruit after an extended period of storage

Disease Comparison

Table: Comparison between blue mold and Mucor rot

Characteristics	Blue mold	Mucor rot
Texture	soft, watery	very soft, juicy
Color of decayed area	light tan to dark brown	light brown to brown
Signs of pathogen	white mycelium, blue or blue-green spore masses	gray mycelium with dark sporangia
Color of internal flesh	brown	light brown to brown
Odor	earthy, musty	sweet

SPHAEROPSIS ROT

Sphaeropsis rot is a newly reported postharvest fruit rot disease of apple and pear. This disease first was found on d'Anjou pears, and later it was determined that Sphaeropsis rot can cause even more serious problems on apples. In one instance, 24% of the apple fruit in storage bins was rotted by this disease after several months of storage. Sphaeropsis rot has been observed to occur on d'Anjou pear and on apple varieties including Red Delicious, Golden Delicious, Fuji, and Granny Smith.

Symptoms

The primary symptoms of Sphaeropsis rot are stem-end rot and calyx-end rot originating from infections at the stem and calyx of fruit, respectively. The decayed tissue is firm or spongy, and the decayed areas appear brown. On Red Delicious apples, decay resulting from infection at the calyx of fruit may appear irregular in shape. As the disease advances, the fungus may form pycnidia in the decayed areas, usually starting from infection sites. The pycnidia are black, superficial or partially embedded in the decayed tissue. The skin of decayed areas generally remains brown or dark brown but may appear dark in aged decayed areas. The internal decayed flesh appears brown. Decay in the fruit flesh develops along the vascular tissue of the decayed fruit, originating from infection of the stem or calyx of fruit. Sphaeropsis rot can be misdiagnosed as gray mold because symptoms of Sphaeropsis rot can be very similar to those caused by gray mold. However, Sphaeropsis rot has a distinct "bandage-like" odor, particularly in the decayed flesh when the fruit is cut. This can be used as a preliminary diagnostic indication of Sphaeropsis rot. A further examination of fruiting bodies, if present, or isolation of the causal agent from decayed tissue may be necessary for confirmation of diagnosis.

Causal Organism

Sphaeropsis pyriputrescens Xiao & J. D. Rogers is the causal agent of Sphaeropsis rot. The fungus grows readily on agar media. Mycelium grows at -3 to 25°C and does not grow at 30°C but survives at this temperature. Optimum temperature for mycelial growth is between 15 and 20°C. Conidia germinate at 0 to 30°C, and a minimum wetness duration of 5-6 h is required for germination at optimum temperature. Conidia can germinate at relative humidity as low as 92%.

On potato dextrose agar at 20°C, *S. pyriputrescens* initially appears colorless but starts to turn light yellow to yellow after 5 to 7 days of incubation. This yellow pigmentation is a useful characteristic in the diagnosis of Sphaeropsis rot.

Occurrence

S. pyriputrescens causes twig dieback and cankers on apple and crabapple trees. 'Manchurian' crabapple trees are very susceptible to twig dieback and cankers caused by this fungus. *S. pyriputrescens* also appears to be associated with dead bark or twigs on trees. Fruiting bodies (pycnidia) of the fungus are often present in these diseased tissues on the trees. Pycnidia are apparently the main type of inoculum for fruit infection. Infection of fruit occurs in the orchard, but symptoms develop after a period of time during storage. The fungus primarily infects the stem and calyx of fruit and causes stem-end rot and calyx-end rot, respectively.

Control

The inoculum of *S. pyriputrescens* is apparently present in the orchard. Removal of cankers and twigs with dieback symptoms would help reduce inoculum of the fungus in the orchard. For orchards with a high percentage of crabapple trees as pollinizers, pruning diseased twigs is particularly important to disease control. Research on chemical control of this disease is currently in progress.

Photo Plate: Sphaeropsis Rot

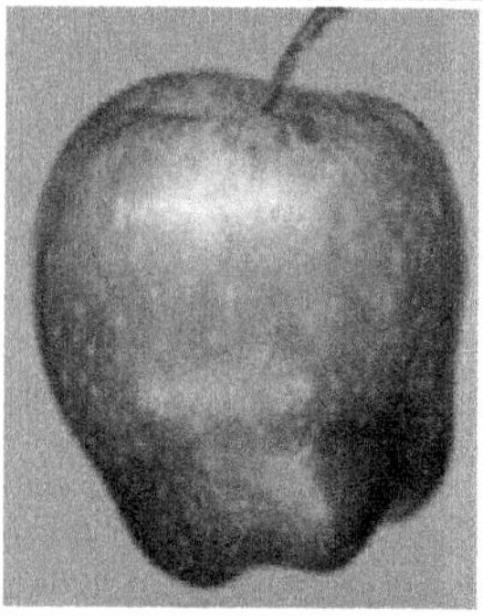

A: Stem-end Sphaeropsis rot originating from infection at the stem of a Red Delicious fruit; decayed area brown, firm

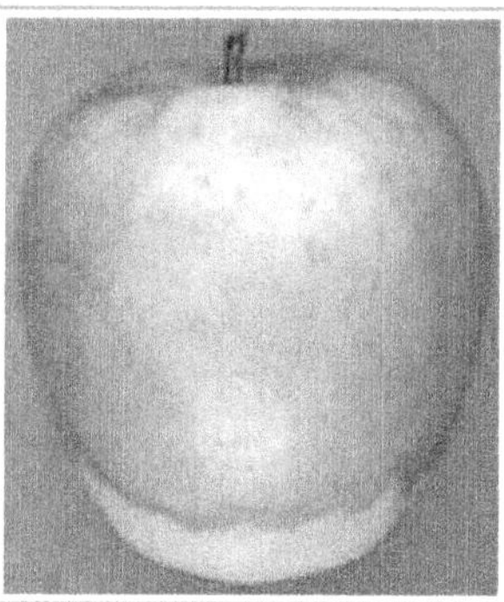

B: Advanced stage of stem-end Sphaeropsis rot on a Golden Delicious fruit

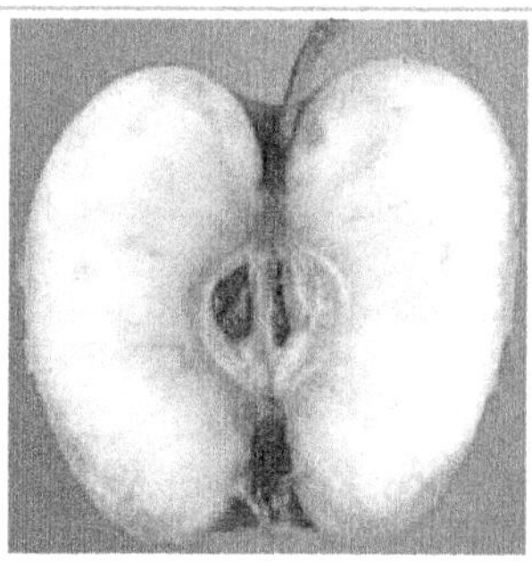

C: Internal flesh light tan to brown; strong distinct "bandage-like" odor commonly associated with it

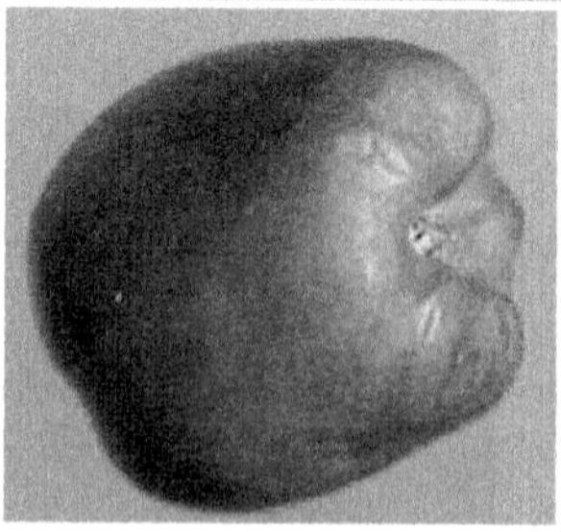

D: Calyx-end Sphaeropsis rot originating from calyx infection on a Red Delicious fruit

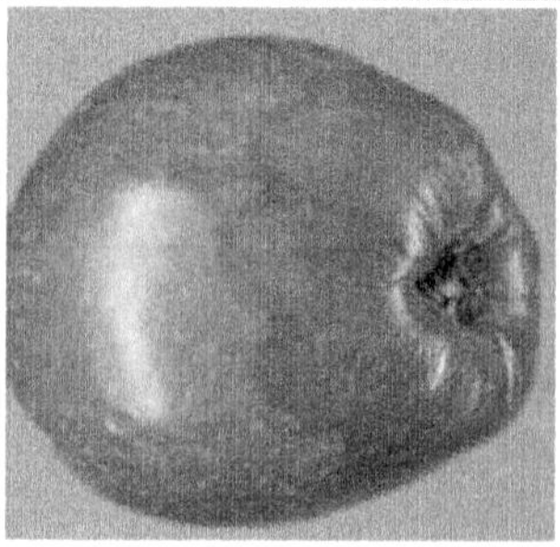

E: Calyx-end Sphaeropsis rot on a Fuji fruit; lesion with irregular margin

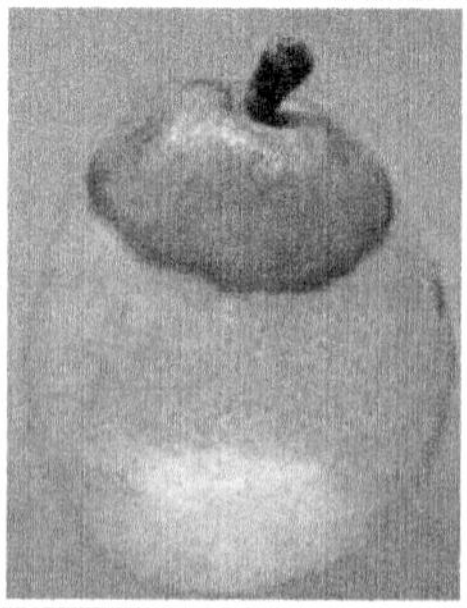

F: Stem-end Sphaeropsis rot on a d'Anjou pear fruit; decayed area brown, firm

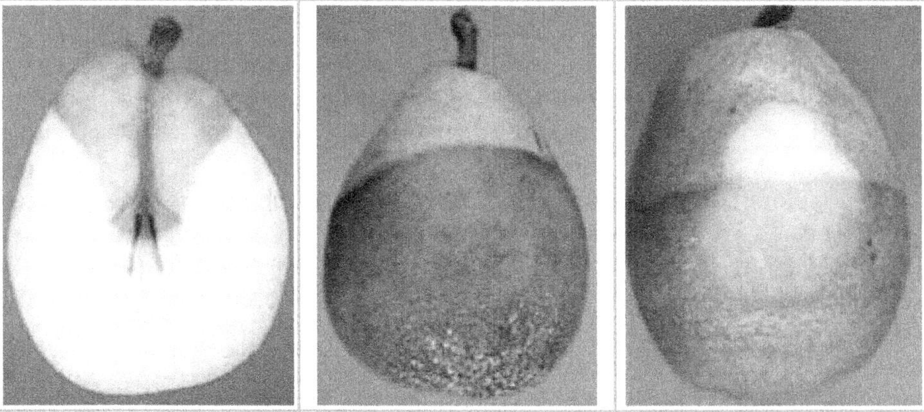

G: Decay in the fruit flesh advances along the vascular tissue of the decayed fruit	H: Calyx-end Sphaeropsis rot; decayed area brown, firm; black pycnidia formed at aged area	I: Internal decayed flesh brown; strong distinct "bandage-like" odor commonly associated with it

Fig. : Symptoms and signs of Sphaeropsis rot (*Sphaeropsis pyriputrescens*) on apples and pears.

Disease Comparison

Table: Comparison between Sphaeropsis rot and gray mold.

Characteristics	Sphaeropsis rot	Gray mold
Texture	spongy to very firm; decayed tissue not separable from the healthy tissue	spongy or firm; decayed tissue not separable from the healthy tissue
Color of decayed area	brown to dark brown; advanced decayed area may turn black	light brown to dark brown; color similar across the decayed area
Signs of pathogen	white mycelia under high humidity; pycnidia may form on decayed fruit at advanced stages	fluffy white to gray mycelia; sporulation under high humidity; gray to brown spore masses; black sclerotia may form at advanced stages
Color of internal flesh	light tan to brown; decay advances along the vascular tissue; decayed vascular tissue brown	light brown to brown
Odor	strong, distinct "bandage-like" odor	generally not detectable

PHACIDIOPYCNIS ROT

Phacidiopycnis rot has previously been reported only in Europe and India and has recently been reported in North America. In Washington State this disease is common on pears, particularly on d'Anjou pears, and is much less common on apples.

Symptoms

Phacidiopycnis rot causes three types of symptoms on pears: stem-end rot, calyx-end rot, and wound-associated rot originating from infection at the stem, calyx and wound on the skin of fruit, respectively. Decayed area is spongy. In the early stage of symptom development, the decayed area appears water-soaked. Color of the decayed area varies with age. As the disease progresses, the aging decayed area turns brown and then black, but the margin of decayed area continues to have a watersoaked appearance. Under high relative humidity, the fungus forms white mycelium, and pycnidia (fruiting bodies) of the fungus are often formed on the decayed area at advanced stages, starting from the infection sites. Pycnidia formed on decayed fruit under commercial fruit-storage conditions are usually immature, and only microconidia are present in immature pycnidia.

In the early stage of symptom development, Phacidiopycnis rot can be misdiagnosed as gray mold because of the similarity in symptoms, but the fruit flesh at the margin of Phacidiopycnis rot appears translucent and watersoaked, whereas internal decayed flesh of gray mold usually appears brown. Phacidiopycnis rot can also be mistaken as Sphaeropsis rot.

Causal Organism

The causal agent of Phacidiopycnis rot is the discomycete *Potebniamyces pyri* Dennis, anamorph *Phacidiopycnis piri* (Fuckel) Weindlmayr. The fungus has two types of conidia, macroconidia and microconidia. Microconidia are considered spermatia in the life cycle of the fungus. Mycelium grows at temperatures from -3 to 25°C but not at 30°C. Conidia germinate in two different manners, by developing germ tubes and budding to produce secondary conidia. Low nutrient levels favor budding, whereas high nutrient levels favor germ-tube development. Conidia germinate at 0 to 30°C but not at 35°C, with optimum temperature between 20 and 25°C. Four to 5 and 6 to 8 h of wetness durations at optimum temperature are required for budding and developing germ tubes, respectively. Secondary conidia germinate at 0 to 25°C in the same manner as mother conidia. Secondary conidia do not germinate at 30°C. Relative humidity affects the mode of conidial germination. At 20°C, conidia germinate only by germ tubes at 98% relative humidity. Conidia germinate in both manners at 100% relative humidity, but germination is dominated by germ tubes. In free water the majority of conidia germinate by budding.

Occurrence

The fungus *Potebniamyces pyri* is a weak canker-causing pathogen on pear trees. It is commonly associated with dead bark and cankers on pear trees. *P. pyri* is present in major pear-producing areas in the U.S. Pacific Northwest and is widespread in pear orchards in the Wenatchee River Valley, the primary production area for d'Anjou pears in Washington State.*P. pyri* survives as mycelium and pycnidia in diseased twigs and dying or dead bark on the trees in the orchard. Apothecia also occur in pear orchards in north-central Washington but at a low frequency. Pycnidia are the main type of inoculum for fruit infection, and viable pycnidia are available throughout pear fruit-growing season in north-central Washington.

Infection of fruit by *P. pyri* occurs in the orchard, and symptoms develop during storage. *P. pyri* infects the stem and calyx of pear fruit as well as skin wounds such as limb rubs and stem punctures created at harvest. The stem and calyx of fruit near harvest are more susceptible to infection by the fungus. Wound infections develop decay symptoms within two months after harvest. Stem-end rot and calyx-end rot symptoms are usually first seen three months after harvest. In north-central Washington, d'Anjou pears are packed soon after harvest and stored in cardboard boxes. At packing, fruit that are infected at the stem and calyx may not have decay symptoms, but symptoms appear on packed fruit after a period of time during storage. Unacceptable levels of decay on packed fruit in cardboard boxes may result in repacking.

Under high relative humidity, the fungus produces fluff mycelium on decayed fruit, which is able to infect surrounding healthy fruit, causing secondary infection through fruit-to-fruit contact.

Control

Ziram applied within two weeks before harvest provides some control of stem-end and calyx-end Phacidiopycnis rot. A postharvest drench treatment with Mertect (thiabendazole) applied prior to storage is effective to control Phacidiopycnis rot originating from infection of wounds on the fruit skin. Research has been undertaken to develop relevant measures for control of stem-end and calyx-end Phacidiopycnis rot.

Photo Plate: Phacidiopycnis Rot

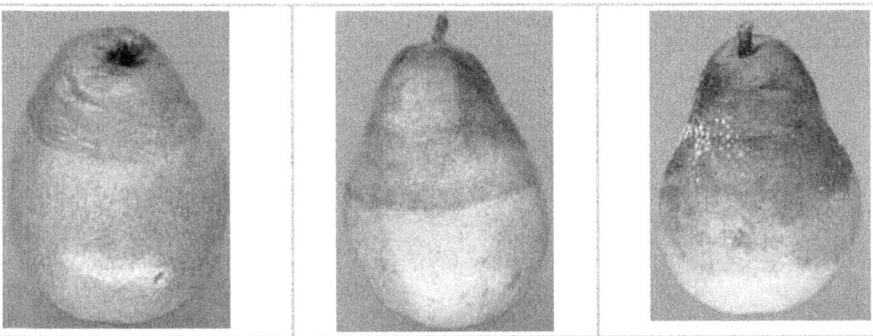

A: Early stage of stem-end Phacidiopycnis rot on a d'Anjou fruit; watersoaked appearance	B: The color of decayed area varies with age; watersoaked at the margin, black at aged area	C: Advanced stage of stem-end rot showing the change in color across the decayed area

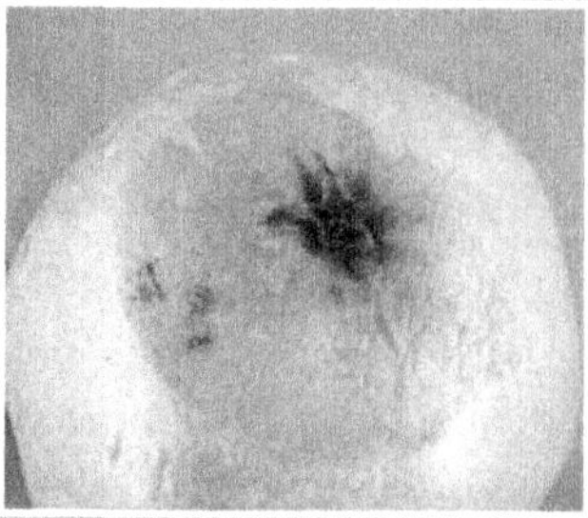

D: Early stage of calyx-end Phacidiopycnis rot; watersoaked appearance

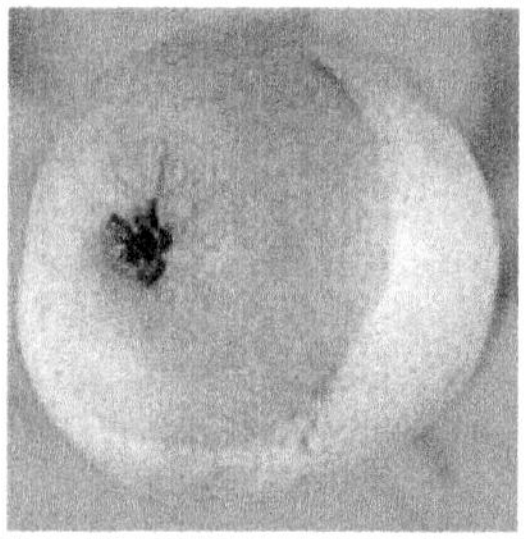

E: As the decay advances, the aging decayed area turns brown to black

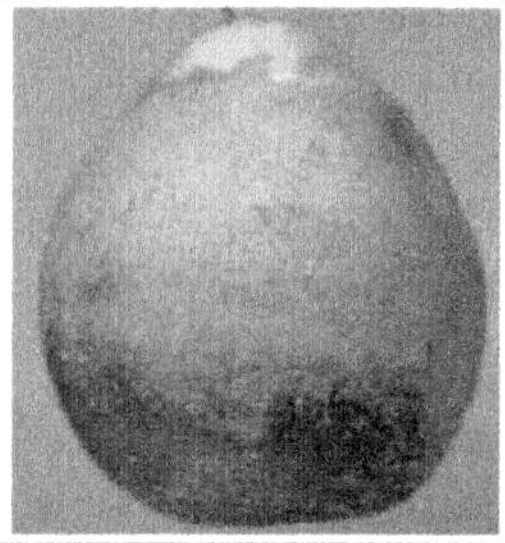

F: Advanced stage of calyx-end rot; black at the calyx-end area; pycnidia may form

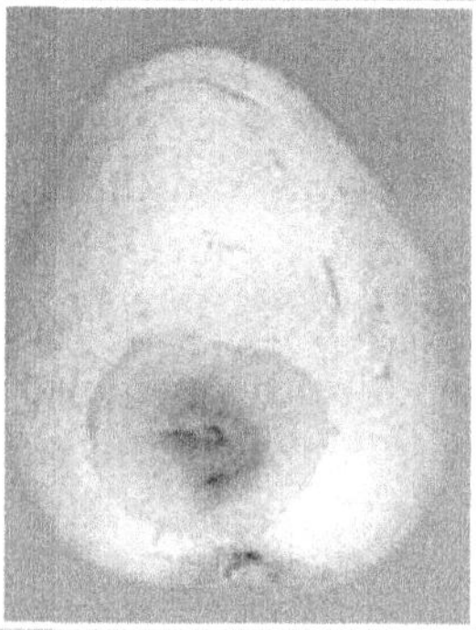

G: Early stage of wound infection commonly associated with limb rubs or punchtures on the fruit created at harvest

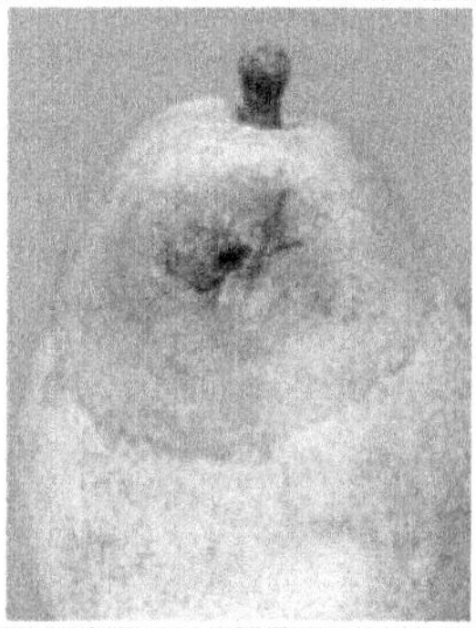

H: Phacidiopycnis rot from wound infection; the infection site turns brown to black as the decay advances

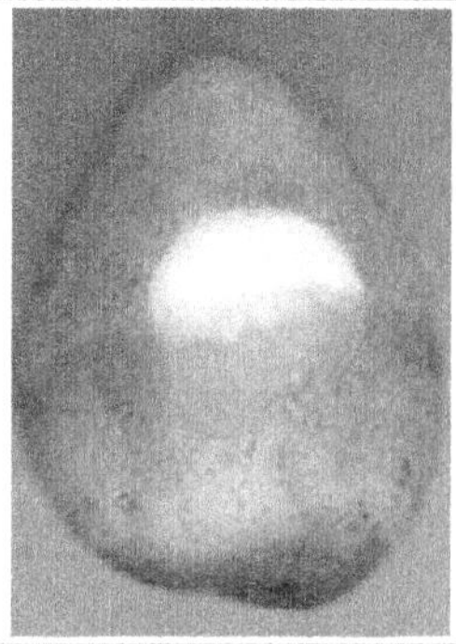

I: Internal decayed flesh translucent, clear at the margin

Fig. : Symptoms and signs of Phacidiopycnis rot caused by *Potebniamyces pyri* on pears.

Disease Comparison

Table: Comparison between Phacidiopycnis rot and gray mold on pears

Characteristics	Phacidiopycnis rot	Gray mold
Texture	spongy; decayed tissue not separable from the healthy tissue	spongy to firm; decayed tissue not separable from the healthy tissue
Color of decayed area	initially watersoaked, then light brown to brown, later black; color varies with age	light brown to dark brown; color similar across the decayed area
Signs of pathogen	white mycelia under high humidity; black pycnidia may form on decayed fruit at advanced stages	fluffy white to gray mycelia; sporulation under high humidity; sclerotia may form
Color of internal flesh	translucent, clear at the margin	light brown to brown at the margin
Odor	mild, distinct	generally not detectable

Table: Comparison between Phacidiopycnis rot and Sphaeropsis rot on pears.

Characteristics	Phacidiopycnis rot	Sphaeropsis rot
Texture	spongy	Firm
Color of decayed area	initially watersoaked, then light brown to brown, later black; color varies with age	brown to dark brown, advanced decayed area may turn black
Signs of pathogen	white mycelia under high humidity; black, immature pycnidia form on decayed fruit at advanced stages	white mycelia under high humidity; pycnidia may form on decayed fruit at advanced stages
Color of internal flesh	translucent, clear at the margin of decayed area; advanced internal decayed flesh may appear dark	brown; decay advances along the vascular tissue of the fruit; decayed vascular tissue brown
Odor	mild, distinct odor	strong, distinct "bandage-like" odor

BULL'S EYE ROT

Bull's eye rot occurs on both apples and pears in the U.S. Pacific Northwest. In Washington State, Bull's eye rot is more commonly seen on Golden Delicious, particularly on apples from orchards with perennial canker problems on trees. Bull's eye rot also occurs in Europe and some other fruit-growing regions.

Symptoms

Bull's eye rot lesion is circular, flat to slightly sunken and appears light brown to dark brown with a lighter brown to tan center. Decayed tissue is firm. Cream-colored spore masses in the aged decayed area may appear. Bull's eye rot commonly originates from infection at lenticels on the fruit skin, but stem-end Bull's eye rot is also commonly seen on Golden Delicious and Gala apples, particularly on the fruit from orchards with over-tree evaporative cooling or irrigation. Calyx-end Bull's eye rot has also been observed on Golden Delicious fruit.

Causal Organism

Four species of *Neofabraea* are known to cause Bull's eye rot on apples. Three species, *N. alba*, *N. perennans*, and a *Neofabraea* sp., have been reported to cause Bull's eye rot on pears in the Pacific Northwest.

Occurrence

N. perennans causes perennial canker on apple trees. On pear trees, *Neofabraea* spp. may survive on the dead bark. It has been reported that the Bull's eye rot fungus may infect fruit anytime between petal fall and harvest but remains latent. Symptoms of Bull's eye rot may appear after only a few months in storage. Fruit become more susceptible to infection by the fungus as the growing season approaches harvest. This disease is more common on fruit from orchards with over-tree irrigation or evaporative cooling, or in years or areas with frequent rains near or during harvest.

Control

In apple orchards with perennial canker, removal of branches with cankers helps reduce inoculum of *Neofabraea* spp. in the orchard. Ziram applied within two weeks before harvest is recommended for control of Bull's eye rot in the Pacific Northwest.

Photo Plate: Bull's Eye Rot

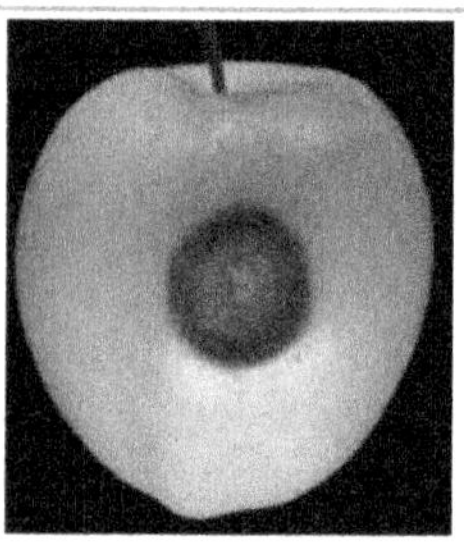

A: Bull's eye rot on a Golden Delicious fruit; lesion flat to slightly sunken, brown to dark brown with lighter brown to tan center

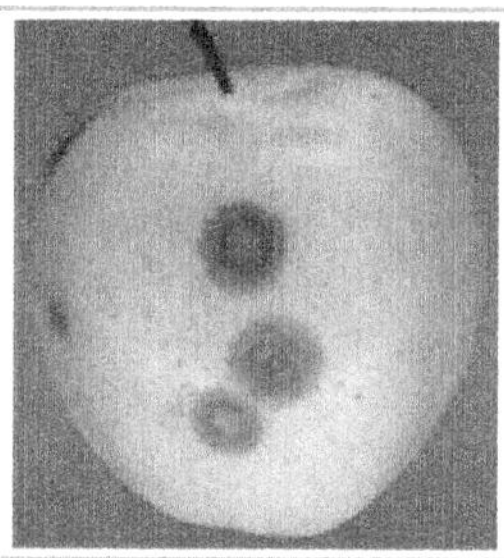

B: Multiple Bull's eye rot lesions on a Golden delicious fruit

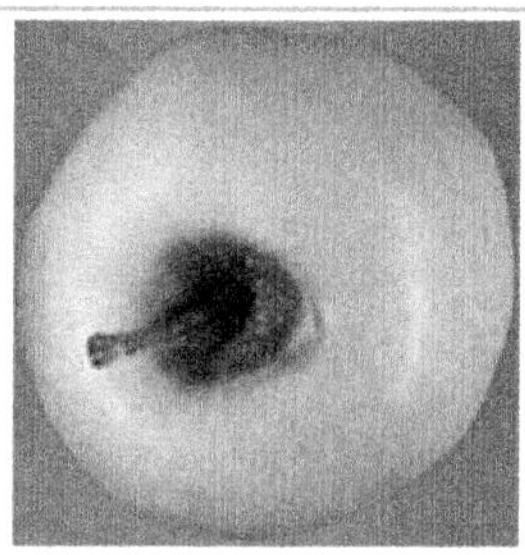

C: Bull's eye rot originating from infection at the stem-bowl area of a Golden delicious fruit

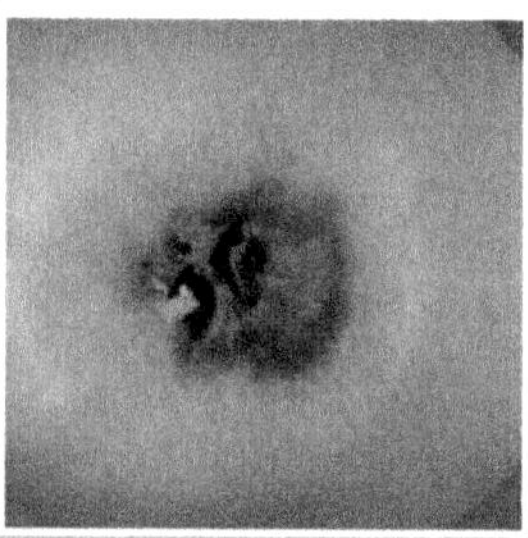

D: Bull's eye rot originating from infection at the calyx end of a Golden Delicious fruit

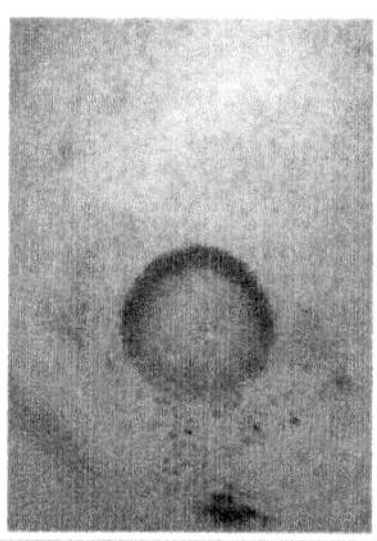

E: Bull's eye rot on a d'Anjou pear fruit

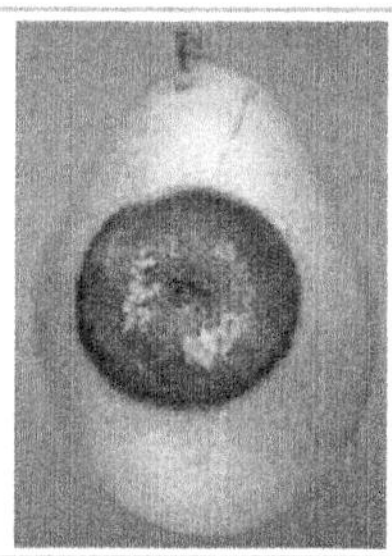

F: Advanced stage of Bull's eye rot on a d'Anjou pear fruit; white mycelium and cream-colored spore masses present at the

Fig. : Symptoms and signs of Bull's eye rot (*Neofabraea* spp.) on apples and pears.

MUCOR ROT

Mucor rot occurs on both apples and pears. Although it can cause significant losses of fruit, Mucor rot is generally not a major problem, particularly when good harvest management and water-sanitation practices at packing are implemented.

Symptoms

Mucor rot originates primarily from infection of wounds on the skin of fruit. Stem-end Mucor rot is also seen on d'Anjou pears. The decayed area appears light brown to brown with a sharp margin. The decayed tissue is very soft and juicy and can be readily separated from the healthy tissue. Gray mycelium with dark sporangia may appear on the decayed area. Mucor rot fruit has a sweet odor. Without the signs of the pathogen present on decayed fruit, Mucor rot can be mistaken as blue mold, particularly in the early stage of symptom development.

Causal Organism

Mucor rot is caused primarily by *Mucor piriformis* E. Fischer.

Occurrence

Mucor piriformis is a soilborne pathogen and survives in the orchard soil. The pathogen enters packing facilities through infested soils or organic debris adhering to field bins. Drench solutions and dump-tank water are the primary source of inoculum for fruit infection at drenching and packing. *Mucor piriformis* may infect the stem, calyx or wounds on the skin of fruit.

Control

Orchard sanitation to remove fallen fruit on the orchard floor helps reduce inoculum levels in the soil. Good harvest management is important for control of Mucor rot. Field bins should be positioned on the orchard floor to minimize contamination of the undersides of bins by infested soil or organic debris. Fruit that fall to the ground during harvest can be contaminated by infested soil and should not be put back in fruit bins.

Photo Plate: Mucor Rot

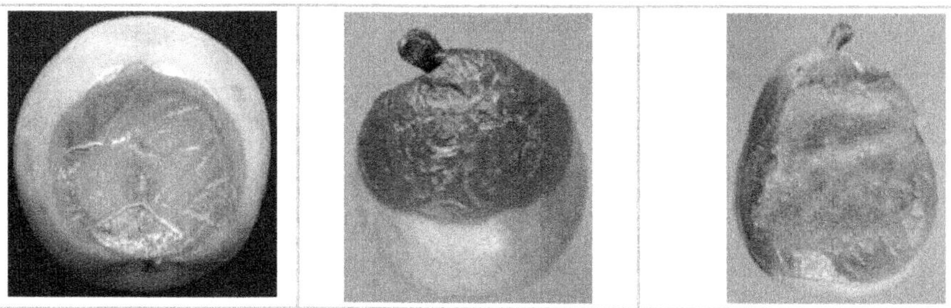

A: Mucor rot on a Golden Delicious fruit; decayed tissue very soft, juicy, often with a sweet odor; lesion with a sharp margin	B: Advanced stage of Mucor rot originating from stem infection on a d'Anjou pear fruit; gray mycelium with dark sporangia	C: Advanced stage of Mucor rot on a d'Anjou pear fruit; gray mycelium with dark sporangia.

Fig. : Symptoms and signs of Mucor rot (*Mucor piriformis*) on apples and pears.

Disease Comparison

Table: Comparison between Mucor rot and blue mold.

Characteristics	Mucor rot	Blue mold
Texture	very soft, juicy	soft, watery
Color of decayed area	light brown to brown	light tan to dark brown
Signs of pathogen	gray mycelium with dark sporangia	white mycelium, blue or blue-green spore masses
Color of internal flesh	light brown to brown	Brown
Odor	sweet	earthy, musty

POST-HARVEST DISEASES OF TREE FRUIT

Diagnosis of Post-harvest Rots

Apples and Pears

Many storage rots are actually initiated in the orchard and can be controlled at this level. There are often no pre-harvest symptoms, so identification is possible only after symptoms develop in storage. In some cases the symptoms are internal and discovered only after the fruit is cut open.

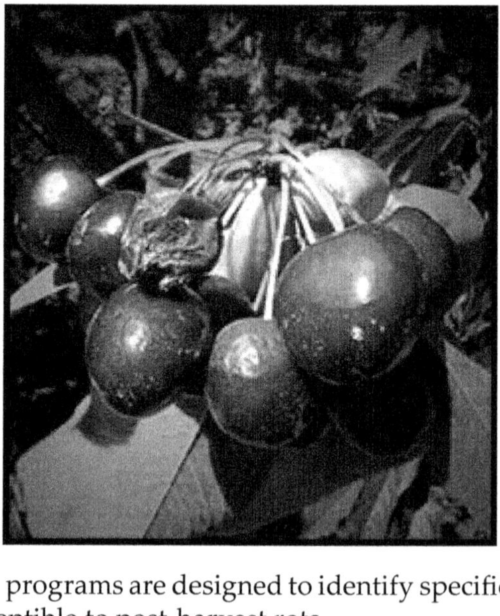

Packinghouse programs are designed to identify specific orchards or areas that are most susceptible to post-harvest rots.

Post-harvest diseases of apple and pear fall into three general categories:

1. Blossom infections
2. Fall lenticel infections

3. Infections that occur at harvest as the result of injury to the fruit, such as stem punctures.

All of these are controllable to some degree by orchard management programs.

Blossom Infections

A number of fungi cause fruit infection by first colonizing or infecting the blossoms. The incidence of infection is much more prevalent in a wet spring. There are often no visible symptoms of disease until after harvest. In some cases infected fruit may colour and fall prematurely in the orchard.

Three common rots in this category are:

1. Moldy core – ***Alternaria*** spp.
2. Symptoms include mold growth in the core region of the fruit. The rot does not spread into the flesh, and can be seen only when the fruit is cut open. Varieties with an open sinus extending from the calyx to the core are more susceptible to moldy core.
3. **Dry eye rot / Blossom end rot - *Botrytis cinerea***
4. Initial symptoms include a reddish discoloration at the calyx end. A soft, water soaked lesion forms which expands to several mm and dries out forming a sunken brown lesion. If infected fruit are harvested, they will have an increased incidence of Botrytis rot in storage.
5. **Core rot – identification unconfirmed**
6. A wet core rot that spreads into the flesh of the apple has been a recent problem in 'Ambrosia', 'Gala' and 'Red Delicious' apple.

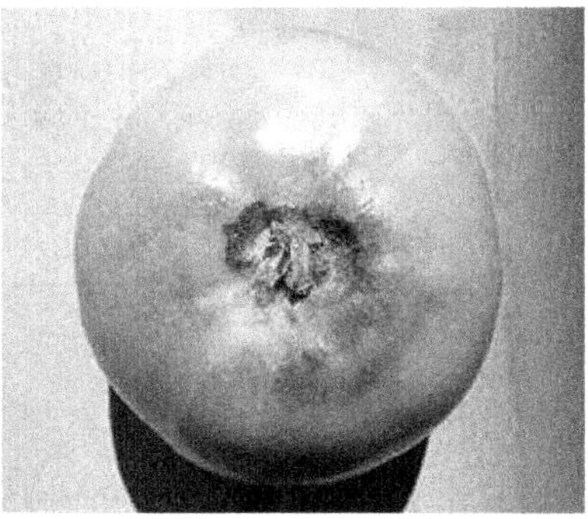

Dry eye rot symptoms on 'MacIntosh' apple, caused by Botrytis cinerea

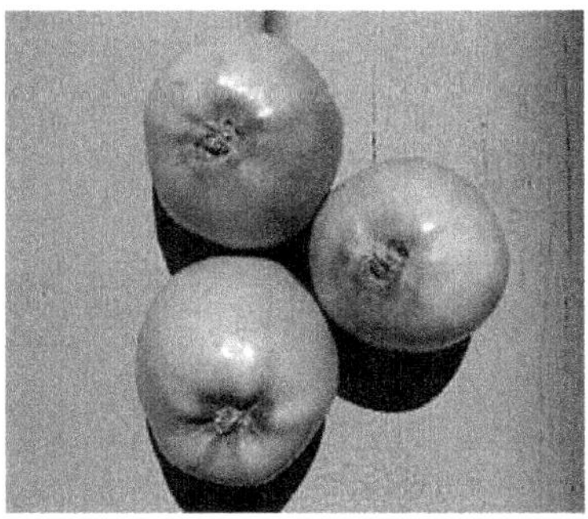

Dry eye rot symptoms on 'MacIntosh' and 'Ambrosia' apple, caused by Botrytis cinerea

Cultural Controls

General sanitation in the orchard helps reduce inoculum. The spores of these fungi are prevalent in the ground litter and in the tree. Therefore, mowing and clearing of leaves, fallen fruit and prunings can help reduce potential sources of infection.

Chemical Controls

These fungal pathogens can be reduced with spring sprays. Although there are no chemicals registered specifically for these fungi, carefully selected scab fungicides at the pink to petal-fall stages will provide some control. See table below for more information.

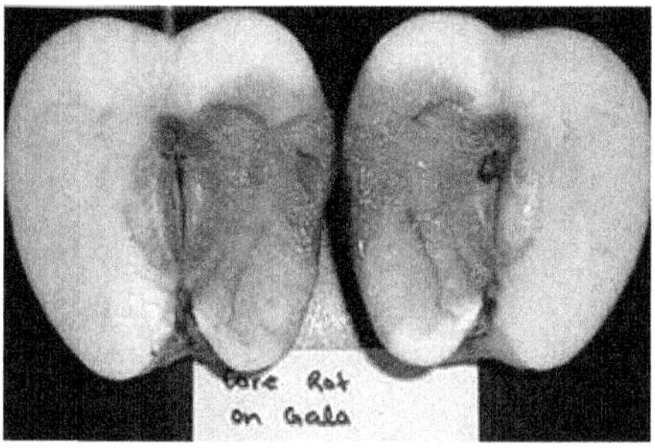

Core rot on 'Gala' apple

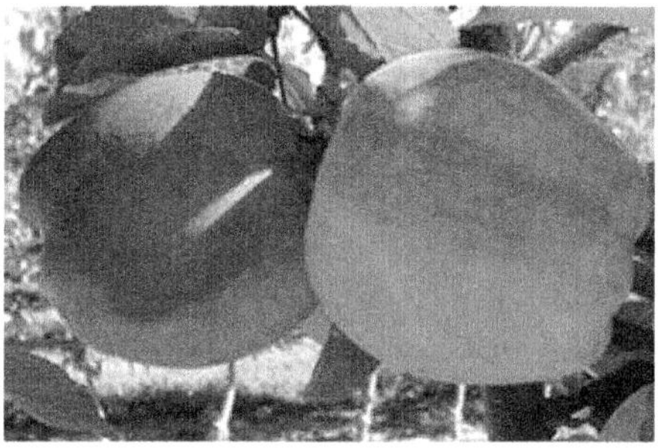

External symptoms of core rot on apple

Fall Lenticel Infections

Several fungi can enter the fruit through weak lenticels and wounds, and may not show symptoms until after a period of storage. The most common in B.C. are:

- **Bull's eye rot** – *Neofabraea* spp.
- Symptoms include circular, brown lesions with a target-like appearance on fruit. This rot is caused by the same fungi that cause perennial and anthracnose cankers. Spores may be spread from cankers to fruit any time during the growing season, but the risk of infection is greatest close to harvest. This disease is more severe in years with frequent rains at harvest.

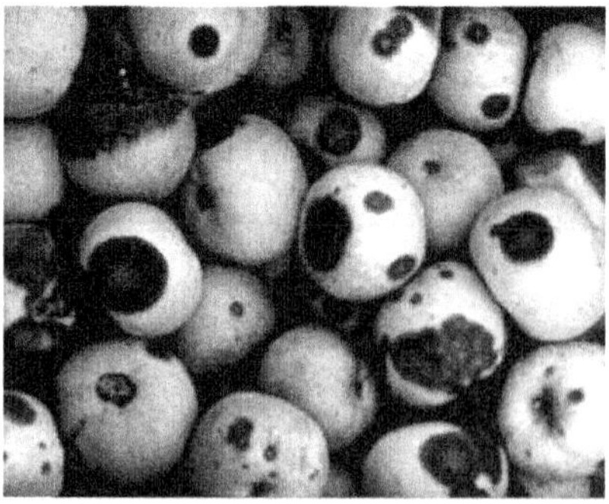

Bull's Eye Rot on apple.

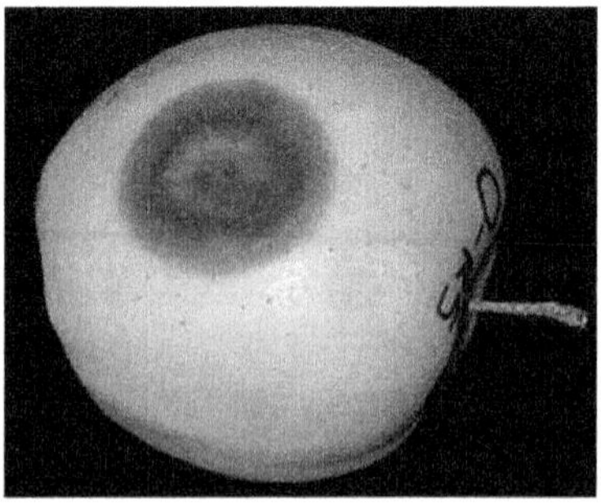

Bull's Eye Rot on apple. Photo courtesy Agriculture & Agri-Food Canada

- **Phacidiopycnis rot** – *Potebniamyces pyri.* (Pears only)
- This fungus is a weak canker-causing pathogen on pear trees. Spores produced in cankers cause infection at the stem and calyx end of pear fruit as well as in skin wounds

Other fungi that occasionally cause fruit infection in the orchard include:

- **Snow mold rot** – low temperature basidiomycete
- Formerly called *Coprinus* rot, this disease is associated with poor handling of bins in the orchard. Clumps of dirt and grass transferred to the bin carry fungi which contaminate dump water and cause fruit infections.
- **Blue mold rot** (*Penicillium* spp.), and Mucor rot (*Mucor* sp.)
- Penicillium spores are ubiquitous in orchards and are capable of infecting via weak lenticels or through skin punctures that occur during harvest. Mucor does not appear to attack through lenticels but is most commonly associated with stem end infections.
- **Sphaeropsis rot** (*Sphaeropsis pyriputrescens*) This fungus produces a rot similar to *Botrytis* spp. It is found on pear and apples but is more important on apples.

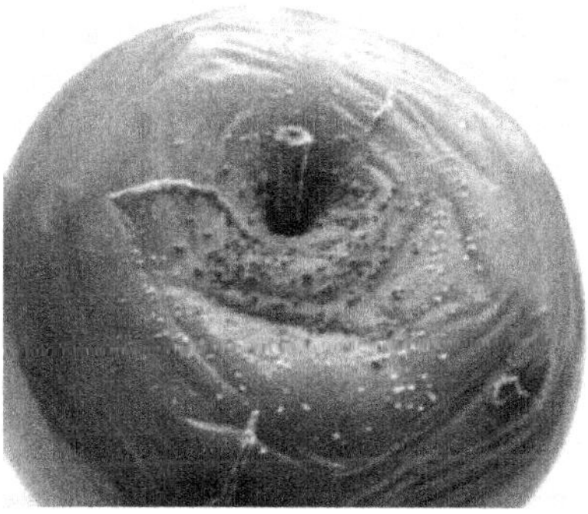

Blue mold rot on apple. Photo courtesy Agriculture & Agri-Food Canada

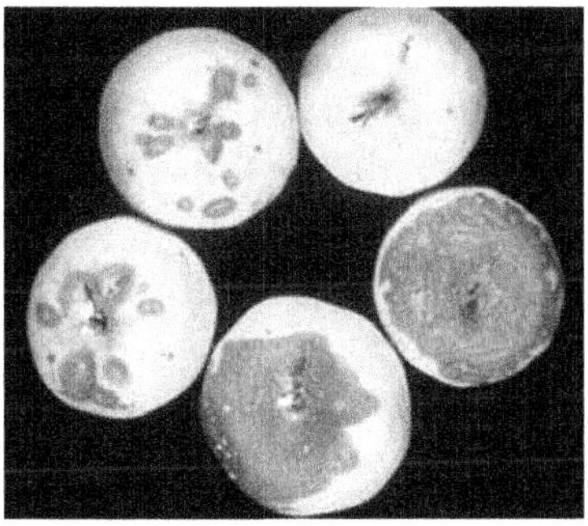

Snow mold rot on apple. Photo courtesy Agriculture & Agri-Food Canada

Cultural Controls

Cleaning up the cankers in the trees is the most effective control for Bull's Eye and Phacidiopycnis rot. Since snow mold rot has been associated with soil and grasses or weeds being transferred from the orchard floor to the bin, careful handling of bins in the orchard should help reduce this problem.

Chemical Control

There are no fungicides registered specifically for these diseases. Some pin point scab sprays will reduce the incidence of these diseases in storage.

Infections that Occur During Harvest as a Result Of Injury

Fruit in the orchard supports a variety of fungal spores on the surface. These spores are opportunistic and will enter the fruit through wounds. *Botrytis* and *Penicillium* spp. are two fungi that commonly infect through wounds at harvest. These pathogens continue to develop and spread in storage, and are responsible for significant post-harvest losses in the packinghouse.

At the time of harvest it is important to handle the fruit gently. Fruit that is transferred to the bins roughly will experience puncture wounds, bruising and other skin breaks. One common wound that is associated with stem end rot is a loosening of the stem-to-apple connection during picking. This results from pulling the apple away from the tree rather than using a twisting motion while holding the stem firmly and steady. This type of injury is most common in the variety Gala.

Cultural

Train pickers to handle the fruit very carefully while harvesting. Remove the fruit by gripping the apple and stem together and separating the apple with pressure on the abscission layer between stem and spur, not by pulling down on the fruit.

Carefully transfer fruit from the picking bag to the bin. This requires gently placing the full bag on the fruit already in the bin and gently pulling the bag up, releasing the apples.

Transporting fruit within the orchard and to the packinghouse should be done carefully to avoid bruising.

Never pick windfalls into the bin.

Stone Fruit

As with apple orchards, stone fruit orchards are full of orchard-generated fungal inoculum. Unlike apples, however, stone fruit are much more vulnerable to breakdown and fungal attack immediately after harvest.

Important post-harvest disease problems of stone fruit in B.C. are:

1. Brown rot – *Monilinia* spp.
2. Botrytis rot – *Botrytis cinerea*
3. Alternaria rot – *Alternaria* spp.
4. Rhizopus rot – *Rhizopus* sp.
5. Mucor rot – *Mucor* sp.

Correct identification of stone fruit decay problems is critical to selecting the correct preventative control strategies in the orchard. If you are unsure which diseases are present, contact your field advisor or submit a sample for diagnosis.

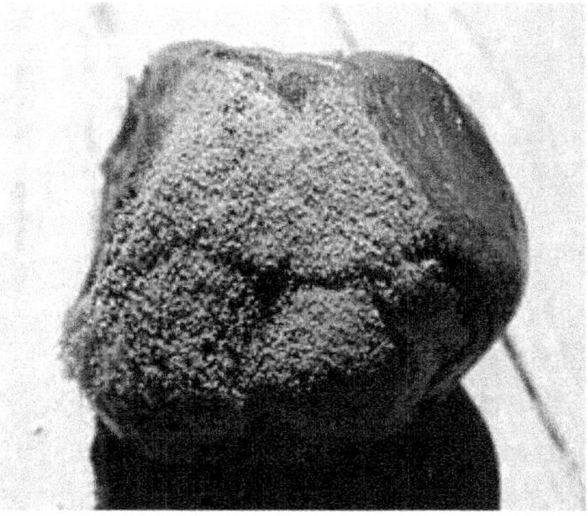

Botrytis on cherry fruit

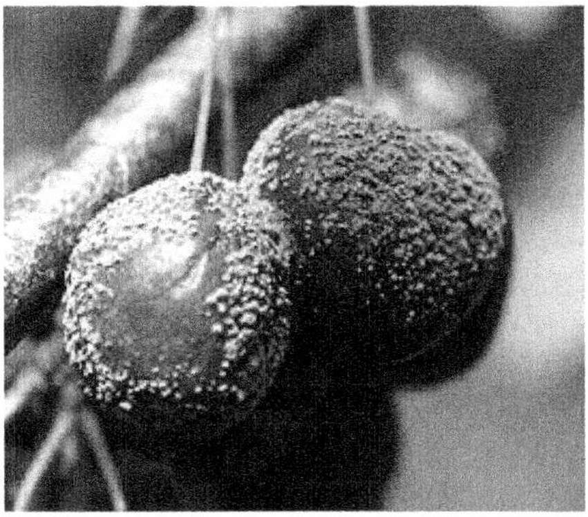

Brown rot on cherry fruit

Brown Rot

Brown rot is common on all stone fruit in the orchard. Blocks of fruit delivered to the market from orchards with heavy fruit infections in the tree will be very prone to post-harvest rot at the market. Inoculum can be found in the litter beneath the tree and in mummies left in the tree from the previous year.

Botrytis Rot

Botrytis rot of mature cherry fruit occurs commonly in the Okanagan and Creston, and can be mistaken for brown rot. Fruit develop a firm brown decay and become covered with light brown spores. *Botrytis* will grow on aborted fruit and trapped leaf debris within the clusters, which then spreads to adjacent fruit. Fruit from infected blocks has a high potential for causing fruit rot problems during post-harvest transport and marketing.

Alternaria Rot

Alternaria is found mainly on cherries and apricots. It is common in split pits of fruit, rain-split cherries and in insect feeding injuries. It also infects aborted cherry fruit, and can spread into attached healthy fruit. On cherry, Alternaria usually shows up post-harvest as a dark green to black growth on the nose of the fruit. Early infections on apricot appear as red rings on the fruit and can occur when the fruit is still green. As the fruit matures the centre of these rings becomes sunken and brown. In these cases fruit is culled at packing or not picked.

Rhizopus Rot

Rhizopus is most common on peaches and nectarines. This rot appears as large masses of black-grey fungus extending outward from the fruit in a whisker-like effect. Symptoms rarely develop in the orchard unless fruit is left to tree ripen. Because this fungus is a wound parasite, breaks in the skin favour disease development. Fruit picked and shipped to market becomes vulnerable to decay as the fruit ripens and sugar concentrations increase, especially if temperature during shipping is allowed to go above 10°C. One diseased peach in a container can infect many other fruits in a few days.

Mucor Rot

Mucor rot is less common on stone fruit. It has the appearance of stiff salt and pepper coloured whiskers on the fruit surface after the rot has progressed deep into the fruit. The spores are spread primarily by water, making the dump tank an important source once it has been contaminated by soil. The fungus is able to grow at 0°C and will rot fruit in cold storage.

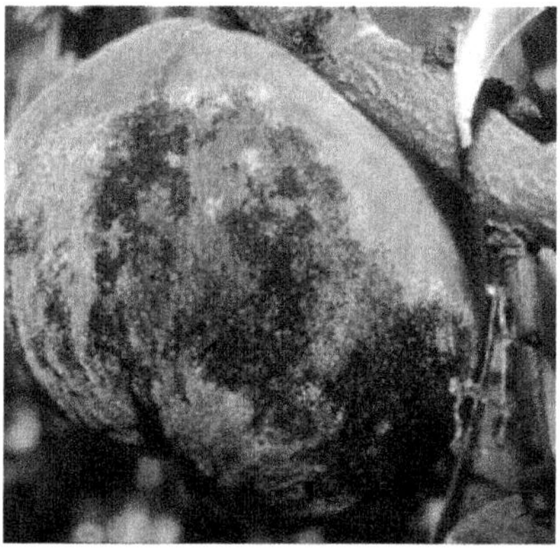

Rhizopus rot on peach. Photo courtesy Agriculture & Agri-Food Canada

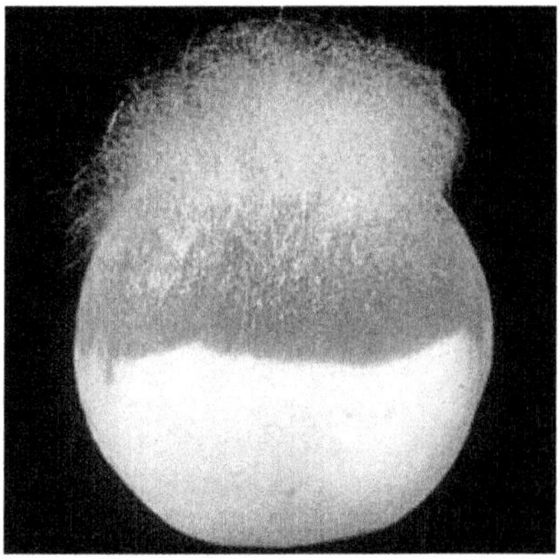

Mucor rot on peach. Photo courtesy Agriculture & Agri-Food Canada

General Management Techniques to Reduce Post Harvest Fruit Rots

Consider post-harvest decay control as an integrated pest management strategy. Cultural controls are important, and options for chemical control are often limited. The following are general management techniques that will help to reduce the occurrence of rots in both pome and stone fruits.

1. Focus on nutrient programs that improve fruit quality. Apply calcium sprays to improve fruit resistance to disease.
2. Control insects that may act as vectors of disease spores.
3. Prune out cankers in varieties or blocks prone to bull's eye rot.
4. Prune for increased air flow to promote faster drying and better spray penetration.
5. Shorten current season's growth to reduce fruit cluster formation on cherry varieties prone to heavy clusters.

6. Ensure good spray coverage into heavy cherry clusters before they close up.
7. Practice orchard sanitation and "soft handling" of fruit.
8. Harvest fruit at proper maturity.
9. Do not put windfalls into bins.
10. Use clean bins. As much as practical, minimize the amount of soil and plant debris brought into the packinghouse on the bottom of bins.
11. Keep fruit cool after harvest, *i.e.* keep bins in the shade.
12. Minimize time between harvest and delivery of fruit to the packing-house.

Management Strategies for Common Post-harvest Rots

In addition to following general management techniques to reduce post-harvest rots, the following practices may be helpful for specific diseases. Follow label instructions and observe pre-harvest intervals for all fungicides. *Rotate fungicides for resistance management.*

Crop	Disease	Season	Source	Control
Apples and Pears	Moldy core	Blossom	Orchard Litter	A reduced incidence of moldy core in blocks sprayed for scab in the spring has been observed.
	Dry eye rot/ blossom end rot	Blossom	Leaf litter	80% bloom to petal fall spray with scab fungicides that also controlBotrytis will reduce this problem (*e.g.* Vangard, Scala on apple)
	Core rot	Blossom	Unknown	Unknown. Spring scab sprays may help to reduce incidence.
	Bull's eye rot	Fall	Tree cankers	Pre-harvest spray with ziram for pinpoint scab will also help to control Bull's eye rot. Remove cankers.
	Snow mold rot	Fall	Soil, orchard grasses	Avoid contaminating bins with grasses or soil
	Phacidiopycnis rot	Fall	Pear tree cankers	Identify and remove cankers.
	Penicillium, Botrytis	Summer/ Fall	Orchard litter	Apply Scala 14 days before harvest for control of post-harvest rots. Late season pinpoint scab sprays with ziram or captan may also help to reduce the incidence in storage.
	Penicillium, Botrytis	Post-harvest		Scholar (fludioxonil) is registered for use on harvested pome fruit in the packinghouse as a post-harvest dip or drench.

Stone Fruit	Brown rot	Spring/ summer	Mummified fruit and wood infections	Control brown rot in the field to reduce post-harvest losses.
	Botrytis, Alternaria	Spring/ summer	Orchard litter	Fungicides applied for control of brown rot will also help to reduce other post-harvest rots. *E.g.* Pristine, Rovral, Lance, Elevate and Vangard have good activity against Botrytis and Alternaria. Observe pre-harvest intervals.
	Rhizopus	Spring/ summer	Orchard litter	Pristine is registered for control of rhizopus rot. Apply the day before harvest for best results.
	Brown rot,Rhizopus, Botrytis	Post-harvest		Scholar (fludioxonil) is registered for use on harvested stone fruit in the packinghouse as a post-harvest dip or drench.

Diagnosis of Post-harvest Rots

Correct identification of fruit rot problems is critical to selecting the correct preventative control strategies in the orchard. Contact your fieldman or crop advisor to assist with identification of any new or unusual disease problems, or submit a sample to the provincial Plant Health Laboratory in Abbotsford. Information on the incidence of post-harvest rot in your crop should also be available through the packinghouse staff and/or feed back from the market.

CHLORINATION AND

Postharvest Disease Control

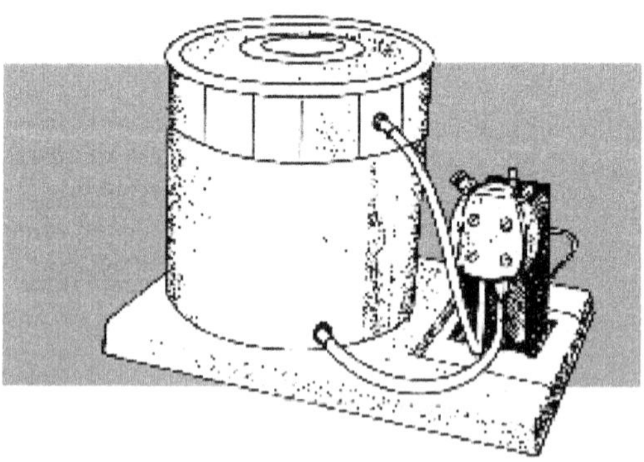

In a hungry and increasingly competitive world, reducing postharvest food losses is a major agricultural goal. For highly perishable commodities, such as tomatoes, squash, and peaches, as much as 30 percent of the harvested crop may be lost to postharvest diseases before it reaches the consumer. Investments made to save food after harvest are usually less costly for the grower and the consumer and less harmful to the environment than efforts to increase production. Even a partial reduction in postharvest losses can significantly reduce the overall cost of production and lessen our dependence on marginal land and other scarce resources.

Many factors contribute to postharvest losses in fresh fruits and vegetables. These include environmental conditions such as heat or drouth, mechanical damage during harvesting and handling, improper postharvest sanitation, and poor cooling and environmental control. Efforts to control these factors are often very successful in reducing the incidence of disease. For example, reducing mechanical damage during grading and packing greatly decreases the likelihood of postharvest disease because many disease-causing organisms (pathogens) must enter through wounds.

Chemicals have been widely used to reduce the incidence of postharvest disease. Although effective, many of these materials have been removed from the market in recent years because of economic, environmental, or health concerns.

Increased interest in the proper postharvest handling of fresh fruits and vegetables in North Carolina has prompted the widespread use of flumes, water dump tanks, spray washers, and hydrocoolers. To conserve water and energy, most postharvest processes that wet the produce recirculate the water after it has passed over the produce. This recirculated water picks up dirt, trash, and disease-causing organisms. If steps are not taken to prevent their spread, these organisms can infect all the produce that is subsequently processed. In the past, various fungicides and bactericides have been used (alone or in combination with chlorination) to prevent the transmission of diseases. These materials have often been favored over chlorination because they provide some residual protection after treatment.

At present, chlorination is one of the few chemical options available to help manage postharvest diseases. When used in connection with other proper postharvest handling practices, chlorination is effective and relatively inexpensive. It poses little threat to health or the environment. This publication has been prepared to acquaint growers, packers, and shippers with the proper use of chlorination.

Postharvest Diseases

Many types of postharvest disorders and infectious diseases affect fresh fruits and vegetables. *Disorders* are the results of stresses related to excessive heat, cold, or improper mixtures of environmental gases such as oxygen, carbon dioxide, and ethylene. Some disorders may be caused by mechanical damage, but all are abiotic in origin (not caused by disease organisms) and cannot be controlled by chlorination or most other postharvest chemicals. However, abiotic disorders often weaken the natural defenses of fresh produce, making it more susceptible to biotic diseases those that are caused by disease organisms. Further, in many cases injuries caused by chilling, bruising, sunburn, senescence, poor nutrition, and other factors can mimic biotic diseases.

Table: Common Postharvest Diseases of North Carolina Fruits and Vegetables.

Commodity and Disease	Pathogena*
Apples	
Blue mold	*Penicillium expansum* (f)
Gray mold	*Botrytis cinerea* (f)
Black rot	*Physalospora obtusa* (f)
Bitter rot	*Glomerella cingulata* (f)
Grapes and small fruit	
Blue mold	*Penicillium* sp. (f)
Gray mold	*Botrytis cinerea* (f)
Rhizopus rot	*Rhizopus stolonifer* (f)
Potatoes	
Fusarium tuber rot	*Fusarium* spp. (f)
Wet rot	*Pythium* sp. (f)
Bacterial soft rot	*Erwinia* spp. (b)
Slimy soft rot	*Clostridium* spp. (b)
Peaches and plums	
Brown rot	*Monilinia fructicola* (f)
Rhizopus rot	*Rhizopus stolonifer* (f)
Gray mold	*Botrytis cinerea* (f)
Blue mold	*Penicillium* sp. (f)
Alternaria rot	*Alternaria* sp. (f)
Gilbertella rot	*Gilbertella persicaria* (f)
Sweetpotatoes	
Bacterial soft rot	*Erwinia chrysanthemi* (b)
Black rot	*Ceratocystis fimbriata* (f)
Ring rot	*Pythium* spp. (f)
Java black rot	*Diplodia gossypina* (f)
Fusarium surface rot	*Fusarium oxysporum* (f)
Fusarium root and stem rot	*Fusarium solani* (f)
Rhizopus soft rot	*Rhizopus nigricans* (f)
Charcoal rot	*Marcrophomina* sp. (f)
Tomatoes and peppers	
Alternaria rot	*Alternaria alternata* (f)
Buckeye rot	*Phytophthora* sp. (f)
Gray mold	*Botrytis cinerea* (f)
Soft rot	*Rhizopus stolonifer* (f)
Sour rot	*Geotrichum candidum* (f)
Bacterial soft rot	*Erwinia* spp. (b) or *Pseudomonas* spp. (b)
Ripe rot	*Colletotrichum* sp. (b)
Vegetables in general	
Watery soft rot	*Sclerotinia* sp. (f)
Cottony leak	*Pythium butleri* (f)
Fusarium rot	*Fusarium* sp. (f)
Bacterial soft rot	*Erwinia* sp. (b) or *Pseudomonas* spp. (b)

* f = fungus, b = bacterium

The control of biotic postharvest diseases depends on understanding the nature of disease organisms, the conditions that promote their occurrence, and the factors that affect their capacity to cause losses. Postharvest diseases may be caused by either fungi or bacteria, although fungi are more common than bacteria in both fruits and vegetables. Postharvest diseases caused by bacteria are rare in fruits and berries but somewhat more common in vegetables. Viruses seldom cause postharvest diseases, although they, like postharvest disorders, may weaken the produce.

Most postharvest fungal diseases (rots) are caused by the dispersion of tiny dustlike spores formed by the actively growing pathogen. Spores have adaptations that allow them to survive in hot, cold, or very dry conditions. They may be carried great distances by wind or water and can cover most exposed surfaces in great numbers.

Spores may remain dormant for long periods until the correct conditions for their germination and growth occur. These conditions include the presence of water (in liquid form or as high relative humidity), warm temperatures, low light levels, adequate levels of oxygen and carbon dioxide, and the presence of nutrients in the form of sugars, starches, or other organic compounds. Many immature fruits and vegetables contain compounds that inhibit the growth of some disease organisms. These compounds and the resistance they provide are often lost during ripening. Therefore, a fresh wound on the surface of a warm, wet, ripened fruit or vegetable enclosed within a shipping container provides an ideal site for postharvest pathogens to colonize and develop. Gentle handling to prevent wounding and thorough cooling immediately after harvest can significantly reduce the incidence of postharvest disease. Figure illustrates the effects of temperature on the development of brown rot (*Monilinia fructicola*) in peaches.

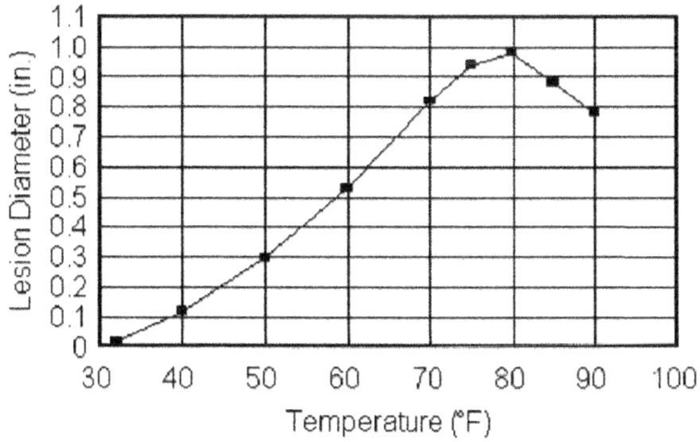

Fig. : The effects of temperature on the growth of ***Monolinia fructicola*** (causal agent of brown rot) in ripe peaches.

Spores of postharvest fungal pathogens are most susceptible to chemical control while they are germinating to produce actively growing mycelium. Under the right circumstances, germination can be rapid, often taking only a few hours. Once active growth is under way and the organism moves below the surface of the fruit or vegetable, chemical control becomes very difficult.

The potential for inoculation and infection by postharvest pathogens is always present when handling fresh produce. Understanding the several different ways by which these organisms can come into contact with the produce can be helpful in formulating control measures.

Soil and Field Conditions

Soil and decaying plant material in the field can contain postharvest pathogens in great abundance. Hard rains and wind can splash and distribute these materials onto unharvested produce. In addition, warm rainy conditions greatly favor the development of diseases in the field.

Contaminated Water

Water from ponds and streams should not be used for postharvest cooling, fluming, and washing. Water from these sources is often contaminated by runoff from fields and packing houses and may therefore contain large concentrations of postharvest pathogens. Using pond or stream water for irrigation can also contaminate produce in the field. Muddy water or water taken from the bottom of the pond is especially likely to be contaminated. Always use potable water from a well or other reliable supply.

Poor Packing House Sanitation

Pathogens brought into the packing house along with the produce will quickly contaminate all working surfaces. Disease-causing organisms will remain viable for months on surfaces such as tank walls, grading belts, and brushes. Wash all produce-handling equipment daily to remove dirt and decayed produce, and disinfect it with a strong chlorine solution on a regular schedule. Keep the packing house and the immediate vicinity clear of any overripe or rotting produce. Remove culls from the packing house and its vicinity immediately.

Air

Even the most meticulous attention to sanitation may not completely prevent contamination of fresh produce by disease organisms. Pathogens are present in the air and will infect produce under suitable circumstances. The best defense against airborne pathogens is sanitation, consistent chlorination, proper handling of the commodity, and quick and thorough cooling.

Although the skin of fruits and vegetables offers considerable protection against infection, pathogens can enter produce through a variety of openings when the produce is wetted. Various wounds, such as punctures, cuts, and abrasions, as well as stems and stem scars provide potential points of entry.

The probability of a pathogen entering the produce increases with the size of the opening, the depth of submergence, the length of time the produce is in the water, and the water temperature. Even tiny natural openings (such as stomata and lenticels) can serve as pathways for disease organisms. A small amount of detergent added to the solution lowers the surface tension, increasing the ability of the chlorine to move into the small openings and destroy the pathogens.

A chlorine concentration of about 55 to 70 ppm at a pH of 7.0 is recommended for sanitizing most fruits and vegetables. A higher concentration may be needed if the pH is higher or if the temperature of the solution is more than 80 F. In actual practice, concentrations of up to 150 ppm of free chlorine have been recommended.

The Chemistry of Chlorination

Chlorine is a very irritating, heavy, greenish yellow gas with a strong, pungent odor. Free chlorine is very reactive, combining with any chemical that will react with oxygen, and is never found uncombined in nature. Chlorine in the gaseous form is a very potent disinfectant, although it is seldom used in that form. It is much safer and easier to use when dissolved in water. Disinfection of produce using chlorine or some other chemical is nearly always done during hydrocooling or during the process of washing the produce to remove soil. Chlorine for disinfection may be obtained from one of three sources: pressurized chlorine gas, calcium hypochlorite (a soluble solid), or a solution of sodium hypochlorite.

Chlorine Gas

Chlorine gas is produced by the electrolysis of salt solutions (principally NaCl) and is furnished commercially in pressurized metal cylinders. Chlorination is accomplished by bubbling a metered amount of the gas into the supply water. Because of dangers involved with the use of chlorine gas and the expense of the metering equipment, the use of the gaseous form for chlorination is usually limited to large applications. Most municipal water supplies are disinfected with chlorine gas. The limited amount of chlorination required by most postharvest fruit and vegetable operations makes the use of chlorine gas impractical.

Calcium Hypochlorite

The most common source of chlorine used in postharvest chlorination is calcium hypochlorite. It is available commercially in the form of either a granulated powder or large tablets. Most commercial formulations are 65 percent calcium hypochlorite, with the balance consisting of stabilizers and inert materials. Calcium hypochlorite is relatively stable as long as it is kept dry, and it may be stored for extended periods. The property that makes it stable also makes it difficult to dissolve completely in water. Adding granulated calcium hypochlorite directly to the water often results in undissolved particles that adhere to the produce, causing undesirable bleaching and chlorine burns.

This problem is particularly common in hydrocoolers because calcium hypochlorite is very slow to dissolve in cold water. Therefore, always dissolve granulated calcium hypochlorite in a small quantity of tepid water before adding it to the wash tank or hydrocooler. Calcium hypochlorite may be obtained in tablets that are added directly to the hydrocooler or wash tank to eliminate the problem of chlorine burns. Properly used, the tablets will dissolve slowly to yield a continuous supply of chlorine to the water. However, the tablets must be positioned carefully to ensure proper mixing of the chemical with the water.

Sodium Hypochlorite

The active ingredient of most liquid household bleaches, sodium hypochlorite is commonly used when the scale of postharvest chlorination is limited. Sodium hypochlorite is not generally available in solid form because it is difficult to store. It absorbs moisture readily from the atmosphere, causing it to release chlorine gas.

Household bleach is usually marketed as a solution of water and 5.25 percent sodium hypochlorite. Larger containers of 12.75 percent or 15 percent sodium hypochlorite solutions are also available through some laundry and swimming pool chemical suppliers. For the same amount of chlorination, a sodium hypochlorite solution is generally more expensive than granular calcium hypochlorite because of the additional shipping and handling costs associated with the water it contains.

Chlorination chemicals can be added to the water manually, or concentrated solutions of sodium or calcium hypochlorite can be injected into the wash tank or hydrocooler at a continuous and measured rate. Commercial chlorine injector systems are particularly useful in operations where a continuous supply of clean, chlorinated water is required. Injector systems, such as the one shown in Figure, consist of a feed tank and an electrically operated pump with a variable output. Chlorine injectors should always be isolated from water supply lines with an approved check valve to prevent backflow into the fresh water system.

Chlorination Effectiveness

Chlorination is a dynamic chemical process. Its effectiveness is influenced by a number of factors. Proper chlorination requires frequent monitoring of the solution and a thorough understanding of the factors involved. These factors include the pH of the solution, chlorine concentration, water temperature, amount of organic matter present, exposure time, and the growth stage of the pathogens present.

Solution pH

The pH of a solution is a measure of its acidity or alkalinity. A solution that is neutral (neither acid nor alkaline) has a pH of 7.0. Solutions with pH numbers less than 7.0 are acid; the lower the number, the greater the acidity. On the other hand, the greater the number above 7.0, the more alkaline the solution. A change of one pH unit indicates a ten-fold change in acidity or alkalinity.

The pH of the solution has a significant effect on the level of chlorination activity. When chlorine gas or one of the hypochlorite salts is added to water, each will generate chlorine gas (Cl2), hypochlorous acid (HOCl), or hypochlorite ions (OCl) in various proportions, depending on the pH of the solution. The form desired for chlorination is hypochlorous acid (HOCl). Hypochlorite ions are relatively inactive, and chlorine gas quickly bubbles out of the solution, causing worker discomfort and serving no useful purpose.

At a pH slightly above neutral, half of the chlorine will be in the form of hypochlorous acid and the other half in the form of hypochlorite ions. Very little will be in the gaseous form. Solutions that are more acid have a higher percentage of hypochlorous acid but are very unstable, allowing more of the chlorine to escape from the solution as a gas. To maximize the proportion of hypochlorous acid and hence the effectiveness of the solution, the pH should be kept in the practical range between 6.5 and 7.5.

Because well water in North Carolina varies from moderately acid to moderately alkaline, the pH should be checked with a pH meter or test papers before and after the chemicals are added and frequently during operation. Furthermore, even if the water initially has a near-neutral pH, the addition of hypochlorites will change the pH.

Different sources of chlorine have different effects on pH:

- Chlorine gas decreases pH
- Sodium hypochlorite increases pH
- Calcium hypochlorite increases pH slightly.

It may be necessary to add a common acid (like vinegar) to lower the pH. Small amounts of sodium hydroxide (lye) may be used to raise the pH. Inexpensive test papers for checking both the chlorine level and pH may be obtained from most swimming pool and chemical supply houses.

Chlorine Concentration

The concentration of a small amount of chemical in a solution is measured in units of parts per million (ppm). In the case of chlorination, this unit of measure indicates the number of parts of available chlorine, by weight, that there are in a million parts of solution. The quantity of calcium or sodium hypochlorite that must be added to a certain quantity of water to obtain a given concentration depends on :

- the available chlorine content of the compound
- the concentration of the compound
- the volume of water to be treated.

Table shows the minimum chlorine concentrations needed to kill all pathogens within one minute at two different temperatures, assuming a neutral pH. Table gives the amount of 5.25 percent solution hypochlorite solution that must be added to 100 gallons of water to obtain various chlorine concentrations from 25 to 150 ppm. Table gives the same information fo 65 percent calcium hypochlorite granules.

Table: Minimum Chlorine Concentration Necessary to Kill All Pathogens Within 1 Minute at Two Temperatures at Neutral pH

	Chlorine Concentration (ppm)	
	77 F	104 F
Fungi	30-40	10
Bacteria	20	10

Table: Amounts of 5.25 Percent Sodium Hypochlorite (NaOCl) Solution Required to Obtain a Specified Concentration of Chlorine in 100 Gallons of Water at Neutral pH

Pints of solution per 100 gallons of water	Approximate chlorine concentration (ppm)
0.4	25
0.8	50
1.2	75
1.6	100
2.0	125
2.4	150

Table: Amount of 65 Percent Calcium Hypochlorite ($Ca(OCl)_2$) Granules Required to Obtain a Specified Concentration of Chlorine in 100 Gallons of Water at Neutral pH

Ounces of granules per 100 gallons of water	Approximate chlorine concentration (ppm)
0.5	25
1.0	50
1.5	75
2.0	100
2.5	125
3.0	150

Temperature

The activity of chlorine increases with the temperature of the solution. Unnecessarily warm solutions should be avoided, however, because the chlorine escapes into the air more rapidly as the temperature increases. On the other hand, in hydrocooling the combined effects of low temperature and high pH values reduce chlorination efficiency.

Organic Matter

Chlorine has a particular affinity for soil particles and organic matter. Chlorinating dirty producc therefore depletes the chlorine supply much faster than relatively clean produce. The amount of chlorine constantly decreases with chlorination reactions. The more organic matter (such as fruit, leaves, or soil) in the tank, the more chlorine will be lost. As a result, the chlorine level should be checked and adjusted hourly, especially when large loads of produce are being processed. Extremely dirty produce (such as sweetpotatoes) is commonly washed with clean water before it is placed into the chlorination tank.

Exposure Time

The effectiveness of chlorination depends greatly on the length of time the produce is exposed to the chlorine solution. Quick dips are much less effective than longer exposures. However, most of the sanitizing action of the chlorine is accomplished within the first several minutes of exposure. Prolonged exposure to strong chlorine solutions has been known to cause surface bleaching. Experience is the best guide to the correct combination of treatment time and chlorine concentration for the crop being processed.

Growth Stage of the Pathogen

Disease organisms may be either in the active vegetative form or in the form of spores. Chlorine will readily kill the vegetative form, but fungal spores are 10 to 1,000 times more difficult to kill. Therefore, chlorine treatment rarely eliminates all pathogens and sterilizes the surface of the produce. Many spores may remain on the surface to develop later should the opportunity arise. Further, chlorine kills only on contact, not systemically, and is effective only on exposed pathogens such as those suspended in water or those on the surface of produce; chlorine does not kill pathogens below the skin because it cannot contact them. Chlorination leaves no residual effect. Therefore, produce exposed to pathogens after treatment is susceptible to reinfection.

Wastewater Disposal

Chlorinated wastewater is customarily drained at the end of each work day or more often if circumstances dictate. This wastewater often contains sediment, pesticides, and other suspended matter. If it is discharged to a municipal wastewater treatment plant or to surface waters (canals, creeks, or ponds), regulatory agencies may consider it to be industrial wastewater. Land application of this material is normally permitted, but a nondischarge permit may be required. Operators may be required to obtain wastewater discharge permits.

If you plan to use chlorination, check with the local office of the North Carolina Department of Environment, Health, and Natural Resources to determine whether a permit is required. Illegal disposal of hydro-cooler wastewater may result in a substantial penalty.

The Environmental Protection Agency (EPA) regulation of May 8, 1991, (40 CFR part 180) exempts calcium hypochlorite and chlorine gas from residue tolerance requirements when it is used before or after harvest on any raw agricultural commodity. Any amount can legally be used; however, excessive chlorination can damage equipment, injure the surface of fruits and vegetables, and waste money. In addition, the use of excessive amounts of chlorine may pose a worker health and safety hazard that is regulated by the North Carolina Department of Labor.

PRACTICAL RULES FOR SUCCESSFUL CHLORINATION

- **If water is not necessary in the packing process, do not use it.** Wetting the produce greatly increases the likelihood of damage by postharvest diseases. If the produce must be washed to remove soil, there is no alternative to wetting. Hydrocooling also necessitates wetting the produce, although other methods, such as forced-air cooling, may be a viable option in some cases. When water is necessary in packing lines (for example, in dumping tanks, flumes, or a hydrocooler), always treat it to reduce the risk of disease.
- **Monitor the chlorine concentration and the condition of the water.** Check the chlorine concentration and pH frequently using test papers or electronic equipment. Automatic chlorination equipment is available that will continually monitor the condition of the solution, add chlorine, and correct the pH. Also, monitor the water temperature.
- **Avoid overexposure.** Do not allow the produce to remain in contact with the solution longer than necessary. Check circulation patterns in chlorination tanks to eliminate dead spots.
- **Change the water frequently.** Chlorination efficiency is poor in dirty water. If necessary, wash very dirty produce with clean water before it comes into contact with the chlorinated water.
- **Dispose of wastewater properly**. Before installing chlorination equipment, plan how you will dispose of the wastewater. Land application of wastewater is normally allowed, but check to see if a permit is needed. Illegal disposal of chlorinated water could result in a substantial fine.
- **Practice good sanitation.** Hose off the packing equipment and floors daily; remove any dirt and trash that has settled in the chlorination tank. Sanitize the equipment with a spray solution composed of four pints of 5.25 percent sodium hypochlorite solution in 10 gallons of water. As an alternative, steam clean the equipment with an approved detergent. Do not allow culls or decayed produce to remain in or around the packing house.

- **Protect workers.** For their safety and comfort, workers must be protected from the chlorine fumes associated with excessively high levels of chlorine. If the amount of chlorine gas in the work area is great enough to cause worker discomfort, the amount of chlorine being used is well above that required for proper postharvest sanitation. If air monitoring equipment is not available, chlorine concentrations can be checked by asking a person who has not been desensitized by the odor to enter the work area. If he or she can smell the chlorine, the level is probably adequate. The concentration is too high if workers are continually irritated by the odor.
- **Remember that chlorination will not solve all your problems.** Even the best chlorination program may not be sufficient to prevent all postharvest decay. Prompt handling, proper sanitation, and rapid cooling should all be part of your postharvest disease management program. Produce infected in the field or otherwise damaged cannot be saved by chlorination.

CITRUS POSTHARVEST DISEASES AND INJURIES RELATED TO IMPACT ON PACKING LINES

Brazil is the first world producer of citrus fruit, with 20 million tons for each year. São Paulo State is responsible for 81% of the Brazilian production, from which 82% goes to juice production, 17% goes to the domestic market and only 1% is destined to fresh fruit exports. This low exporting/production ratio is due to the cultivars, taxes, phytosanitary barriers, and quality demands by the international market. European Union, for example, classifies citrus black spot as a quarantine disease. Therefore, a careful selection process is necessary at the packinghouse in order to avoid exports of injured fruits. Other typical postharvest diseases such as green mold and stem-end rot also reduce fruit quality. Therefore, fruit has to be treated in order to minimize problems with the importer.

In Spain, the percentage of the fruit rotting after harvest in a typical season is 3 to 6%. However, under favorable disease conditions, losses up to 50% can occur during marketing. In Northeastern Brazil, a survey of postharvest diseases in citrus indicated a 21.9% incidence of fungal rots. Several kinds of injuries may affect fruit's quality after harvest. The damaged tissue becomes susceptible to infections caused by pathogenic microorganisms, besides showing scars visually detrimental to the fruits, reducing their market value. Miller & Wagner (1991) found that 80% of impacts on fruits along a citrus sorting line varied from 25 and 150 G. These impacts can injure the fruits.

The purposes of this work were to characterize injuries in 'Valencia' orange and 'Murcott' tangor, destined for foreign markets, at different stages in the packinghouse and to identify critical points, as well as the extent of impact on fruits along the citrus processing lines in packinghouse.

Material and Methods

The terminology proposed by Zadoks (1985) was adopted in this paper: any visible and measurable symptoms caused by a harmful organism or mechanical impact are called an injury; any reduction in the quantity and/or quality of yield is called damage; any reduction in financial return due to harmful organisms is called loss.

Characterization of Injuries and Quantification of Postharvest Damages in Citrus

Samples were collected at five points in the processing line of a packinghouse located in Matão, São Paulo State, Brazil, every 14 day. Fruits of 'Valencia' sweet orange (*Citrus sinensis* L. Osbeck) were destined to the European fresh market and came from Gavião Peixoto, São Paulo State (21°50′ S; 48°29′ W, 515 m altitude), harvested in July-October 2004 (eight sampling dates); from Matão, São Paulo State (21°60′ S; 48°36′ W, 585 m altitude), harvested in August 2005 (one sampling date), and from Uberlândia, Minas Gerais State (18°53′ S; 48°17′ W, 890 m altitude), harvested in August-October 2005 (five sampling dates). The 'Murcott' tangor [*Citrus reticulata* Blanco × *C. sinensis* (L.) Osb.] fruits were destined to the Asian fresh market and came from Pratânia, São Paulo State (22°48′ S; 48°39′ W, 685 m altitude), harvested in October-November 2004 (four sampling dates).

One hundred fruits were collected at each of the following stages commonly found in the packinghouses processing fruits to foreign markets:

i) from plastic field bins at the arrival at the packinghouse;
ii) after pre-washing with sodium hypochlorite at 200 mg L^{-1} active chlorine and neutral detergent;
iii) after degreening with ethylene at 1-5 $\mu L\ L^{-1}$ (from three to five days);
iv) at the packing table (after another washing with sodium hypochlorite and detergent and application of wax + thiabendazole and imazalil fungicides, both at 1000 mg L^{-1});
v) at the fiberboard cartons (19 kg) on pallets.

Because there was no need to degreened 'Murcott' tangor, the stages assessed for then were (i), (iv), and (v) only. Fruits were placed separately inside trays and stored at 25°C and 85% RH for 21 days. During the first 24 hours of storage, fruits were stored in a humid chamber in a plastic bag to induce rots. The incidence of injuries was visually assessed at the retrieval of fruits from the humid chambers and every three days. The fungi pathogenicity was confirmed by the inoculation of healthy fruits.

The incidences of postharvest diseases were compared at different processing stages by a non-parametric test of multiple proportions comparison at

5% probability, according to Zar (1999). This non-parametric test uses untransformed data.

Impact Critical Points in Citrus Processing Lines

An instrumented sphere (70 mm) was used to assess the extent of impacts at transference points along the citrus processing line. The 110-meter long processing line (Maf-Roda), with electronic sorter had 19 automated transference points. The instrumented sphere, with an impact range between 15 and 500 G, was placed with the fruits at the reception stage and followed the flow of the fruit until the sorting stage. The time of residence of the instrumented sphere in each stage, as well as the time to complete the entire cycle, was monitored by a precision chronometer. Assessments were repeated five times in the total cycle of the processing line.

The acceleration force (G) in the processing line at different transference points was compared using ANOVA followed by multiple comparisons by Tukey's test ($p < 0.05$). The incidence of oleocellosis in three stages of processing fruits (reception, after pre-washing with sodium hypochlorite/detergent and table) was correlated to the impact (sum of acceleration in different transference points) in the respective stages and analyzed by F test ($p < 0.05$).

Results and Discussion

Postharvest Diseases

Mean rot incidences (2004/05) at fruit arrivals, pre-washing, degreening, packing table, and pallet stages for 'Valencia' sweet orange, destined to the European fresh market, were 1.8, 1.6, 2.6, 2.1 and 1.9%, respectively , which were considered low in comparison to the incidences in varieties of citrus destined to the domestic market: 21% higher under similar experimental storage conditions. This lower rotting incidence is probably due to the intensive management of diseases in those orchards. The most frequent pathogens identified from lesions on symptomatic fruit were *Lasiodiplodia theobromae* (Pat.) Griffon & Maubl., *Penicillium digitatum* Sacc., *Alternaria citri* Ellis & N. Pierce, and *Phomopsis citri* Fawcett. *Colletotrichum gloeosporioides* Penz, *Fusarium* spp, *Geotrichum citri-aurantii* (Ferraris) E.E. Butler,*Penicillium italicum* Wehmer and *Rhizopus stolonifer* (Ehrenb.) Vuill were also detected but low in frequency.

Rot incidences among 'Murcott' tangors, destined to the Asian fresh market, tended to decrease with processing from 2.6% at arrival of fruits in the packinghouse to 1.6% at the packing table and 1.1% at pallet stage. The most frequent pathogens were *C. gloeosporioides, L. theobromae,* and *P. digitatum. P. citri., G. citri-aurantii* and *Fusarium* spp. were also found.

L. theobromae was the most frequent pathogen among sweet oranges, which is consistent with Dantas *et al.* (2003) results for 'Pera' variety commercialized at Distribution Centre of Recife, Brazil, where the stem-end rot incidence was 53% of total fruit rot. *Lasiodiplodia* stem-end rot is an important postharvest disease in warm and humid citrus growing regions such as Florida and Caribbean. *L. theobromae* usually infects the fruit from its button (calyx and disk) at the stem-end of the fruit, leading to development of soft brown to black decay symptoms at both fruit ends. Decay develops rapidly during and after excessive degreening and can be observed in the fruits at the packinghouse. It is often observed at a market arrival or shortly thereafter. In this paper, the incidence of stem-end rot by *Lasiodiplodia* tended to be higher among degreened sweet oranges, including that receiving fungicides application in 2005.

Green mould is one of the most important fruit decays in Florida and in the Mediterranean region. In a survey with 'Pera', 'Lima' and 'Natal' oranges and 'Murcott' tangor for the São Paulo State domestic market, green mould was the most frequent disease found with a 15.9% average incidence in all these citrus varieties. The extensive spores production by this pathogen, which is easily dispersed by air, ensures its presence wherever fruit is handled, including in the field, packinghouse, transit containers, and the in marketplace. Infection takes place only through wounds, where nutrients are available to stimulate spore germination, where fruit decay begins.

The reduction of *P. digitatum*inocula through washing citrus with detergent and sodium hypochlorite, as well as with additional protection by the fungicides thiabendazole and imazalil, reduced green mold incidence to below 0.5% in orange and to zero in tangor. *P. digitatum* can develop resistance to some fungicides, particularly to the benzimidazoles. Problems of resistance can be minimized using two or three fungicides with different modes of action, in addition to the monitoring of the resistant populations in air. Fludioxonil, a newly registered and classified as reduced-risk fungicide in the United States, effectively reduced both *Lasiodiplodia*stem-end rot and green mold when applied through a simulated commercial packing line system. This fungicide was also effective for the control of *P. digitatum* isolates resistant to thiabendazole on oranges.

Anthracnose generally occurs on previously injured fruits by sunburn, wind, pests and phytotoxicity by pesticides. However, anthracnose symptoms can be observed even in non-injured fruits of some varieties of tangerine and their hybrids. This was the case of 'Murcott' tangor, in which anthracnose was the most prevalent disease. In Florida, anthracnose is a major cause of decay in tangerines that are harvested early in the fall, when long periods of degreening are required to enhance fruit appearance. Washing of the fruit prior to degreening removes many appressoria and reduces disease incidence. In this study, it was difficult to isolate the effect of washing due to the low field incidence of this disease.

Control of postharvest diseases in citrus is vital for maintaining quality and shelf-life in a market where transport from producer to consumer may take several weeks. Postharvest infections can be reduced by sanitation of plastic field bins and packinghouse facilities (machines, chambers, floor tiles) with chlorine and quaternary ammonium containing products. Fungicides such as thiabendazole and imazalil are important in order to minimize postharvest decay, as observed in sweet orange and tangor in 2004, although in some cases, such as in sweet orange in 2005, stem-end rots increased after fungicides application. Without using fungicides, sales of fresh Florida citrus would be reduced by at least 50%. The environment control by using cold rooms in storage and the use of refrigerated containers during transport of processed fruits delay the development of rots considerably and are important practices to reduce postharvest diseases, especially when combined with chemical treatment.

Processing effectively removed fruit with sooty mold, caused by *Capnodium citri* Berk. & Desm., and other biotic injuries from pathogens which infect fruits in the field to incidence below or near 1.0% in 'Valencia' sweet orange. However, in 'Murcott' tangor, 13.0% of fruit collected in the pallet had alternaria brown spot (*Alternaria alternata* (Fr.) Keissler) due to the high field incidence of this disease.

The incidences of citrus black spot (*Guignardia citricarpa* Kiely) in 'Valencia' orange and 'Murcott' tangor at the intial stages of processing were 1.6 and 1.0%, respectively. However, after packing, even after 21 days of storage, no fruit showed citrus black spot symptoms. The reduction in citrus black spot incidence results from the careful selection process at arrivals of the fruits in the packinghouse and after degreening. Waxing treatment was found to reduce postharvest development of citrus black spot in studies in Australia, whereas the postharvest fungicides treatment had no effect on postharvest symptom development.

Associations between titratable acidity and soluble solids content with rot incidence, both on 'Valencia' orange and 'Murcott' targor fruits, were generally variable and not significant (dates not presented). Although associations between the increase of the infections with the rise of the soluble solids or the reduction of the acidity were found to be related in cherry fruits , apple and peach , these characteristics had not assisted in the understanding of the causes of resistance/susceptibility on the citric fruits, thus having to be considered only in the planning of the harvest. During citrus fruit maturation, drastic variations in the chemical composition do not occur.

Postharvest Mechanical Injuries

Oleocellosis increased with processing in 'Valencia' sweet orange, from 28.8% at arrival in the packinghouse to 53.4% in the packing table. The reduction in mechanical injuries at the pallet is probably due to the removal of these kinds of defecting fruits during the grading process. Oleocellosis symptoms in 'Murcott' tangor also increased, during processing, from 7.8% at arrival in packinghouse to 20.8% in pallet.

Incidences of dehydrated 'Valencia' orange and 'Murcott' tangor fruits decreased with the processing. Among 'Valencia' oranges, 38% were dehydrated at the arrival to the packinghouse, while only 2% were dehydrated in the pallets. Similarly, among 'Murcott' tangor fruits 19%, were dehydrated at the arrival in the packinghouse, while less than 1% were dehydrated in the pallet. A better preservation during storage of fruits from the packing table and pallet stages, related to water loss, is due to the covering of fruits with carnauba wax. Besides reducing the dehydration, the wax sealing improved its visual quality. According to Waks *et al.* (1985) the waxes minimize stem-end rind breakdown and others collapses of rind tissue and can protect the fruit of the entrance of pathogen.

Other fruit injuries, such as healed lesions from pest attacks or wind injury, as well as pesticide spray injuries, were reduced with the sorting and culling processes. Yet, non-healed lesions from the harvest increased up to the processing table stage. Non-healed lesions are due to fruit impacts in sorting line surfaces, and they could be related to incidence of some postharvest diseases, like Penicillium rots, that enter fruit only through wounds. Other imperfections observed in orange and tangor fruits were pressure bruises, chimeras, ethylene damage, and wax damage. These imperfections are not presented due to their very low incidence.

Impact Critical Points in Citrus Processing Lines

Most impacts (94.7%) were within 30-95 G and caused by drops against hard surfaces. Maximum acceleration was observed at the placement of washed fruits into empty bins destined to degreening. Mean acceleration in this point reached 272 G and differed from other points ($p < 0.05$). A positive correlation was observed between the impacts in the processing line and the incidence of oleocellosis in 'Valencia' fruits (r = 0.97).

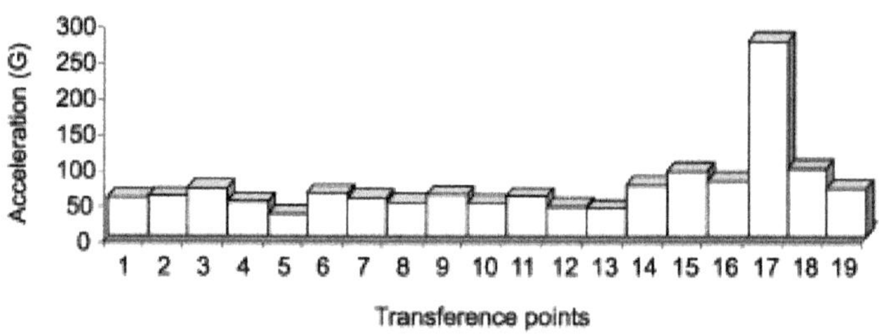

Fig. : Mean accelerations in the citrus processing line at different transference points: 1 - Reception, 2- canvas belt entrance, 3- distribution conveyor, 4- end of distribution conveyor, 5- stall of washing, 6- balsa entrance, 7- conveyor (main selection), 8- first exit of conveyor (main selection), 9- fall, 10- second exit (main selection), 11- classification entrance, 12- classification exit, 13- return from classifier, 14- return from conveyor, 15- first Fall (return conveyor), 16- Second fall (return conveyor), 17- entrance to bins destined to deyreening, 18- entrance of classification table, 19- exit of classification table.

The 272 G acceleration at the return of washed fruits to bins destined to degreening corresponds to the most critical situation favoring fruit injuries. The first falling fruits are exposed to higher drops (80 cm) and, consequently, greater impacts when in comparison to the last falling fruits. Despite the high acceleration value observed in this study, Miller & Wagner (1991) reported that acceleration values may reach 300 G during fruit reception stages. Points that favor the occurrence of mechanical injuries along the sorting line include: fruit drops at the entrance of the processing line; transference of fruits among the line components; hard surfaces; high speed of fruits; inadequate maintenance of equipment; and excessive wax residue on the line.

Mechanical injuries caused by processing of grapefruits (*Citrus paradise* Macf.), observed by Skaria *et al.* (2003), were followed by increased rot incidences. Sorted fruits, even treated with thiabendazole and sodium o-phenylphenate (SOPP), presented higher incidence of green mold than carefully packed fruits in the field, without postharvest treatment with fungicides. 'Tahiti' lime fruits submitted to impacts showed great depreciation in external appearance, with symptoms of blossom-end rots and reduction of shelf-life by six days, when in comparison to the 15-day shelf-life of non-impacted fruits. Sargent *et al.*, (1992) commented that modifications in some points of classification lines of table tomatoes, such as surface protections, reduced impact pressure by up to 50%. Timm & Brown (1991) also observed decreased impacts in classification lines of avocado, papaya and pineapple when using rubber covers. Impact decreases can be promoted by using cushioning materials, reducing rough surfaces, such as concrete or metal, which are the most likely cause of fruit damage.

POST-HARVEST LOSSES (VEGETABLES)

The post-harvest sector includes all points in the value chain from production in the field to the food being placed on a plate for consumption. Postharvest activities include harvesting, handling, storage, processing, packaging, transportation and marketing.

Losses of horticultural produce are a major problem in the post-harvest chain. They can be caused by a wide variety of factors, ranging from growing conditions to handling at retail level. Not only are losses clearly a waste of food, but they also represent a similar waste of human effort, farm inputs, livelihoods, investments and scarce resources such as water. Post-harvest losses for horticultural produce are, however, difficult to measure. In some cases everything

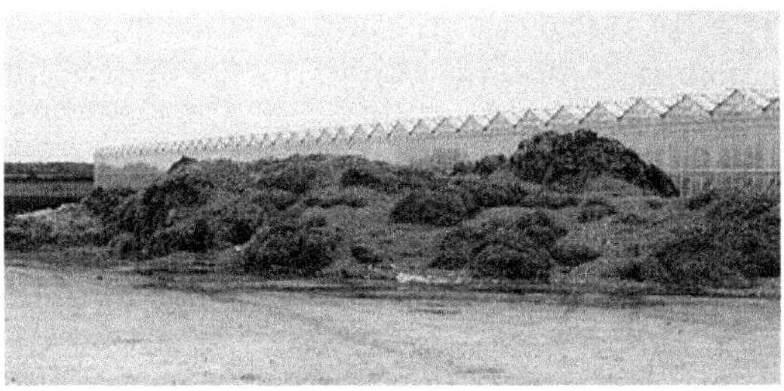

Fig. : Discarded tomatoes on a compost heap at nurseries.

harvested by a farmer may end up being sold to consumers. In others, losses or waste may be considerable. Occasionally, losses may be 100%, for example when there is a price collapse and it would cost the farmer more to harvest and market the produce than to plough it back into the ground. Use of average loss figures is thus often misleading. There can be losses in quality, as measured both by the price obtained and the nutritional value, as well as in quantity.

On-farm Causes of Loss

There are numerous factors affecting post-harvest losses, from the soil in which the crop is grown to the handling of produce when it reaches the shop. Pre-harvest production practices may seriously affect post-harvest returns. Plants need a continuous supply of water for photosynthesis and transpiration. Damage can be caused by too much rain or irrigation, which can lead to decay; by too little water; and by irregular water supply, which can, for example, lead to growth cracks. Lack of plant food can affect the quality of fresh produce, causing stunted growth or discoloration of leaves, abnormal ripening and a range of other factors. Too much fertilizer can harm the development and post-harvest condition of produce. Good crop husbandry is important for reducing losses. Weeds compete with crops for nutrientsand soil moisture. Decaying plant residues in the field are also a major loss factor.

Causes of Loss After Harvest

Fruits and vegetables are living parts of plant and contain 65 to 95 percent water. When food and water reserves are exhausted, produce dies and decays. Anything that increases the rate at which a product's food and water reserves are used up increases the likelihood of losses. Increases in normal physiological changes can be caused by high temperature, low atmospheric humidity and physical injury. Such injury often results from careless handling, causing internal bruising, splitting and skin breaks, thus rapidly increasing water loss.

Respiration is a continuing process in a plant and cannot be stopped without damage to the growing plant or harvested produce. It uses stored starch or sugar and stops when reserves of these are exhausted, leading to ageing. Respiration depends on a good air supply. When the air supply is restricted fermentation instead of respiration can occur. Poor ventilation of produce also leads to the accumulation of carbon dioxide. When the concentration of carbon dioxide increases it will quickly ruin produce.

Fresh produce continues to lose water after harvest. Water loss causes shrinkage and loss of weight. The rate at which water is lost varies according to the product. Leafy vegetables lose water quickly because they have a thin skin with many pores. Potatoes, on the other hand, have a thick skin with few pores. But whatever the product, to extend shelf or storage life the rate of water loss must be minimal. The most significant factor is the ratio of the surface area of the fruit or vegetable to its volume. The greater the ratio the more rapid will be the loss of water. The rate of loss is related to the difference between the water vapour pressure inside the produce and in the air. Produce must therefore be kept in a moist atmosphere.

Diseases caused by fungi and bacteria cause losses but virus diseases, common in growing crops, are not a major post-harvest problem. Deep penetration of decay makes infected produce unusable. This is often the result of infection of the produce in the field before harvest. Quality loss occurs when the disease affects only the surface. Skin blemishes may lower the sale price but do not render a fruit or vegetable inedible. Fungal and bacterial diseases are spread by microscopic spores, which are distributed in the air and soil and via decaying plant material. Infection after harvest can occur at any time. It is usually the result of harvesting or handling injuries.

Ripening occurs when a fruit is mature. Ripeness is followed by senescence and breakdown of the fruit. The category "fruit" refers also to products such as aubergine, sweet pepper and tomato. Non-climacteric fruit only ripen while still attached to the parent plant. Their eating quality suffers if they are harvested before fully ripe as their sugar and acid content does not increase further. Examples are citrus, grapes and pineapple. Early harvesting is often carried out for export shipments to minimise loss during transport, but a consequence of this is that the flavour suffers.

Climacteric fruit are those that can be harvested when mature but before ripening has begun. These include banana, melon, papaya, and tomato. In commercial fruit marketing the rate of ripening is controlled artificially, thus enabling transport and distribution to be carefully planned. Ethylene gas is produced in most plant tissues and is important in starting off the ripening process. It can be used commercially for the ripening of climacteric fruits. However, natural ethylene produced by fruits can lead to in- storage losses. For example, ethylene destroys the green colour of plants. Leafy vegetables will be damaged if stored with ripening fruit. Ethylene production is increased when fruits are injured or decaying and this can cause early ripening of climacteric fruit during transport.

Damage in the Marketing Chain

Fruits and vegetables are very susceptible to mechanical injury. This can occur at any stage of the marketing chain and can result from poor harvesting practices such as the use of dirty cutting knives; unsuitable containers used at harvest time or during the marketing process, *e.g.* containers that can be easily squashed or have splintered wood, sharp edges or poor nailing; over packing or under packing of containers; and careless handling of containers. Resultant damage can include splitting of fruits, internal bruising, superficial grazing, and crushing of soft produce. Poor handling can thus result in development of entry points for moulds and bacteria, increased water loss, and an increased respiration rate.

Produce can be damaged when exposed to extremes of temperature. Levels of tolerance to low temperatures are importance when cool storage is envisaged. All produce will freeze at temperatures between 0 and -2 degrees Celsius. Although a few commodities are tolerant of slight freezing, bad temperature control in storage can lead to significant losses.

Some fruits and vegetables are also susceptible to contaminants introduced after harvest by use of contaminated field boxes; dirty water used for washing produce before packing; decaying, rejected produce lying around packing houses; and unhealthy produce contaminating healthy produce in the same packages.

Losses directly attributed to transport can be high, particularly in developing countries. Damage occurs as a result of careless handling of packed produce during loading and unloading; vibration (shaking) of the vehicle, especially on bad roads; and poor stowage, with packages often squeezed into the vehicle in order to maximise revenue for the transporters. Overheating leads to decay, and increases the rate of water loss. In transport it can result from using closed vehicles with no ventilation; stacking patterns that block the movement of air; and using vehicles that provide no protection from the sun. Breakdowns of vehicles can be a significant cause of losses in some countries, as perishable produce can be left exposed to the sun for a day or more while repairs are carried out.

At the retail marketing stage losses can be significant, particularly in poorer countries. Poor-quality markets often provide little protection for the produce against the elements, leading to rapid produce deterioration. Sorting of produce to separate the saleable from the unsaleable can result in high percentages being discarded, and there can be high weight loss from the trimming of leafy vegetables. Arrival of fresh supplies in a market may lead to some existing, older stock being discarded, or sold at very low prices.

Avoiding Loss

Losses can be avoided by following good practices as indicated above. There is also a wide range of post-harvest technologies that can be adopted to improve losses throughout the process of pre-harvest, harvest, cooling, temporary storage, transport, handling and market distribution. Recommended technologies vary depending on the type of loss experienced. However, all interventions must meet the principle of cost-effectiveness. In theory it should be possible to reduce losses substantially but in practice this may be prohibitively expensive. Especially for small farms, for which it is essential to reduce losses, it is difficult to afford expensive and work-intensive technologies.

There are no reliable methods for evaluating post-harvest losses of fresh produce. Any assessment can only refer to a particular value chain on a particular occasion and, even then, it is difficult to account for quality loss or to differentiate between unavoidable moisture loss and losses due to poor postharvest handling and other factors described above. Accurate records of losses at various stages of the marketing chain are rarely kept, particularly in tropical countries where losses can be highest, making reliable assessment of the potential cost-effectiveness of interventions at different stages of the chain virtually impossible. The lack of such information may lead to misplaced interventions by governments and donors.

MANAGEMENT OF POST-HARVEST MANGO DISEASES

In Australia, the most serious diseases of mango are caused by fungi. The majority of these diseases infect fruit in the field and the critical phases for disease control in mango production are during flowering and fruit set, and after harvest.

Post-harvest diseases can ruin an otherwise high-quality crop. With some post-harvest diseases, the damage occurs when fruit is ripening and ready for display to buyers. Other diseases show symptoms before harvest, so affected fruit can be culled during grading and packing.

Post-harvest disease control in mango is currently achieved through a combination of pre- and post-harvest fungicide application, orchard hygiene and post-harvest temperature management.

Inoculum of post-harvest diseases occurs on leaves, stems and flowers, so field control is important to reduce post-harvest losses. Post-harvest temperature management is also important because post-harvest diseases are favoured by temperatures above 25 °C during ripening.

During long-term storage of mangoes in a controlled atmosphere (5 per cent oxygen and 2 per cent carbon dioxide at 13 °C for three or more weeks), diseases appear that are not encountered in fruit stored for shorter periods. The cool temperature, high humidity and fact that fruit begins to reach the end of its storage life favour these pathogens.

Anthracnose

Anthracnose (caused by *Colletotrichum gloeosporioides* and occasionally *C. acutatum*) is the major fungal disease of mangoes. Major losses occur from flowering to fruit set and again after harvest. Sunken black spots appear on the surface of the fruit during ripening. The disease is most severe following periods of wet weather. Pre- and post-harvest control measures are required.

Pre-harvest Management

Spray with mancozeb at the label rate weekly during flowering and then monthly until harvest. Stop spraying 14 days before harvest. During dry weather, you can reduce flower sprays to fortnightly intervals.

If rain occurs during flowering, apply prochloraz at the label rate in a tank mix with mancozeb. Prochloraz needs to be applied only every 3-4 weeks.

Copper oxychloride sprays at the label rate used for bacterial black spot control also control anthracnose, however copper oxychloride should not be used during flowering. Where bacterial black spot is serious, copper oxychloride can be substituted for mancozeb sprays after flowering.

Post-harvest Management

Unheated prochloraz spray can be used to control anthracnose. Prochloraz is not effective against stem end rot.

Apply prochloraz as per label recommendations at ambient temperature. Prochloraz has been approved for use only as a non-recirculated spray over fruit. Complete coverage of the fruit is essential for effective control.

Post-harvest treatments will not provide complete disease control.

Stem End Rot

Stem end rot (*Botryosphaeria* spp., *Lasiodiplodia theobromae* and other fungi) is a soft, watery rot that develops from the stem end as fruit ripens after harvest.

Post-harvest Management

The chemical fludioxonil is currently permitted for the management of stem end rot in mangoes. The permit specifies three use rates. A low rate for use in heated dip systems to be used when disease pressure is suspected to be low, a double rate for heated dip or flood spray systems when disease pressure is expected to be high and a higher rate that can be used as an ambient flood spray. Agitation and recharge is essential to maintain chemical levels in the treatment so consult the label for directions of use in your situation.

Do not use fruit from orchards with a history of stem end rot losses for long-term storage. The severity of stem end rot can be assessed in the following way.

Harvest 100 mature fruit at random from throughout the orchard. Leave them untreated and store at 25 °C until they are fully ripe. Ideally, less than one-tenth - and certainly no more than one-third - of the fruit should develop symptoms of stem end rot by the time they are fully ripe.

Transit Rot

Transit rot (*Rhizopus stolonifer*) appears after harvest and can cause sporadic fruit losses under high-humidity conditions. Pale, watery lesions appear and fungal growth may cover affected fruit. The disease can spread from fruit to fruit and from contaminated packaging material such as wood wool.

Post-harvest Management

Remove reject fruit that may harbour Rhizopus from the packing shed. Spray packing equipment and the shed with a sanitising agent. Steam cleaning or high-pressure hot water cleaning should precede the use of a sanitiser.

No fungicides are approved for transit rot control in mangoes.

Other Post-harvest Diseases

A few other fungi (*Aspergillus niger, Mucor* spp. *etc.*) occasionally cause fruit losses during storage. They cause rots on the sides or at the stem end of fruit. Rough harvesting and handling can encourage these diseases. Careful handling and good hygiene measures will help manage these problems.

Export Requirements

Choose fruit from orchards with low disease levels. Do not export fruit from orchards with a history of stem end rot or other diseases.

Where prochloraz is not acceptable to the importing country, export is not recommended unless fruit is treated with a combination of hot water and vapour heat. Vapour heat treatment was developed for export markets that require quarantine security against fruit fly but do not accept chemical disinfestation treatments. Vapour heat treatment will control anthracnose during short-term storage, but will not give adequate control of stem end rot. Dipping fruit in hot water at 48-52 °C for 5 minutes 24 hours before vapour heat treatment will improve stem end rot control.

Sanitation of Packing Equipment

Clean equipment daily with a steam cleaner or high-pressure hot water applicator. Otherwise, use a hose with a suitable detergent. Clean any equipment on which dirt and sap collects at lunch time as well as at the end of the day. If you are having a problem with fruit rots other than anthracnose or stem end rot, you may need to sanitise the equipment using a solution that is approved for use in a food business.

A GLIMPSE INTO THE WORLD OF POSTHARVEST DISEASE RESEARCH

There is a need to figure out novel disease management strategies necessary for staying one step ahead of postharvest decay pathogens. Evaluating wild apples for resistance to postharvest diseases, understanding fungi causing decay in storage, and finding alternatives for controlling rots are briefly discussed.

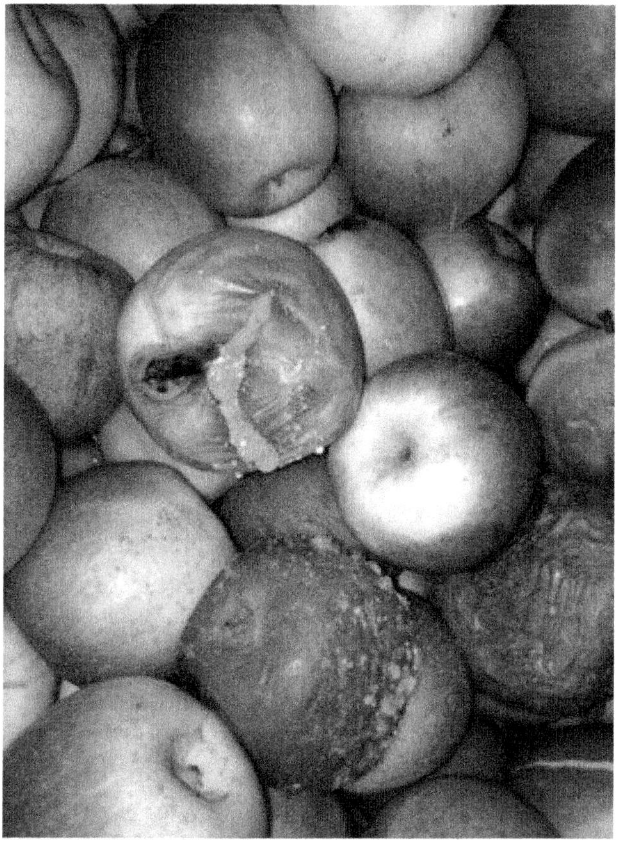

My introduction in the world of tree fruit research was not out in the orchard, but in cold storages. When folks are opening up their storages to examine fruit bins this time of year, it's hard not to reflect on my roots.

Postharvest diseases caused by fungi are economically important and a consistent problem for the pome fruit and packing industry. Some of the most problematic diseases occurring in storage include blue mold (#1 offender), gray mold, bitter rot, Mucor rot, as well as common rot summer diseases (white rot, black rot, bitter rot). The fungi causing blue mold (*Penicillium* spp.) are especially important since these fungi are capable of producing harmful mycotoxins, such as patulin, that contaminate processed fruit products like juices, sauces, and butters.

Resistant cultivars are available for problematic preharvest diseases, such as apple scab, fire blight, and bacterial spot; however, resistance to postharvest fungal diseases, like blue mold, is lacking in commercial cultivars since breeding has been focused on disease resistance for field production. Consequently, the need to figure out novel disease management strategies is necessary for staying one step ahead of these decay pathogens. Here, I want to share a few areas of research where I have been involved to give you an idea of what scientists are up to these days in fighting the war against postharvest diseases.

Evaluating wild Apples from Kazakhstan for Resistance to Postharvest Fruit Diseases

Central Asia has been described to be the origin for the domestic apple. In the 1990s, samples from wild forest stands of the distant ancestor of the domesticated apple located in Kazakhstan were collected by USDA scientists that resulted in the "Kazak" collection of apple germplasm seedlings. The Kazak orchard is located at the USDA-ARS Plant Genetic Resources Unit on the campus of Cornell's New York Agricultural Research Station in Geneva, New York and represents a wide variety of horticultural traits, such as forms and colors.

Consequently, the "Kazak Orchard" has been utilized to search for disease resistance, particularly to postharvest diseases. My former lab spent several years evaluating apple fruit from 83 different apple accessions for resistance to the most important apple storage diseases: blue mold (*Penicillium expansum*) and bitter rot (*Colletotrichum acutatum*). Two accessions were discovered to be "immune" to both diseases: despite apple wounds inoculated with spores, no disease developed. When we examined if the spores were still viable in the inoculated wound, the spores grew on artificial media indicating the apple was somehow shutting down the process of the spores' ability to germinate and cause infection. Such data provide a foundation for future research, such as serving as a source of resistance for conventional breeding programs and understanding the molecular mechanisms and genetics of resistance against postharvest diseases.

Evaluating Fungi Causing Postharvest Decay in Storage

When managing decay pathogens in storage, it is important to determine which pathogens are responsible for causing decay and to evaluate the efficacy of the current fungicide chemistries controlling these diseases. Only three chemistries are available to control postharvest diseases: Scholar® (fludioxonil), Mertect® (thiabendazole), and Penbotec® (pyrimethanil). We obtained infected fruit from three different packinghouses using different preharvest and postharvest decay management practices, including organic.

Over the course of two years, we consistently observed a variety of decay-causing fungi like *Penicillium* spp. (blue mold), *Botrytis* spp. (gray mold), *Alternaria* spp. (Alternaria rot), *Mucor* spp. (Mucor rot), *Colletotrichum* spp. (bitter rot), and *Botryosphaeria* spp. (white and black rots). As expected, the most prevalent fungi isolated were *Pencillium* spp. causing blue mold. In addition, we discovered several *Penicillium* spp. not previously described as pathogens of apple fruit. These species were especially important since both are capable of producing the patulin mycotoxin. Mertect® has been around a very long time and has well known documented resistance, especially for *Penicillium* spp.

Not surprisingly, we found *Penicillium* spp. isolates from all three packinghouses that were resistant to Mertect®, with the lowest number of resistant isolates found on organic apples. The importance for the apple industry is that we now have specific information concerning the decay-causing fungi in storage, thereby allowing for the selection of the most appropriate chemical controls. (Special thanks to the State Horticultural Association of Pennsylvania for funding this work!)

Alternatives for Controlling Postharvest Rots on Stone Fruit

The microbial community living on fruit is vast and some organisms can be beneficial to plants for controlling plant pathogens. Consequently, these microbes can be a source for finding alternatives to control diseases. Much research has been conducted evaluating the microbial community on pome and citrus fruits due to searching for biocontrol agents, some of which have become commercial products. Unfortunately, little is known about the microbial community on stone fruits and their ability to be exploited to control diseases. In collaboration with Dr. Wojciech Janisiewicz at the USDA-ARS Appalachian Fruit Research Station in Kearneysville, WV, we found the occurrence of yeasts on plum that were able to control brown rot on plums after harvest. These results suggest that the surfaces of plums harbor yeasts that may be able to serve as a potential biocontrol for brown rot on stone fruits.

What Can You do to Prevent Storage Disease this Season?

Preventing disease in storage begins in the orchard: practicing orchard sanitation, properly timing fungicides, harvesting at optimum maturity, and using methods to prevent bruising. Development of disease is dependent upon the initial amount of inoculum in the orchard. The goal of sanitation is to reduce the number of fungal spores available to cause fruit rots for the coming season. Pre-season practices include removing and destroying infected plant tissue where fungal spores overwinter, such as mummified fruit (black rot, white rot, bitter rot, brown rot) and dead branches or twigs. Applying urea in the fall for scab management will also help knock back the spores of summer disease fruit rots that also infect leaves (black rot, bitter rot, Alternaria rot).

During the season, it's important to be mindful of the fallen fruit underneath the trees. Fallen fruit encourages fungal spores residing in the soils (blue mold, gray mold, Mucor rot) to make copious amounts of spores, potentially contaminating bins in the orchard and thereby providing a vehicle for entering the packinghouse. Consequently, it's very important to minimize the soil on the bins during harvest time. This is especially important for controlling Mucor rot. Postharvest fungicides exist to control blue mold and gray mold in storage; however, no commercial fungicides are available to control Mucor rot. Although you may have done your best to minimize the amount of spores lurking around to cause fruit decay, fungicide applications are still important for cover sprays.

Captan and Topsin® M are the workhorses for summer disease control, but strobilurins (Flint®, Merivon®, Pristine®) need to be considered, as well. Although strobilurins may have lost their efficacy to control apple scab as a result of resistance, strobilurins are still excellent controlling fruit rots and should not be overlooked when figuring out one's cover spray programs. In addition to controlling the typical summer diseases, the fungicides Merivon® and Pristine® are labeled to control blue mold and gray mold and are very helpful keeping disease in check after fruit are harvested. Although resistance to strobilurins has not been described or noted for the summer rots, resistance management strategies should still be followed, such as rotating and tank mixing with a broad spectrum fungicide.

POST HARVEST AND STORAGE DISEASE OF ONIONS AND GARLIC

Onions are approaching maturity, but should not be pulled and cured until temperatures drop. Bulbs will sunburn if they are uprooted and laid on soil when temperatures exceed 90 degrees and will not dry well in high humidity. Wait until cooler, less humid weather before pulling unless they are to be marketed immediately. Most garlic crops have been curing for the past several weeks and may be ready to top and store. The number of growers producing these crops seems to be increasing, and in many cases growers plan to store bulbs into fall and winter. It is also important to be aware of possible pathogens that affect onions and garlic in storage.

Fusarium basal rot of onion & garlic (caused by *Fusarium oxysporum f. sp. cepae*) can affect onion, garlic, and other Allium spp. The fungus lives a long time in soil. Infection often is associated with mechanical or insect injury. Foliar symptoms appear as a unilateral or general wilt and a yellow to tan dieback of leaf tips during mid to late season. Foliage death may occur over several weeks. In onions, red-brown rot appears where roots are attached to the basal plate. Rot and discoloration usually affect all of the base and up into the bulb scales; affected tissue appears brown and watery when bulbs are cut open. Sometimes, a white moldy growth develops on the stem plate or between affected scales. Bulbs may appear normal at harvest, but rot may progress in storage.

In storage, garlic bulbs appear spongy or sunken. Infected bulbs are softened, brown and watery when cut open. There may be a white, light pink or reddish fungal growth (mycelium) covering the cloves, or in the rot cavities. Deep cracks form in the cloves, followed by break down of the tissue, which will eventually dry down to a portion of its original size, the cloves becoming crinkled and small.

There are resistant onion cultivars available. In a Malheur County, OR test, these cultivars had less basal rot: 'Golden Cascade', 'Cima', 'Oro Grande', 'Valient', and 'Cashe'. Your mileage may vary in the Northeast. Other onion cultivars described as tolerant are 'Bronze Reserve', 'North Star', 'Sassy Brassy', and 'Sentinel'.

Cultural control:

- Use 4-year or longer crop rotations.
- Protect plants from insect, fertilizer, or other injury.
- Storing bulbs at 39°F minimizes postharvest losses.

Botrytis Neck Rot of onions and garlic is caused by fungi that overwinter on plant debris in soil, on infected bulbs, and as sclerotia in soil. The pathogen can also be seed-borne. Botryitis neck rot is caused by a different pathogen from Botrytis leaf spot. Botryitis neck rot is seen primarily in storage. Most common causes of severe losses are excessive nitrogen, which delays crop maturity; irrigation and/or rain late in the season; inadequate or improper curing; and improper storage. Infection is through neck tissue or wounds in bulbs. Some bulb infections may arise from symptomless leaf infections, but often the fungus directly enters the neck via airborne spores when onions are topped before soft, susceptible top tissues dry properly.

Symptoms

In onion, the disease is more apparent after harvest, while bulbs are in storage. At first, soft neck tissue looks water soaked, and a yellow discoloration moves down the neck into the scales. Bulbs break down to a soft mass. A gray mold develops between the onion scales, later producing small to large black bodies (sclerotia) which develop as a solid layer around the neck.

In garlic, the disease usually appears first on necks near the soil line at any time after spring greenup when weather conditions permit. The disease becomes worse when it starts early in the season. Extensive development of sclerotia is best seen on maturing bulbs just before and during harvest. The fungus moves rapidly into the succulent garlic bulb's neck region, producing a water-soaked appearance. A gray mold develops on the surface of or between garlic scales, later producing black bodies (sclerotia) which develop around the neck. Before bulbing, plants may die or recover if weather permits. Bulbs infected late break down to a soft mass, and secondary infections by other organisms follow.

Cultural Control: Bulb Onions

- Allow tops to mature well (at least half of leaves brown), then lift or undercut the onions. In dry weather, cure onions on the ground 6 to 10 days.
- Be sure onions are well dried and necks tight (*i.e.* the tissue does not slide when you roll your neck between your fingers) before topping. Bacterial diseases and Botrytis Neck rot can move through green tissue into bulbs. These diseases do not move in dry tissue.
- Minimize bruising and mechanical injury in topping and storing.

- Store in well-ventilated houses at 32°F or slightly higher. Use higher temperatures if humidity cannot be held below 75%.
- Practice a crop rotation of at least three years.

Cultural Control: Garlic

- Allow the tops to mature well, then lift or undercut the garlic. - If normal dry weather prevails, cure garlic on the ground for 6 to 10 days. Otherwise, cure in a well ventilated area at 70-80 °F.
- When topping, minimize bruising and mechanical injury.
- Store garlic in well-ventilated houses at temperatures of 32°F, or just slightly higher.
- Avoid frequent and excessive irrigation. - Visual inspection of seed garlic with a hand lens may help. Examine the basil plate or bottom of garlic for gray mold.

Purple Blotch of onion & garlic is caused by Alternaria porri, a fungus that overwinters on infected bulbs and debris in the field, and can be seedborne in onion. Another fungus, Stemphylium vesicarium, can also cause purple blotch in onions. These are most common as leaf diseases, but can affect bulbs in storage. Infection often follows injury caused either by other fungi, such as Botrytis, by thrips feeding injury or by sand or dust during windstorms. Older leaves are more susceptible, and plants approaching maturity are more susceptible than young plants. Spores require rain or persistent dew to cause infection. Optimum temperatures are 77 to 81°F. Almost no infection occurs below 55°F. 'Sweet Spanish' onions are very susceptible.

Symptoms: A small, water-soaked lesion develops on leaves and soon develops a white center. Zones may appear as the spot enlarges and the lesion turns brown to purplish. Red or purple margins often encircle the purplish centers and are surrounded by yellowish tissue. In moist weather, the spot's surface usually becomes covered with a brownish black, powdery fungus growth. Leaves with large spots turn yellow and are blown over by the wind. Bulbs may decay during and after harvest. Decay shows first as a watery rot around the neck and is particularly noticeable because of the yellowish to wine-red discoloration in the neck region. As the fungus moves through onion scales, the tissue turns yellow then a wine-red and dries to a papery texture.

Cultural control: - Practice long rotations with unrelated crops. - Reduce hours of leaf wetness with wider plant spacing and good weed control. - Destroy old onion cull piles and bury debris. - Avoid injuring bulbs during production. - Properly cure bulbs in the field; lift and allow onions to dry several days before topping. - Plant tolerant or resistant varieties. Avoid Sweet Spanish onion because they are extremely susceptible.

White Rot of garlic & onion is caused by a fungus, *Sclerotium cepivorum*, that produces hardy *sclerotia* which may live in soil for years. Only *Allium spp.* such as onion, leek, and shallot are attacked. *Sclerotia* can infect plants from 12 inches below the soil surface. One sclerotium can infect a group of 20 to 30 adjacent plants. Fungal activity is favored by cool soils and is restricted above 75°F. Once the disease is in a field, it is very difficult to grow *Allium spp.*successfully. Disease spreads with infected sets or transplants, grazing animals, and movement of infested soil.

Symptoms: Leaves decay at the base, turn yellow, wilt, and topple over. Older leaves collapse first. Roots rot,and the plant can be pulled up easily. Fluffy mycelium may be on remaining roots and bulb. Affected bulbs may become watery, and outer scales crack as they dry and shrink. Small sclerotia (0.02 inch, or about the size of a poppy seed) form in and on the surface of affected bulb parts, often around the neck. White rot can continue to decay infected bulbs in storage if humidity is not kept low. - Plant only disease-free material in disease-free soil. - When working in infected fields avoid moving soil contaminated with sclerotia into new fields by washing equipment before it enters another field. - If practical, dig out all plants in infested spots in the field and the healthy plants growing next to diseased ones. Remove some soil with both diseased and healthy plants. Dispose of the material in a landfill or hot composting operation.

Blue Mold of garlic, caused by Penicillium spp. fungi, is a harvest and storage disease and may be particularly prevalent on fresh garlic. Several Penicillium species cause Blue Mold and these species are common in soil, on plant and animal debris, and in senescing plant tissues. Some species may be borne in infected cloves or seed. Invasion is usually through wounds, mechanical bruises, or freezing injury.

Cultural Control

- Control other diseases in the field to prevent avenues for infection.
- Harvest bulbs with a minimum or bruising and wounding and dry promptly.
- Store bulbs at 5° C (40° F) with low relative humidity.

YEAST APPLICATION FOR CONTROLLING APPLE POSTHARVEST DISEASES ASSOCIATED WITH PENICILLIUM EXPANSUM

Apple is one of the most important fruits produced in China. To provide fruit throughout the year, fresh apples are stored after harvest. Postharvest losses caused by fungal diseases are the major factor limiting the storage life of apples. Postharvest fungal diseases of apple are mainly caused by *Penicillium expansum*. Traditionally, this disease is controlled by the application of synthetic fungicides. However, the potential impact on environment as well as human health largely limits their application. It is reported that some microbes become fungicide-resistant, and thus a fungicide's effect on controlling fungal growth may be greatly reduced. Considering the human health and pollution risks, some fungicides are prohibited from use in many developed countries (*e.g.* America, England).

Recently, biological control has been developed as an alternative to synthetic fungicides , and considerable success has been achieved by utilizing antagonistic microorganisms for controlling postharvest diseases. Ever since Guter reported that *Bacillus subtilis* was antagonistic toward fruit pathogens, many studies involving antagonistic microorganisms to control postharvest diseases of fruits and vegetables have been done. Roberts (1990) discovered that *Cryptococcus laurentii* has antagonistic activity against many postharvest pathogens. The competition for nutrients may play a role in the antagonism of *Cryptococcus laurentii.*

Decay caused by *Rhizopus* sp. is reduced 70% when strawberries are treated with *Aureobasiduim pullulans* before storage. Calvente found that *Rhodotorula glutinis* produced rhodotorulic acid, which enhanced biocontrol activity of *Rhodotorula glutinis* against *Penicillium expansum* in postharvest apples. When applied to wounds 12 h after inoculation, *Pseudomondia corragata* significantly reduced brown rot in nectarines and peaches caused by *Monilinia fructicola.* Studies on postharvest biocontrol of fruits and vegetables have become an important new area of research.

Advantages of utilizing antagonistic microorganisms include reducing environmental pollutation, effectively controlling postharvest diseases, and producing high quality and safe food. Unfortunately, only a few studies in this field were reported in China. There is a new opportunity to study biological control of postharvest diseases of fruits and vegetables in our country.

In this paper the capability of some kinds of yeast to control apple postharvest diseases caused by *Penicillium expansum* was studied, and the one with the best effect was identified. The antagonistic conditions of that yeast and the methods of antagonism were also discussed.

Materials and Methods

Microorganisms: Antagonists—*Saccharomyces* sp., *Candida* sp., *Cryptococcus* sp., and *Rhodotorula*sp. Pathogen --- *Penicillium expansum,* isolated from microorganisms of rotten fruits.

Reagents: pH 7.0, 0.05 M phosphate buffer; 70% alcohol; 0.5% NaCl; $FeSO_4$.

Instruments: haemocytometer; microscope; incubator; rotary shaker.

Antagonist Isolation

5 g soil under a pear tree was added to 50 mL sterile water and mixed well. The solution and its diluents were streaked onto wort agar plates. The plates

were incubated at 28°C for a few days. Red yeast and other yeast single-cell colonies were re-cultured on wort agar plates and stored in test tubes.

Pathogen Isolation

5 g rotten apple tissue was dissolved in 50 mL sterile water. The solution and the diluents were streaked on to PDA plates containing 1.6 μg/mL penicillin and incubated at 28°C.

In Vitro Test

Four yeasts, *Saccharomyces* sp.,*Candida* sp., *Cryptococcus* sp. and *Rhodotorula* sp. were cultured in wort liquid culture medium on a rotary shaker (100 mL culture medium volume in a 500 mL Erlenmeyer flask, 200 rpm) for 72 h at 28°C. *Penicillium expansum* spores and washed yeast cells were counted with a haemocytometer and adjusted to a concentration of 4.4×10^6 spores/mL and 6×10^6cells/mL, respectively. Aliquots of 0.2 mL *Penicillium expansum* spore suspension were added to PDA plates and then stored in the refrigerator for 2 h. Washed yeast cell suspension was injected into holes (diameter 6.5 mm; four holes per plate) on PDA plates and cultured at 28°C. Plates were checked once or twice a day.

In Vivo Test

Washed *Cryptococcus* sp. cells were cultured in wort liquid culture medium amended with Fe^{2+} (0 μmol/L, 2.5 μmol/L, 5 μmol/L or 20 μmol/L) on a rotary shaker (25 mL culture volume in a 100 mL Erlenmeyer flask, 200 rpm) for 72 h at 28°C. The yeast cells were harvested by centrifugation at 2,000 rpm for 10 min, washed twice with sterile phosphate buffer (pH = 7.0), and suspended in phosphate buffer. The suspension was adjusted to concentrations of 3.5×10^8 cells/mL, 3.5×10^7 cells/mL, or 7×10^6cells/mL.

Apple fruits (no wound or scar in the surface, fresh) were selected for experiment. They were surface-sterilized with 0.5% NaCl for 5 min and then washed with tap water. After air-drying, apples were treated with 70% ethanol. Each fruit was wounded four times at its equator with the sterile head of a finishing nail (6 mm diameter × 4 mm deep). To obtain the best biocontrol efficacy of the yeast, apple fruits were divided into four groups to study four factors (the concentration of the yeasts, inoculatation time, concentration of Fe^{2+} in culture medium, and storage temperature) that influence the antagonism. 10 fruits were arranged in a randomized complete block for each group, and the experiment was repeated three times. All data presented in the figure are averages.

The Antagonistic Effect of the Different Yeast Concentrations

Wounds were inoculated with 20 μL yeast (grown in culture medium that contained 2.5 μmol/L Fe^{2+}) cell suspension at the concentrations of 7.0×10^6, 3.5

× 10^7, and 3.5 × 10^8cells/mL. After 12 h, a 20 μL*Pencillium expansum* suspension at a concentration of 1 × 10^4 spores/mL was added to each wound. The apples were then stored at 15°C.

The Antagonistic Effect of Different Inoculation Times

Wounds were inoculated with 20 μL yeast (cultured in culture medium that contained 2.5 μmol/L Fe^{2+}) cell suspension at a concentration of 3.5 × 10^7 cells/mL. After 1 h, 12 h, 36 h, a 20 μL *Pencillium expansum*suspension in a concentration of 1 × 10^4 spores/mL was added to each wound, and the apples were stored at 15°C.

Antagonistic Effect of Fe2 +concentration in Culture Medium

Wounds were inoculated with a 20 μL yeast (grown in culture medium that contained 0 μmol/L, 2.5 μmol/L, 5 μmol/L, and 20 μmol/L Fe^{2+}, respectively) cell suspension at concentration of 3.5 × 10^7cells/mL. After 12 h, a 20 μL*Pencillium expansum* suspension in a concentration of 1 × 10^4 spores/mL was added to each wound, and the apples were stored at 15°C.

Antagonistic Effect Of Different Storage Temperatures

Wounds were inoculated with a 20 μL yeast (grown in culture medium that contained 2.5 μmol/L Fe^{2+}) cell suspension at 3.5×10^7 cells/mL. After 12 h, a 20 μL *Pencillium expansum* suspension in 1 × 10^4spores/mL was added to each wound. The apples were stored at 10°C, 15°C, and 20°C, respectively.

Controls

CK1 control means nothing was added to wounds at 15°C, and CK2 control means only*Pencillium expansum* was added to wounds at 15°C.

Statistical Analysis

The incidence and severity of decay were analyzed by an analysis of variance (ANOVA) with SAS Software (SAS Institute, version 6.08, Cary, NC). Statistical significance was judged at the level $p < 0.01$. When the analysis was statistically significant, Duncan's Multiple-Range Test (SSR Test) was used to test mean separations among mean values of each treatment.

Results

Effects of the Antagonist

An antagonistic effect was observed on PDA plates which had*Pencillium expansum* for 2 h followed by *Saccharomyces* sp.,*Candida* sp., *Cryptococcus* sp., or*Rhodotorula* sp. *Candida* sp. and*Cryptococcus* sp. had greater antagonistic efficacy than *Rhodotorula* sp. and *Saccharomyces* sp. Since *Cryptococcus* sp. had the greatest effects, it was selected for further study.

Effects of Yeast Concentration on Decay

High, middle and low concentrations of yeast (LC = 7.0×10^6 cells/mL; MC = 3.5×10^7cells/mL; HC = 3.5×10^8 cells/mL) were inoculated into wounds of apple. The apple fruits were kept at 15°C. Lesion diameters were recorded every 24 h after the treatment.

It was obvious from Figure that the antagonism was related to the concentration of antagonist. When concentration of the yeast reached 3.5×10^7 cells/mL and spore suspension of the pathogens was 1×10^4 spores/mL, distinct antagonism was observed. The postharvest biocontrol capability of the apple fruits was enhanced with the increase of the antagonist concentration.

Effects of Different Inoculation Times on Decay

The pathogens were added 1 h, 12 h, 36 h or 60 h after the inoculation of the antagonist. Then the fruits were stored at 15°C, and the disease severity was observed every 24 h after inoculation.

It can be seen that the later the pathogens were introduced after the inoculation of the yeast, the better the antagonism was. The reason may be that with enough time the yeast had reproduced and used the nutrient (we call it competitive antagonism), or the yeast had secreted enough antagonistic substances. So we should apply the antagonist early after the harvest to prevent the pathogens from infecting.

Effects of Concentrations of Fe2+ in Culture Medium on Decay

Others have reported that the Fe^{2+}concentration in the cells has a great effect on the antagonism. We added different concentrations of Fe^{2+} into the culture medium to make the yeast cells contain different concentration Fe^{2+} and then inoculated the wounds with these yeasts. We stored the fruits at 15°C and observed the results 35d after the inoculation.

The experimental results showed that the yeast had its greatest antagonism when cultured in a medium containing 2.5 μmol/L Fe^{2+}, but when too much Fe^{2+} was in the culture medium, the yeast accelerated the decay. The antagonism appears to be related to the competition for Fe^{2+}. Yeast cells became enriched in the wounds by chelated Fe^{2+} to put off the germination of pathogens and then strengthened biocontrol efficacy. Charlang *et al.* (1981) reported that spores could not germinate without taking up large quantities of Fe^{2+}.

Effect of Different Storage Temperatures on Decay

Apple fruits were stored at 10°C, 15°C, 20°C after the inoculation of the antagonist and the pathogen. Lesion diameters were recorded every 24 h after challenge inoculation.

The higher the storage temperature, the faster the apple decay. The antagonistic effects were inversely related to the storage temperature; the antagonism was better when the storage temperature was lower. It is important to keep apple fruit at a low temperature.

Summary of the Analysis of Variance for the Effects of Each Treatment

The effects of *Cryptococcus laurentii*'s concentration, inoculation time of *Pencillium expansum*, concentration of Fe^{2+} in culture medium, and storage temperature on decay be adjusted to an appropriate concentration to achieve the best antagonism.

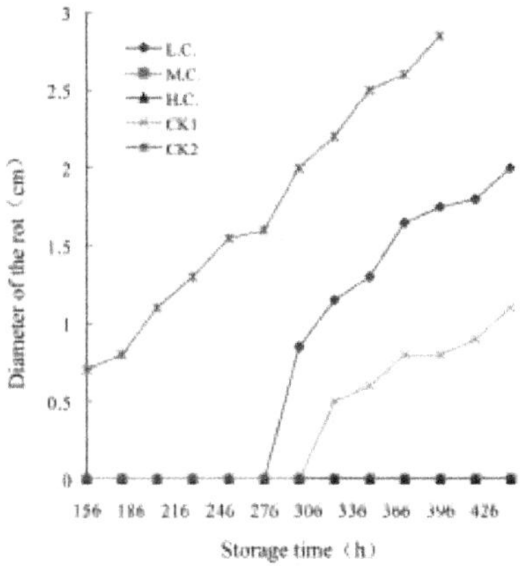

Fig.: Effects of *Cryptococcus laurentii* concentration on apple decay.

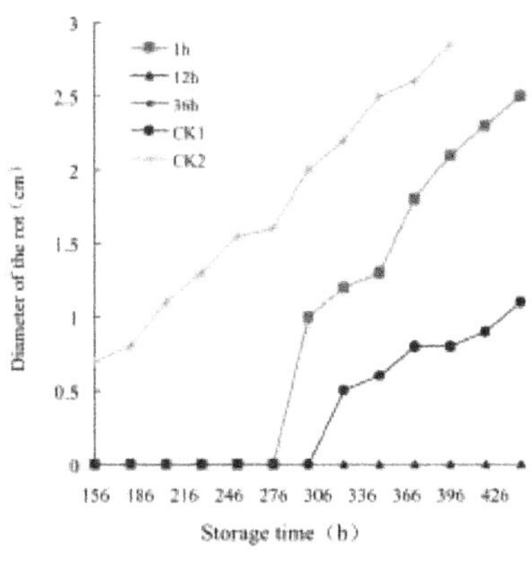

Fig.: Effect of inoculation time of *Pencillium expansum* on apple decay.

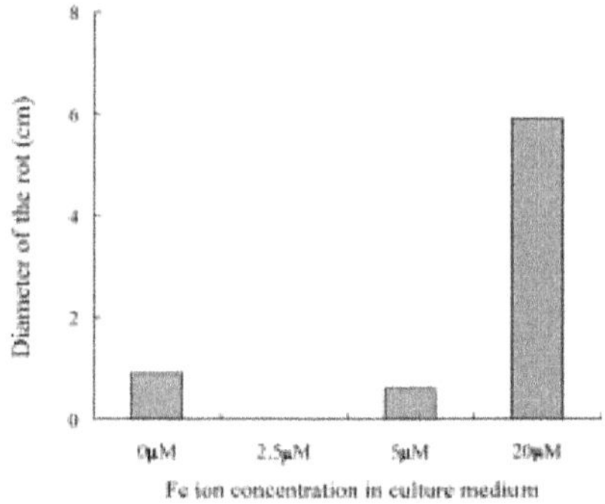

Fig.: Effects of concentration of Fe^{2+} in culture medium on apple decay.

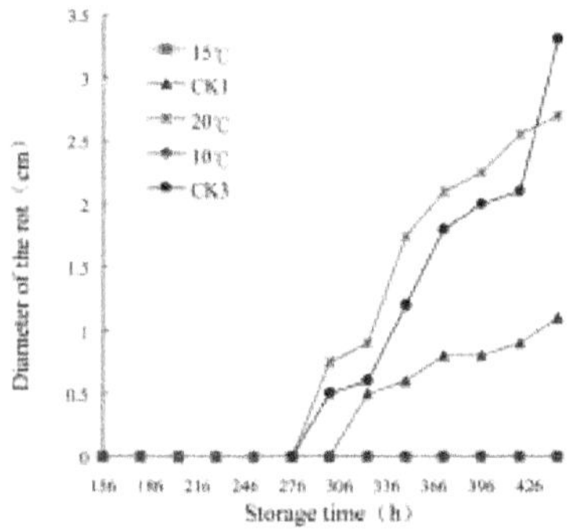

Fig.: Effect of storage temperature on apple decay.
CK3 control means nothing was added to wounds at 20°C.

Discussion

The results of this study indicate that *Pencillium expansum* causes postharvest decay of apple and that *Cryptococcus* sp. and *Candida* sp. have antagonistic biocontrol efficacy. The biocontrol efficacy of *Cryptococcus* sp. was superior. Biocontrol of postharvest decay caused by different epiphytes may need different antagonists.

Yeast was the most common biological control agent. This study showed a valuable method to control epiphyte pathogen caused postharvest decay by using antagonistic yeasts. However, the methods of yeast biocontrol include many aspects and need further study. Better methods

ARBUSCULAR MYCORRHIZAL FUNGI IN PLANT DISEASE CONTROL ARBUSCULAR MYCORRHIZA

An arbuscular mycorrhizal fungus (plural *mycorrhizae* or *mycorrhizas*, a.k.a. *endomycorrhiza, AM fungi,* or AMF) is a type of mycorrhiza in which the fungus penetrates thecortical cells of the roots of a vascular plant. (Not to be confused with ectomycorrhiza or ericoid mycorrhiza.)

Arbuscular mycorrhizas are characterized by the formation of unique structures, arbuscules and vesicles by fungi of the phylum Glomeromycota. AM fungi help plants to capture nutrients such as phosphorus, sulfur, nitrogen and micronutrients from the soil. It is believed that the development of the arbuscular mycorrhizal symbiosis played a crucial role in the initial colonisation of land by plants and in the evolution of the vascular plants.

It has been said that it is quicker to list the plants that do not form mycorrhizae than those that do. This symbiosis is a highly evolved mutualistic relationship found between fungi and plants, the most prevalent plant symbiosis known, and AM is found in 80% of vascular plant families in existence today.

The tremendous advances in research on mycorrhizal physiology and ecology over the past 40 years have led to a greater understanding of the multiple roles of AMF in the ecosystem. This knowledge is applicable to human endeavors of ecosystem management, ecosystem restoration, and agriculture.

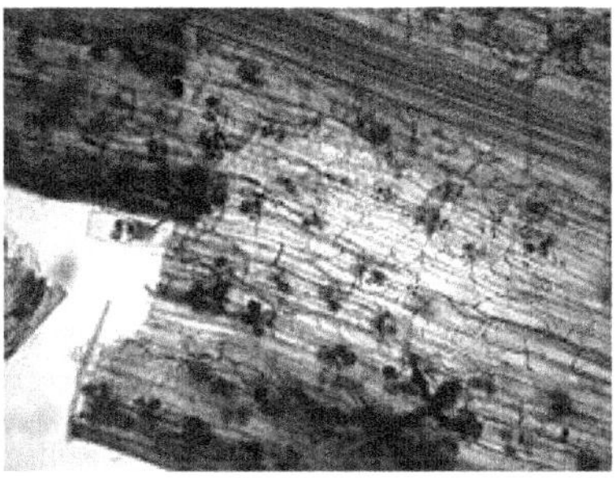

Fig. : Flax root cortical cells containing paired arbuscules.

Evolution of Mycorrhizal Symbiosis

Paleobiology

Both paleobiological and molecular evidence indicate that AM is an ancient symbiosis that originated at least 460 million years ago. AM symbiosis is ubiquitous among land plants, which suggests that mycorrhizas were present in the early ancestors of extant land plants. This positive association with plants may have facilitated the development of land plants.

The Rhynie chert of the lower Devonian has yielded fossils of the earliest land plants in which AM fungi have been observed. The fossilized plants containing mycorrhizal fungi were preserved in silica.

The Early Devonian saw the development of terrestrial flora. Plants of the Rhynie chert from the Lower Devonian (400 m.yrs ago) were found to contain structures resembling vesicles and spores of present *Glomus* species. Colonized fossil roots have been observed in *Aglaophyton major* and *Rhynia,* which are ancient plants possessing characteristics of vascular plants and bryophytes with primitive protostelic rhizomes.

Intraradical mycelium was observed in root intracellular spaces, and arbuscules were observed in the layer thin wall cells similar to palisade parenchyma. The fossil arbuscules appear very similar to those of existing AMF. The cells containing arbuscules have thickened walls, which are also observed in extant colonized cells.

Mycorrhizas from the Miocene exhibit a vesicular morphology closely resembling that of present Glomerales. The need for further evolution may have been lost due to the readily available food source provided by the plant host. However, it can be argued that the efficacy of signaling process is likely to have evolved, which could not be easily detected in the fossil record. A finetuning of the signaling processes would improve coordination and nutrient exchange between symbionts while increasing the fitness of both the fungi and the plant symbionts.

The nature of the relationship between plants and the ancestors of arbuscular mycorrhizal fungi is contentious. Two hypotheses are:

- Mycorrhizal symbiosis evolved from a parasitic interaction that developed into a mutually beneficial relationship.
- Mycorrhizal fungi developed from saprobic fungi that became endosymbiotic.

Both saprotrophs and biotrophs were found in the Rhynie Chert, but there is little evidence to support either hypothesis.

There is some fossil evidence that suggests that the parasitic fungi did not kill the host cells immediately upon invasion, although a response to the invasion was observed in the host cells. This response may have evolved into the chemical signaling processes required for symbiosis.

In both cases, the symbiotic plant-fungi interaction is thought to have evolved from a relationship in which the fungi was taking nutrients from the plant into a symbiotic relationship where the plant and fungi exchange nutrients.

Molecular Evidence

Increased interest in mycorrhizal symbiosis and the development of sophisticated molecular techniques has led to the rapid development of genetic evidence. Wang *et al.* (2010) investigated plant genes involved in communication with order Glomales fungal partners (DMI1, DMI3, IPD3). These three genes could be sequenced from all major clades of modern land plants, including liverworts, the most basal group, and phylogeny of the three genes proved to agree with then current land plant phylogenies. This implies that mycorrhizal genes must have been present in the common ancestor of land plants, and that they must have been vertically inherited since plants colonized land.

Physiology

Presymbiosis

The development of AM fungi prior to root colonization, known as presymbiosis, consists of three stages: spore germination, hyphal growth, host recognition and appressorium formation.

Spore Germination

Spores of the AM fungi are thick-walled multi-nucleate resting structures. The germination of the spore does not depend on the plant, as spores have been germinated under experimental conditions in the absence of plants both *in vitro* and in soil. However, the rate of germination can be increased by host root exudates. AM fungal spores germinate given suitable conditions of the soil matrix, temperature, carbon dioxide concentration, pH, and phosphorus concentration.

Hyphal Growth

The growth of AM hyphae through the soil is controlled by host root exudates known as strigolactones, and the soil phosphorus concentration. Low-phosphorus concentrations in the soil increase hyphal growth and branching as well as induce plant exudation of compounds that control hyphal branching intensity.

The branching of AM fungal hyphae grown in phosphorus media of 1 mM is significantly reduced, but the length of the germ tube and total hyphal growth were not affected. A concentration of 10 mM phosphorus inhibited both hyphal growth and branching. This phosphorus concentration occurs in natural soil conditions and could thus contribute to reduced mycorrhizal colonization.

Host Recognition

Root exudates from AMF host plants grown in a liquid medium with and without phosphorus have been shown to affect hyphal growth. Spores of *Gigaspora magarita* were grown in host plant exudates. Hyphae of fungi grown in the exudates from roots starved of phosphorus grew more and produced tertiary branches compared to those grown in exudates from plants given adequate phosphorus. When the growth-promoting root exudates were added in low concentration, the AM fungi produced scattered long branches. As the concentration of exudates was increased, the fungi produced more tightly clustered branches. At the highest-concentration arbuscules, the AMF structures of phosphorus exchange were formed.

This chemotaxic fungal response to the host plants exudates is thought to increase the efficacy of host root colonization in low-phosphorus soils. It is an adaptation for fungi to efficiently explore the soil in search of a suitable plant host.

Further evidence that arbuscular mycorrhizal fungi exhibit host-specific chemotaxis, that enable hyphal growth toward the roots of a potential host plant: Spores of *Glomus mosseae* were separated from the roots of a host plant, nonhost plants, and dead host plant by a membrane permeable only to hyphae. In the treatment with the host plant, the fungi crossed the membrane and always emerged within 800 μm of the root, but not in the treatments with nonhost plants and dead plants.

Molecular techniques have been used to understand the signaling pathways between arbuscular mycorrhizae and plant roots. In 2003 it was shown how the AM undergoes physiological changes in the presence of exudates from potential host plant roots, to colonize it. Host plant root exudates trigger and turn on AM fungal genes required for the respiration of spore carbon compounds. In experiments, transcription rate of 10 genes increased half-hour after exposure and at an even greater rate after 1 hour. after 4 hours exposure AM respond with morphological growth. Genes isolated from that time are involved in mitochondrial activity and enzyme production. The fungal respiration rate, measured by O_2 consumption rate, increased by 30% 3 hours after exposure to root exudates, indicating that host plant root exudates stimulate AMF spore mitochondrial activity. It may be part of a fungal regulatory mechanism that conserves spore energy for efficient growth and the hyphal branching upon receiving signals from a potential host plant.

Appressorium

When arbuscular mycorrhizal fungal hyphae encounter the root of a host plant, an appressorium or 'infection structure' forms on the root epidermis. From this structure hyphae can penetrate into the host's parenchyma cortex. AM need no chemical signals from the plant to form the appressoria. AM fungi could form appressoria on the cell walls of "ghost" cells in which the protoplast had been removed to eliminate signaling between the fungi and the plant host. However, the hyphae did not further penetrate the cells and grow in toward the root cortex, which indicates that signaling between symbionts is required for further growth once appressoria are formed.

Symbiosis

Once inside the parenchyma, the fungus forms highly branched structures for nutrient exchange with the plant called "arbuscules". These are the distinguishing structures of arbuscular mycorrhizal fungus. Arbuscules are the sites of exchange for phosphorus, carbon, water, and other nutrients. There are two forms: *Paris* type is characterized by the growth of hyphae from one cell to the next; and *Arum* type is characterized by the growth of hyphae in the space between plant cells. The choice between *Paris* type and*Arum* type is primarily determined by the host plant family, although some families or species are capable of either type.

The host plant exerts a control over the intercellular hyphal proliferation and arbuscule formation. There is a decondensation of the plant's chromatin, which indicates increased transcription of the plant's DNA in arbuscule-containing cells. Major modifications are required in the plant host cell to accommodate the arbuscules. The vacuoles shrink and other cellular organelles proliferate. The plant cell cytoskeleton is reorganized around the arbuscules.

There are two other types of hyphae that originate from the colonized host plant root. Once colonization has occurred, short-lived runner hyphae grow from the plant root into the soil. These are the hyphae that take up phosphorus and micronutrients, which are conferred to the plant. AM fungal hyphae have a high surface-to-volume ratio, making their absorptive ability greater than that of plant roots. AMF hyphae are also finer than roots and can enter into pores of the soil that are inaccessible to roots. The third type of AMF hyphae grows from the roots and colonizes other host plant roots. The three types of hyphae are morphologically distinct.

Nutrient Uptake and Exchange

AM fungi are obligate symbionts. They have limited saprobic ability and depend on the plant for their carbon nutrition. AM fungi take up the products of the plant host's photosynthesis as hexoses.

Carbon transfer from plant to fungi may occur through the arbuscules or intraradical hyphae. Secondary synthesis from the hexoses by AM occurs in the intraradicalmycelium. Inside the mycelium, hexose is converted to trehalose and glycogen. Trehalose and glycogen are carbon storage forms that can be rapidly synthesized and degraded and may buffer the intracellular sugar concentrations. The intraradical hexose enters the oxidative pentose phosphate pathway, which produces pentose for nucleic acids.

Lipid biosynthesis also occurs in the intraradical mycelium. Lipids are then stored or exported to extraradical hyphae where they may be stored or metabolized. The breakdown of lipids into hexoses, known as gluconeogenesis, occurs in the extraradical mycelium. Approximately 25% of the carbon translocated from the plant to the fungi is stored in the extraradical hyphae. Up to 20% of the host plant's carbon may be transferred to the AM fungi. This represents the host plant's considerable carbon investment inmycorrhizal network and contribution to the below-ground organic carbon pool.

Increasing the plant's carbon supply to the AM fungi increases uptake and transfer of phosphorus from fungi to plant Likewise, phosphorus uptake and transfer is lowered when the photosynthate supplied to the fungi is decreased. Species of AMF differ in their abilities to supply the plant with phosphorus. In some cases, arbuscular mycorrhizae are poor symbionts, providing little phosphorus while taking relatively high amounts of carbon.

The main benefit of mycorrhizas to plants has been attributed to increased uptake of nutrients, especially phosphorus. This may be due to increased surface area in contact with soil, increased movement of nutrients into mycorrhizae, a modified root environment, and increased storage. Mycorrhizas can be much more efficient than plant roots at taking up phosphorus. Phosphorus travels to the root or via diffusion and hyphae reduce the distance required for diffusion, thus increasing uptake. The rate of phosphorus flowing into mycorrhizae can be up to six times that of the root hairs. In some cases, the role of phosphorus uptake can be completely taken over by the mycorrhizal network, and all of the plant's phosphorus may be of hyphal origin. Less is known about the role of nitrogen nutrition in the arbuscular mycorrhizal system and its impact on the symbiosis and community. While significant advances have been made in elucidating the mechanisms of this complex interaction, much investigation remains to be done.

Mycorrhizal activity increases the phosphorus concentration available in the rhizosphere. Mycorrhizae lower the root zone pH by selective uptake of NH_4^+ (ammonium-ions) and by releasing H^+ ions. Decreased soil pH increases the solubility of phosphorus precipitates. The hyphal NH_4^+ uptake also increases the nitrogen flow to the plant as the soil's inner surfaces absorb ammonium and distribute it by diffusion.

Ecology

Biogeography

Arbuscular mycorrhizal fungi are most frequent in plants growing on mineral soils, and are of extreme importance for plants growing in nutrient-deficient substrates such as involcanic soil and sand dune environments. The populations of AM fungi is greatest in plant communities with high diversity such as tropical rainforests and temperate grasslands where they have many potential host plants and can take advantage of their ability to colonize a broad host range. There is a lower incidence of mycorrhizal colonization in very arid or nutrient-rich soils. Mycorrhizas have been observed in aquatic habitats; however, waterlogged soils have been shown to decrease colonization in some species. Arbuscular mycorrhizal fungi are found in 80% of plant species and have been surveyed on all continents except Antarctica. The biogeography of glomeromycota is influenced by dispersal limitation, environmental factors such as climate, soil series and soil pH and plant community. While evidence from 2000 suggests that AM fungi are not specialists on their host species, studies as of 2002 have indicated that at least some fungi taxa are host specialists.

Response to Plant Communities

The specificity, host range, and degree of colonization of mycorrhizal fungi are difficult to analyze in the field due to the complexity of interactions between the fungi within a root and within the system. There is no clear evidence to suggest that arbuscular mycorrhizal fungi exhibit specificity for colonization of potential AM host plant species as do fungal pathogens for their host plants. This may be due to the opposite selective pressure involved.

In pathogenic relations, the host plant benefits from mutations that prevent colonization, whereas, in a mutualistic symbiotic relationship, the plant benefits from mutation that allow for colonization by AMF. However, plant species differ in the extent and dependence on colonization by certain AM fungi, and some plants may be facultative mycotrophs, while others may be obligate mycotrophs. Recently, mycorrhizal status has been linked to plant distributions, with obligate mycorrhizal plants occupying warmer, drier habitats while facultative mycorrhizal plants occupy larger ranges of habitats.

The ability of the same AM fungi to colonize many species of plants has ecological implications. Plants of different species can be linked underground to a common mycelial network. One plant may provide the photosynthate carbon for the establishment of the mycelial network that another plant of a different species can utilize for mineral uptake. This implies that arbuscular mycorrhizae are able to balance below-ground intra–and interspecific plant interactions.

Since Glomeromycota fungi live inside plant roots, they can be influenced substantially by their plant host and in return affect plant communities as well. Plants can allocate up to 30% of their photosynthate carbon to AM fungi and in return AM fungi can acquire up to 80% of plant phosphorus and nitrogen. The diversity of AM fungal communities has been positively linked to plant diversity, plant productivity and herbivory. Arbuscular mycorrhizal fungi can be influenced by small scale interactions with the local plant community. For example, the plant neighborhood around a focal plant can alter AM fungal communities as can the order of plant establishment within sites.

AM Fungi and Plant Invasion

During invasions of plant species, the AM fungal community and biomass can be drastically altered. In the majority of cases AM fungal biomass and diversity decrease with invasions. However, some mycotrophic plant species may actually increase AM fungal diversity during invasion.

The mycorrhizal status of invasive plant species often varies between regions. For example, in the United Kingdom and central Europe recently invasive plants are more frequently obligately mycorrhizal than expected, while invasive plants in California were found to be less frequently mycorrhizal than expected.

Interactions Between AM Fungi and Other Plant Symbionts

All symbionts within a plant host interact, often in unpredictable ways. A 2010 meta-analysis indicated that plants colonized by both AM fungi and vertically transmitted endophytes often are larger than plants independently colonized by these symbionts. However, this relationship is context-dependent as AM fungi can interact synergistically with fungal endophytes inhabiting the leaves of their host plant, or antagonistically,. Similar ranges of interactions can occur between AM fungi and ectomycorrhizal fungi and dark septate endophytes.

Response to Environmental Gradients

Arbuscular mycorrhizal fungi vary across many environmental gradients. Their tolerance to freezing and drying is known to shift between AM fungal taxa. AM fungi become less prevalent and diverse at higher soil nutrient and moisture concentrations, presumably because both plants allocate less carbon to AM fungi and AM fungi reallocate their resources to intraradical hyphae in these environmental conditions. Over the long term, these environmental conditions can even create local adaptation between plant hosts, AM fungi and the local soil nutrient concentrations. AM composition often becomes less diverse on mountain tops than at lower elevations, which is driven by the composition of plant species.

AM fungi have been shown to improve plant tolerance to abiotic environmental factors such as salinity. They alleviate salt stress and benefit plant growth and productivity.

Rhizosphere Ecology

The rhizosphere is the soil zone in the immediate vicinity of a root system.

Arbuscular mycorrhizal symbiosis affects the community and diversity of other organisms in the soil. This can be directly seen by the release of exudates, or indirectly by a change in the plant species and plant exudates type and amount.

Mycorrhizae diversity has been shown to increase plant species diversity as the potential number of associations increases. Dominant arbuscular mycorrhizal fungi can prevent the invasion of non-mycorrhizal plants on land where they have established symbiosis and promote their mycorrhizal host.

Recent research has shown that AM fungi release an unidentified diffusional factor, known as the myc factor, which activates the nodulation factor's inducible gene MtEnod11. This is the same gene involved in establishing symbiosis with the nitrogen fixing, rhizobial bacteria. When rhizobium bacteria are present in the soil, mycorrhizal colonization is increased due to an increase in the concentration of chemical signals involved in the establishment of symbiosis. Molecules similar to Nod factors were isolated from AM fungi and were shown to induce MtEnod11, lateral root formation and enhance mycorrhization. Effective mycorrhizal colonization can also increase the nodulations and symbiotic nitrogen fixation in mycorrhizal legumes.

The extent of arbuscular mycorrhizal colonization and species affects the bacterial population in the rhizosphere. Bacterial species differ in their abilities to compete for carbon compound root exudates. A change in the amount or composition of root exudates and fungal exudates due to the existing AM mycorrhizal colonization determines the diversity and abundance of the bacterial community in the rhizosphere.

The influence of AM fungi on plant root and shoot growth may also have indirect effect on the rhizosphere bacteria. AMF contributes a substantial amount of carbon to the rhizosphere through the growth and degeneration of the hyphal network. There is also evidence to suggest that AM fungi may play an important role on mediating the plant species' specific effect on the bacterial composition of the rhizosphere.

Glomeromycota and Global Climate Change

Global climate change is affecting AM fungal communities and interactions between AM fungi and their plant hosts. While it is generally accepted that interactions between organisms will affect their response to global climate change, we still lack the ability to predict the outcome of these interactions in future climates. In recent meta-analyses, AM fungi were found to increase plant biomass under drought conditions and decrease plant biomass under simulated nitrogen deposition studies,. Arbuscular mycorrhizal fungi themselves have been shown to increase their biomass in response to elevated atmospheric CO_2.

Plants Lacking Arbuscular Mycorrhizae

Members of the mustard family (Brassicaceae), such as cabbage, cauliflower, canola, and crambe, do not establish arbuscular mycorrihizal fungi on their roots.

Molecular Genetic Analyses of Arbuscular Mycorrhizal Fungi

In the past ten years there have been spectacular advances in molecular genetic technologies and tools. These advances allow microbial and mycorrhizal ecologists to ask new and exciting questions about the ecological and evolutionary roles of arbuscular mycorrhizal (AM) fungi as individuals, in communities and ecosystems. Genetic analyses of AM fungi have been used to explore the genetic structure of single spores using multilocus genotyping, AM fungal diversity and adaptation across multiple grassland communities, all the way up to a global investigation of AM fungal diversity, which greatly increased the described molecular diversity within the phylum Glomeromycota.

All the recent advances in molecular genetics clearly permit the analysis of microbial communities at much finer and functional scales and potentially with more confidence than previous methods. The classical AM fungal identification method of spore extraction from soil and further spore morphological analysis is fraught with complicating issues due to the various strategies and forms of AM fungi, *e.g.*, lack of sporulation in certain species, seasonality, high unculturability, possible misidentification (human error), and new evidence of multi-nucleate spores and high genetic variation within clonal AM species. Because of these various problems, in the past researchers likely misrepresented the true composition of AM fungal communities present at any one point in time or place.

Additionally, by following the traditional extraction, culture and microscopic identification methods, there is no way to determine active/functioning AM fungal populations, which are likely the most important when attempting to relate plant-AM symbiotic interactions and mechanisms to ecological or ecosystem function. This is especially true in the case of root colonization analyses, which can determine percentage of roots colonized by AM fungi. The major problem with this analysis is in field soils, which contain multiple species of AM fungi in association with a target plant at the same time.

The identification of the associated fungal symbionts is impossible without the use of molecular methods. Though genetic analysis of AM fungal communities has advanced a great deal in the past decade, the methodology is not yet completely refined. Below is an overview of the methods used in molecular genetic analyses of AM fungi, along with applications to research, future directions and some of their problems.

Overview of Methods

DNA/RNA

Genetic analyses of AM fungi from soil and root samples range in their applicability to answer ecological or phylogenetic questions. DNA analyses utilize various nuclear markers to describe AM fungi and represent different regions of the nuclear ribosomal operon (18S rRNA) found in all eukaryotic organisms. The DNA analysis of AM fungi using these markers began in the early 1990s and are continuing to be developed today. The small subunit (SSU) rRNA gene, the internal transcribed spacer (ITS) gene, and the large subunit (LSU) rRNA gene are currently the most common DNA markers used. The SSU region has been used most frequently in ecological studies, while the ITS and LSU regions have been predominantly used in taxonomic constructions of the phylum Glomeromycota.

General Procedure

The first step of all molecular genetic analyses is the preparation and/or preservation of a sample. In the case of AM fungi, samples typically come in the form of soil or roots that will contain AM spores, hyphae and/or various AM colonization structures. Sample preservation will vary depending on the desired analysis (DNA or RNA). For analysis of DNA, samples should either be processed immediately or kept frozen prior to nucleic acid extraction. For analysis of RNA, samples should be cryogenically frozen (−196 °C) almost immediately upon collection, or stored in an RNA stabilization and preservation reagent (*e.g.* RNAlater).

The next step is to extract the desired nucleic acids from the sample, which can be performed manually using various published extraction methods or by using one of the many commercially available DNA/RNA extraction kits. Due to the labile nature of RNA, synthesis of complementary DNA (cDNA) using extracted RNA as a template is performed for further analysis. For most molecular genetic sequencing methods of AM fungi a PCR step is required to increase the total amount of target DNA/RNA/cDNA. There are many PCR conditions proposed for analysis of AM fungi and some of the most accessible are briefly summarized below.

PCR Methods

- Reaction mixture:
- 20 μl Qiagen HotStarTaq Master mix
- 0.23 μM of each primer (NS31 and AM1, more on AM fungal specific primers below)
- 2 μl template DNA
- PCR:
- Run on an MWG AG Biotech Primus 96 Plus thermocycler
- 15 minutes at 99 °C
- 5 cycles of 30 seconds at 42 °C
- 60 seconds at 72 °C
- 45 seconds at 92 °C
- 35 cycles of 30 seconds at 65 °C
- 60 seconds at 72 °C
- 45 seconds at 92 °C
- 30 seconds at 65 °C
- 10 minutes at 72 °C
- PCR products then separated by gel electrophoresis on 1.5% agarose gel in 0.5 x TBE

- Separated PCR products were then purified using the Qiagen QIAquick Gel Extraction kit

From Krüger *et al.* 2009:

- Reaction mixture:
- 0.02 U μl^{-1} Phusion polymerase
- 1X Phusion buffer with 1.5 mM $MgCl_2$
- 200 μM of each dNTP
- 0.5 μM of each primer : SSUmAf-LSUmAr and SSUmCf-LSUmBr
- PCR:
- Thermal cycling was performed in an Eppendorf Mastercycler Gradient
- 5 minute initial denaturation at 99 °C
- 40 cycles of 10 second denaturation at 99 °C
- 30 seconds annealing at 60 °C
- 1 minute elongation at 72 °C
- 10 minute final elongation
- To visualize PCR product, load onto 1% agarose gel with 1x sodium borate buffer at 220 V, and stain with ethidium bromide (1 $\mu g\ ml^{-1}$)

Primer Selection for Arbuscular Mycorrhizal Fungi

One difficulty with the genetic analysis of arbuscular mycorrhizal fungi has been the selection of ideal, comprehensive, and repeatable primers or primer sets. Currently there are four common AM fungal specific markers/primers used in genetic sequencing to describe AM fungal communities in a sample, ideally to species level identification. These sequence markers are designed for the nuclear ribosomal RNA (rRNA) in the 18S region and are either used individually or in some combination.

The partial small subunit (SSU), the partial large subunit (LSU), and the internal transcribed spacer (ITS1, 5.8S, ITS2) are the regions used for genetic sequencing of AMF. Additionally, there are 'primer sets' that incorporate a combination of these different regions into one target primer for AMF, these include the "Krüger " and the "Redecker" primers. The "Krüger" primer utilizes the partial SSU, the ITS, and the partial LSU regions, while the "Redecker" primer utilizes the partial SSU and the ITS.

Currently, there is no consensus as to which primers or primer sets, being used with varying degrees of success, repeatability and species-level resolution, are best for molecular genetic analysis of AMF. Additionally, the current advances and coming changes in genetic sequencing technology, *e.g.* Sanger, to 454 pyrosequencing, to IlluminaHiSeq/MiSeq, can force researchers to only use certain primers. The large size of the "Krüger" (~1500bp) and "Redecker"

(~900bp) primer sets prohibit use with newer sequencing technology (*e.g.* Illumina MiSeq) as opposed to 454 pyrosequencing that is capable of these long read lengths.

Though Roche Diagnostics has announced the discontinuation of the 454 platform for 2016, it is still commonly used in genetic analyses. Perhaps new 'all-inclusive' AM specific primers should be created to support the new technologies for as descriptive a molecular analysis from the "Kruger" primer set using 454 pyrosqeuncing, as shown below. The reverse may also be true, where molecular technologies should be developed with both long read lengths (which would allow for large primer sets) as well as sequencing depth.

Kohout *et al.* present a study using all of the aforementioned primers/primer sets on identical plant samples using 454 sequencing analysis. Results of their experiment are summarized below.

- "Kruger" primers yielded relatively higher diversity parameters than other comparable primers (LSU, ITS2)
- "Kruger" primers showed significantly higher Shannon diversity measures than did SSU primer
- "Redecker" primers yielded the most different, but maybe most descriptive community composition of all primers tested. This may be explained by the ability of the "Redecker" primers to find less abundant AMF lineages such as the Claroideoglomeraceae or the Paraglomeraceae
- LSU primers had a strong bias towards Glomeraceae, excluding other families
- SSU primers had a bias towards Glomeraceae and underestimated the presence of different families within the Glomeromycota, including the Claroideoglomeraceae, Diversisporaceae and Paraglomeraceae.

Family	MOTUs from SSU	MOTUs from LSU	MOTUs from ITS2	MOTUs from "Kruger"	MOTUs from "Redecker"
Glomeraceae	14	19	17	21	8
Claroideoglomeraceae	1	2	2	3	5
Archaeosporales	1	1	1	1	2
Diversisporaceae	1	0	2	3	2
Paraglomeraceae	0	0	0	0	2
Acaulosproraceae	0	0	1	1	1

MOTU = Molecular operational taxonomic unit, synonymous with OTU or phylotype.

qPCR and qRT-PCR

Real-time PCR or quantitative PCR (qPCR), is becoming a well-established method to quickly amplify and simultaneously quantify targeted AM fungal

DNA from biological samples (plant roots or soils). Fairly recent developments in qPCR markers allow researchers to explore the relative abundance of AM fungal species within roots in greenhouse experiments as well as in the field to identify local AM fungal communities.

qPCR markers for arbuscular mycorrhizal fungi will consist of AM specific primers and fluorescently labeled hydrolysis probes. These AM specific primers can be chosen by the researcher and this decision is typically guided by the question at hand, resources available, and willingness to troubleshoot in the lab.

Microarray

DNA microarray analysis is currently being used in AM fungal research to simultaneously measure the expression of many genes from target species or experimental samples. The most common tool or method is to use functional gene array (FGA) technology, a specialized microarray that contains probes for genes that are functionally important in microbial processes such as carbon, nitrogen or phosphorus cycling. FGAs have the ability to simultaneously examine many functional genes. This technique is typically used for general analysis of functional microbial genes, but when complemented with genetic sequencing, inferences can be made about the connection between fungal community composition and microbial functionality.

PLFA/NLFA

Specific organismal chemical signatures can be used to detect biomass of more cryptic organisms, such as AM fungi or soil bacteria. Lipids, more specifically phospholipids and neutral lipids, contain fatty acids connected to a glycerol backbone. The fatty acid composition of organisms varies, and the proportions of specific fatty acids can be organism specific. For example, in AM fungi the proportion of the fatty acids, 16:1ω5 and 18:1ω7, in the phospholipid portion account for approximately 58% of total fatty acid composition. The fatty acid, 16:1ω5 is the most commonly used acid to characterize AM fungi in soils and can be used as a strong indicator of mycelial biomass in soil sample.

Neutral lipid fatty acid analysis of AM fungi is typically looked upon as a method to indicate energy storage, but most importantly, the ratio of NLFA (16:1ω5) to PLFA (16:1ω5) can potentially be used to indicate nutritional status of AM fungi in soils. Energy is mainly stored in AM fungi as neutral lipids in storage structures like spores and vesicles. Because of this NLFA correlates quite well with the number of spores in a given volume of soil. The ratio of NLFA concentration to PLFA concentration (active mycelia) can then give the proportion of carbon allocated to storage structures.

Problems with lipid fatty acid analyses include the incomplete specificity of fatty acids to AM fungi, the species- or genera-specific variation in fatty acid composition can complicate analysis in systems with multiple AM fungal species (*e.g.* field soil), the high background levels of certain fatty acid concentration in soils, and that phospholipids are correlated to an organism's membrane area, and the surface to volume ratio can vary widely between organisms such as bacteria and fungi. More work must be done to identify the efficacy of this method in field soils with many genera and species of AM fungi to discern the methods ability to discriminate between many varying fatty acid compositions.

Future Research Directions with AM Fungi

An exciting prospect for future analysis of AM fungi is the use of stable isotope probes. Stable isotope probing (SIP) is a technique that can be used to determine the active metabolic function of individual taxa within a complex system of microbes. This level of specificity, linking microbial function and phylogenetics, has not been achieved previously in microbial ecology. This method can also be used independently of classical culture methods in microbial ecology, allowing for *in situ* analysis of functional microbes.

Stable Isotope Probing (SIP)

SIP, more explicitly DNA/RNA-based SIP, uses stable-isotope enriched substrates, such as ^{13}C, ^{15}N, or $H_2{}^{18}O$, and then analyzes the 'labeled' markers using species specific DNA or RNA markers. The analysis of labeled DNA is performed by separating unlabeled and labeled DNA on a cesium chloride gradient formed in an ultra centrifuge.Because all microbial organisms are capable of importing water into their cells, the use of $H_2{}^{18}O$ stable isotope probing is a very exciting new method that can shed light on questions microbial ecologists and biologists have struggled with answering for years, in particular, what are the active microbial organisms in my system? The $H_2{}^{18}O$, or heavy water method will target all organisms that are actively growing, and induce little influence on growth itself. This would be especially true with most greenhouse experiments with arbuscular mycorrhizas because plants must be watered anyway, and water does not directly select for organisms with specific metabolic pathways, as would happen when using ^{13}C and ^{15}N.

Little has been done with this method in arbuscular mycorrhizal experiments, but if proven to work in a controlled experiment, and with further refinement of DNA/RNA fungal community analyses techniques, this may be a viable option to very specifically determine the actively growing portion of AM fungal species across growing seasons, with different plant hosts or treatments, and in the face of climate change.

Phytoremediation

Disturbance of native plant communities in desertification-threatened areas is often followed by degradation of physical and biological soil properties, soil structure, nutrient availability, and organic matter. When restoring disturbed land, it is essential to replace not only the above ground vegetation but also biological and physical soil properties.

A relatively new approach to restoring land is to inoculate soil with AM fungi when reintroducing vegetation in ecological restoration projects (phytoremediation). It has enabled host plants to establish themselves on degraded soil and improve soil quality and health. Soils' quality parameters were significantly improved long-term when a mixture of indigenous arbuscular mycorrhizal fungi species was introduced compared to noninoculated soil and soil inoculated with a single exotic species of AM fungi. The benefits were increased plant growth, increased phosphorus uptake and soil nitrogen content, higher soil organic matter content, and soil aggregation, attributed to higher legume nodulation in the presence of AMF, better water infiltration, and soil aeration due to soil aggregation.

Agriculture

Many modern agronomic practices are disruptive to mycorrhizal symbiosis. There is great potential for low-input agriculture to manage the system in a way that promotes mycorrhizal symbiosis.

Conventional agriculture practices, such as tillage, heavy fertilizers and fungicides, poor crop rotations, and selection for plants that survive these conditions, hinder the ability of plants to form symbiosis with arbuscular mycorrhizal fungi.

Most agricultural crops can perform better and are more productive when well-colonized by AM fungi. AM symbiosis increases the phosphorus and micronutrient uptake and growth of their plant host.

Management of AM fungi is especially important for organic and low-input agriculture systems where soil phosphorus is, in general, low, although all agroecosystems can benefit by promoting arbuscular mycorrhizae establishment.

Some crops that are poor at seeking out nutrients in the soil are very dependent on AM fungi for phosphorus uptake. For example, flax, which has poor chemotaxic ability, is highly dependent on AM-mediated phosphorus uptake at low and intermediate soil phosphorus concentrations.

Proper management of AMF in the agroecosystems can improve the quality of the soil and the productivity of the land. Agricultural practices such as reduced tillage, low phosphorus fertilizer usage, and perennialized cropping systems promote functional mycorrhizal symbiosis.

Tillage

Tillage reduces the inoculation potential of the soil and the efficacy of mycorrhizaes by disrupting the extraradical hyphal network.

By breaking apart the soil macro structure, the hyphal network is rendered non-infective. The disruption of the hyphal network decreases the absorptive abilities of the mycorrhizae because the surface area spanned by the hyphae is greatly reduced. This, in turn, lowers the phosphorus input to the plants that are connected to the hyphal network.

In reduced-tillage system, heavy phosphorus fertilizer input may not be required as compared to heavy-tillage systems. This is due to the increase in mycorrhizal network, which allows mycorrhizae to provide the plant with sufficient phosphorus.

Phosphorus Fertilizer

The benefits of AMF are greatest in systems where inputs are low. Heavy usage of phosphorus fertilizer can inhibit mycorrhizal colonization and growth.

As the soil's phosphorus levels available to the plants increases, the amount of phosphorus also increases in the plant's tissues, and carbon drain on the plant by the AM fungi symbiosis become non-beneficial to the plant.

A decrease in mycorrhizal colonization due to high soil-phosphorus levels can lead to plant deficiencies in other micronutrients that have mycorrhizal-mediated uptake such as copper.

Perennialized Cropping Systems

Cover crops are grown in the fall, winter, and spring, covering the soil during periods when it would commonly be left without a cover of growing plants.

Mycorrhizal cover crops can be used to improve the mycorrhizal inoculum potential and hyphal network.

Since AM fungi are biotrophic, they are dependent on plants for the growth of their hyphal networks. Growing a cover crop extends the time for AM growth into the autumn, winter, and spring. Promotion of hyphal growth creates a more extensive hyphal network. The mycorrhizal colonization increase found in cover crops systems may be largely attributed to an increase in the extraradical hyphal network that can colonize the roots of the new crop. The extraradical mycelia are able to survive the winter, providing rapid spring colonization and early season symbiosis. This early symbiosis allows plants to tap into the well-established hyphal network and be supplied with adequate phosphorus nutrition during early growth, which greatly improves the crop yield.

Soil Quality

Restoration of native AM fungi increases the success of ecological restoration project and the rapidity of soil recovery. AM fungi enhance soil aggregate stability is due to the production of extraradical hyphae and a soil protein known as glomalin.

Glomalin-related soil proteins (GRSP) have been identified using a monoclonal antibody (Mab32B11) raised against crushed AMF spores. It is defined by its extraction conditions and reaction with the antibody Mab32B11.

There is other circumstantial evidence to show that glomalin is of AM fungal origin. When AM fungi are eliminated from soil through incubation of soil without host plants, the concentration of GRSP declines. A similar decline in GRSP has also been observed in incubated soils from forested, afforested, and agricultural land and grasslands treated with fungicide.

Glomalin is hypothesized to improve soil aggregate water stability and decrease soil erosion. A strong correlation has been found between GRSP and soil aggregate water stability in a wide variety of soils where organic material is the main binding agent, although the mechanism is not known. The protein glomalin has not yet been isolated and described, and the link between glomalin, GRSP, and arbuscular mycorrhizal fungi is not yet clear.

ARBUSCULAR MYCORRHIZAS

Mycorrhizal associations produced by *Glomeromycotan* fungi are known as arbuscular mycorrhizas, or vesicular-arbuscular mycorrhizas (formerly also endomycorrhizas, or endotrophic mycorrhizas) and are abbreviated as VAM here. There is disagreement about whether arbuscular mycorrhizas or vesicular-arbuscular mycorrhizas is the most appropriate name to, because some fungi do not produce vesicles, but arbuscules are not consistently used to identify associations (*i.e.* they are absent in myco-heterotrophs and older roots). These associations involve primitive fungi in the Glomeromycota and a wide diversity of plants, as explained at the end of this Section. Components of associations are illustrated below with details in the following sub-sections.

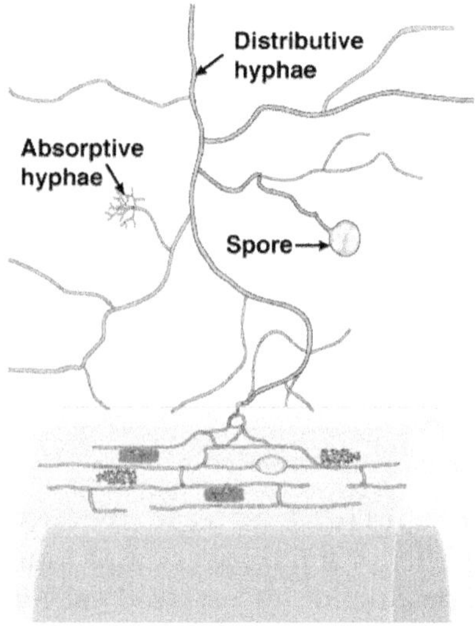

Structures in Soil

- **Hyphae :** A network of hyphae forms in the soil with thicker hyphae which function as conduits.
- **Absorptive hyphae:** Thin highly branched hyphae which are thought to absorb nutrients.
- **Spores:** Large (for a fungus) asexual spherical structures (20-1000+ μm diameter) formed on hyphae in soil, or in roots.

Structures in Roots

- **Hyphae:** these are non-septate when young and ramify within the cortex.
- **Arbuscules :** intricately branched haustoria in cortex cells.
- **Vesicles:** storage structures formed by many fungi.

Looking at Associations

These associations are introduced by 3 diagrams showing increasing magnification views. Methods used to see mycorrhizal structures.

Level 1. Whole Plants

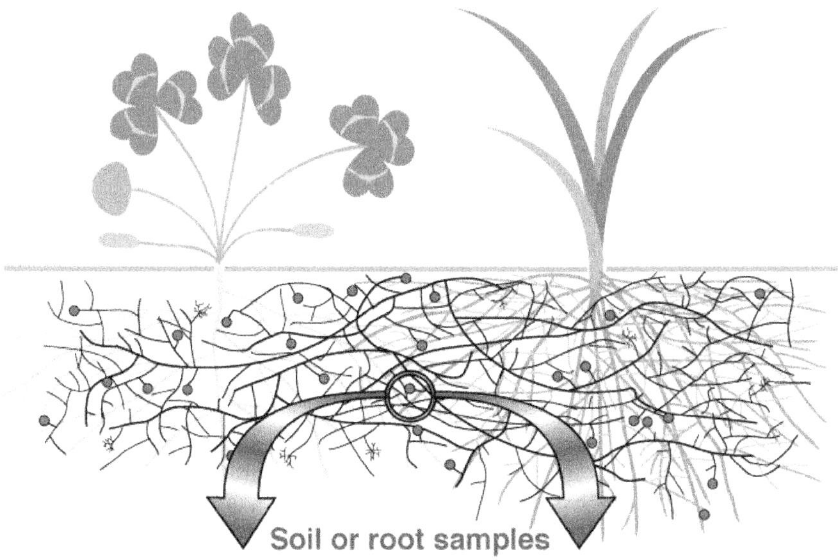

Mycorrhizal roots and the associated networks of hyphae are a major component of most soils, but cannot normally be seen with the naked eye (they are greatly exaggerated in this diagram and soil is omitted).

Samples of roots or soil can be collected, processed and viewed with a microscope as described below to detect mycorrhizas.

Level 2. The Dissecting Microscope

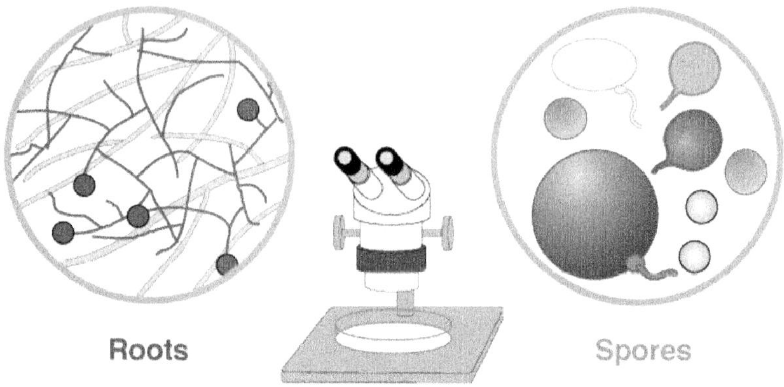

The medium level of magnification provided by a dissecting microscope allows hyphae and spores to be seen after extraction from soil.

Mycorrhizal structures within root are normally not visible, unless roots have been cleared in hot alkali to make them transparent and stained with a dye that binds to fungal hyphae. Observations at this level are normally used to quantify mycorrhizal associations.

Level 3. The Compound Microscope

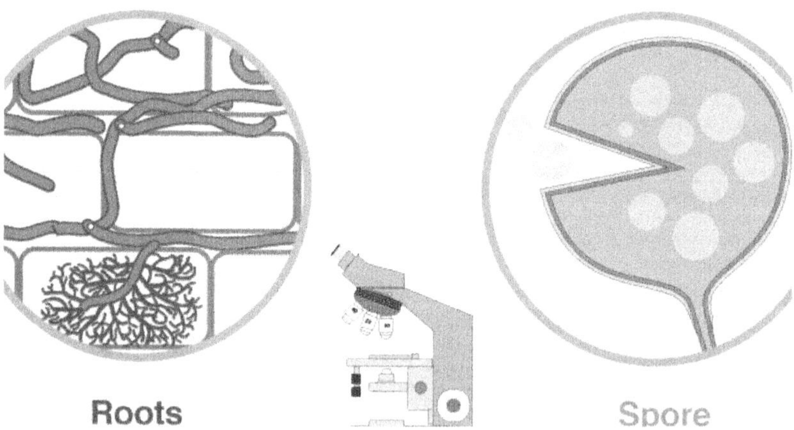

The higher level of magnification of a compound microscope allows mycorrhizal structures within roots to be *confirmed* and details of spore structure to be seen.

Root fragments that have been cleared and stained are examined to reveal structural details of mycorrhizas. Variations in these structures are sometimes used to *identify fungi*.

Observations at this scale allow fungi to be identified by spore structure. Spores are *separated from soil* by washing soil through very fine sieves, then concentrated by centrifugation in a dense medium such as a sugar solution.

Structures and Developmental Stages

VAM associations form when host roots and compatible fungi are both active in close proximity and soil conditions are favourable. The stages in root colonization by VAM fungi are illustrated below. Associations formed by species of *Glomus, Gigaspora, Scutellospora* and *Acaulospora* are shown. Terminology is explained in the *glossary*.

Soil Hyphae

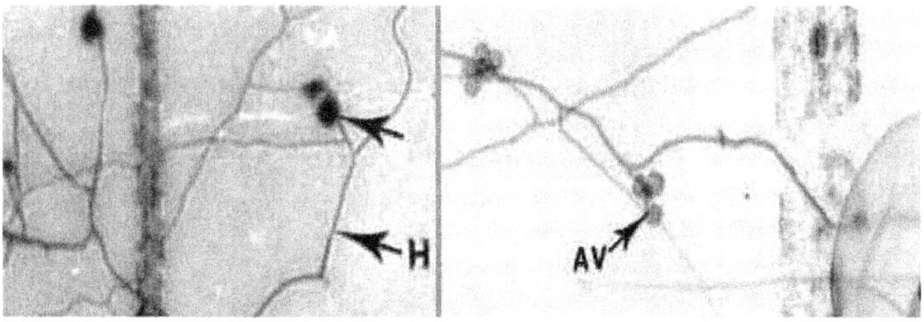

Mycorrhizal associations may be initiated by spore germination as illustrated here. Hyphae may also originate from fragments of roots. In many cases there already is a pre-existing network of hyphae resulting from previous root activity. Hyphae resulting from spore germination have a limited capacity to grow and will die if they do not encounter a susceptible root within a week or so. Hyphae emerge from a germination shield within the spore in *Scutellospora* and *Acaulospora*species.

Germinating hyphae which emerged several days after spores were extracted from dry soil. These spores are of *Gigaspora decipiens* (left) and *Scutellospora cerradensis* (right).

Spores germinated on membrane filters in contact with soil solution. Hyphae in the picture on the left were stained with Trypan blue. AV = accessory vesicles. GS = germination shield. Bars = 100 μm.

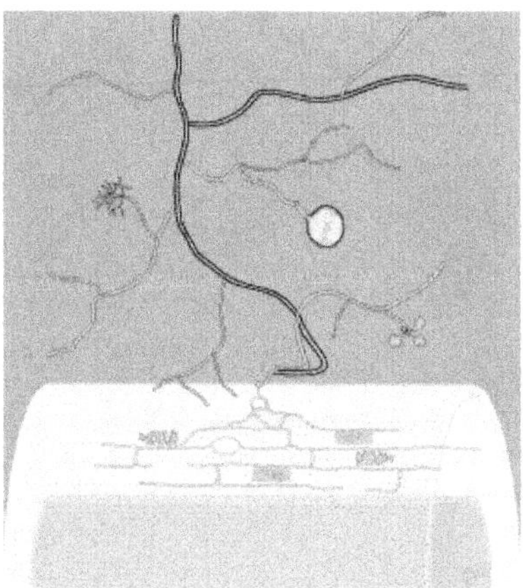

Soil hyphae, also known as extraradical or external hyphae, are filamentous fungal structures which ramify through the soil. They are responsible for nutrient acquisition, propagation of the association, spore formation, *etc.* VAM fungi produce different types of soil hyphae including thick "runner" or "distributive" hyphae as well as thin "absorptive" hyphae. The finer hyphae can produce "branched absorptive structures" (BAS) where fine hyphae proliferate (Bago *et al.* 1998). Hyphae of *Scutellospora* and*Gigaspora* species produce clustered swellings with spines or knobs called auxiliary cells. *Roles of hyphae* are discussed in another section.

The network of hyphae in the soil is only connected to roots by the entry points that initiate mycorrhizal associations, as hyphae do not grow out of living roots.

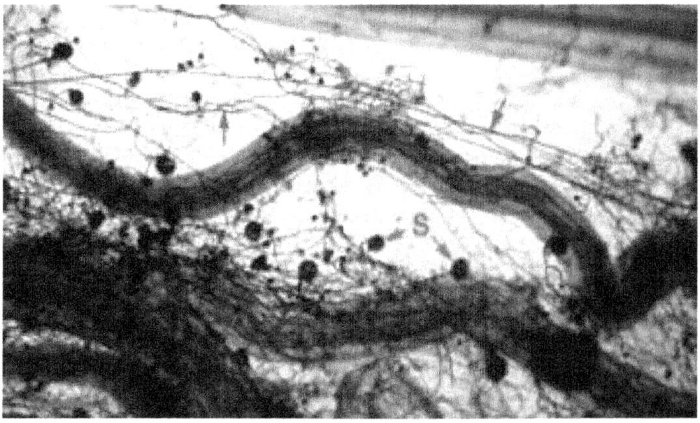

Mycorrhizal root system washed carefully from coarse sand to reveal the intact network with external hyphae (arrow) with spores (S) produced by *Glomus mosseae.*

bar = 100 μm

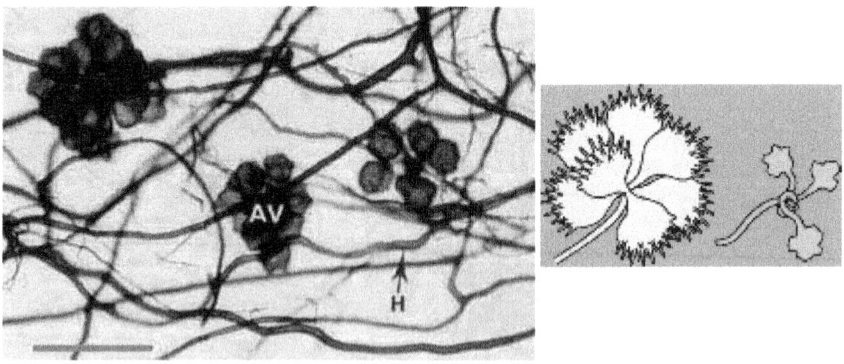

Darkly pigmented soil hyphae (H) of a*Scutellospora*species with auxiliary cells (AV).

Bar = 100 μm

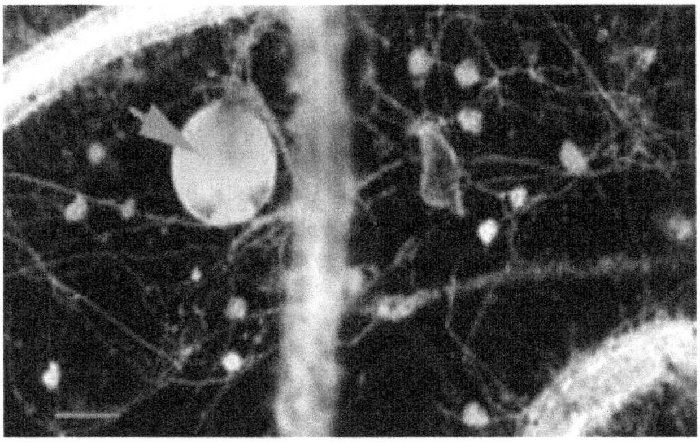

Soil hyphae produced by a single germinated spore of *Gigaspora* (arrow) used to start a mycorrhizal association. The hyphal network has produced accessory vesicles and is spreading through the root system of a clover plant.

Bar = 100 μm

Root Contact and Penetration

Mycorrhizal associations start when soil hyphae respond to the presence of a root by growing towards it, establishing contact and growing along its surface. Next, one or more hyphae produce swellings called appressoria between epidermal cells. Root penetration occurs when hyphae from the appressoria penetrate epidermal or cortical cells to enter the root. These hyphae cross the hypodermis (through *passage cells* if these are present in an exodermis) and start branching in the outer cortex.

The images in the following sections are of cleared roots which have been stained with Chlorazol black. Many images are from a study of the time course of VAM formation by *Glomus versiforme* in leek (*Allium porrum*) roots.

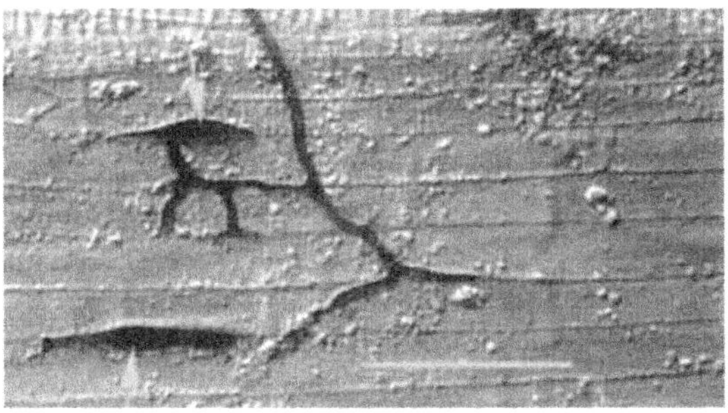

Soil hyphae have produced 2 appressoria between epidermal cells (arrows). These are seen here in a surface view of a root with attached hyphae.

Whole mount focused on surface.

Bar = 100 μm

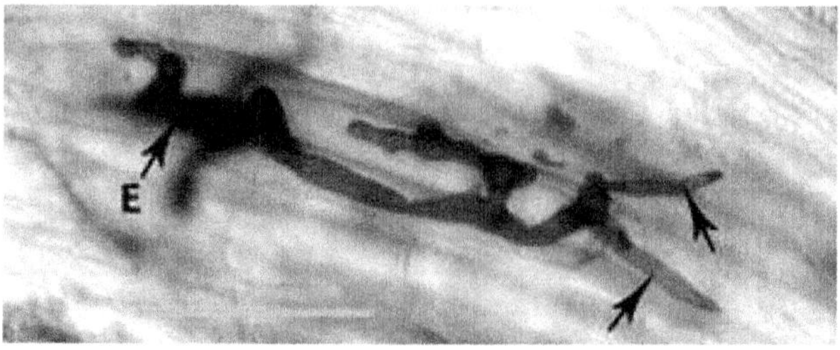

Hyphae at an entry point (E) penetrating cortex cells (arrows) approximately 1 day after contact with the root.

Whole mount focused on outer cortex.

Bar = 100 μm

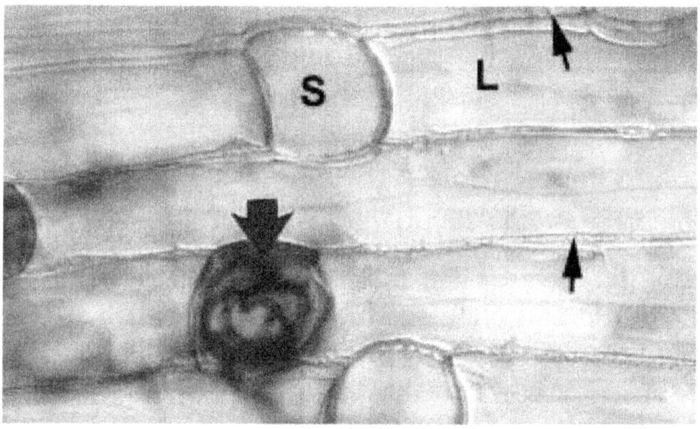

Alternating long (L) and short (S) cells in the dimorphic *exodermis* of a *Smilacina racemosa* root. Hyphae of VAM fungi have penetrated unsuberised short cells (arrows).

Surface view of whole root.

Bar = 100 μm

Hyphal Proliferation in the Cortex

Aseptate hyphae spread along the cortex in both directions from the entry point to form a colony. Hyphae within root are initially without cross walls, but these may occur in older roots. Gallaud (1905) observed that VAM associations in different species formed two distinctive morphology types, which he named the *Arum* and *Paris* series after host plants. These are now known as linear (1.1) and coiling (1.2) associations respectively. The differences between these two modes of spread within the root cortex are illustrated below. Both types of associations are important in ecosystems.

1. **Linear** (Arum) series associations where hyphae proliferate in the cortex by growing longitudinally between host cells. This occurs because

hyphae grow through longitudinal intercellular air spaces that are present.

2. **Coiling** (Paris) series where hyphae spread by forming coils within cells because there are no continuous longitudinal air spaces.

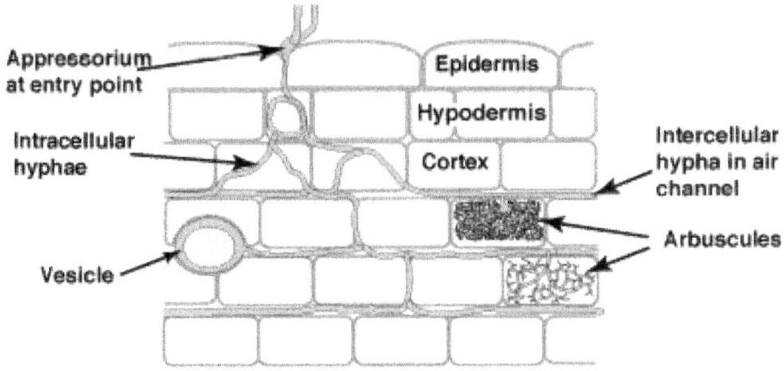

A colony refers to hyphal growth within a root resulting from the same external hyphae (1 or more connected entry points). These are also called infection units.

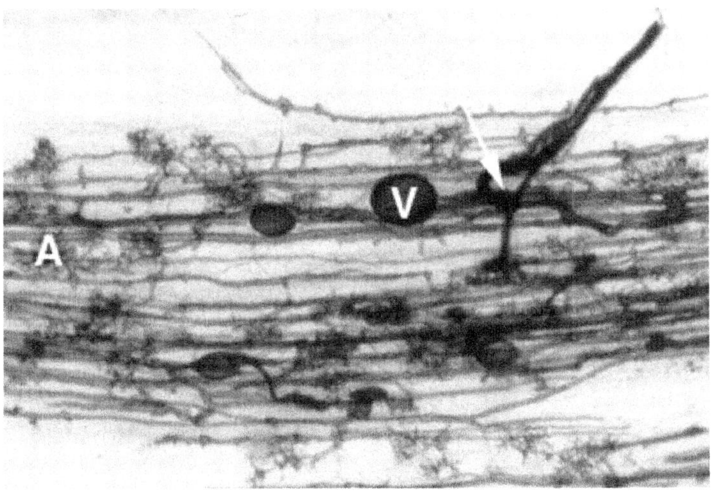

Part of a colony of a VAM fungus (*Glomus* sp.) with hyphae, arbuscules (A) and vesicles (V) growing from an entry point (arrow).

Highly magnfied views of a cleared and stained root

Coiling (Paris) Arbuscular Mycorrhizas

These are associations where hyphae spread primarily by intracellular growth following a convoluted path through cortex cells. The resulting colonies of VAM fungi generally have a coiled appearance, but may have more digitate branching patterns. Arbuscules may be restricted to a single layer of cells in the inner cortex.

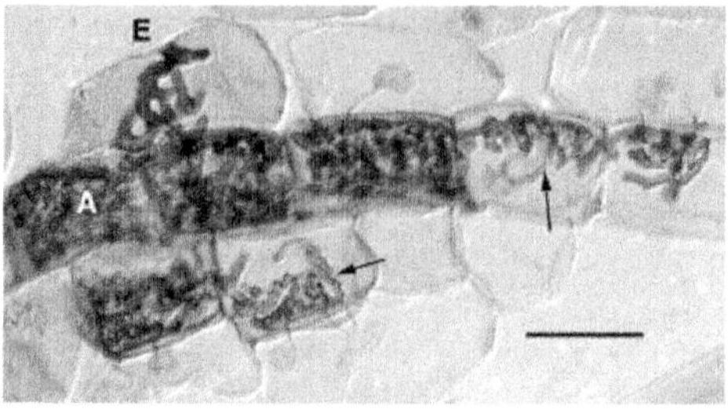

Colony of a VAM fungus spreading from the entry point (E) by convoluted hyphae (arrows) in the cortex of an *Erythronium americanum*root. Hyphal coils tend to occur in roots without prominent air channels.

A = arbuscule

Bar = 100 μm

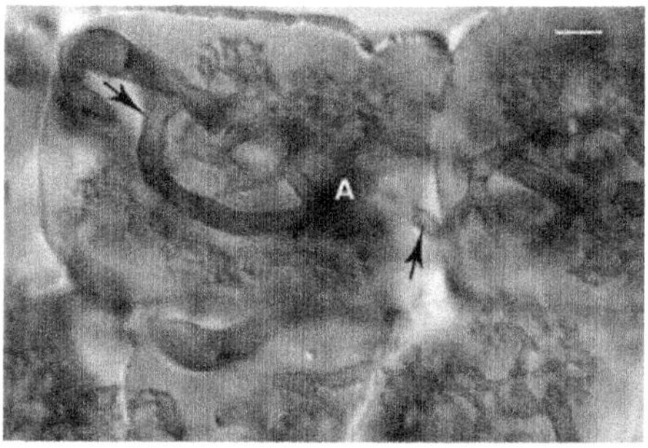

Higher magnification view of two arbuscules (A) growing directly from coils in adjacent root cortex cells of *Erythronium americanum*. Note how arbuscular branches arise from the same hyphae (arrows) which connect hyphae in adjacent cells.

Bar = 10 μm

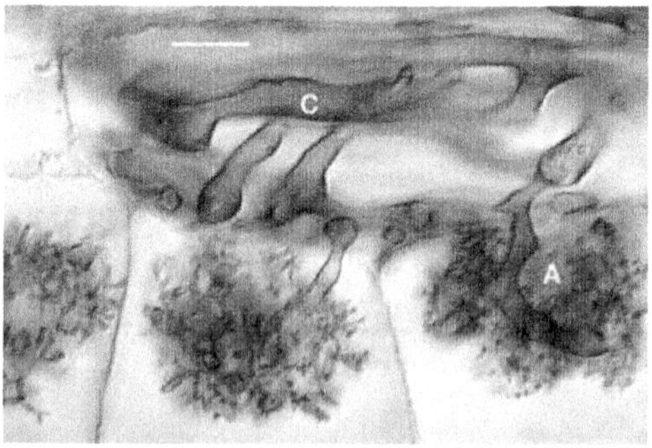

Arbuscules (A) and convoluted hyphae (arrow) in the inner cortex of an *Asarum canadense* root. Arbuscules only form in the innermost cortex cell layer next to the endodermis in this species.

Bar = 10 μm

Linear (Arum) Arbuscular Mycorrhizas

These are associations where hyphae grow along longitudinal intercellular air channels between the walls of root cells. A relatively rapid parallel spread of intercellular hyphae may occur along these *channels*. The resulting colonies of VAM fungi have a linear appearance.

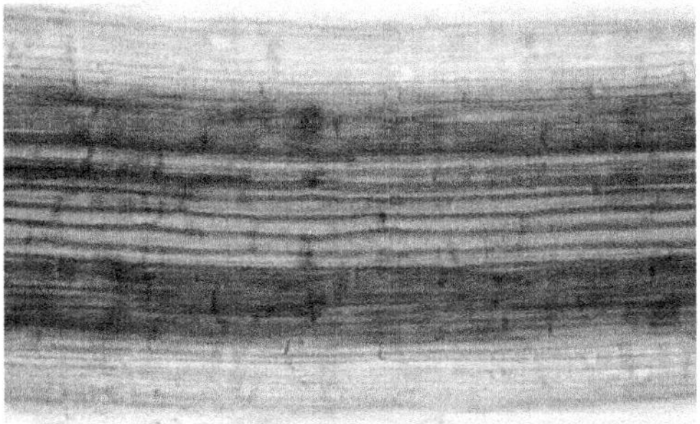

Intercellular air channels (arrows) in a whole mount of a living leek root (*Allium porrum*), shown for comparison with mycorrhizal development in the same host. These channels run continuously from the apex to the base of roots.

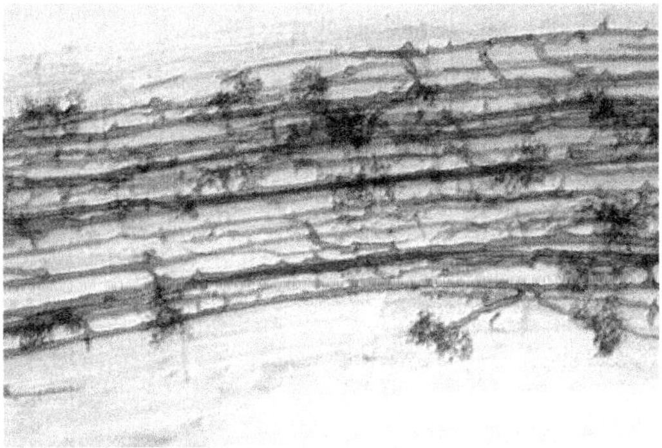

Longitudinal growth of hyphae of a VAM fungus (*Glomus versiforme*) along cortex air channels in a leek root. Note progressive development of arbuscules with increasing distance from the growing tips of hyphae.

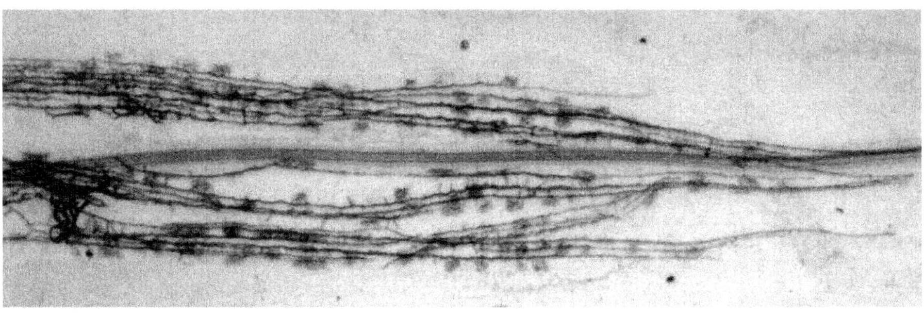

Arbuscules

Arbuscules are intricately branched haustoria that formed within a root cortex cell. They were named by Gallaud (1905), because they look like little trees. Arbuscules are formed by repeated dichotomous branching and reductions in hyphal width, starting from an initial trunk hypha (5-10 um in diameter) and ending in a proliferation of fine branch hyphae (< 1 um diameter).

Arbuscules start to form approximately 2 days after root penetration. They grow inside individual cells of the root cortex, but remain outside their cytoplasm, due to invagination of the plasma membrane. Arbuscules are considered the major site of exchange between the fungus and host. This assumption is based on the large surface area of the arbuscular interface, but has not been confirmed. Arbuscule formation follows hyphal growth, progressing outwards from the entry point. Arbuscules are short-loved and begin to collapse after a few days, but hyphae and vesicles can remain in roots for months or years.

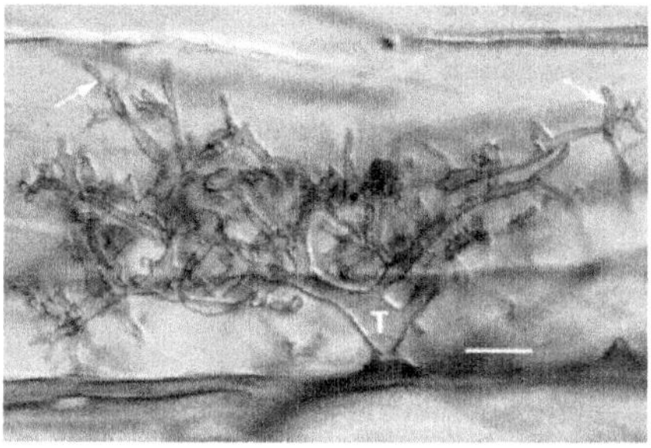

Developing arbuscule of *Glomus mosseae* in a root cell with fine branch hyphae (arrows). The trunk (T) of this arbuscule branched from an intercellular hyphae.

Nomarski Interference Contrast Image

Bar = 10 μm

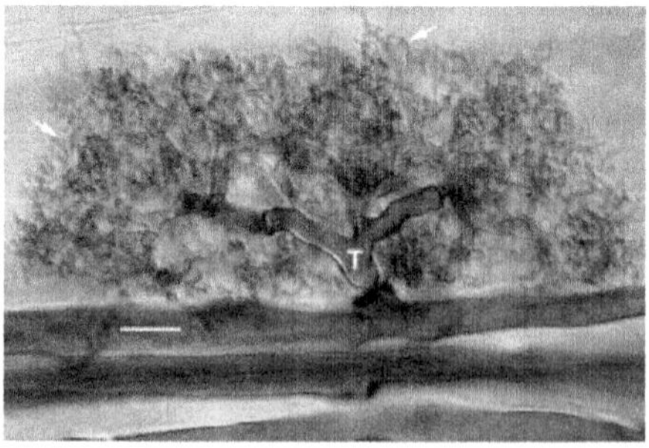

Mature arbuscule of *Glomus* showing trunk (T) and numerous fine branch hyphae (arrows).

Nomarski Interference Contrast Image

Bar = 10 μm

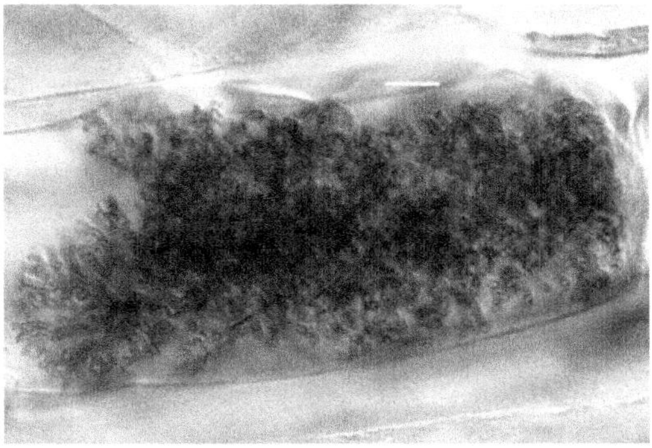

An arbuscule of *Glomus versiforme* in a root cortex cell with branch hyphae densely packed in the cortex cell of the host.

Nomarski Interference Contrast Image

Bar = 10 μm

Zoom in with roll-over

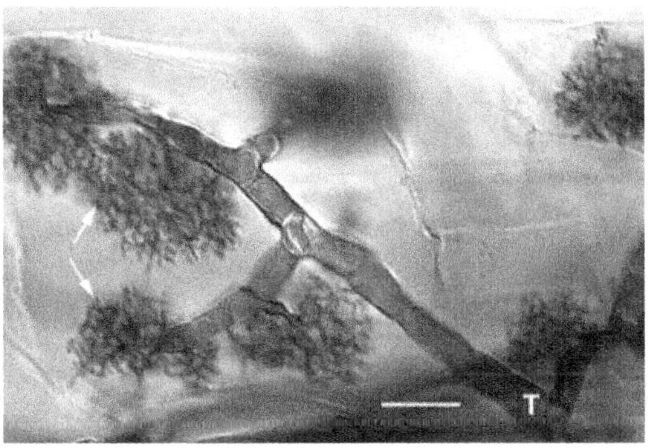

Arbuscule of *Gigaspora margarita* with an elongated trunk hypha (T) and tufts of fine branch hyphae (arrows). Note how this arbuscule differs from the *Glomus* arbuscules above.

Nomarski Interference Contrast Image

Bar = 10 μm

Zoom in with roll-over

Vesicles

Vesicles develop to accumulate storage products in many VAM associations. Vesicles are initiated soon after the first arbuscules, but continue to develop when the arbuscules senesce. Vesicles are hyphal swellings in the root cortex that contain lipids and cytoplasm. These may be inter- or intracellular. Vesicles can develop thick walls in older roots and may function as propagules. Some fungi produce vesicles which are similar in structure to the spores they produced in soil, but in other cases they are different.

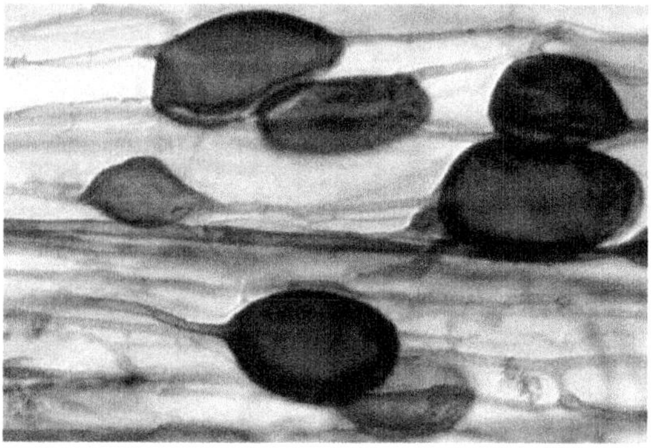

Vesicles (V) produced by a *Glomus* species in a leek root. This root also contains many intercellular hyphae.

Approx. 50 μm long

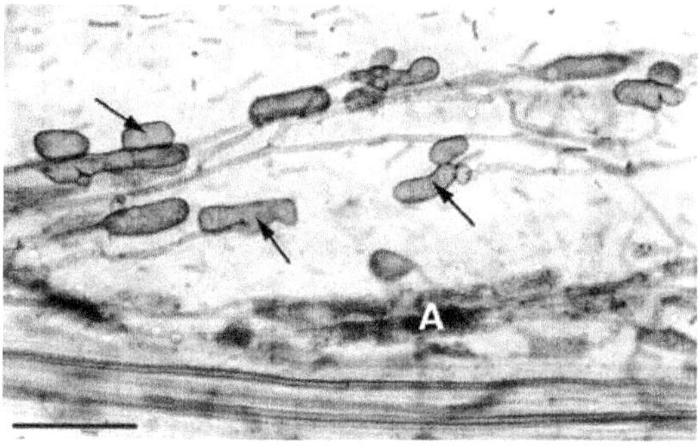

Lobed vesicles of an *Acaulospora* species in a clover root.

Arrows = vesicles, A = arbuscules

Bar = 100 μm

Structural Diversity

It is possible to identify individual Glomeromycotan fungi by recognising characteristic root morphology patterns in roots. Identification of endophytes within roots is important for culture quality control, because contaminating fungi can be identified months before they sporulate. This procedure can also be used to determine the mycorrhizal inoculum potential of different fungi by growing trap plants in a soil.

It is also possible to identify Glomeromycotan fungi by colonization patterns in roots, but it is difficult to separate species. Morphological features that are important include variations in vesicles (size, shape, wall thickness, wall layers, position and abundance), hyphal branching patterns, the diameter and structure of hyphae (especially near entry points), and the staining intensity of hyphae (dark or faint).

Mycorrhizas Produced by Glomus Species

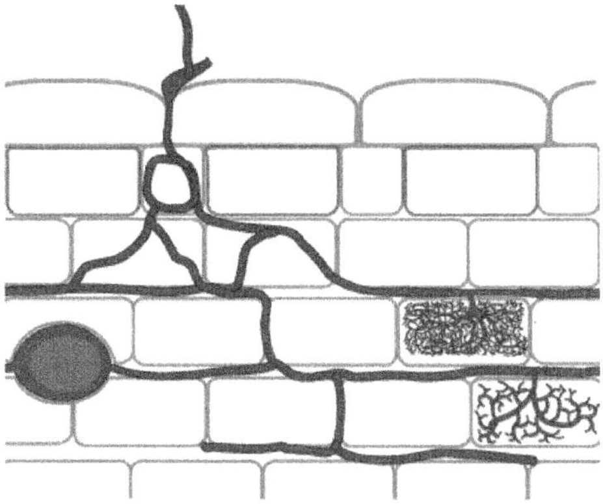

- Relatively straight hyphae ramify along the root cortex (if root anatomy permits), often producing "H" branches which result in simultaneous growth in 2 directions. Staining of these hyphae is usually relatively dark.
- *Arbuscules* can be dense and compact.
- Oval *vesicles,* which usually form between root cortex cells, are present in many cases. These vesicles persist in roots and often develop thickened and/or multi-layered walls.

Mycorrhizas Produced by Scutellospora and Gigaspora Species

- *In Scutellospora* VAM looping hyphae are often present near entry points. This genus has similar root colonisation patterns to *Acaulospora,* but hyphae in the cortex are generally thick-walled and stain darkly.

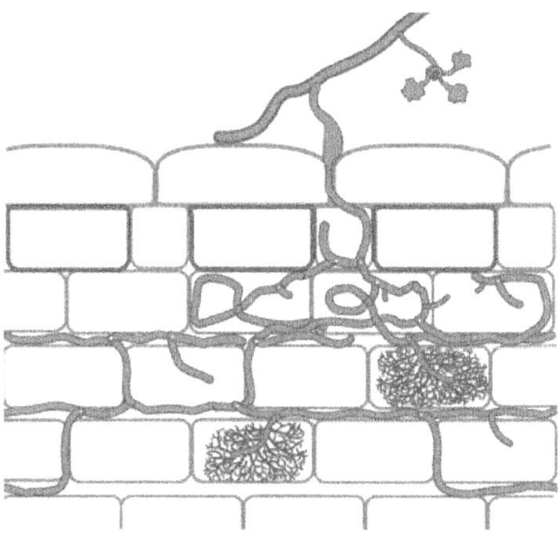

- Internal vesicles are not present.
- *Arbuscular trunk hyphae* normally are much longer and thicker than

those of*Glomus*. Arbuscules appear wispy due to relatively long curving branches.

- The root colonization pattern for *Gigaspora* is very similar to that for*Scutellospora,* with wide hyphae.

Mycorrhizas Produced by Acaulospora Species

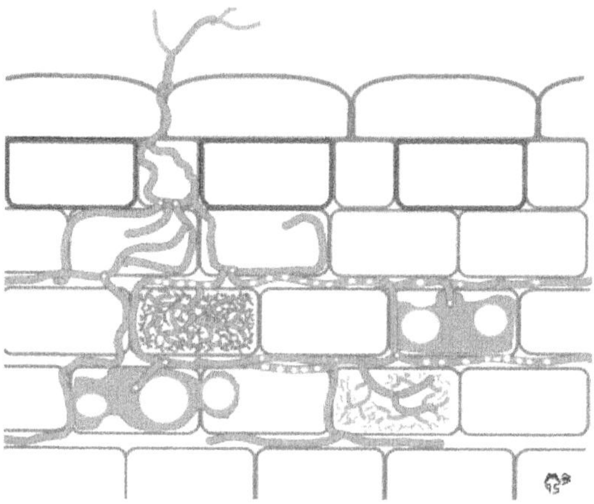

- Entry point hyphae have characteristic branching patterns. Hyphae in the outer cortex generally are more irregularly branched, looped or coiled than for*Glomus*. Colonies in roots are often relatively small.
- Internal hyphae are thin walled, often stain weakly and thus may be very hard to see, but may be visible due to rows of lipid droplets. External hyphae are usually also very hard to see.
- Intracellular oil-filled *vesicles,* that are initially rectangular, but often become irregularly lobed due to expansion into adjacent cells, are a characteristic feature. These have thin walls and do not persist in roots.

Mycorrhizas Produced by Fine Endophytes

- These unusual fungi have been called*Glomus tenue,* but are substantially different from other *Glomus* species.
- Fine endophytes can easily be distinguished by their very narrow hyphae (< 1 um in diameter) and net-like growth pattern in roots.
- Small hyphal swelling (< 5 um) can occur near entry points and may be analogous to vesicles.

Spores

Spores form as swellings on one or more subtending hypha in the soil or in roots. These structures contain lipids, cytoplasm and many nuclei. Spores usually develop thick walls with more than one layer and can function as propagules. Spores may be aggregated into groups called sporocarps. Sporocarps may contain specialised hyphae and can be encased in an outer layer (peridium). Spores apparently form when nutrients are remobilised from roots where associations are senescing. They function as storage structures, resting stages and propagules. Spores may form specialised *germination structures*, or hyphae may emerge through the subtending hyphae or grow directly through the wall.

Diversity of Spores in a Single Soil Sample

The following pictures show some of the fungi isolated into pot cultures started using spores or soil collected from Kakadu National Park in tropical northern Australia.

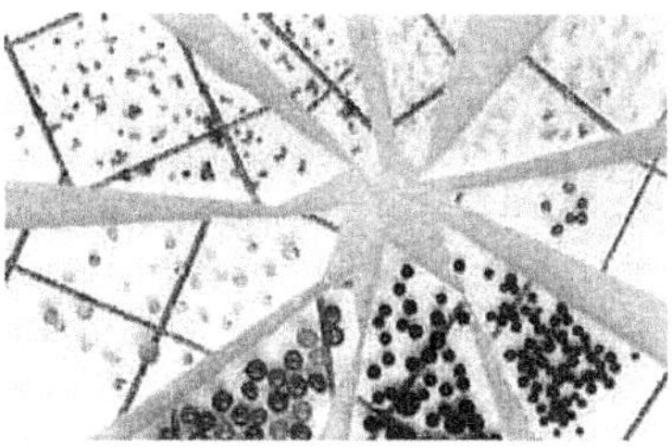

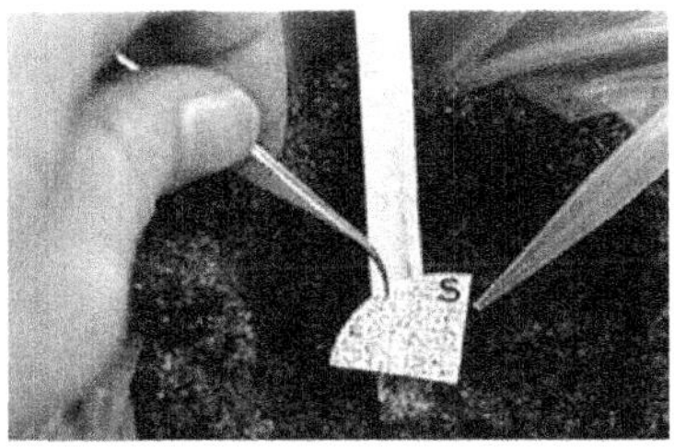

Left: Spores *separated from soil* and sorted into categories based on size and colour. **Right:** Spores (S) on a piece of filter paper used to start a "pot culture" using pasteurised soil in which a host plant was grown.

The images below show some of these *spores viewed with a compound microscope*. Inner wall layers have been revealed by crushing spores and by staining with Melzer's reagent (I,KI). All scale bars = 100 μm.

Spores of *Glomus*

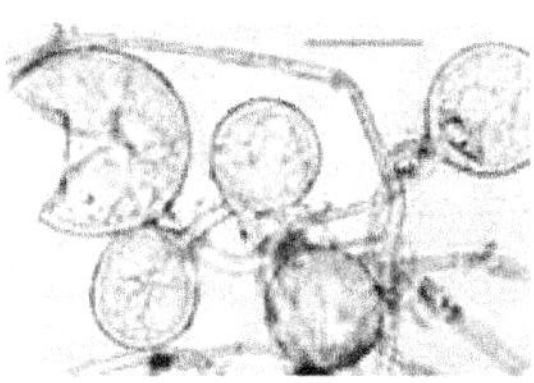

Relatively small white spores of a *Glomus species*.

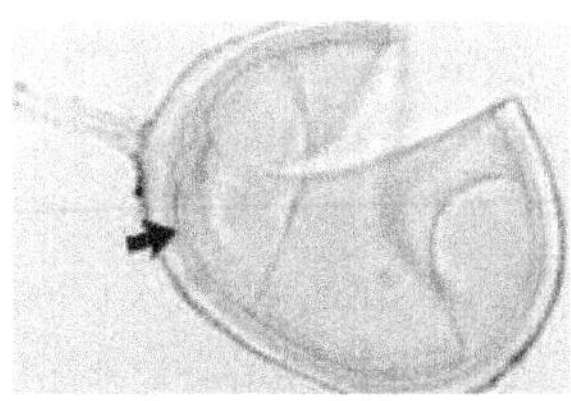

Spore of *Glomus clarum* which has a visible inner wall layer (arrow).

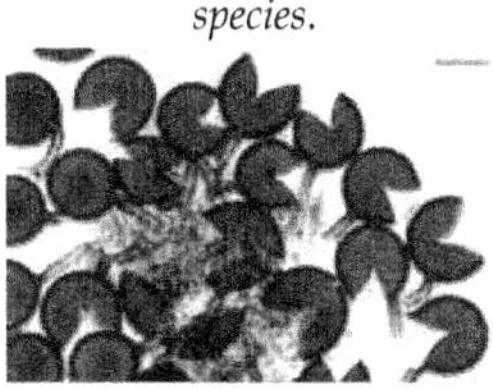

Sporocarp of *Glomus invermaium* typical of the dead spores often found in field-collected soil.

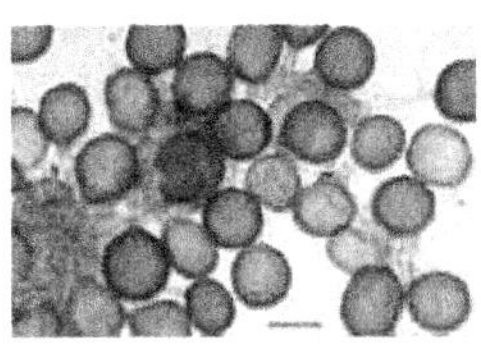

Living spores of *Glomus invermaium*from a pot culture.

Spores of *Acaulospora*

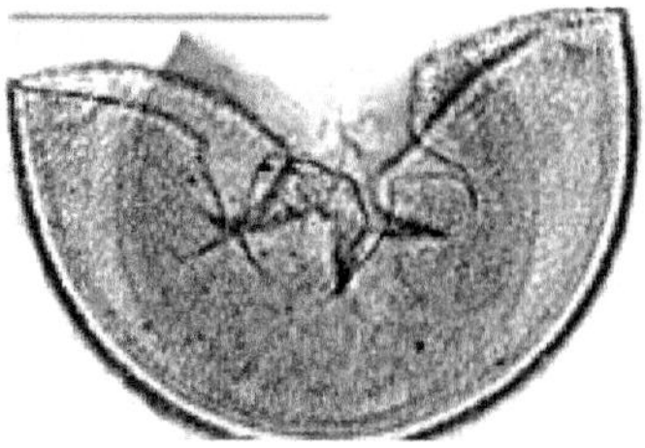

Acaulospora spore with deep pits in the outer wall and inner wall layers stained by Melzer's reagent.

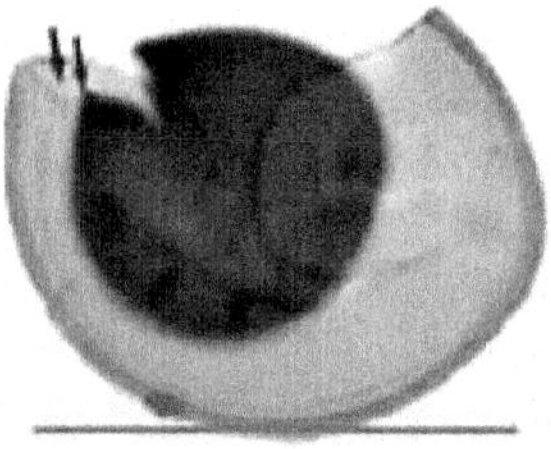

*Acaulospora*spore with several inner wall layers (arrows). One layer has stained darkly with Melzer's reagent.

Spores of *Scutellospora* Species

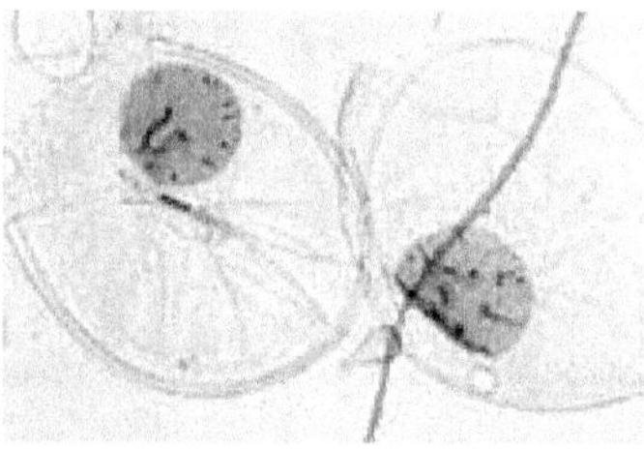

White *Scutellospora cerradensis* spores with prominent browngermination shields.

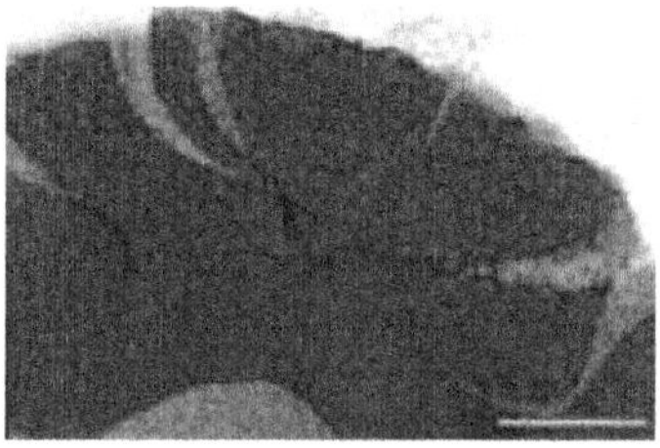

Large black spore with deep pits of*Scutellospora reticulata*.

Arbuscular Mycorrhizal Fungi

Populations of arbuscular mycorrhizal fungi in the Glomeromycota are thought to have occupied the same soil habitats for millions of years, slowly adapting to changes in site conditions. Many of these fungi have worldwide distribution patterns, but soil factors such as pH restrict the distribution of other taxa. Consequently, habitat information is as important as knowledge of the taxonomic identity of fungi, for comparing the results of experiments, or the selection of isolates for practical use.

The classification of the Glomeromycota is based on the structure of their soil-borne spores and DNA sequences. Accurate identification of these fungi often requires them to be isolated in cultures with host plants, to observe developmental stages, avoid the loss of diagnostic features and obtain healthy spores for DNA extraction. There now are many DNA sequences of Glomeromycotan fungi in databases such as *Genbank*. Information on Glomeromycotan systematics is available via the links below.

Arbuscular Mycorrhizal Fungi

- VAM fungi belong to the Glomeromycota.

- They are primitive fungi at the base of the tree for higher fungi (basidiomycetes).
- They associated with *first land plants* and appear to have evolved very slowly since then.
- They have no known sexual state.
- They produce microscopic structures, or relatively small sporocarps (truffle-like).
- Just over 200 species of these fungi are described, yet they are capable of forming mycorrhizal associations with the majority of plants.

Classification Scheme for Glomeromycotan Taxa

This classification follows Schüßler *et al.* (2001) with emendations of Oehl and Sieverding (2004), Walker and Schüßler (2004), Sieverding and Oehl (2006), and Spain *et al.* (2006).

<table>
<tr><th>Taxonomic Hierarchy</th><th>List of genera</th></tr>
<tr><td>
• Archaeosporales
o Ambisporaceae
o Archaeosporaceae
o Geosiphonaceae
• Diversisporales
o Acaulosporaceae
o Entrophosporaceae
o Diversisporaceae
o Gigasporaceae
o Pacisporaceae
• Glomerales
o Glomeraceae
• Paraglomerales
o Paraglomeraceae
</td><td>
<table>
<tr><td>Family</td><td>Genera</td></tr>
<tr><td>Acaulosporaceae</td><td>Acaulospora, Kuklospora</td></tr>
<tr><td>Ambisporaceae</td><td>Ambispora</td></tr>
<tr><td>Archaeosporaceae</td><td>Archaeospora, Intraspora</td></tr>
<tr><td>Diversisporaceae</td><td>Diversispora</td></tr>
<tr><td>Entrophosporaceae</td><td>Entrophospora</td></tr>
<tr><td>Geosiphonaceae</td><td>Geosiphon (not a mycorrhizal fungus)</td></tr>
<tr><td>Gigasporaceae</td><td>Gigaspora, Scutellospora</td></tr>
<tr><td>Glomeraceae</td><td>Glomus (Sclerocystis)</td></tr>
<tr><td>Pacisporaceae</td><td>Pacispora</td></tr>
<tr><td>Paraglomeraceae</td><td>Paraglomus</td></tr>
</table>
</td></tr>
</table>

MYCORRHIZAS

Symbiosis and Mutualism

The terms symbiotic and mutualistic have been used interchangeably to describe mycorrhizal associations and parasitic fungi have also been called symbiotic, but many scientists now only call beneficial associations symbiotic. Symbiosis is defined broadly as "two or more organisms living together" and in most cases both partners benefit. There are many types of symbiosis evolving different combinations of plants, fungi, microbes and animals. Only plant-fungus associations are considered in detail here, but several others are illustrated below.

Fungal symbioses have been defined as "all associations where fungi come into contact with living host from which they obtain, in a variety of ways, either metabolites or nutrients". However, this definition excludes mycorrhizal associations of myco-heterotrophic plants, where plants are nutritionally dependant on fungi. Only the broadest definition of symbiosis - "living together of two or more organisms", applies universally to mycorrhizal associations.

Mutualistic associations are a subset of symbioses where two or more different living organisms receive mutual benefits, as illustrated in the diagram below.

Examples of Symbiotic Associations Between Animals and Algae

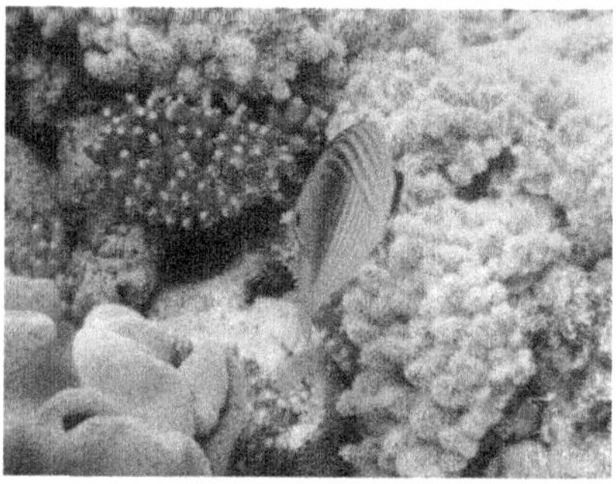

Diverse coral reef community on the Great Barrier Reef (Heron Island). Corals are symbiotic associations between an animal (coral polyp) and photosynthetic algae (zooxanthellae) inside the polyp.

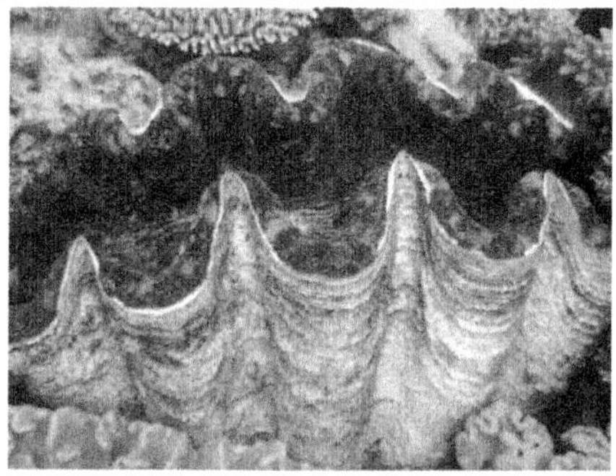

Giant Clam in the Great Barrier Reef. Note algae in mantle (rollover image shows greater detail).

Other Examples of Symbiotic Associations

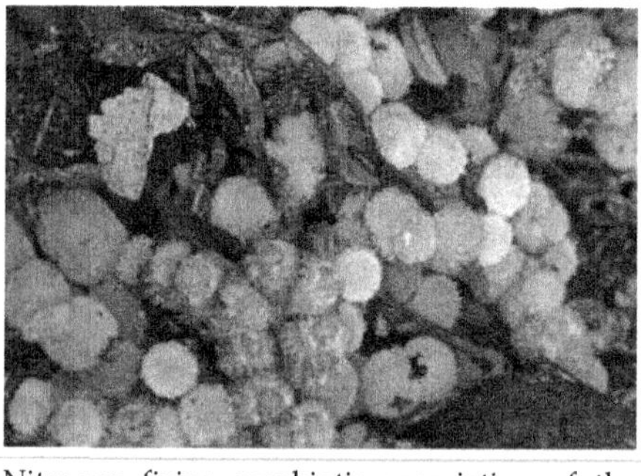

Nitrogen fixing symbiotic association of the cycad *Macrozamia riedlei*. These coralloid roots at the soil surface which contain cyanobacteria that fix nitrogen. Other nitrogen fixing associations include bacteria in nodules of peas (the Fabaceae) and actinomycete nodules in several other plant families.

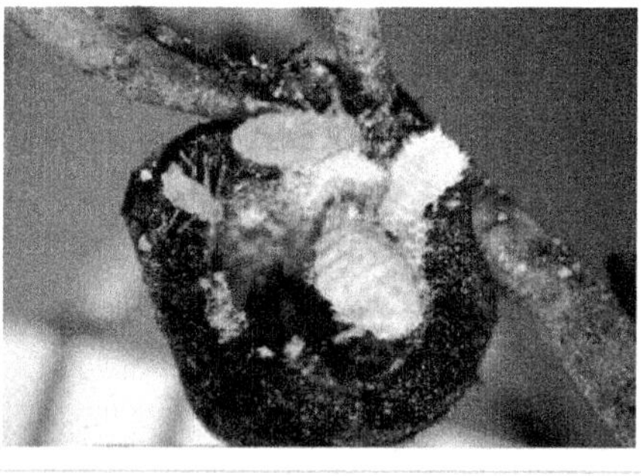

The ash bolete (*Gyrodon merulioides*), which occurs under ash trees in North America (*Fraxinus americana*), has a symbiotic association with aphids (seen in cross section).

Plant-Fungal Symbioses

Mycorrhizas are the most important type of symbiotic plant-fungus associations, but there are a wide diversity of other associations between plants and fungi, as illustrated in the diagram below. The relationship between mycorrhizas and other types of plant-fungus associations, such as parasitic or endophytic associations, are also shown below.

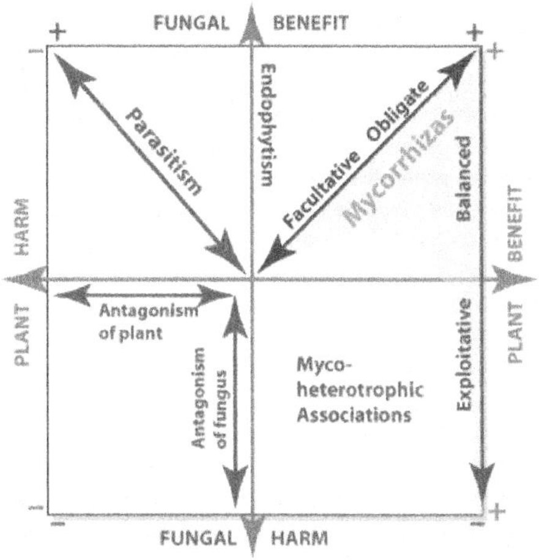

This diagram compares types of plant-fungus interactions and each is explained separately below.

Mutualistic associations occupy the mutual benefit (+ +) quadrant in diagrams contrasting the relative benefits (+) or harm (-) to two interacting organisms. This is a phase plane diagram that describes biological interactions according to a cost-benefit model, where mutualism is an isocline showing both partners are more successful together than they are alone.

Explanations

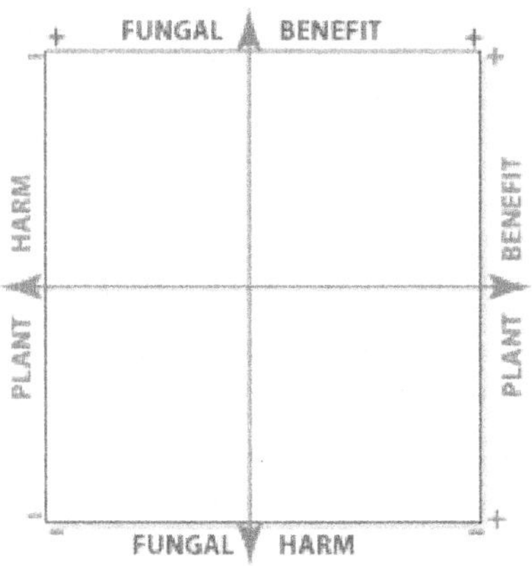

The vertical axis is a continuum of fungal harm or benefits.
The horizontal axis is a plant harm-benefit continuum.

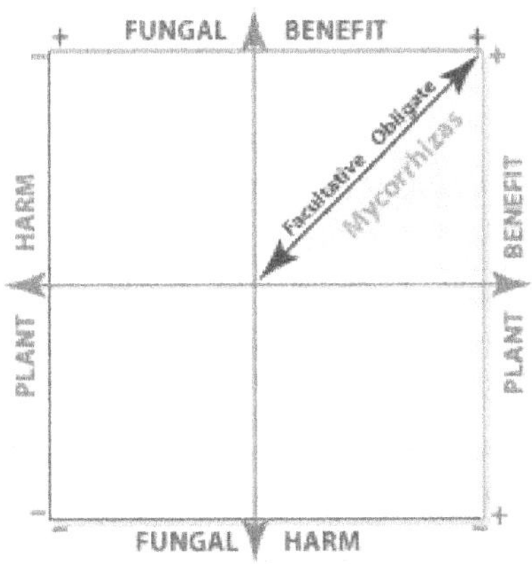

Fungus benefits are linked to plant benefits in balanced mycorrhizas.
Obligate associations require greater investment from both partners than facultative mycorrhizas.

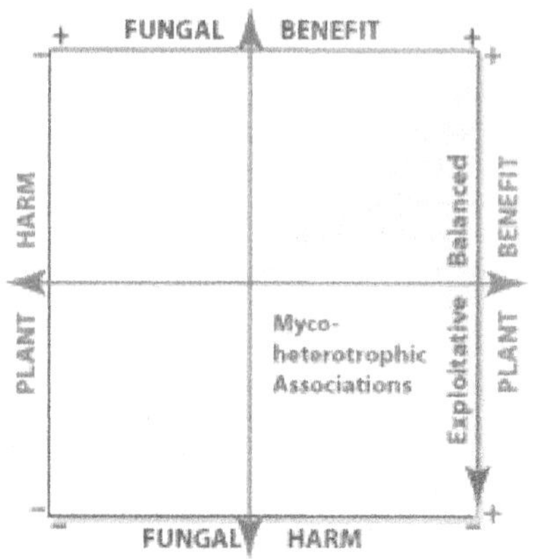

Exploitative mycorrhizas (myco-heterotrophs) are parallel to the vertical axis - plant benefit occurs at expense of fungi.

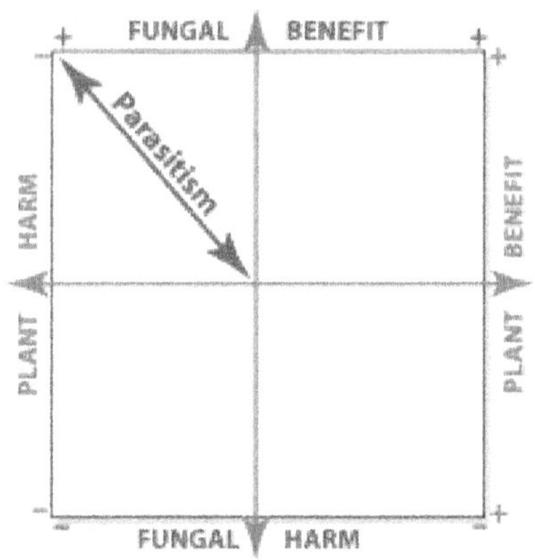

Parasitic plant-fungal associations are those where fungal benefits are linked to plant harm.

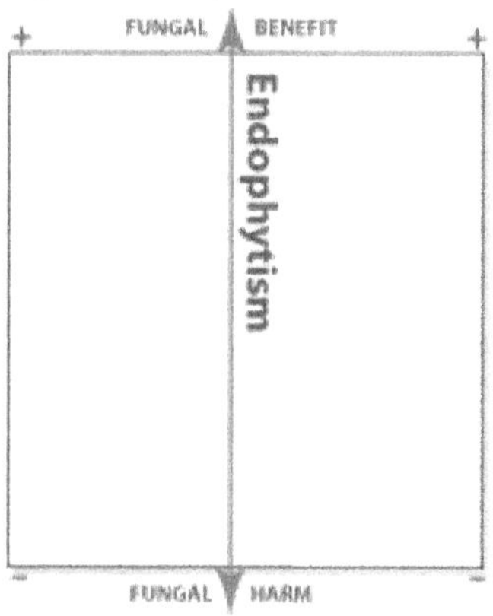

Endophytic plant-fungus associations (no plant harm or benefit).

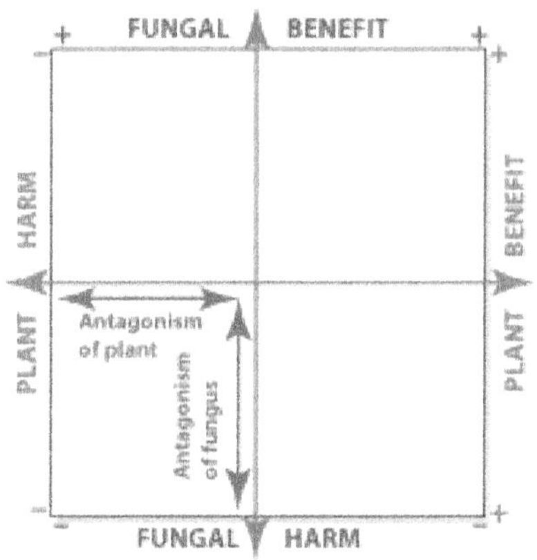

Other categories of plant-fungus interactions include antagonism of fungi by plants or plants by fungi (causing harm to another organism without gaining direct benefits).

Definition of Mycorrhizas

The name mycorrhizas, which literally means fungus-root, was invented by Frank (1885) for non-pathogenic symbiotic associations between roots and fungi. A revised definition that includes non-mutualistic mycorrhizas and excludes other plant-fungus associations is provided below.

Mycorrhizas are symbiotic associations essential for one or both partners, between a fungus (specialised for life in soils and plants) and a root (or other substrate-contacting organ) of a living plant, that is primarily responsible for nutrient transfer. Mycorrhizas occur in a specialised plant organ where intimate contact results from synchronised plant-fungus development.

Key Characteristics of Mycorrhizas		
Fungus	Symbiosis	Plant
Soil inhabitant	Intimate contact at interface for nutrient transfer	Control of association
Plant inhabitant	Essential for one or both partners	Specialised organ
Specialised hyphae	Synchronised development	Root or stem

Notes

1. The structure and development of mycorrhizal fungus hyphae is substantially altered in the presence of roots of host plants. These root-borne hyphae are distinct from hyphae which are specialised for growth in soil.
2. All mycorrhizas have intimate contact between hyphae and plant cells in an interface where nutrient exchange occurs.
3. The primary role of mycorrhizas is the transfer of mineral nutrients from fungus to plant. In most cases there also is substantial transfer of metabolites from the plant to fungus.
4. Mycorrhizas require synchronised plant-fungus development, since hyphae only colonise young roots (except orchid mycorrhizas and exploitative VAM).
5. Plants control the intensity of mycorrhizas by root growth, digestion of old interface hyphae in plant cells (AM, orchid), or altered root system form (ECM).
6. Roots evolved as habitats for mycorrhizal fungi. Mycorrhizas normally occur in roots, but can be hosted in stems in some cases (*e.g.* some orchids).

Categories of Mycorrhizal Associations

Consistent definitions of mycorrhizal associations are required for accurate communication of data. The flowchart below groups similar types of mycorrhizas together using categories regulated by the host and morphotypes caused by different fungi. Categories and subcategories are defined in the subsequent table.

Association	Categories	Morphotypes

Arbuscular Mycorrhizal Associations	VAM
	– Coiling
	  – Beaded
	  – Inner cortex
	  – Exploitative
	  – Others
	– Linear
		VAM
		– Glomus Types
		– Scutellospora & Gigaspora
		– Acaulospora types
		– Fine endopytes
		– Medium endopytes
		– Others

Ectomycorrhizal Associations	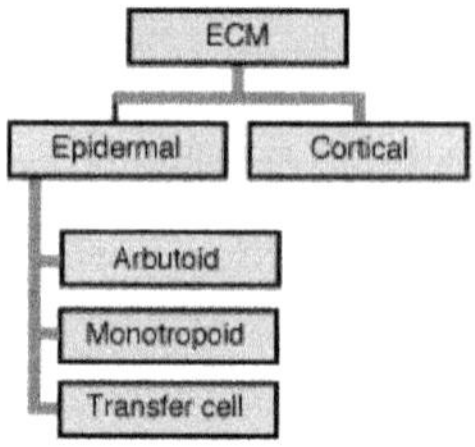	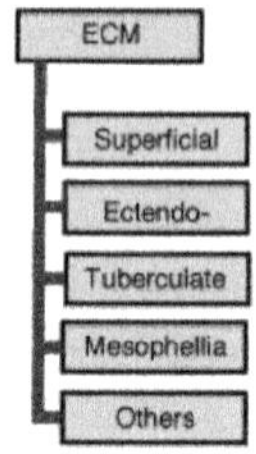

Hierarchical Classification Scheme for Mycorrhizal Associations

No.	Category	Definition	Hosts	Fungi
1	*Arbuscular mycorrhizas*	Associations formed by Glomeromycotan fungi in plants that usually have arbuscules and often have vesicles (also known as vesicular-arbuscular mycorrhizas, AM, VAM).	Plants	Glomeromycota
1.1	*Linear VAM*	Associations that spread predominantly by longitudinal intercellular hyphae in roots (formerly known as Arum series VAM).	Plants	As above
1.2	*Coiling VAM*	Associations that spread predominantly by intracellular hyphal coils within roots (formerly known as Paris series VAM).	Plants	As above
1.2.1	*Beaded VAM*	Coiling VAM in roots, where interrupted root growth results in short segments divided by constrictions.	Woody plants	As above
1.2.2	*Inner cortex VAM*	Coiling VAM with arbuscules in one layer of cells of the root inner cortex.	Plants	As above
1.2.3	*Exploitative VAM*	Coiling VAM of myco-heterotrophic plants, usually without arbuscules.	Achlorophyllous plants	As above
2	*Ecto-mycorrhiza*(ECM)	Associations with a hyphal mantle enclosing short lateral roots and a Hartig net of labyrinthine hyphae that penetrate between root cells.	Hosts	Higher fungi (asco-, basidio- and zygo- mycetes)
2.1	*Cortical*	Hartig net hyphae penetrate between multiple cortex cell layers of short roots	Most are gymnosperm trees	As above
2.2	*Epidermal*	Hartig net fungal hyphae are confined to epidermal cells of short roots	Angiosperms (most are trees)	As above
2.2.1	Transfer cell	Epidermal Hartig net with transfer cells (plant cells with wall ingrowths)	*Pisonia* (Nyctaginaceae).	*Tomentella* spp. in*Pisonia*
2.2.2	*Monotropoid*	Exploitative epidermal ECM of myco-heterotrophic plants in the Ericaceae where individual hyphae penetrate epidermal cells.	Ericaceae (Monotropa, Pterospora, Sarcodes)	Basidiomycetes

2.2.3	*Arbutoid*	ECM of autotrophic plants in in the Ericaceae where multiple hyphae penetrate epidermal Hartig net cells.	Ericaceae (part only)	Basidiomycetes
3	*Orchid*	Associations where coils of hyphae (pelotons) penetrate within cells in the plant family Orchidaceae.	Hosts	Most are basidiomycetes in*Rhizoctonia* alliance.
3.1	*Orchid Root*	Associations within a root cortex.	Orchidaceae	As above
3.2	*Orchid Stem*	Associations within a stem or rhizome.	Orchidaceae	As above
3.3	*Exploitative Orchids*	Associations of myco-heterotrophic orchids.	Orchidaceae (fully or partially achlorophyllous)	Orchid, ectomycorrhizal, or saprophytic fungi
4	*Ericoid*	Coils of hyphae within very thin roots (hair roots) of the Ericaceae.	Ericaceae (most genera)	Most are Ascomycetes
5	Sub- epidermal	Hyphae in cavities under epidermal cells, only known from an Australian monocot genus.	*Thysanotus* spp. (Laxmaniaceae)	Unknown

Morphology of Mycorrhizal Associations

The mycorrhizal association types defined in the table above are briefly described and illustrated below. More information on *arbuscular mycorrhizas* and *ectomycorrhizas* are provided in dedicated Sections of this site.

Arbuscular Mycorrhizas

Arbuscular mycorrhizas (Vesicular-Arbuscular Mycorrhizas, VAM or AM) are associations where Glomeromycete fungi produce arbuscules, hyphae, and vesicles within root cortex cells. These associations are defined by the presence of arbuscules. Fungi in roots spread by linear hyphae or coiled hyphae.

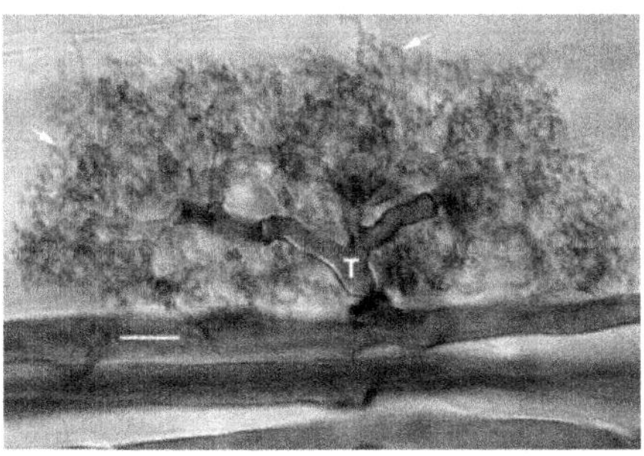

Arbuscule of a *Glomus* species in a root cortex cell. *More information.*

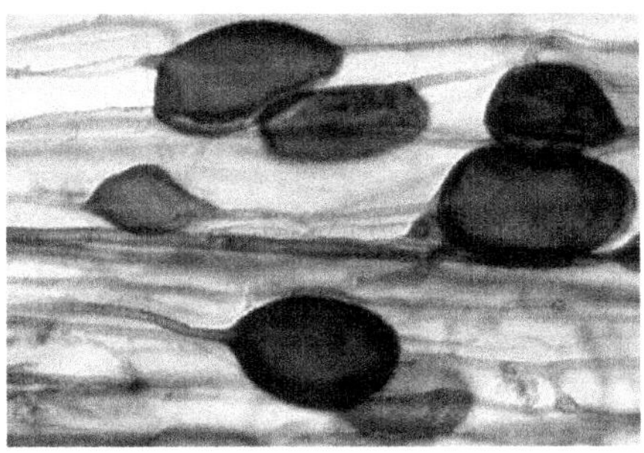

Vesicles of a *Glomus* species in a root cortex. *More information.*

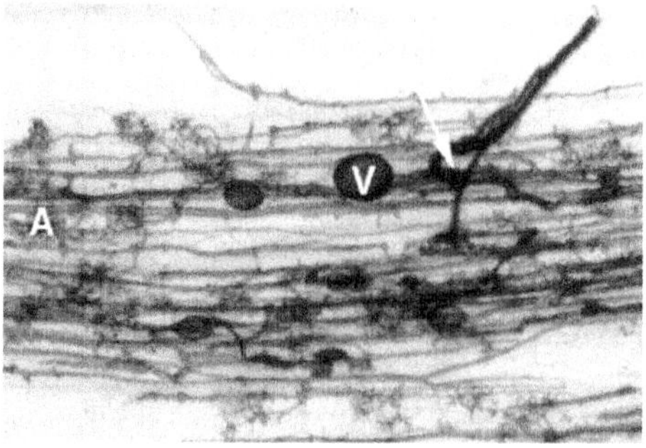

Fig. : Linear association in root of *Allium porrum*-with arbuscules (A) and vesicles (B) on longitudinal hyphae near entry point (arrow).

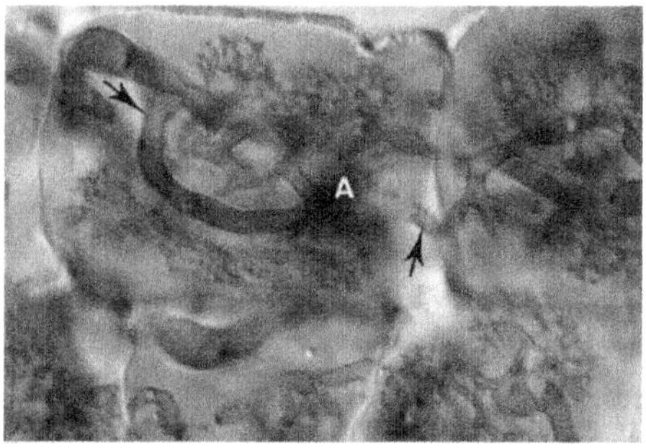

Fig. : Coiling association with arbuscules (A) on coiling hyphae (arrow) in a root of *Erythronium americanum*. *More information*.

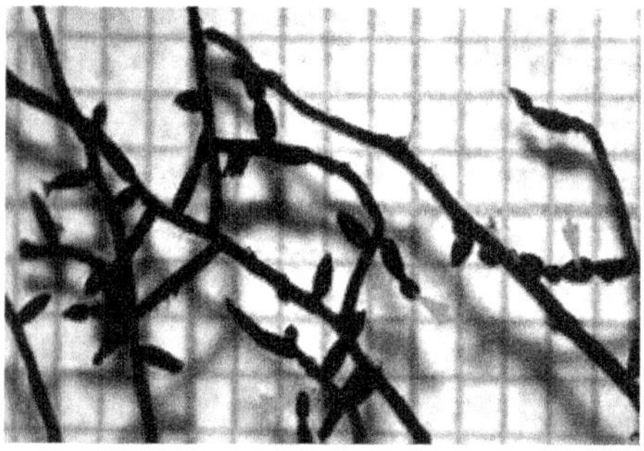

Fig. : Beaded roots (arrows) of Sugar Maple (*Acer saccharum*) - a VAM host. *More information*

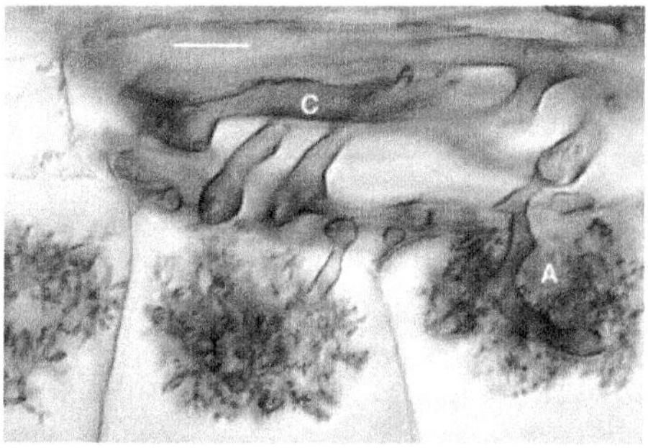

Fig. : Coiling association with arbuscules (A) only in the inner cortex layer of *Asarum canadense* roots.

Myco-heterotrophic "Arbuscular" Mycorrhizas

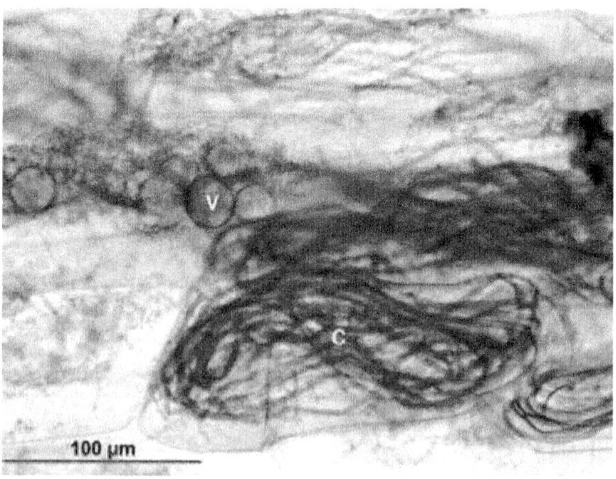

Coils of hyphae in the rhizome of*Psilotum nudum* a *whisk fern*. This is a type of VAM association without arbuscules from a young sporophyte with green shoots that is not fully myco-heterotrophic.
Cleared and stained rhizome
C = coil, V = vesicle
Roll-over animation shows a single coil.

Ectomycorrhizas

Ectomycorrhizas (ECM) are associations where fungi form a mantle around roots and a Hartig net between root cells. These associations are defined by Hartig net hyphae which grow around cells in the epidermis or cortex of short swollen lateral roots. *ECM associations*. The former category of ECM is a *morphotype* (defined by fungi not hosts). Characteristics of this ECM morphotype are summarised by Yu *et al.* (2001).

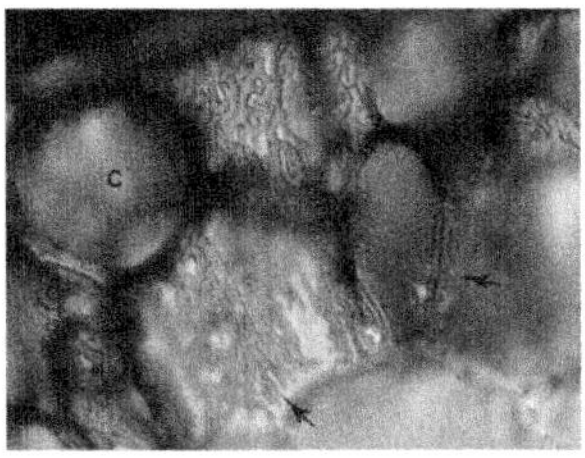

Cortical Hartig net of *Pinus* ECM root.

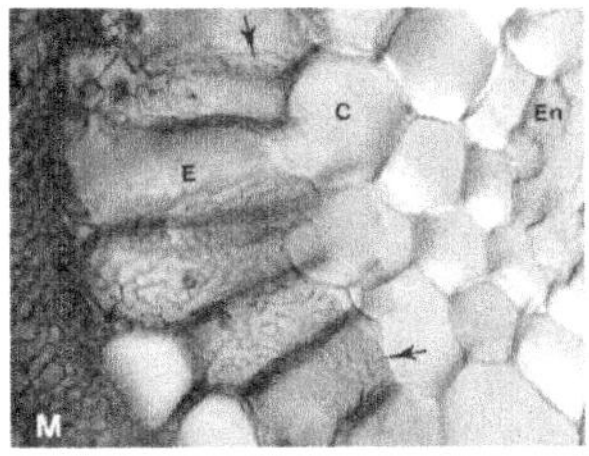

Epidermal Hartig net of *Populus* ECM.

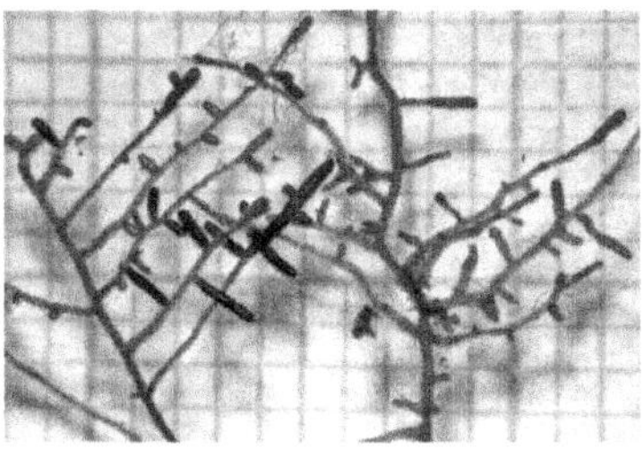

Betula root system showing thicker branched or unbranched ECM roots borne on thinner lateral roots.

Monotropoid

Monotropoid mycorrhizas are ECM associations of a few genera of myco-heterotrophic plants in the Ericaceae. These associations are characterised by limited hyphal penetration into epidermal cells. Information on structure of associations and the identity of mycorrhizal fungi in *Monotropa, Pterospora, Sarcodes, etc.*.

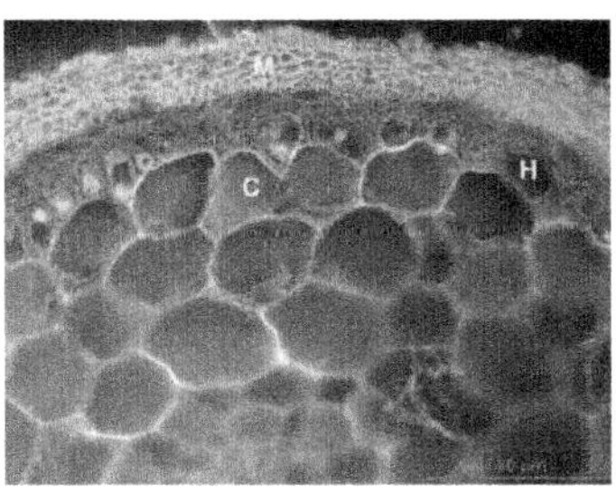

Monotropa root with epidermal Hartig net (H) and mantle (M) in a cross-section viewed with UV light. Roll-over - hypha projecting into an epidermal cell (arrow) in stained root section.

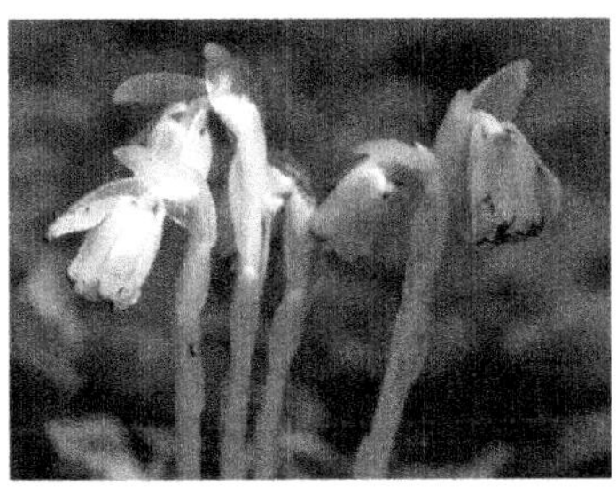

Monotropa uniflora (Canada) is a myco-heterotrophic plant lacking chlorophyll that is entirely dependant on ECM fungi linked to nearby trees.

Arbutoid Mycorrhizas

Arbutoid mycorrhizal associations are variants of ECM found in certain plants in the Ericaceae characterised by hyphal coils in epidermal cells. These mycorrhizal roots are described by Largent *et al.* (1980), Molina & Trappe (1982) and Massicotte *et al.*. *Gaultheria* and *Kalmia* have ericoid mycorrhizas as well as arbutoid associations.

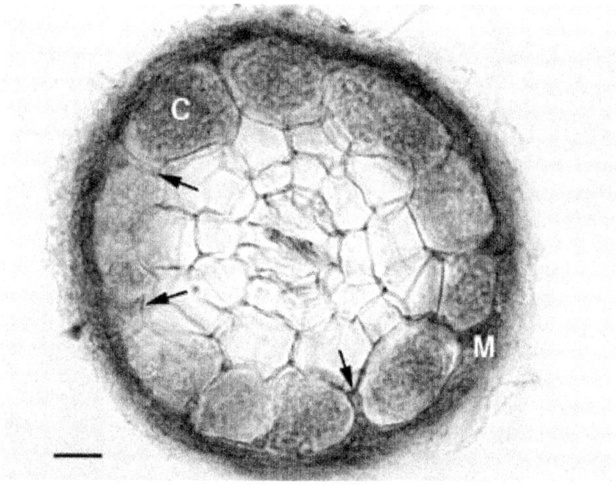

Arbutus unedo root with Hartig net (arrows), coils (C) and mantle (M) of stained or unstained hyphae. Highly magnified views of sectioned and cleared roots (Bar = 20 μm)

Orchid Mycorrhizas

Orchid mycorrhizas consist of coils of hyphae within roots or stems of orchidaceous plants. Details of Orchid mycorrhizal associations are not provided here, but *Australian Orchids* found to have mycorrhizas are listed.

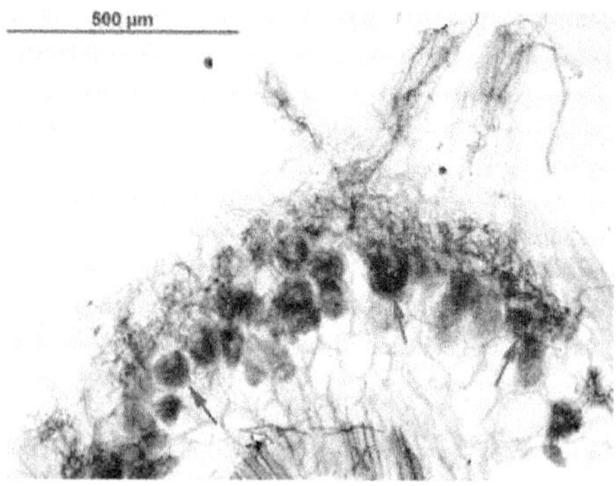

Orchid mycorrhizas with hyphae in trichomes and hyphal coils in stem of *Pterostylis vittata* Highly magnified views of cleared and stained hand section.

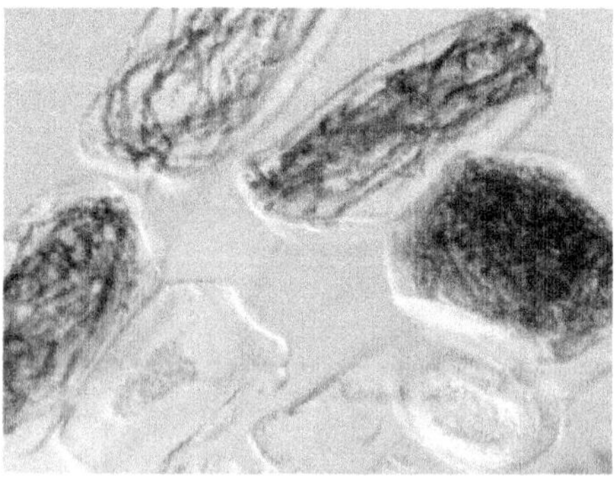

Hyphal coils from orchid mycorrhizas in*Epipactis helleborine* root.

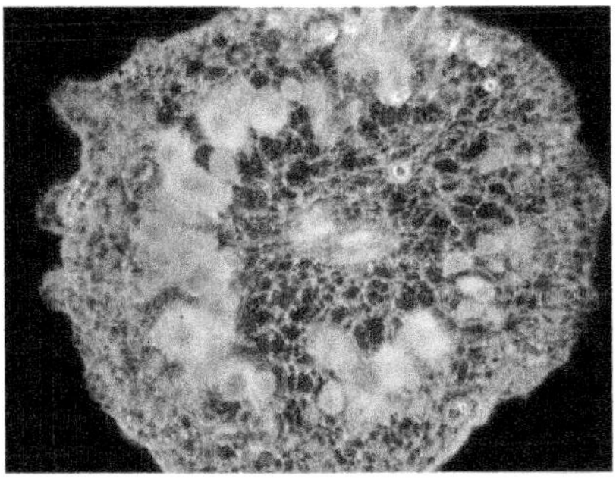

Hyphal coils (pelotons) of an exploitative mycorrhizal association in a myco-heterotrophic orchid. Coils are white of brown fuzzy balls. This rhizome of the Western Underground Orchid (*Rhizanthella gardneri*) is 5 mm wide.

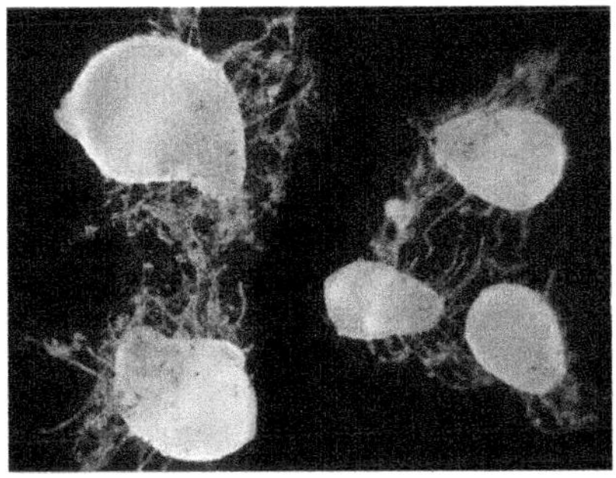

Seedlings of *Rhizanthella gardneri* germinated by a mycorrhizal fungus linked to ECM roots of a shrub (*Melaleuca* sp.). These subterranean seedlings are 2-10 mm long with a zone of brown hyphal coils clearly visible at their base.

Ericoid Mycorrhizas

Ericoid mycorrhizas have hyphal coils in outer cells of the narrow "hair roots" of plants in the family Ericaceae. These associations are not described in detail here, but *Australian plants* with these mycorrhizas are listed.

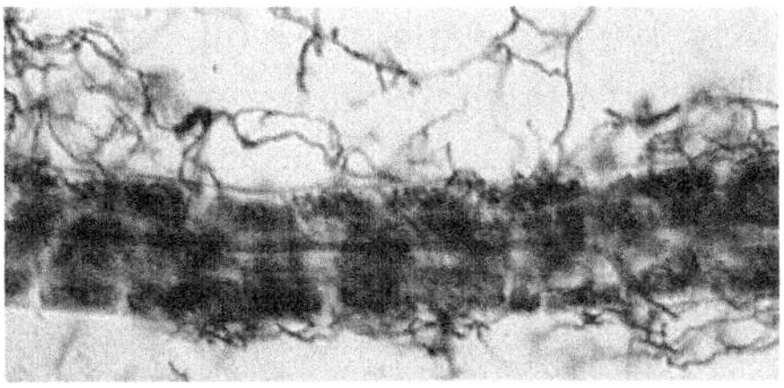

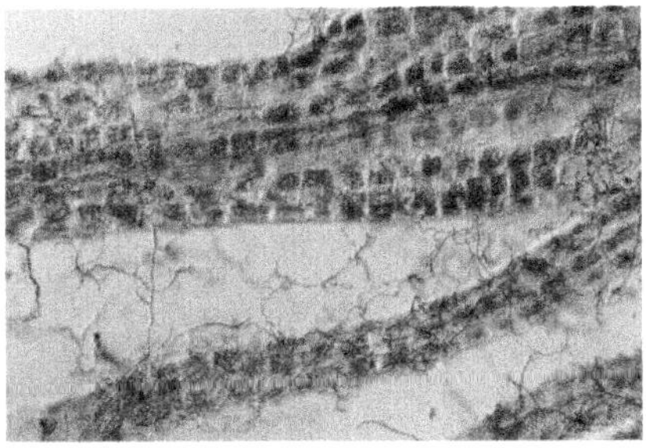

Ericoid mycorrhizas with hyphal coils in hair roots of *Leucopogon verticillatus*
Highly magnified views of cleared and stained roots.

Subepidermal Association of Thysanotus

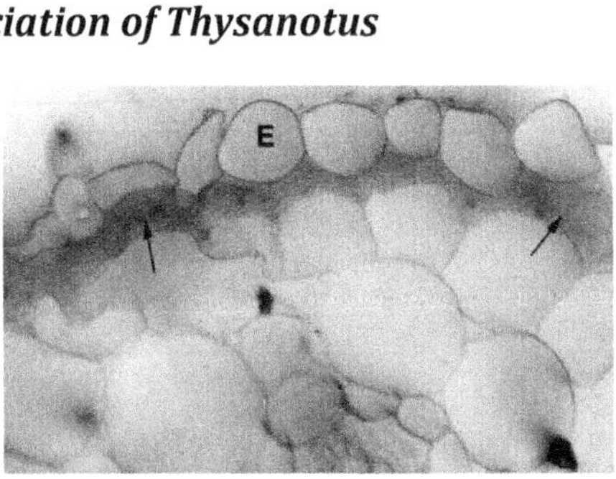

The Australian lilies in the genus *Thysanotus*(Laxmaniaceae) have *unique mycorrhizas*where fungus hyphae grow in a cavity under epidermal cells. Other members of this family have VAM or have NM roots.
Highly magnified views of a sectioned and stained root of *Thysanotus* sp.
Arrows point to hyphae under the epidermis (E)

Host Plants

Field surveys have found that plants with mycorrhizal associations predominate in most natural ecosystems, as summarised the in table below.

Association	Occurrence
Vesicular Arbuscular Mycorrhizal (VAM) Plants	• Plants with VAM are common in most habitats • It is easier to say where they are not found
Ectomycorrhizal (ECM) Plants	• Trees with ECM are dominant in coniferous forests, especially in cold boreal or alpine regions • ECM trees and shrubs common in many broad-leaved forests in temperate or mediterranean regions • ECM trees also occur in some tropical or subtropical savanna or rain forests habitats
Nonmycorrhizal (NM) Plants	• NM plants are most common in disturbed habitats, or sites with extreme environmental or soil conditions • NM plants appear to be more common in Australia than in other continents.

Mycorrhizal Fungi

Members of the fungus kingdom obtain nutrition from many sources, including decomposition of organic substrates, predation and parasitism, and involvement in mutualistic associations. Mycorrhizal fungi are a major component of the soil microflora in many ecosystems, but usually have limited saprophytic abilities. They are considered to have many *important roles* in natural and managed ecosystems. These fungi are introduced in the table below.

Mycorrhiza	Phylum	Families	Anamorphs	Teliomorphs
Arbuscular	Glomeromycota	Glomaceae, Acaulosporacae, *etc.*	*Glomus, Scutellospora, Acaulospora, etc.*	none
Ecto- mycorrhiza (ECM)	Basidiomycota, Ascomycota, Zygomycota	Many families including Amanitaceae, Cortinariaceae, Boletaceae, *etc.*	Most ECM fungi lack anamorphs, but*Cenococcum* is one example	Many genera including*Amanita, Cortinarius, Russula, etc.*
Monotropoid ECM	Basidiomycota	Russulaceae, *etc.*	NA	*Russula, Tricholoma, Rhizopogon, etc.*
Orchid: not myco-heterotrophic	Basidiomycota (Ascomycete)	*Ceratobasidiaceae, Tulasnellacea, Sebacinaceae* (related to Chanterellaceae?) (also many others are reported)	Sterile hyphae:*Rhizoctonia* alliance:*Epulorhiza, Ceratorhiza, Tulasnella, etc.* as well as *Fusarium, etc.*	*Ceratobasidium, Thanatophorus, Sebacina, etc.*
Orchid: myco-heterotrophic	Basidiomycota	Russulaceae, Telephoraceae, *etc.*	NA	unrelated clades of ECM, orchid and saprophytic fungi
Ericoid	Ascomycota (Basidiomycota)	Helotiaceae (Sebacinaceae)	NA	*Hymenoscyphus, Rhizoscyphus, (Sebacina)*

EVOLUTION OF MYCORRHIZAS

Mycorrhizal evolution is hypothesised to have progressed from endophytic to balanced symbiotic associations where both partners are interdependent due to the exchange of limiting resources. Fossil Glomeromycotan associations and spores, as well as DNA-based phylogenies suggest mycorrhizas evolved with thc first land plants. This evidence is used to identify mycorrhizal plant lineages and estimate their age in a summary provided below. Information on mycorrhizas of primitive plants has been updated with new references and more recent phylogenetic schemes for these plants.

Topic	Major Sources
Evolution of mycorrhizas	Brundrett 2002
Lower plant mycorrhizas	Many sources including: Maeda 1954, Read *et al.* 2000, Kottke & Nebel 2005 and Ligrone *et al.* 2007
Fossil mycorrhizas	Pirozynski & Dalpé 1989, Tayor & Krings 2005

Examples of Primitive Conifers from Western Australia

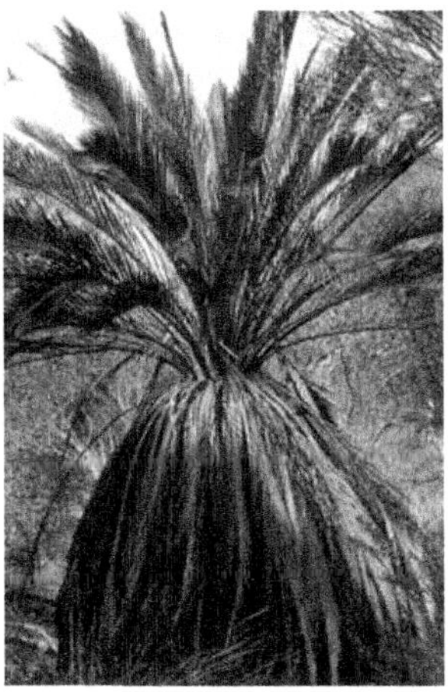

Macrozamia dyeri
Zamia
Zamiaceae

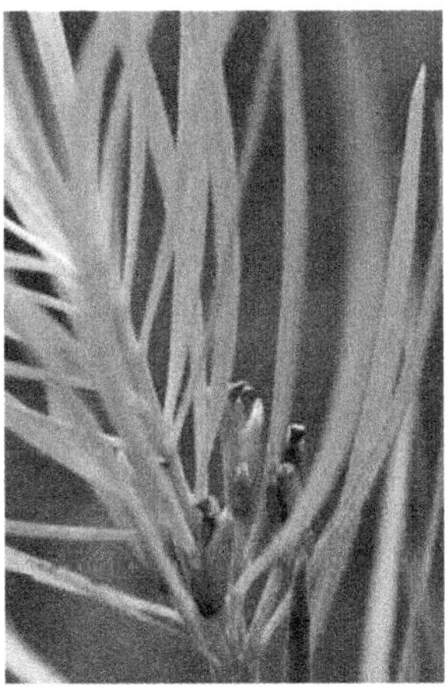

Podocarpus drouynianus
Emu Plum
female plant

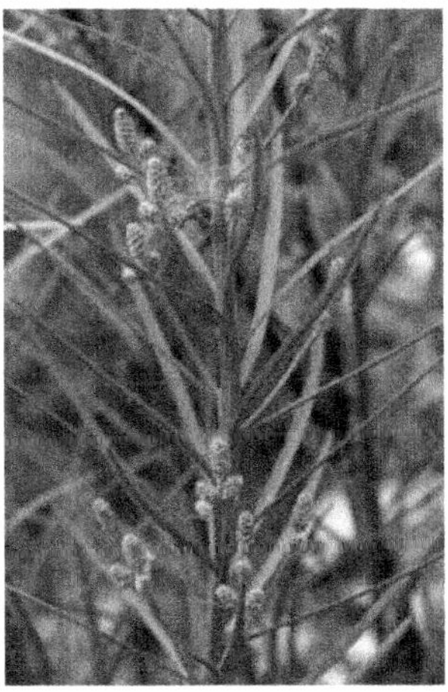

Podocarpus drouynianus
Podocarpaceae
male plant

Fossil History of Mycorrhizas

Fossils of early land plants contain fungal associations which are similar to arbuscular mycorrhizas in extant plants, as shown below, and ancient fossil spores of Glomeromycotan fungi are also known. This story is described in detail in recent reviews, so only summary tables of land plant and fungus history are presented here. Photos of 400 Myr old fossil mycorrhizas are reproduced below courtesy of Professor Thomas Taylor at the *University of Kansas*.

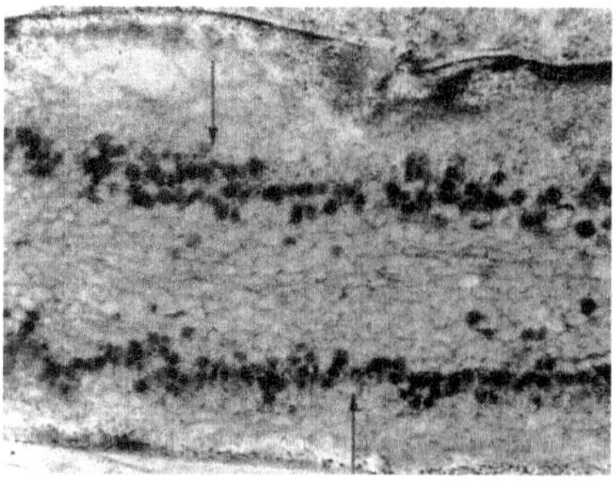

400 MY old fossil mycorrhiza-like association inAglaophyton major rhizome.

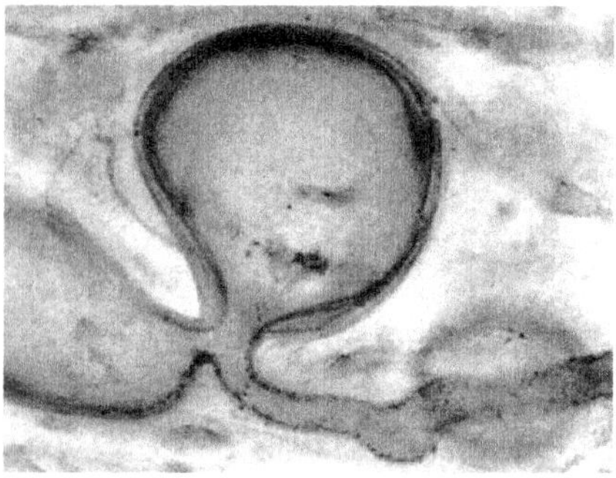

Vesicle-like structures in *Aglaophyton major*rhizome.

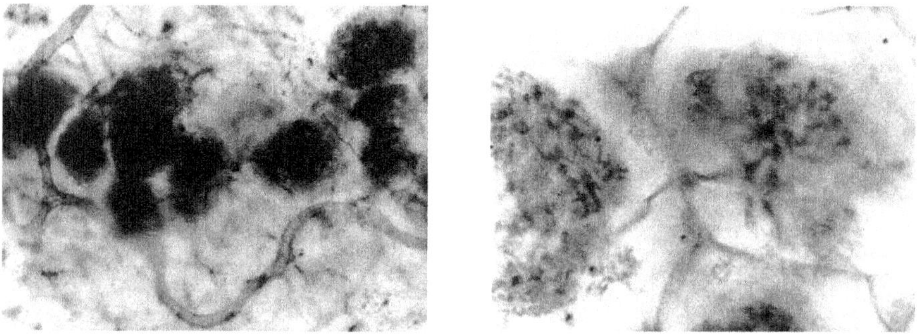

Arbuscule-like structures in *Aglaophyton major* major rhizome.

Table. The mycorrhizal status of major plant lineages, with their approximate ages estimated usingfossil evidence and molecular plant phylogenetics.

Period	Plant	Mycorrhizas
Ordovician to Silurian (476-432 Myr)	The first bryophyte-like land plants	Limited fossils without roots, mycorrhizas unknown.
Ordovician (476 Myr) to Present	Liverworts and hornworts	Currently have VAM-like associations (with or without arbuscules), other associations, or are NM
Silurian (415-425 Myr) to Present	Mosses	No roots. Currently NM. Some have endophytic Glomeromycotan fungi (see below)
Devonian (400 Myr)	*Aglaophyton major, Nothia aphylla* and other early land plants of uncertain affinities	Fossils with VAM-like hyphae and arbuscules in rhizome or thallus (Taylor *et al.* 1995, Taylor & Krings 2005, Krings *et al.* 2007)
Devonian (395 Myr) to Present	Lycopods: Lycopodium, Selaginella, *etc.*	VAM in sporophyte, underground gametophyte MH VAM
Devonian (395 Myr) to Present	Sphenophytes: Equisetum, *etc.*	Living *Equisetum* species have VAM

Mid Devonian (385 Myr)	Sphenophytes, Lycopods, Pteridophytes	First plants with roots resembling those of modern plants
Mid Devonian (385 Myr) to Present	Pteridophytes (Ferns)	Most living ferns have VAM
Mid Devonian to Present	Cycads	VAM
Triassic (215-235 Myr)	Cycad from Antarctica (Antarcticycas sp.)	Earliest fossil VAM association in roots?.
Permian (265 Myr) to Present	Ginkgoales: Ginkgo, *etc.*	Extant tree has VAM
Triassic (235 Myr) to Present	Southern Hemisphere conifers: Araucariaceae, Podocarpaceae	Living trees have VAM
Early Jurassic (190 Myr) to Present	Northern Hemisphere conifers (except Pinaceae): Cupressaceae, Taxodiaceae	Trees with VAM
Early Jurassic (190 Myr) to Present	Gnetales: *Ephedra, Gnetum, Welwitschia*	VAM or ECM (Gnetum)
Early Cretaceous? (>120 Myr) to Present	Conifers in Pinaceae: Larix, Picea, Pinus, Tsuga, *etc.*	ECM trees with heterorhizic roots
Early Cretaceous (~120 Myr) to Present	Angiosperms	VAM, NM or other mycorrhiza type
Cretaceous (100 Myr) to Present	Fagales: Betulaceae, Casuarinaceae, Fagaceae, Juglandaceae, Myricaceae, Nothofagaceae	Single lineage of ECM trees or shrubs with heterorhizic roots (some VAM also)
Late Cretaceous, Oligocene, or Eocene to present (90-30 Myr)	Fabaceae (Caesalpiniaceae, Mimosaceae), Myrtaceae, Salicaceae, Tiliaceae, *etc.*	Several separate ECM lineages (see *Section 5*)
Cretaceous (100 Myr) to Present	Orchidaceae	Orchid mycorrhizas. Orchids age estimate based on biogeography and phylogenetics
Cretaceous (100 Myr) to Present	Mainly NM families: Proteaceae, Cyperaceae, Restionaceae, *etc.*	Oldest known fossils of plants likely to have true NM roots
Late Cretaceous (80 Myr)	Ericaceae	Oldest known fossils likely to have had ericoid mycorrhizas

This table was originally published in Brundrett (2002). The mycorrhizal status of lineages is primarily derived from observation of living descendants with some fossil evidence. Data on plant lineages and fossil histories are from Stewart & Rothwell (1993), Taylor & Taylor (1993), Kenrick & Crane (1997), Wing & Boucher (1998), Hill *et al.* (1999), Renzaglia *et al.* (2000) and Barrett & Wills (2001). All dates are approximate. Abbreviations: VAM = arbuscular mycorrhizal, ECM = ectomycorrhizal, NM = nonmycorrhizal.

Table. The early history of fungi in the Glomeromycota thought to form vesicular-arbuscular mycorrhizas based on fossil evidence and DNA sequence data.

Period	Evidence
Cambrian? >500 Myr	Fungi probably colonised land long before plants, but left scant early fossil evidence
Proterozoic? >570Myr	DNA sequences show Glomeromycotan fungi are the monophyletic sister group to the higher fungi
400-500 Myr	Approximate origin of Glomeromycota according to a DNA molecular clock estimates
Ordovician (460 Myr)	Fossil spores identified as Glomeromycotan fungi
Devonian (400 Myr)	Fossil spores identifiable as *Scutellospora*, a genus within the Glomeromycota
Devonian (400 Myr) to present	Many reports of fossil spores likely to be Glomeromycotan
Unknown to present	Geosiphon – a soil crust fungus with endosymbiotic cyanobacteria that rRNA sequence data suggests is a primitive Glomeromycotan fungus

Mycorrhizal Status of Primitive Plants

Bryophytes

Bryophytes might be expected to be nonmycorrhizal since they lack roots - the organ which is normally required to host these associations and have a dominant gametophyte stage. However, some of these primitive plants have associations with glomeromycete fungi, or other fungi in their thallus.

Liverworts (Hepatophyta: 15 orders or families)

Associations with Glomeromycotan fungi in a thallus considered to be like VAM. Examples include*Treubia* sp. which has a coiling associations without arbuscules , as well as*Conocephalum*, *Haplomitrium* and *Pellia* spp. which have arbuscules.

Liverworts (Hepatophyta: 11 families or orders)

Liverworts where mycorrhizas were not seen in samples examined include members of the families Metzgeriaceae, Pleuroziaceae, Sphaerocarpales, Ricciaceae and some Aneuraceae (*Riccardia* spp.)

Liverworts (Hepatophyta: Makinoaceae, Metzgeriaceae)

Mycorrhiza-like associations with basidiomycetes that include members of the *Rhizoctonia* alliance.

Mosses (Bryophyta)

Saprophytic activity of both VAM and ECM fungi occurs in senescent moss tissue. Zhang & Guo (2007) observed hyphae and vesicles of Glomeromycotan fungi in a wide diversity of moss species.

Hornworts (Anthocerotophyta)

Anthoceros and *Phaeoceras* have associations by Glomeromycotan fungi with arbuscules.

Fern Allies

Selaginella, Isoetes and *Lycopodium* are primitive plants that belong to a separate clade between the ferns and hornworts (bryophytes). Living relatives of some of the earliest land plants typically have VAM associations, but some of these relationships are hard to decipher, especially in myco-heterotrophic gametophytes.

Huperzia, Lycopodium, Phylloglossum (Huperziaceae, Lycopodiaceae)

Lycopodiaceae typically have VAM in sporophytes and myco-heterotrophic VAM in gametophytes, but some samples are nonmycorrhizal.

Selaginella (Selaginellaceae)

Most species examined had roots with VAM.

Isoetes (Isoetaceae)

These submerged aquatic plants often contain VAM associations, even when growing at considerable depths

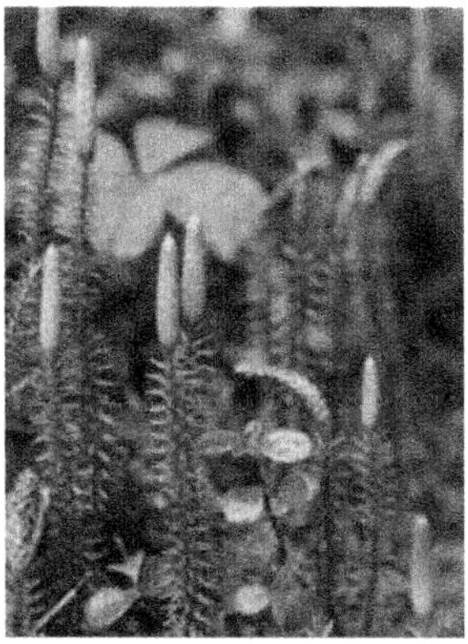

Lycopodium sp.
Club Moss
a fern ally

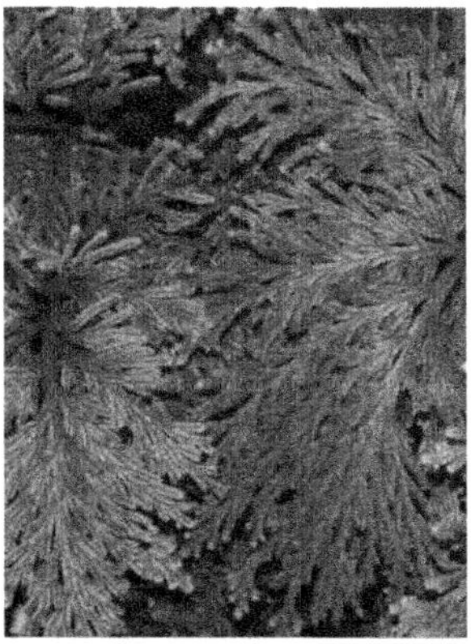

Selaginella sp.
Selaginella
a fern ally

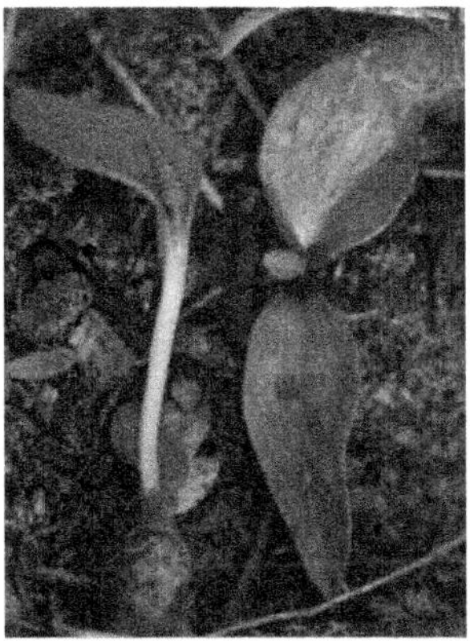

Ophioglossum lusitanicum
Adders Tongue
a primitive fern

Primitive Ferns

Phylogenetic analyses based on gene sequence data have shown that the "fern allies" such as *Equisetum* and *Psilotum*are in a monophyletic clade with other ferns. Subterranean gametophytes of primitive ferns have myco-heterotrophic VAM associations with *hyphal coils*, but no arbuscules.

Psilotum, Tmesipteris (Psilotaceae)

Coiling VAM in subterranean gametophyte. Often NM when growing as an epiphyte or in potting mix - Gemma & Koske 1995, Brundrett unpublished. Sporophytes and gametophytes have *VAM with hyphal coils* that lack arbuscules in their rhizomes.

Botrychium, Helminthostachys, Ophioglossum (Ophioglossaceae)

Botrychium and *Ophioglossum* sporophytes have VAM and subterranean gametophytes of these plants have myco-heterotrophic VAM.

Equisetum (Equisetaceae)

Equisetum sporophytes can have VAM or NM roots and probably have facultative associations, as some individuals are NM. In one study 11 of 12 samples had VAM, while another found VAM in 21 of 23 samples.

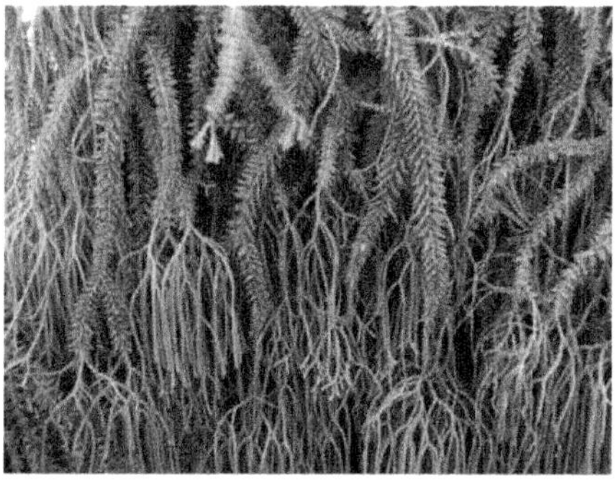

Huperzia (Lycopodium) phlegaria(Lycopodiaceae) - Tassel fern, a Lycopod.

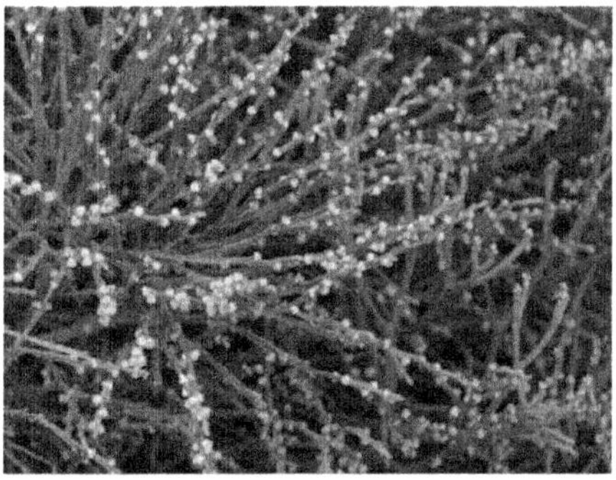

*Psilotum nudu*m (Psilotaceae) - Whisk Fern, Fork Fern.

Ferns

Comprehensive studies of ferns listed below in different regions of the world found the majority to be mycorrhizal with VAM associations. However, most ferns have fine roots with long root hairs, suggesting they have*facultative mycorrhizas*. More advanced Leptosporangiate ferns are considered to be less dependent on mycorrhizas than more primitive ferns. Photosynthetic gametophytes of ferns are usually not mycorrhizal, but *Pellaea* and *Adiantum* had VAM in their gametophytes. Substrates strongly influence mycorrhizal status of ferns, since aquatic ferns, such as *Azolla, Marsilea* and*Salvinia,* and epiphytes are less likely to be mycorrhizal than terrestrial species.

Mycorrhizal Surveys of Ferns and Fern Allies

- VAM in 80 out of 85 species from Japan (aquatic ferns NM).
- VAM in 94 of 101 species in 21 families from New Zealand by Cooper (1976).
- VAM in 21 out of 35 species from Europe in data summarised by Harley & Harley (1987).
- VAM in 17 out of 33 species in Canada.
- VAM in 66 out of 89 species (83% of terrestrial vs. 55% of epiphytic ferns) in Hawaii by Gemma *et al.* (1992).
- VAM in 13 out of 22 epiphytic species in Hawaii by Gemma & Koske (1995).
- 20 species of epiphytic ferns in Malaysia were not mycorrhizal, but many of the same species had VAM in soil.
- VAM in 88 out of 256 species in Yunan Province, China by Zhao (2000).
- VAM in 31 out of 34 species sampled in southern China by Zhang *et al.* (2004).
- VAM in 19 out of 85 species in sothern Ecuador by Lehnert *et al.* (2009).

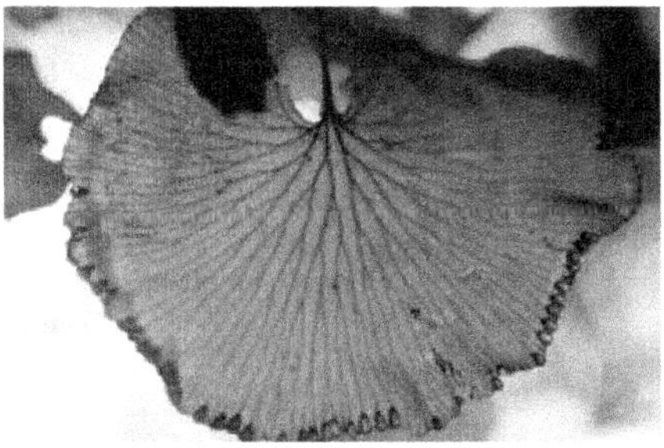

Trichomanes reniforme a filmy fern from New Zealand.

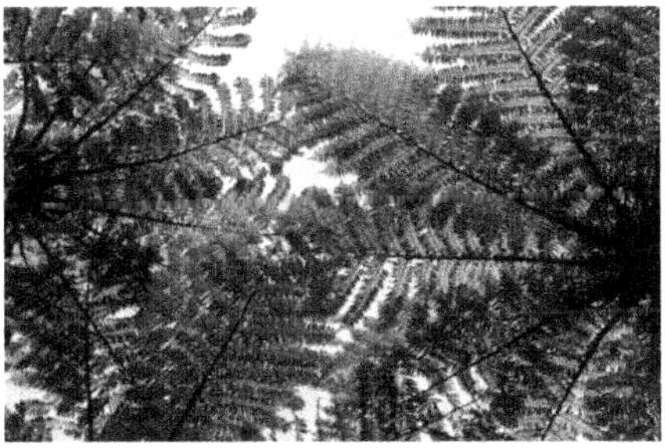

Dicksonia squarrosa a tree fern from New Zealand.

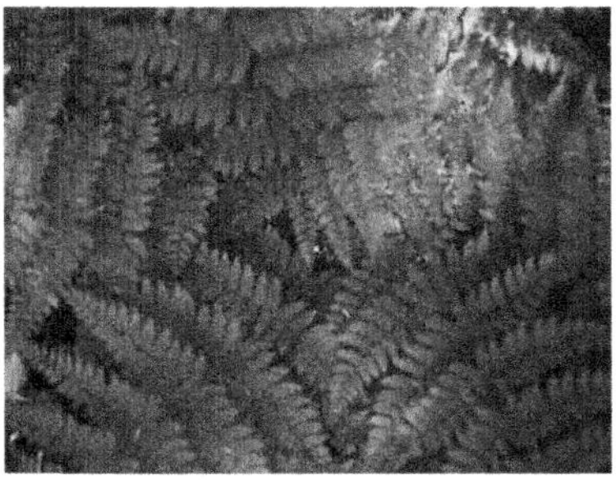

Dryopteris sp. a woodland fern from Canada.

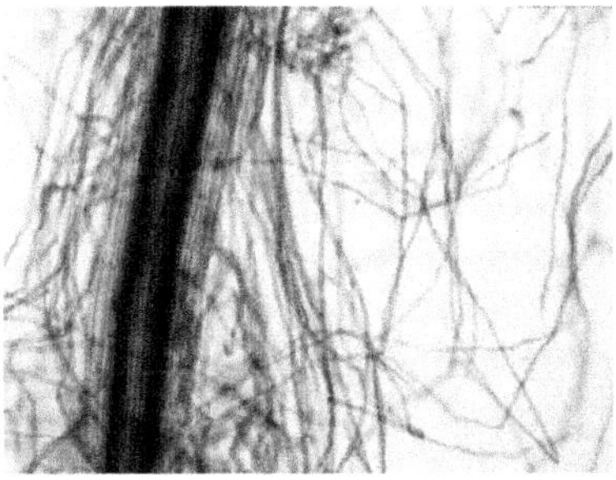

Dryopteris has typical fern roots with very long root hairs.

Gymnosperms

The gymnosperms are the next major group to evolve after the ferns and gave rise to the angiosperms. There are two major mycorrhizal strategies of gymnosperms: VAM and ECM that follow major clades. The genus *Gnetum* and genera in the Pinaceae have *ECM roots*, while all the rest have VAM. Gymnosperms are classified into groups below from the most to least primitive following molecular phylogenies.

Ceratozamia, Cycas, Dioone, Lepidozamia, Macrozamia, Zamia - Cycads (Cycadaceae, Zamiaceae, Stangeriaceae)

VAM occur in cycad roots sampled from India, Florida USA, Australia, *etc.* (Brundrett & Abbott 1991, Reddell *et al.* 1996, Muthukumar & Udaiyan 2002, Fisher & Vovides 2004). See the *Cycad Pages* for information on cycad identification, ecology, propagation, Nitrogen fixing associations, *etc.*

Ginkgo Biloba - Maidenhair Tree (Ginkgoaceae)

The only living member of this clade is a tree with VAM.

Ephedra (Ephedraceae)

VAM hosts.

Gnetum (Gnetaceae)

Tropical climbers with ECM association.

Welwitschia Mirabilis (Welwitschiaceae)

One remaining species in family. VAM - Jacobson *et al.* (1993).

Abies, Cathaya, Cedrus, Keteleeria, Larix, Picea, Pinus, Pseudolarix, Pseudotsuga, Tsuga (Pinaceae)

A great many studies of plants in this family (that includes pine trees and other conifers important to forestry) have shown them to be ECM hosts, as summarised by Harley & Harley (1987) and Brundrett *et al.* (1990). These trees and their associated fungi dominate boreal forests in the Northern Hemisphere. Roots of these trees also contain hyphae of VAM fungi, presumably as non-functional (endophytic) associations.

Agathis, Araucaria, Phyllocladus, Podocarpus, Wollemia, etc. *(Araucariaceae, Podocarpaceae)*

These conifers were once dominant in forests worldwide, but now confined to relatively cool and wet areas in the Southern Hemisphere. They consistently have VAM associations in *beaded roots*. *Parasitaxus usta* is an unusual parasitic plant in the Podocarpaceae from New Caledonia that apparently has both mycorrhizal and physical connections with its host.

Actinostrobus, Austrocedrus, Callitris, Calocedrus, Cephalotaxus, Chamaecyparis, Cryptomeria, Cunninghammia, Cupressus, Juniperus, Metasequoia, Sequoia, Sequoiadendron, Taxodium, Taxus, Thuja, Thujopsis, Torrea, Widdringtonia (Cupressaceae)

These conifers include the largest and some of the longest-lived trees in the world. The genera listed above have VAM associations. VAM associations also occur in fossil *Metasequoia* roots. Older reports of ECM in this group are probably incorrect. The current definition of Cupressaceae includes the Cephalotaxaceae, Taxaceae, *etc.*.

Sciadopitis (Sciadopityaceae)

Roots of plants in this family, outgroup to the Cupressaceae, also have VAM.

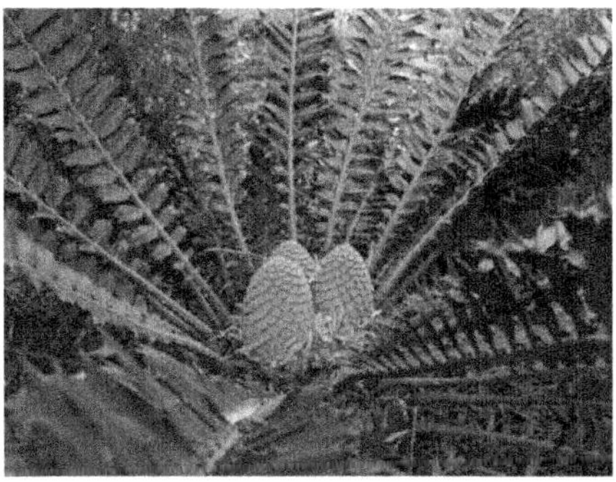

Encephalartos ferox a Cycad (Zamiaceae) with female cones.

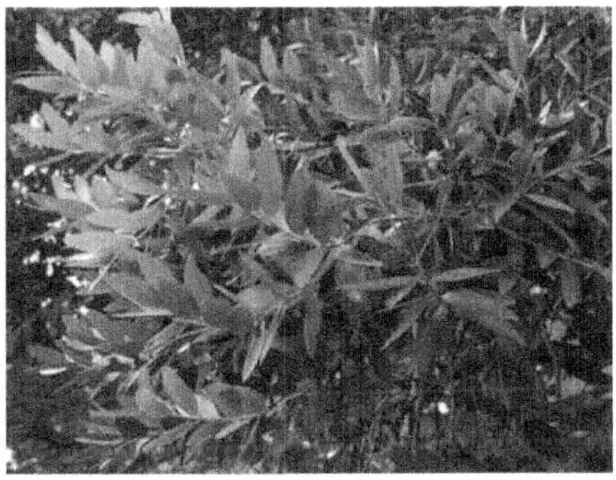

Agathis robusta (Araucariaceae) - a southern hemisphere conifer.

Conifers with VAM

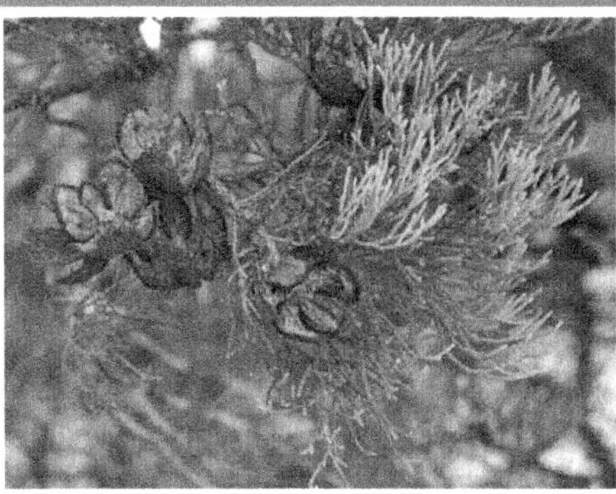

Callitris columellaris (Cupressaceae) - Cyperus Pine (Western Australia).

Conifers with ECM

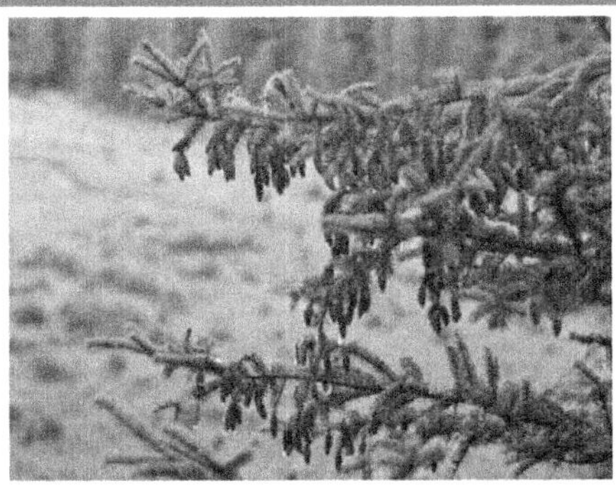

Picea glauca (Pinaceae) - White Spruce (Ontario, Canada).

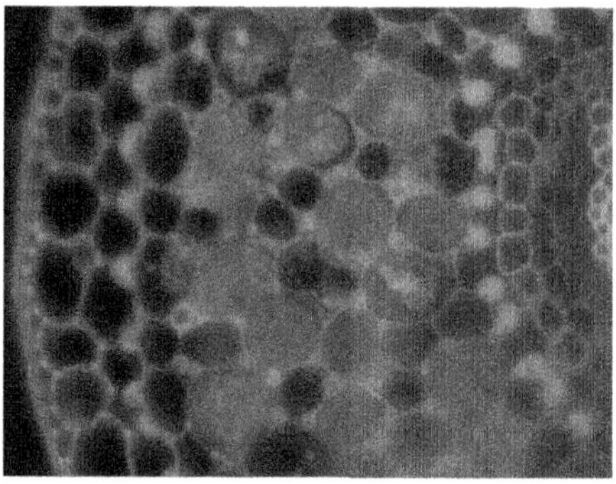

Thuja occidentalis (White Cedar, Cupressaceae) autofluorescence image of *VAM* in root cross section (Ontario, Canada).

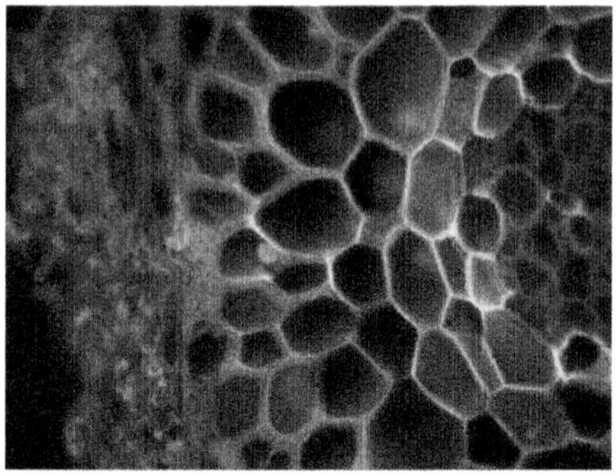

Larix laracina (Larch, Pinaceae) autofluorescence image of *ECM* cross section (Ontario, Canada).

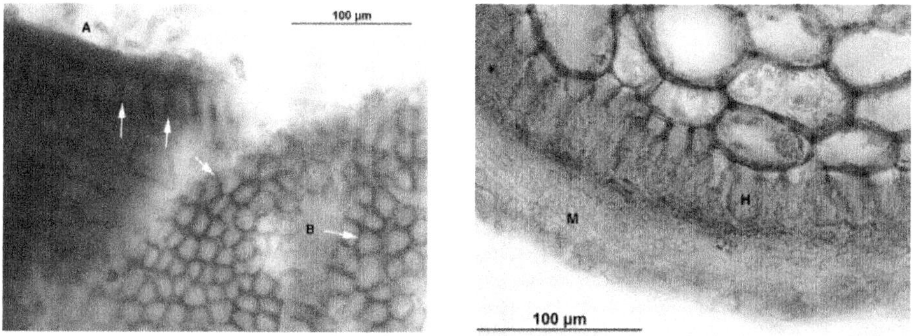

Squashed cleared and stained root (left) and section (right) of *Gnetum* sp. short root showing Hartig net in (A) side and (B) top views. Arrows point to Hartig net hyphae. This association has a thick mantle (M) and an epidermal *Hartig net* (H) which differs from the cortical Hartig net of ECM conifers.

Mycorrhizas of Flowering Plants

The strongest evidence that VAM is the ancestral condition for angiosperms is provided its near- ubiquitous occurrence in these plants today. Trappe (1987) compiled data for 6507 angiosperm species, of which 67% had VAM (including 12% considered to be facultative), 15% had another association type and 18% were NM. A more recent survey by Wang & Qui (2006) of 2964 species found very similar results (85% of species mycorrhizal). Other summaries of the UK flora , Hawaiian angiosperms (83% mycorrhizal - Koske *et al.*, 1992) and Australian plants of plants from natural ecosystems have provided similar results.

A brief summary of mycorrhizal lineages in the angiosperms is provided below primarily based on data from my review. These lineages are shown in the Figures below and the dates when they diverged are summarised in the*mycorrhizal history* table above. Comprehensive lists of *ECM hosts, NM plants* and *Australian plants* are provided in Sections 5, 6 and 8 of this site.

Ectomycorrhizal Associations in the Fagales

It is probable that the origin of ECM in the Fagales occurs in the Cretaceous after the Pinaceae already had these associations. Information on the The Fagales lineage (Betulaceae, Casuarinaceae, Juglandaceae, Nothofagaceae and Fagaceae).

ECM in other Families

Other angiosperms that have independently evolved the capacity to host ECM include some members of the Ericaceae and 11 families in 6 orders of the rosids. These clades include many families of VAM plants, so have more recently become ECM hosts. The Dipterocarpaceae, Cistaceae and Sarcolaenaceae are closely related ECM families in the Malvales that share an common ancestor. Several isolated lineages of ECM plants occur in otherwise NM families: the sedges and several genera in the Caryophyllales (Neea, Pisonia: Nyctaginaceae).

Mycorrhizas in the Ericaceae

Members of the Ericaceae have complex evolutionary trends, starting from a VAM ancestor, progressing to ECM, then to arbutoid ECM and culminating in ericoid mycorrhizas, or exploitative ECM in myco-heterotrophs like *Monotropa*.

Orchid Mycorrhizas (Orchidaceae)

Orchids are a single consistent lineage with *orchid mycorrhizas*.

Nonmycorrhizal (NM) Plants

There are at least 10 lineages of *NM plants* in the angiosperms, but most also contain VAM plants. As explained in *Section 6*, NM lineages include many aquatic plants, parasitic plants, insectivorous plants that often occur in harsh habitats, such as wet, saline, or arid soils.

Myco-heterotrophic Plants with Exploitative Mycorrhizas

Plants with exploitative mycorrhizas occur in many separate lineages. These include subterranean gametophytes of *primitive plants*, as well as angiosperms with exploitative mycorrhizas in the dicotyledon's (Diapensiaceae, Ericaceae, Gentianaceae, Polygalaceae) and the monocots (Burmanniaceae, Corsicaceae, Iridaceae, Orchidaceae, Petrosaviaceae, Triuridaceae).

Comparing data on the occurrence of mycorrhizas in Angiosperm families with phylogenies based on multiple gene sequence data.

- Clades with multiple plant families with fairly consistent mycorrhizal associations are shown in bold.
- Other clades have several types of mycorrhizal associations.

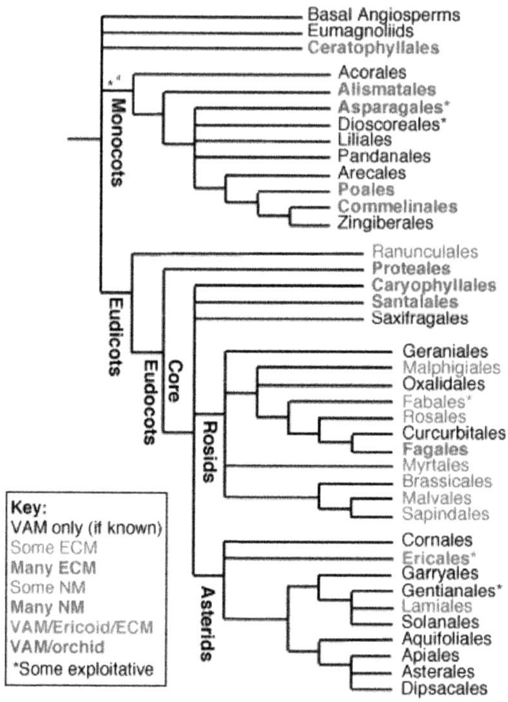

- The simplified angiosperm family tree follows APG (1988) and Soltis *et al.* (2000).
- Information about the mycorrhizal status of families is compiled from many sources.

Evolutionary lineages of plants with different types of mycorrhizas or loss of mycorrhizal dependency.

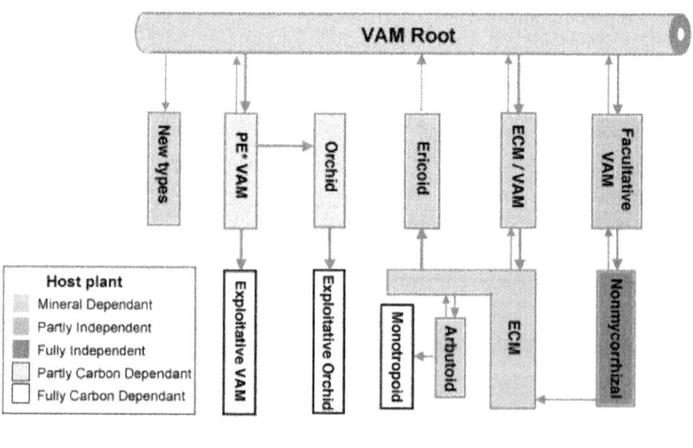

- Boxes and arrows are not to scale
- PE = partially exploitative

Evolution of Roots

Hypothetical events in the evolution of roots are summarised in the diagram below. As plants colonised the land they would have faced powerful selection pressure to increase the surface area of their absorptive systems in soil in parallel with increases in their photosynthetic organs (fine branches and leaves). The evolution of roots most likely started by specialisation of underground stems (rhizomes) as shown in the diagram below, but there is little fossil evidence of intermediate stages in this process.

Roots continued to progressively evolve into a hierarchy of structures (branch orders), with the finest elements becoming progressively thinner and more diffuse, culminating in the very complex roots in angiosperms with NM *cluster roots*. Fine roots tend to be short-lived, while relatively coarse roots form a more-permanent network used for mechanical support, production of fine roots and transport. Structural evidence that roots evolved with the need to provide habitats for mycorrhizal fungi as a primary function.

Diagram summarising probable stages in the evolution of stems, rhizomes, leaves and roots from the thallus of an early bryophyte-like land plant, using a hypothetical final example with a woody trunk. These 5 stages are listed below.

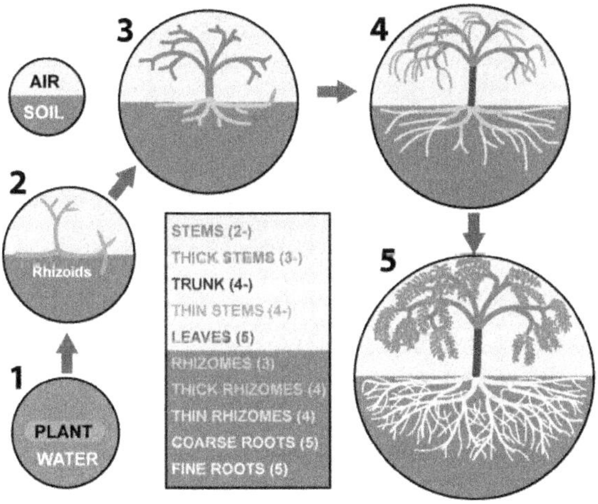

Hypothetical Stages in Root Evolution

1. Aquatic plants without roots or leaves colonise land
2. Early land plants with upright branches to intercept sunlight and horizontal stems (rhizomes) for propagation.
3. Complex shoot and rhizome branching allows more effective use of resources and competition with other plants.
4. Dimorphic subterranean rhizome systems evolve in response to allow more optimal nutrient and water uptake, mycorrhiza formation, and

mechanical support. This occurs in parallel with the evolution of specialised aerial branches that are more efficient at photosynthesis.

5. Some rhizomes become thinner and longer to increase contact with the soil and absorptive capacity, eventually becoming specialised as roots. These host mycorrhizal associations, grow faster and have a relatively short lifespan. In contrast, the thick rhizomes evolve protective features to limit permeability and facilitate long-term survival in soil.

ARBUSCULAR MYCORRHIZAL SYMBIOSIS

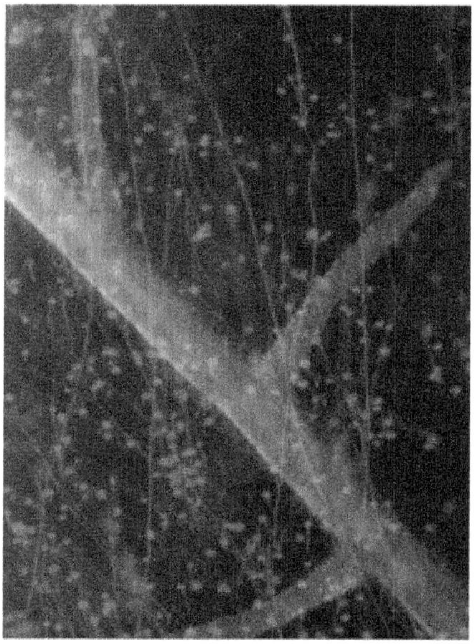

Over 80% of plants obtain their mineral nutrients with the help of symbiotic fungi that grow in and around their roots. These root-fungal partnerships (mycorrhizas) evolved around the time that plants first invaded the land, over 400 million years ago and today are found across all major taxa of plants and in ecosystems across the planet. Billions of tons of carbon flow annually from leaves to mycorrhizas to support this relationship, which is responsible for huge nutrient uptake fluxes into plants worldwide. Despite the importance of these fluxes to plant ecology, evolution and function, we have only recently begun to describe the pathways involved.

Our work has traced the routes by which carbon flows from roots to fungi and how nitrogen flows in the other direction. Using isotopic labeling, spectroscopic analysis, and gene characterization and expression measurements we have developed working models for carbon and nitrogen metabolism and transport (illustrated below). We are working to extend these models, which are now largely accepted, to address unanswered questions about their mechanisms and to include regulation.

Photo of One of Our Mycorrhizal Cultures Taken by Stacy LaClair

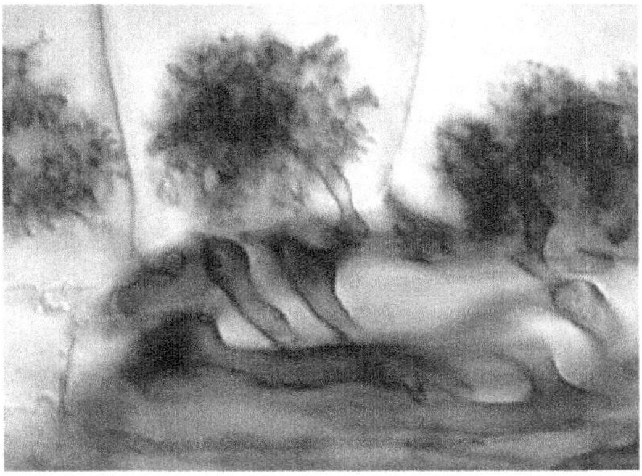

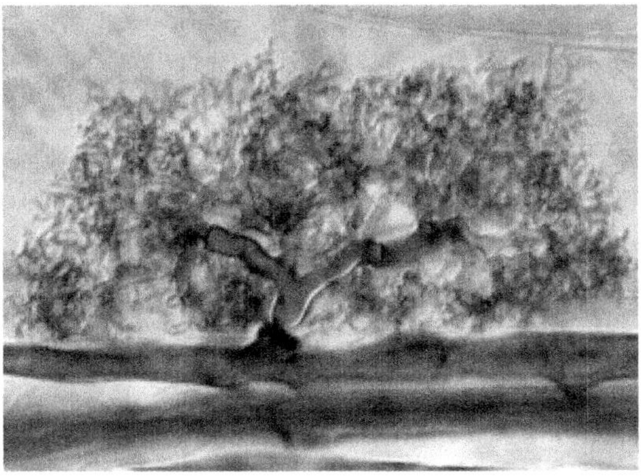

These images show the fungal arbuscules growing inside living plant root cells. These structures give the symbiosis its name and function as an interface for the exchange of nutrients between the partners.

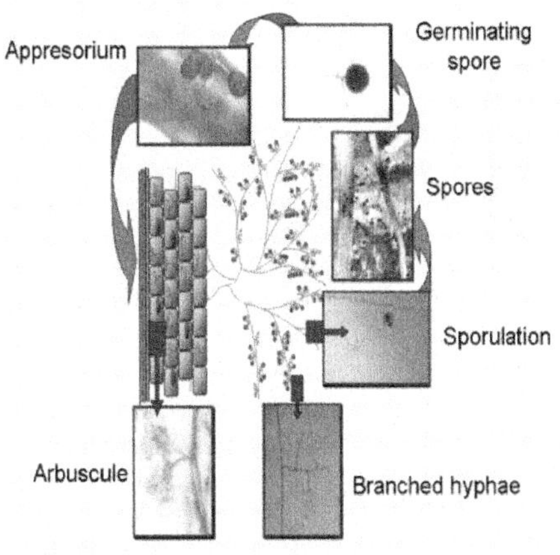

Fig. : The life cycle of an Arbuscular mycorrhizal fungus.

The germination and hyphal growth from asexual spores in the soil is stimulated by signaling compounds released by roots. These hyphae form infection structures (appresoria) on the surface of host roots the fungus grows into the root forming hyphae between cells and arbuscules that penetrate cell walls without killing the plant cells. Hyphae also grow out into the soil forming a branched mycelium that functions to explore the soil and take up mineral nutrients. Spores are formed by this external mycelium, completing the life cycle.

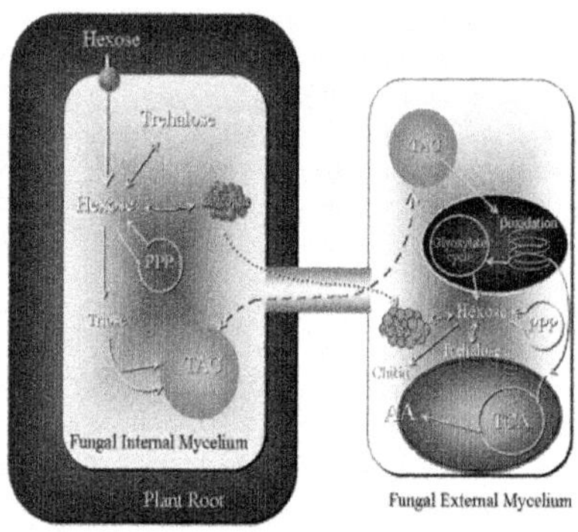

Fig.: The movement of carbon in the AM symbiosis.

We have used stable isotopic labeling experiments supported by gene expression and microscopy measurements to elucidate the processes shown above. Carbon is transferred from the plant to the fungus inside the roots as hexose and this is made into short term fungal storage carbohydrates (glycogen and trehalose).

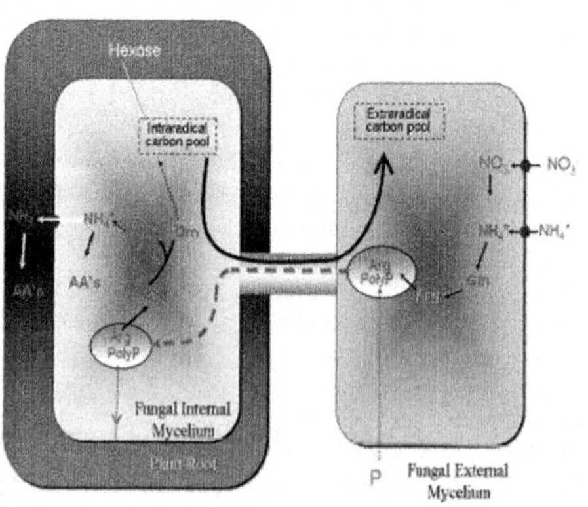

Fig.: The movement of Nitrogen in the AM symbiosis.

The fungus makes triacylglycerol (TAG) in the root which is exported within the fungus to the external mycelium and to fungal mycelium in other roots (but carbon is not transferred to roots). TAG is broken down in the external mycelium and the glyoxylate cycle is used to produce storage and structural carbohydrates (cell walls) as well as energy via the TCA cycle and other products including amino acids (AA).

Based on previous work we proposed the scheme shown here by which nitrogen moves from soil to plant via arbuscular mycorrhizal fungi. We subsequently provided evidence from labeling and gene expression that this is how N moves through the symbiosis. Inorganic N (NO_3^- and NH_4^+) is taken up by the external mycelium, assimilated and converted to arginine, which is transported (probably in association with Polyphosphate) within the fungus to the fungal mycelium inside plant roots. There the arginine is broken down to release ammonium which is transferred to the plant without carbon. More recently we have identified most of the fungal metabolic genes involved in N movement and begun to characterize the regulation of the pathway.

Chapter 12

Commercialization of Arbuscular Mycorrhizal Biofertilizer

BIOFERTILIZER

A Bio fertilizer (also *bio-fertilizer*) is a substance which contains living microorganisms which, when applied to seed, plant surfaces, or soil, colonizes the rhizosphere or the interior of the plant and promotes growth by increasing the supply or availability of primary nutrients to the host plant. Bio-fertilizers add nutrients through the natural processes of nitrogen fixation, solubilizing phosphorus, and stimulating plant growth through the synthesis of growth-promoting substances. Bio-fertilizers can be expected to reduce the use of chemical fertilizers and pesticides. The microorganisms in bio-fertilizers restore the soil's natural nutrient cycle and build soil organic matter.

Through the use of bio-fertilizers, healthy plants can be grown, while enhancing the sustainability and the health of the soil. Since they play several roles, a preferred scientific term for such beneficial bacteria is "plant-growth promoting rhizobacteria" (PGPR). Therefore, they are extremely advantageous in enriching soil fertility and fulfilling plant nutrient requirements by supplying the organic nutrients through microorganism and their byproducts. Hence, bio-fertilizers do not contain any chemicals which are harmful to the living soil.

Bio-fertilizers provide eco-friendly organic agro-input and are more cost-effective than chemical fertilizers. Bio-fertilizers such asRhizobium, Azotobacter, Azospirilium and blue green algae (BGA) have been in use a long time. Rhizobiuminoculant is used for leguminous crops. Azotobacter can be used with crops like wheat, maize, mustard, cotton, potato and other vegetable crops. Azospirilluminoculations are recommended mainly for sorghum, millets, maize, sugarcane and wheat. Blue green algae belonging to a generalcyanobacteria genus, *Nostoc* or *Anabaena* or *Tolypothrix* or *Aulosira,* fix atmospheric nitrogen and are used as inoculations for paddy crop grown both under upland

and low-land conditions. *Anabaena* in association with water fern Azolla contributes nitrogen up to 60 kg/ha/season and also enriches soils with organic matter.

Other types of bacteria, so-called phosphate-solubilizing bacteria, such as Pantoea agglomerans strain P5 or Pseudomonas putida strain P13, are able to solubilize the insoluble phosphate from organic and inorganic phosphate sources. In fact, due to immobilization of phosphate by mineral ions such as Fe, Al and Ca or organic acids, the rate of available phosphate (P_i) in soil is well below plant needs. In addition, chemical P_i fertilizers are also immobilized in the soil, immediately, so that less than 20 percent of added fertilizer is absorbed by plants. Therefore, reduction in P_i resources, on one hand, and environmental pollutions resulting from both production and applications of chemical P_i fertilizer, on the other hand, have already demanded the use of new generation of phosphate fertilizers globally known asphosphate-solubilizing bacteria or phosphate bio-fertilizers.

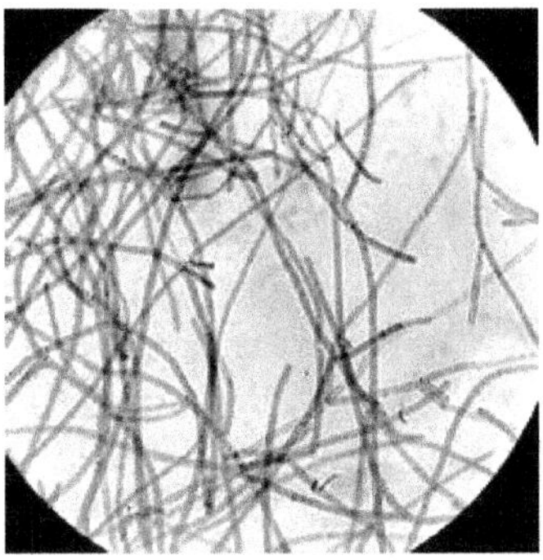

Fig. : *Tolypothrix,* cyanobacteria often used as fertilizer.

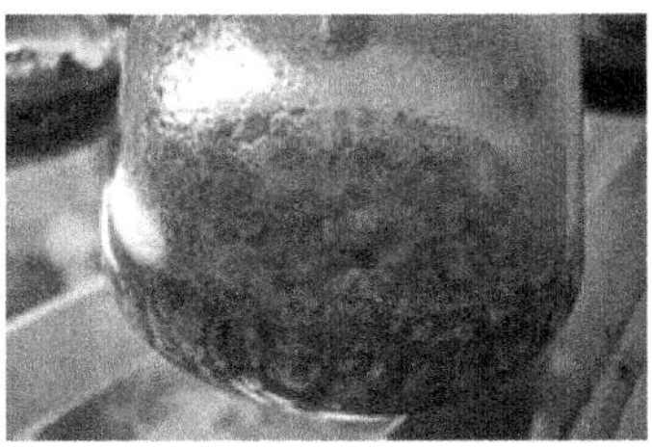

Fig. : Blue-green algae cultured in specific media. Blue-green algae can be helpful in agriculture as they have the -green algae is used it as a bio-fertilizer.

Benefits

A bio-fertilizer provides the following benefits:

1. Since a bio-fertilizer is technically living, it can symbiotically associate with plant roots. Involved microorganisms could readily and safely convert complex organic material in simple compounds, so that plants are easily taken up. Microorganism function is in long duration, caus-

ing improvement of the soil fertility. It maintains the natural habitat of the soil. It increases crop yield by 20-30%, replaces chemical nitrogen and phosphorus by 25%, and stimulates plant growth. It can also provide protection against drought and some soil-borne diseases.

2. Bio-fertilizers are cost-effective relative to chemical fertilizers. They have lower manufacturing costs, especially regarding nitrogen and phosphorus use.

Some important groups of Bio-fertilizers:

1. **Azolla-Anabena symbiosis:** Azolla is a small, eukaryotic, aquatic fern having global distribution.Prokaryotic blue green algae Anabena azolla resides in its leaves as a symbiont. Azolla is an alternative nitrogen source. This association has gained wide interest because of its potential use as an alternative to chemical fertilizers.
2. **Rhizobium:** Symbiotic nitrogen fixation by Rhizobium with legumes contribute substantially to total nitrogen fixation. Rhizobium inoculation is a well-known agronomic practice to ensure adequate nitrogen.

LIQUID BIO-FERTILIZERS

Nitrogen and Phosphorous are two major nutrients for plant growth which are provided through Chemical Fertilizers or Organic Manure. Certain soil microorganisms have an ability to fix part of elementary form of atmospheric nitrogen to the available form for the plants and / or solubilize part of the bound phosphates of the soil and thereby make them available to the plant which increases fertility of the soil and yield. These characteristics make them useful as Biofertilizers.

Thus, the Biofertilizers are the 'Live Fertilizers' of biological origin. GSFC is one of the leading manufacturers of Biofertilizers since 1984. GSFC's Biofertilizers are available under the brand name of 'Sardar Biofertilizers'.

With the expanding horizon of science, GSFC envisage demand of time and decided to switch over from carrier based Biofertilizers and put 'Liquid Biofertilizers' in the service of farming community with the proven production technology from Anand Agricultural University, Anand. The manufacturing of Liquid Biofertilizers will be undertaken shortly under the brand name of 'Sardar'.

Initially following 3 products of Sardar Liquid Biofertilizers would be available :

1. Azotobacter (for Cereals, Cash Crops, Horticultural Crops)
2. Azopirillum (for Cereals, Cash Crops, Horticultural Crops)
3. Phosphate Solubilizing Bacteria (for all Crops)

Advantages of Biofertilizers in General:

1. Harmless and Eco-friendly low cost agro-input.
2. Higher Shelf-life
3. Higher bacterial count
4. Add nutrients to the soil / make them available to the crop & secret certain growth promoting substances.
5. Help the proliferation and survival of beneficial microorganisms in soil.
6. Under certain conditions it exhibit anti-fungal activity and thereby protect the plant from pathogenic fungus.
7. Increase soil fertility and crop production.
8. Varied application methods including Drip Irrigation.

BIOFERTILIZERS AND FREQUENCY OF APPLICATION

Biofertilizers are inoculants that supply living soil microbes that improve soil fertility, soil tilth, and plant mineral nutrition. These include primarily mycorrhizal fungi and beneficial rhizosphere bacteria. Since fungi and bacteria are living organisms capable of reproduction and self perpetuation, why do we recommend repeat applications of them?

The answer to this question is related to the biology of these organisms and their host plants. To be accurate, it is necessary to address mycorrhizal fungi and bacteria separately.

Mycorrhizal Fungi

Background

Mycorrhizal fungi form a very intimate, symbiotic relationship with the plant host. This involves the actual infection, or "colonization", of the root by these fungi. This "colonization" produces a fungus-root complex, called a "mycorrhiza", that improves water and mineral absorption. Once established, a mycorrhizal symbiosis is rather persistent, and less likely to be lost due to ordinary environmental stresses. However, spread of the mycorrhizal fungi to roots is dependent on the spread of new root growth. If stressed soil conditions limit root growth, then spread of mycorrhizal fungi will be seriously limited. Therefore, fertilizer or other treatments to encourage more root growth for colonization by mycorrhizal fungi is particularly helpful under conditions of severe soil stress, such as compaction, drought, low fertility, or competition with turf. A mycorrhizal fungus can be replaced by other fungi through competition over time. Inoculating with mycorrhizal fungi strains adapted to stressed sites will improve upon the overall function of the root system by adding to, or supplanting, the existing native fungal populations.

Trees and Shrubs

So how often should mycorrhizal fungi be reapplied to a tree or shrub?

It depends on the circumstances. Here are some guidelines:

New plantings should be inoculated at planting time to support survival during transplant stress. Mature, urban, suburban, or landscape trees should be inoculated at least once. Reinoculation makes sense if the tree subsequently experiences any of the following since its first inoculation:

- Construction damage, particularly to the root system
- Root pruning of any kind
- Excessive soil compaction
- A very severe drought
- Salt stress
- Any other significant stress (*e.g.*, competition from turf)

If trees are of particularly high value, or if they are subjected to repeated traumas or stresses such as those listed above, then a regular reapplication regime is warranted. Most urban trees and shrubs live in a continuously stressed soil environment characterized by soil compaction, turf competition, and drought. The most frequent reapplication regime would be a yearly application of mycorrhizal fungi, but for most trees, annual reapplications are not needed. In fact, a single inoculation can theoretically last the life of the tree. If repeated mycorrhizal treatments are desired, a more moderate regime would involve reinoculation with mycorrhizal fungi every 5 years. A special inoculation outside of this timetable would be warranted any time the tree or shrub experiences a severe stress incident such as those listed above. As with all treatments, economic considerations will play a part in determining application frequency. But even one inoculation with mycorrhizal fungi can have a profound effect on the health and hardiness of a tree or shrub.

Scheduling of Mycorrhizal Fungal Inoculations for Trees and Shrubs

In summary, the scheduling of mycorrhizal fungal inoculations depends on circumstances and economic considerations. The following guidelines are recommended:

- Transplants should be inoculated with mycorrhizal fungi at planting to promote survival during transplant stress.
- A mycorrhizal treatment should be undertaken if a tree or shrub shows signs of stress due to severe drought, compaction, salt stress, or root damage.
- High value trees or shrubs can be put on a regular mycorrhizal treatment schedule of once every 3 to 5 years.

- Extremely high value trees or shrubs in stressed soils can be inoculated annually as a hedge against stress.

Scheduling of Mycorrhizal Fungal Inoculations for Flower Beds and Gardens

Flowerbeds and gardens that are dug up and replanted periodically should be reinoculated at each planting. The fungi cannot survive without the plant hosts. While some spores and root fragments may carry the fungi over from one planting to the next, there is no way to guarantee the effectiveness of this. Reinoculation at every planting is recommended for flowerbeds.

Rhizosphere Bacteria

What about Rhizosphere bacteria? How often should they be reapplied?

Background

Rhizosphere bacteria are free-living, and do not form a true symbiotic relationship with plants. They do not colonize plant tissue. Instead, these bacteria live in the rhizosphere, or "root zone". This is the space very close to living roots where plants release (or "exude") their nutritious byproducts, called "root exudates". Rhizosphere bacteria thrive on these root exudates, which include sugars and vitamins. In addition, plants drop off old dead root cells, which rhizosphere bacteria readily consume.

The presence of these bacteria so close to the root is beneficial to the plant, because these bacteria perform some very valuable functions. Some will absorb atmospheric nitrogen gas, and incorporate it into bacterial proteins. Incorporating nitrogen gas into useful compounds is called "fixing" nitrogen. When these bacteria die individually, their proteins are decomposed by other bacteria into soluble forms of nitrogen, like ammonia, which the plant can absorb. Other types of rhizosphere bacteria can convert insoluble mineral phosphorus into soluble phosphates, which plants can absorb and use. Remember that soluble forms of nitrogen and phosphates are two of the most important essential mineral nutrients required by plants. By fixing atmospheric nitrogen and solubilizing mineral phosphorus, these rhizosphere bacteria increase soil fertility, hence the term "biofertilizer". By virtue of their tendency to cause aggregation of small soil particles, these bacteria also improve soil structure or "tilth".

So How Often Should Rhizosphere Bacteria be Reapplied to a Tree or Shrub?

Rhizosphere bacteria are free-living, and must establish a lasting foothold in the soil and root environment. The rhizosphere is not typically "empty" when these inoculants are applied (unless they are applied to fumigated soil or new potting soils). Other soil microbes occupy this niche. These other microbes can present significant competition to newly introduced strains. Typically, bacterial inoculants must be applied repeatedly before the bacteria succeed in dominating the rhizosphere. Reapplications ensure that their populations are established and maintained at effective levels.

Fortunately, bacteria are easier to produce than are mycorrhizal fungi. As a result, bacterial inoculants are less expensive than mycorrhizal fungi inoculants, and regular applications of bacteria inoculants are not hindered by the same economic pressures. In addition, unlike endomycorrhizal fungi, bacterial inoculants can be applied by spray or drench. Therefore, the ease and economy of bacterial inoculants helps to compensate for their need to be reapplied.

Because of the great variability among soils from different locations, it is impossible to predict how frequently inoculations must be applied to ensure the continued establishment of the selected bacterial species. Generally, a regular, repeated application regimen should do the trick. Typical application frequencies vary from a maximum of every two weeks, to a minimum of twice a year. Sometimes, bacteria are added to a given inoculation regimen to supplement the other treatments. Considerations of economics and observed results will often influence the final application schedule.

MYCORRHIZA

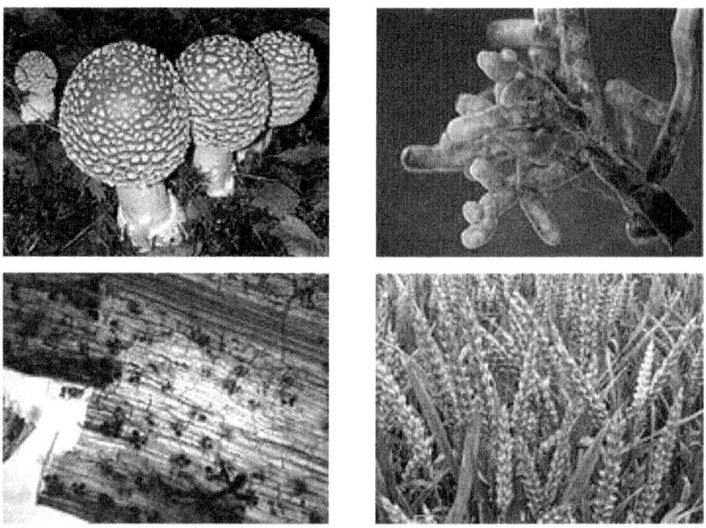

Fig. : Many conspicuous fungi such as the fly agaric form ectomycorrhiza with tree rootlets. Arbuscular mycorrhizais very common in plants, including crop species such as wheat.

A mycorrhiza is a symbioticassociation composed of a fungus and roots of a vascular plant. In a mycorrhizal association, the fungus colonizes the host plant's roots, either intracellularly as in arbuscular mycorrhizal fungi (AMF or AM), or extracellularly as inectomycorrhizal fungi. They are an important component of soil life and soil chemistry. The association is generally mutualistic, but occasionally weakly pathogenic.

Mutualist Dynamics

Fungi in mycorrhizae form a mutualistic relationship with the roots of most plant species. The roots in the relationship, and the plants themselves are referred to as mycorrhizal if mycorrhizae are formed. While only a small proportion of all species has been examined, 95% of those plant families are predominantly mycorrhizal.They are named after their presence in the plant's rhizosphere (root system).

Recent research with ectomycorrhizal plants in boreal forests has indicated that mycorrhizal fungi and plants have a relationship that may be more complex than simply mutualistic. This relationship was noted when mycorrhizal fungi were unexpectedly found hoarding nitrogen from plant roots in times of nitrogen scarcity. Researchersargue that some mycorrhizae distribute nutrients based upon the environment with surrounding plants and other mycorrhizae. Researchers go on to explain how this updated model explains why mycorrhizae do not alleviate plant nitrogen limitation, and why plants can switch abruptly from a mixed strategy with both mycorrhizal and nonmycorrhizal roots to a purely mycorrhizal strategy as soil nitrogen availability declines.

Sugar-water/Mineral Exchange

This mutualistic association provides the fungus with relatively constant and direct access to carbohydrates, such as glucose and sucrose. The carbohydrates are translocated from their source (usually leaves) to root tissue and on to the plant's fungal partners. In return, the plant gains the benefits of the mycelium's higher absorptive capacity for water and mineral nutrients due to the large surface area of fungal hyphae, which are much finer than plant roots, thus improving the plant's mineral absorption capabilities.

Plant roots alone may be incapable of taking up phosphate ions that are demineralized in soils with a basic pH. The mycelium of the mycorrhizal fungus can, however, access these phosphorus sources, and make them available to the plants they colonize. Thus many plants are able to obtain phosphate, without using soil as a source. For example, in some dystrophic forests, large amounts of phosphate are taken up by mycorrhizal hyphae acting directly on leaf litter, bypassing the need for soil uptake. *Inga alley cropping,* proposed as an alternative to slash and burn rainforest destruction, relies upon mycorrhiza within the Inga Tree root system to prevent the rain from washing phosphorus out of the soil. In some cases, the transport of water, carbon, and nutrients could be done directly from plant to plant through mycorrhizal networks that are underground hyphal networks created by mycorrhizal fungi that connect individual plants together.

Suillus tomentosus, a basidiomycete fungus, produces specialized structures known as tuberculate ectomycorrhizae with its plant host lodgepole pine (*Pinus contorta* var. *latifolia*). These structures have been shown to host nitrogen fixing bacteria which contribute a significant amount of nitrogen and allow the pines to colonize nutrient-poor sites.

Mechanisms

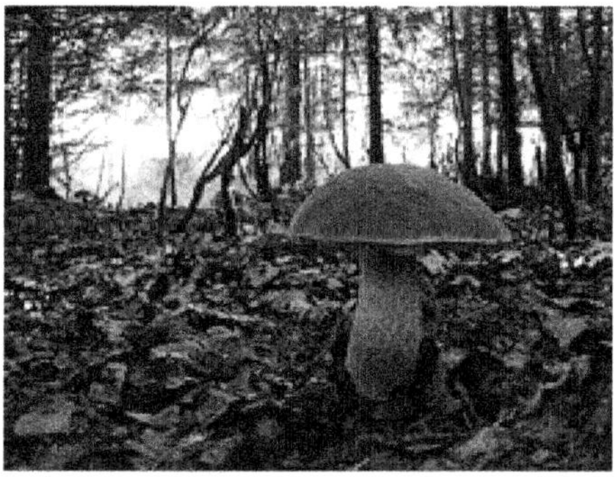

Leccinum aurantiacum, anectomycorrhizal fungus

The mechanisms of increased absorption are both physical and chemical. Mycorrhizal mycelia are much smaller in diameter than the smallest root, and thus can explore a greater volume of soil, providing a larger surface area for absorption. Also, the cell membrane chemistry of fungi is different from that of plants (including organic acid excretion which aids in ion displacement). Mycorrhizas are especially beneficial for the plant partner in nutrient-poor soils.

Disease, Drought and Salinity Resistance and its Correlation to Mycorrhizae

Mycorrhizal plants are often more resistant to diseases, such as those caused by microbial soil-borne pathogens. AMF was also significantly correlated with soil biological fertility variables such as soil fungi and soil bacteria, including soil disease. Furthermore, AMF was significantly correlated with soil physical variable, but only with water level and not with aggregate stability and are also more resistant to the effects of drought. It is known the significance of arbuscular mycorrhizal fungi alleviation of salt stress and their beneficial effects on plant growth and productivity. Although salinity can affect negatively arbuscular mycorrhizal fungi, many reports show improved growth and performance of mycorrhizal plants under salt stress conditions.

Colonization of Barren Soil

Plants grown in sterile soils and growth media often perform poorly without the addition of spores or hyphae of mycorrhizal fungi to colonise the plant roots and aid in the uptake of soil mineral nutrients. The absence of mycorrhizal fungi can also slow plant growth in early succession or on degraded landscapes. The introduction of alien mycorrhizal plants to nutrient-deficient ecosystems puts indigenous non-mycorrhizal plants at a competitive disadvantage.

Resistance to Toxicity

Fungi have been found to have a protective role for plants rooted in soils with high metal concentrations, such as acidic and contaminated soils. Pine trees inoculated with *Pisolithus tinctorius* planted in several contaminated sites displayed high tolerance to the prevailing contaminant, survivorship and growth. One study discovered the existence of *Suillus luteus* strains with varying tolerance of zinc. Another study discovered that zinc-tolerant strains of *Suillus bovinus* conferred resistance to plants of *Pinus sylvestris*. This was probably due to binding of the metal to the extramatricial mycelium of the fungus, without affecting the exchange of beneficial substances.

Occurrence of Mycorrhizal Associations

At around 400 million years old, the Rhynie chert contains an assemblage of fossil plants preserved in sufficient detail that mycorrhizas have been observed in the stems of *Aglaophyton major*.

Mycorrhizas are present in 92% of plant families studied (80% of species), with arbuscular mycorrhizas being the ancestral and predominant form, and the most prevalent symbiotic association found in the plant kingdom. The structure of arbuscular mycorrhizas has been highly conserved since their first appearance in the fossil record, with both the development of ectomycorrhizas, and the loss of mycorrhizas, evolving convergently on multiple occasions.

Types of Mycorrhiza

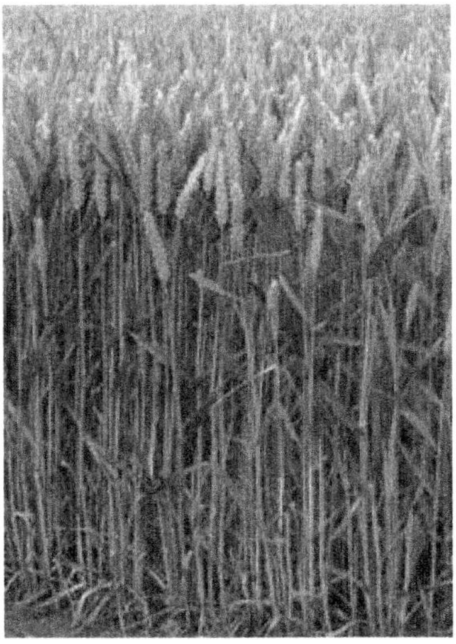

Fig. : Wheat is arbuscular mycorrhizal.

Mycorrhizas are commonly divided into *ectomycorrhizas* and *endomycorrhizas*. The two types are differentiated by the fact that the hyphae of ectomycorrhizal fungi do not penetrate individual cells within the root, while the hyphae of endomycorrhizal fungi penetrate the cell wall and invaginate the cell membrane. Endomycorrhiza includes *arbuscular, ericoid,* and *orchid mycorrhiza,* while *arbutoid mycorrhizas* can be classified as *ectoendomycorrhizas. Monotropoid* mycorrhizas form a special category.

Arbuscular Mycorrhiza

Endomycorrhizas are variable and have been further classified as arbuscular, ericoid, arbutoid, monotropoid, and orchid mycorrhizas. Arbuscular mycorrhizas, or AM (formerly known as vesicular-arbuscular mycorrhizas, or VAM), are mycorrhizas whose hyphae enter into the plant cells, producing structures that are either balloon-like (vesicles) or dichotomously branching invaginations (arbuscules). The fungal hyphae do not in fact penetrate the protoplast (*i.e.* the interior of the cell), but invaginate the cell membrane. The structure of the arbuscules greatly increases the contact surface area between the hypha and the cell cytoplasm to facilitate the transfer of nutrients between them.

Arbuscular mycorrhizas are formed only by fungi in the division Glomeromycota. Fossil evidence and DNA sequence analysis suggest that this mutualism appeared 400-460 million years ago, when the first plants were colonizing land. Arbuscular mycorrhizas are found in 85% of all plant families, and occur in many crop species. The hyphae of arbuscular mycorrhizal fungi produce the glycoprotein glomalin, which may be one of the major stores of carbon in the soil. Arbuscular mycorrhizal fungi have (possibly) been asexual for many millions of years and, unusually, individuals can contain many genetically different nuclei (a phenomenon called heterokaryosis).

Ectomycorrhiza

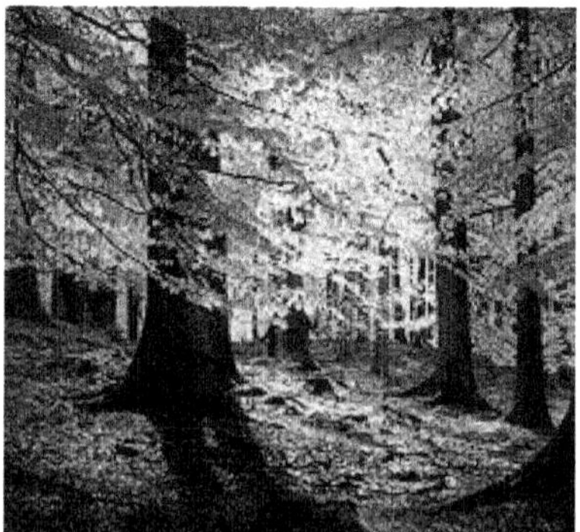

Fig. : Beech is ectomycorrhizal.

Ectomycorrhizas, or EcM, are typically formed between the roots of around 10% of plant families, mostly woody plants including the birch, dipterocarp, eucalyptus, oak, pine, and rose families, orchids, and fungi belonging to the Basidiomycota, Ascomycota, and Zygomycota. Some EcM fungi, such as many *Leccinum* and *Suillus*, are symbiotic with only one particular genus of plant, while other fungi, such as the *Amanita*, are generalists that form mycorrhizas with many different plants. An individual tree may have 15 or more different fungal EcM partners at one time. Thousands of ectomycorrhizal fungal species exist, hosted in over 200 genera. A recent study has conservatively estimated global ectomycorrhizal fungal species richness at approximately 7750 species, although, on the basis of estimates of knowns and unknowns in macromycete diversity, a final estimate of ECM species richness would probably be between 20000 and 25000.

Ectomycorrhizas consist of a hyphal sheath, or mantle, covering the root tip and a Hartig net of hyphae surrounding the plant cells within the root cortex. In some cases the hyphae may also penetrate the plant cells, in which case the mycorrhiza is called an ectendomycorrhiza. Outside the root, Ectomycorrhizal extramatrical mycelium forms an extensive network within the soil and leaf litter.

Nutrients can be shown to move between different plants through the fungal network. Carbon has been shown to move from paper birch trees into Douglas-fir trees thereby promoting succession in ecosystems. The ectomycorrhizal fungus *Laccaria bicolor* has been found to lure and kill springtails to obtain nitrogen, some of which may then be transferred to the mycorrhizal host plant. In a study by Klironomos and Hart, Eastern White Pine inoculated with *L. bicolor* was able to derive up to 25% of its nitrogen from springtails.

The first genomic sequence for a representative of symbiotic fungi, the ectomycorrhizal basidiomycete *Laccaria bicolor*, has been published. An expansion of several multigene families occurred in this fungus, suggesting that adaptation to symbiosis proceeded by gene duplication. Within lineage-specific genes those coding for symbiosis-regulated secreted proteins showed an up-regulated expression in ectomycorrhizal root tips suggesting a role in the partner communication. *Laccaria bicolor* is lacking enzymes involved in the degradation of plant cell wall components (cellulose, hemicellulose, pectins and pectates), preventing the symbiont from degrading host cells during the root colonisation. By contrast, *Laccaria bicolor* possesses expanded multigene families associated with hydrolysis of bacterial and microfauna polysaccharides and proteins. This genome analysis revealed the dual saprotrophic and biotrophic lifestyle of the mycorrhizal fungus that enables it to grow within both soil and living plant roots.

Ericoid Mycorrhiza

Ericoid mycorrhizas are the third of the three more ecologically important types. They have a simple intraradical (grow in cells) phase, consisting of dense coils of hyphae in the outermost layer of root cells. There is no periradical phase and the extraradical phase consists of sparse hyphae that don't extend very far into the surrounding soil. They might form sporocarps (probably in the form of small cups), but their reproductive biology is little understood.

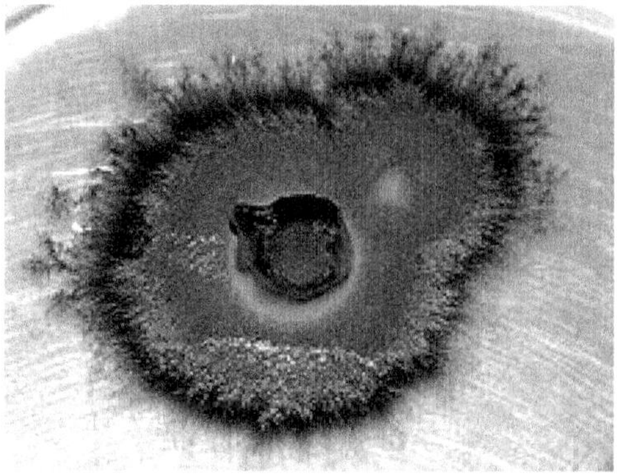

Fig. : An ericoid mycorrhizal fungus isolated from *Woollsia pungens.*

Ericoid mycorrhizas have also been shown to have considerable saprotrophic capabilities, which would enable plants to receive nutrients from not-yet-decomposed materials via the decomposing actions of their ericoid partners.

Arbutoid Mycorrhiza

This type of mycorrhiza involves plants of the Ericaceae subfamily Arbutoideae. It is however different from ericoid mycorrhiza and resembles ectomycorrhiza, both functionally and in terms of the fungi involved. The difference to ectomycorrhiza is that some hyphae actually penetrate into the root cells, making this type of mycorrhiza an *ectendomycorrhiza.*

Monotropoid Mycorrhiza

This type of mycorrhiza occurs in the subfamily Monotropoideae of the Ericaceae. These plants are heterotrophic or mixotrophic and derive their carbon from the fungus partner. This is thus a non-mutualistic, parasitic type of mycorrhizal symbiosis.

Orchid Mycorrhiza

All orchids are myco-heterotrophic at some stage during their lifecycle and form orchid mycorrhizas with a range of basidiomycete fungi. Their hyphae penetrate into the root cells and form typical coils.

Discovery

Associations of fungi with the roots of plants have been known since at least the mid-19th century. However early observers simply recorded the fact without investigating the relationships between the two organisms. This symbiosis was studied and described by Franciszek Kamieński in 1879–1882. Further research was carried out by Albert Bernhard Frank, who introduced the term *mycorrhiza* in 1885.

BENEFITS AND LIMITATIONS OF BIOFERTILIZATION IN AGRICULTURAL PRACTICES

The growing need for supply of agronomic products for food and consumer goods processing by the modern society has caused substantial increases in agrarian activities in recent decades. As a result, the need for implementation of methods that allow, among other things, to improve the efficiency of crops, mitigate adverse impacts on the soil, reduce the use of chemical fertilizers, and increase revenues per cultivated area, have been addressed. For this reason, the implementation of conservative agriculture (CA) models has been a cornerstone of farming practices globally. The CA focuses on reducing adverse impacts on the environment, increasing crop yields and inputs, and implementing sustainable techniques for development of agriculture.

Biological fertilization is based on the use of natural inputs including fertilizers, decaying remains of organic matter, crops excess, domestic sewage, animal manure, and microorganisms such as fungi and bacteria. They are used to improve fixation of nutrients in the rhizosphere, produce growth stimulants for plants, improve soil stability, provide biological control, biodegrade substances, recycle nutrients, promote mycorrhiza symbiosis, and develop bioremediation processes in soils contaminated with toxic, xenobiotic and recalcitrant substances.

Additionally, the use of bio-fertilizers can improve productivity per area in a relatively short time, consume smaller amounts of energy, mitigate contamination of soil and water, increase soil fertility, and promote antagonism and biological control of phytopathogenic organisms. The aforementioned aspects are translated into profitable benefits for farmers as a result of lower costs associated with the process of fertilization and higher crop yields. In this sense, biologic fertilizers application can bring benefits from an economic, social, and environmental point of view. However, the implementation of fertilization techniques requires feasibility studies, monitoring of environment variables involved in metabolic processes, acquisition of biological inputs, capital investment, time, and trained personnel. In order to achieve a sustainable agriculture is necessary the implementation of plans, programs, projects and initiatives directed toward the minimization of environmental impacts and consequent benefits for farmers and producers.

As a result of recent investigations performed to effectively assess the implementation of biological fertilizers under varying conditions, a review on their benefits and limitations is required to provide a valid background for academics, farmers and producers to perform future research complementing current work that deeply assess economical, environmental and social aspects related to the agricultural expansion worldwide. This paper focuses on the review of current research resulting from the use of biological fertilizers in different regions of the world to obtain a framework that facilitates the development of future investigations in the agricultural sector and, consequently, promote the reduction of environmental impacts associated to the permanent use of chemical fertilization. Firstly in this chapter, an introduction to biofertilization techniques is addressed to understand the fundamentals that explain their functioning. Next in the paper, benefits and limitations of biofertilizers are mentioned followed by current researches' results from the use of the techniques in various geographical backgrounds. Finally, conclusions and recommendations complete the paper.

Biological Nitrogen Fixation (BNF)

Biological nitrogen fixation is considered a key process in the biosphere and fundamental constituent of sustainable agriculture. It allows the conversion of gaseous nitrogen (N_2) to the mainly forms of available nitrogen (*e.g.*, nitrite, nitrate, and ammonium) for the development of metabolic processes of plants. The conversion process of gaseous nitrogen and similar products (more available for plant's growth) takes place by the action of microorganisms in the soil. These microorganisms include: Azospirillum, Azotobacter, Beijerinckia (*i.e.*, microorganisms that establish associations with grass plants), Rhizobium, Bradyrhizobium, and Azorhizobium (*i.e.*, bacteria establishing symbiosis with legumes), Frankia (*i.e.*, symbiotic actinomycetes with woody plants), Nostoc (*i.e.*, blue-green algae establishing symbiosis with different plants) or Anabahena (*i.e.*, ferns).

Development of BNF depends on specialized microorganisms, *i.e.*, those who are carriers of nitrogenase enzyme. These are responsible for its production through biological and physicochemical processes. Additionally, BNF has shown minimal environmental impacts. Its usefulness and efficiency for the optimum physical plant development have been extensively recognized. Sessitsch *et al* (2002) estimated that approximately 80% of fixed nitrogen on the planet is due to gram-negative activity of Rhizobium bacteria. The acquisition strategy for reducing atmospheric nitrogen by Rhizobium-legume-association is a complex process. Rhizobium induces the legume to form nodules, thereby establishing metabolic cooperation, in which the bacteria reduce nitrogen (N_2) to ammonia (NH_4). The latter is exported to the plant tissue to be assimilated into proteins and other complex nitrogenous compounds. Simultaneously, the leaves reduce carbon dioxide (CO_2) into sugars through photosynthesis and transport it to the roots. There, Rhizobium provides ATP for the diatomic nitrogen immobilization, taking advantage from that source of energy and facilitating the development of photosynthetic and growth processes of plants.

In addition, it is estimated that Rhizobium-legume association is responsible for setting annually 35 million tons of nitrogen. This amount significantly influences the fertilization of soils globally and favors the development of agriculture and forestry activities in several parts of the world.

Biological Fertilization Techniques

Society must meet its food needs through agricultural resources. Therefore, the use of methods that are effective and feasible to obtain better yields and meet global demand of inputs has become increasingly necessary. Similarly, alternative methods arise to increase the soil fertility. The main scope of these methods is to provide greater efficiencies, increase the quality of agricultural products, minimize crop time, and reduce costs. On the other hand, contamination of soils, extensive and continuous use of chemical inputs and monoculture has led to the need of incorporating fewer invasive fertilization methods. Next in this chapter, some biological fertilization techniques, that represent lower environmental, implementation costs and efficiencies comparable to conventional chemical fertilizers used in the world, are presented and explored.

Benefits and Limitations of Biological Fertilizers

Biological practices can offer a wide range of opportunities for the development of better agrarian practices due to the advantages and benefits provided for the soil, products and farmers. Nevertheless, limitations of these practices are also well studied and recognized, which implies that feasibility studies should be carried out to find out better solutions for each particular case in agricultural activities. Next in this section, some benefits and limitations are mentioned to highlight the need of future research on some issues.

Table: Benefits and limitations of biological fertilizers.

Benefits	Limitations
Biological fertilizers can mobilize nutrients that favor the development of biological activities in soils.	Compost products have highly variable concentrations of nutrients. In addition, implementation costs are higher than those of certain chemical fertilizers.
Maintenance of plant health is enhanced by the addition of balanced nutrients.	Extensive and long-term application may result in accumulation of salts, nutrients, and heavy metals that could cause adverse effects on plant growth, development of organisms of the soil, water quality, and human health.
Food supply is provided and growth of microorganisms and beneficial soil worms is impelled.	Large volumes are required for land application due to low contents of nutrients, in comparison with chemical fertilizers.
As a result of the good structure provided to the soil, root growth is promoted.	Main macronutrients may not be available in sufficient quantities for growth and development of plants.
The content of organic matter in soil is higher than normal levels.	Nutritional deficiencies could exist, caused by the low transfer of micro- and macro-nutrients.
Promotes the development of mycorrhizal associations, which increases the availability of phosphorus (P) on the soil.	
Help to eliminate plantar diseases and provide continuous supply of micronutrients to the soil.	
Contribute to the maintenance of stable nitrogen (N) and phosphorus (P) concentrations.	
Improvements on the capacity of nutrients' exchange in the soil.	

Animal Manure

For several decades, animal manure has been widely used by farmers for soil fertilization, given the low costs associated with its production, transportation, and processing. This wide availability and the nutritional intake of trace elements make it an attractive alternative for the development of fertilization on soils suffering nutritional deficiencies.

Manure has many benefits, which include:

- It is a biological fertilizer with high proportions of nitrogen (N) and potassium (K), medium proportions of calcium (Ca) and phosphorus (P), and low proportion of magnesium (Mg) and sulfur (S). It allows getting favorable effects on physicochemical stability of soils, plants growth, and development of beneficial microbial populations.
- Manure adds organic matter to the soil.
- The composition of organic solids is between 20% and 40%.
- Given the high nitrogen content, decomposition of organic matter is developed more quickly.
- Despite having low content of phosphorus (P), manure prevents blockage of this element, making it available for plants.

Aspects such as the type, age, and health of the animal affect the proportion of macro- and micro-nutrients available in the manure. For example, sheep and poultry manure contain high levels of nitrogen (N), while manure from pigs, cattle, and horses have lower proportions of this element. The type of bedding (*i.e.*, ferns or other plants that serve as disposal of urine and excreta) also affects the quality of manure. Furthermore, the usefulness and usability of the product depends on the proportion of heavy metals and other chemical substances. The application of manure in the soil must be made in quantities or concentrations acceptable by rules and regulations from environmental and health authorities. In most cases, it is recommended that manure is sprayed in a thin layer over large portions of land, rather than stacked on a small portion. The purpose of the technique is to promote soil aeration, maximize the efficiency of agricultural production, and facilitate the development of biological activities that are able to create a medium rich in nutrients for plants growth.

Although animal manure provides improved availability of nutrients and facilitates plants growth, it also has disadvantages and limitations of particular interest. Some of the limitations are referred to possible risks on the safety of consumers, physicochemical, and biological stability of soils. In this regard, high contents of ammonia from manure can burn foliage and roots of plants; the presence of manure could increase the amount of weed flora and costs associated with transportation, and manure application are superior to those of traditional techniques. Besides that, the presence of heavy metals (*e.g.*, mercury, chromium, lead) pose a threat as a result of their carcinogenic potential and their capability of bio-accumulation and bio-magnification in the food chain. For this reason, the use of manure to fertilize soils should be well assessed and considered in order to evaluate the cost-benefit ratio. Also, technical tests must be carried out to verify its safety. Finally, excessive application of manure can generate important reductions of plants growth, extreme levels of nitrogen, ammonia, and salts that could lead to different undesired scenarios for farmers and the soil itself.

Arbuscular Mycorrhiza (AM)

Most plants of agricultural interest are endomycorrhiza and belong to the arbuscular mycorrhiza type (AM). Mycorrhiza is a mutualistic association existing between fungi and most land plants. These partnerships are easy to locate in distinct places, from aquatic to desert, occurring at different altitudes and latitudes. Therefore, its value in terms of availability and ease of use in various geographical conditions is widely recognized.

The fungi that form symbiotic associations are obligated bio trophic, meaning that they can only complete their life cycle by colonizing roots of host plants. This type of symbiotic association has been called bio-fertilizer and crop bio-protector. It is also considered relevant for integrated management programs of soils and crops. Arbuscular mycorrhiza fungi belong to Glomeromycota division. The most abundant and diverse is the genus' Glomus, consisting of fungal inoculants (*i.e.*, mycorrhiza fungi widely used in agricultural activities worldwide). Mycorrhiza inoculants application in soils provides benefits for agricultural and forest crops such as increased growth rate and tolerance of plants to drought and soil salinity. Salinas et al (2005) stated that Glomus (*i.e.*, vesicular-arbuscular) could supplement or replace chemical fertilizers of crops in varying environmental conditions.

The proper selection and application of arbuscular mycorrhiza fungi improves plant nutrition and increases the resistance of plants against pathogens and stress conditions (both biotic and abiotic). Furthermore, the wide range of options and applicability of AM in different regions makes it an attractive technique to replace, partially or completely, chemical fertilization of soils.

Biosolids

Dumping of human waste treated at Sewage Treatment Plants (STP) on the soil has a fairly broad historical trajectory. In the early seventies, application of sewage sludge in soil or sediment began for agricultural and forestry purposes in the United States. In fact, the Environmental Protection Agency (U.S.EPA) estimates that half of the sludge produced in the United States are spread on the soil mostly applied as fertilizer in large portions of land. Biosolids can be applied to the soil through techniques such as dumping, injection, irrigation, among others, depending on the local environmental and financial conditions.

These techniques help to decrease spreading of odors, insects influence on crops, minimize runoff losses, and loss of ammonia in the air.

Several options for using the sludge from STP are suitable, for example: landfill disposal, incineration, and direct application on the soil. The latter requires dissolving the sludge, before being applied to the soil, where it is decomposed by microorganisms and filtered by the soil matrix. Therefore, it is the most promising use from the economic and environmental perspective. Moreover, the composition of biosolids is useful for soil nutrition, which explains the increased rate of use as amendment in several countries. The U.S.EPA classifies biosolids according to their content of heavy metals. Those with lower concentrations can be applied under more flexible security controls. Biosolids with higher concentrations are not likely to be used. Hence, they must be incinerated or landfilled. The acceptance or denial of biosolids used as fertilizers is based on safety parameters, such as hazardous characteristics (*i.e.*, corrosivity, reactivity, explosivity, toxicity, flammability, and biological hazards). If biosolids do not exhibit these characteristics, they can be certified as safe to the soil and may be applied. These sanitary regulations are primarily intended to reduce risks to human health and the environment based on the potential for contamination of water resources, crops, and ecosystems.

Composting

Composting is one of the oldest techniques used for the stabilization of natural wastes and soil biologic fertilization. The main objective of this practice is to obtain a stable, chemical and biological rich product with micro and macro nutrients.

The composting process works as follows: initially, strains of microorganisms break down living waste, generating temperature differentials , while the pH of the medium decreases as a result of the production of natural acids. Once the temperature gets close to 40°C, thermophilic bacteria initiate degradation processes, making the temperature to reach 65°C (under these conditions the metabolism of certain fungi is deactivated). During this stage, biological transformation reactions are developed by actinomycetes and fungi spore forming bacteria. These quickly consume easily degradable compounds such as sugars, proteins, starch, and fat.

In addition, the pH tends to be alkaline due to the release of ammonium ion. Once degraded the organic material, the reactions rate decreases as well as temperature. This stage is known as cooling. Both thermophilic and mesophilic fungi are capable of degrading cellulose during this phase. Finally, ripening process begins, which requires several months at least four to be completed. This stage can lead to the complete degradation of compounds and to obtain stable material. The composting process can lead to obtain: stable humus and humic- and fulvicin-acids; characterized by high nutritional value and potential for fertilization of soils with nutriment deficiencies.

Benefits provided by compost are broad and can be from the physical, chemical, biological and environmental realm. Application of compost depends on the conditions of organic matter, moisture, temperature, the pH and presence of microorganisms in the pile. For example, compost increases drainage and absorption of moisture in soils with structural deficiencies or lack of nutrients. It also permits to:

(1) increase crop productivity,
(2) promote plant growth by incorporation of essential nutrients,
(3) facilitate implementation in different types of soil,
(4) reduce runoff, and
(5) obtain economic benefits for farmers.

Green Manures

Green manures consist of green plant tissue incorporated on the soil to correct or improve physical characteristics or its chemical properties. Fast-growing crops such as oats, vetch, berseem clover, rye, and peas are mainly used as green manures. The use of green manure has a positive influence on certain soil characteristics. For example, soil nutrients susceptible to loss by drainage are retained. On the other hand, certain long-rooted manures capture nutrients from lower soil horizons and have the ability to transport them to the surface, which increases their availability for the development of metabolic processes of plants.

Green manures increase the amount of available organic matter in the soil for development of metabolic processes of native flowers and other plant species. Being in direct contact with the soil matrix, plant material is susceptible to microbial decomposition, which produces humic compounds that are able to increase the adsorption capacity of nutrients, promote drainage, aeration, and soil granulation. In addition, decomposition products serve as a substrate for those microorganisms responsible of biological transformation processes. These processes have a positive impact on the production of carbon dioxide, ammonia, nitrites, nitrates, and other simple compounds that are easily assimilated by plants for growth and development.

Finally, green manure applications can be combined with natural inputs to improve soil structure, minimize erosion, and increase water availability in the soil (*i.e.*, trough evaporation reduction). For example, mineral fertilization with green manure increases yields per hectare and promotes development of mycorrhizal associations. It is also a source of essential nutrients and it is a way to foster development of AMF strains (Arbuscular Mycorrhizal Fungi).

Microbial Inoculants (MI)

Microbial inoculants (MI) are substances or biological aggregates containing microbial populations as fermentation fungi, bacteria, and lactobacilli. Their high nutritional content of salts allows reactions with organic matter in the soil, producing favorable substances for plant nutrition (*e.g.*, vitamins, organic acids, chelated minerals, and antioxidants). Microbial inoculants are capable to modify characteristics of the soil such as micro- and macro-flora and can improve biological balance. In addition, their antioxidant properties promote decomposition of organic matter and increase humus content in the soil matrix. The latter has positive effects on plant growth, quality of harvests, and improvement of chemical, physical and biological stability of soils.

With a rational use of MI, certain physical, chemical, and biological properties can be improved and suppression of biological diseases can be achieved. In this regard:

- On physical conditions: improvement of structure and aggregation of soil particles, reducing compaction, and increasing the pore spaces and water infiltration.
- On chemical conditions: improvement of nutrients availability in the soil, leaving free elements to facilitate their absorption by the root system.
- On soil microbiology: suppression or control through competition of pathogenic populations of microorganisms present on the soil. MI increase microbial biodiversity creating suitable conditions for the development of beneficial microorganisms.

Seaweed Extracts

Seaweeds are mainly composed by trace, and major- and minor-elements helpful for plant nutrition. Other natural substances can also be found, whose effects are similar to those of certain growth regulators, vitamins, carbohydrates, proteins, and biocidal substances. They act against some pests, diseases, and chelating agents such as organic acids and mannitol. The benefits of seaweed use in agriculture (greater efficiency and better fruit quality) may be evident from direct application of pure forms or its derivatives. Species such as Ascophyllum nodosum contain macronutrients and micronutrients needed for cellular nutrition. Recent research has showed that vitamin supplements coming from these species can increase agricultural productivity and revenues. Similarly, it can promote availability of sugars, increase fruit size, minimize the time of cultivation, and help to obtain better shapes and tones of agricultural products.

On the other hand, cyanobacteria (*i.e.*, blue-green algae) are good at obtaining phosphates and micronutrients from media that may contain insoluble minerals. This ability gives them a level of superiority over other species, since they can supply essential nutrients for fertilizing soil and other substrates. The Oregon State University evaluated the effects of applying seaweed extracts in apple orchards (*i.e.*, apple trees). Two treatments were administered in areas of 1 acre. The first consisted on applying half pound of fungicides and herbicides, while the second on applying half pound of seaweed extracts. The authors concluded that 80% of orchards treated with seaweed extract produced fruits with better physical-chemical characteristics. Also 4% increases in yield per cultivated acre were reported by researchers.

Vermicomposting

Vermicomposting is a biological fertilization technique consisting on the use of earthworms' metabolism to produce humus with high nutrients content. To apply it, organic waste is required (*e.g.*, manure, fruit peel, crop residues). The organic material passes through the digestive tract of worms, where it is transformed into a material rich in microorganisms, macro-, and micro-nutrients. Based on that process, a chemical and biological stable fertilizer is obtained. Use, storage, transport, and application of vermicompost in soils are of particular interest for those soils with nutritional shortages. This technique can be developed and applied successfully at both small and large-scale in various environments or under controlled laboratory conditions.

Most common earthworm species used in vermiculture are: Eisenia foetida (*i.e.*, Red California) and Eudrilus eugeniae. The former presents advantages related to its rapid rate of reproduction under conditions of high temperature (above 40 ° C), high densities tolerance (10,000 to 50,000 worm/m2), and resistance to large variations in temperature, pH and moisture. In addition, this species have the ability to thrive in different substrates, under variable conditions. Therefore, Eisenia foetida is the most used worm in vermiculture. On the other hand, Eudrilus eugeniae is a rapidly growing worm, prolific, but difficult to manage, since removal from the substrate is much more complex than that of Eisenia foetida.

Recently, effectiveness of vermicomposting in Havana, Cuba, has been proved. The research study conducted by Berc et al (2005) addressed the need to increase productivity and reduce adverse environmental impacts caused by indiscriminate use of chemical fertilizers and pesticides to get better agrarian practices in their geographical area. The authors emphasized on the need of taking advantage from climatic (*e.g.*, solar radiation, rainfall, multi-year average temperature), economic (*e.g.*, agricultural potential), and social (*e.g.* availability of cheap labor) conditions to incorporate vermicomposting techniques and identify their advantages, disadvantages, and limits. The authors got yields exceeding 1.5 tones of humus in a year (in tropical zones) from three tons of organic substrate and 1 m3 of treated soil. Humus resulting from the process exhibited physical, chemical, and microbiological properties suitable for application on local soils with nutrients deficiencies such as phosphorus (P), nitrogen (N), and potassium (K).

Efficiency and Benefits of Biological Fertilization

In the past two decades, agricultural activities based on conservative agriculture (CA) have become largely visible in the world. These have produced favorable results on agricultural productivity and sustainability of agriculture, for both traditional and extensive techniques.

Development of research focused on biological fertilization has increased over the last ten to fifteen years. The increasing number of annual publications in international journals such as Scientia Horticulturae, Crop Production, Food Science and Technology, and Bioresource Technology indicates a widespread interest on the subject. In addition, professionals and specialized personnel are being required to solve different issues around the globe concerning mass production and environmentally friendly practices.

In the next section of this paper, some national and international researches based on the use of biological fertilizers for agricultural production are presented and discussed.

Alegre and Morales (2000) conducted a research on the East coast of Peru, evaluating the influence of green manures (cowpea and Crotolaria sp.), varying doses of cattle manure and pea compost over performance of potato crops. Yields of 55 t/ha (*i.e.*, tones per hectare), 53 t/ha, 47.7 t/ha, 43.3 t/ha, and 33.3t/ha were obtained. These productivity values were not reported previously for that same species in the study area. The latter mentioned verifies the efficiency of biologic fertilizers in agricultural production. On the other hand, some physicochemical properties of the soil were improved and environmental impacts (caused by the continuous use of chemical fertilizers) were gradually mitigated.

Singh and Sharma (2003) evaluated the growth of kidney bean plots fertilized with vermicompost from degraded municipal solid waste. The researchers used Eisenia foetida earthworms. The study was conducted in New Delhi by the Indian Institute of Technology. The results showed that the combination of microbial inoculants (*i.e.*, P. sajor-caju, T. Haezianum and A. Chrocooccum) with vermicomposting has a positive effect on the growth of kidney bean crops. The significant role played by fungi on the degradation rate of solid waste and the significant involvement of bacteria in the atmospheric fixation of nitrogen and subsequent transformation into more available forms were emphasized by the authors.

Rajendran and Devaraj (2004) reported 40% increases in the levels of phosphorus (P) and nitrogen (N) through the use of the microbial species: Azospirillum, Phosphobacterion, MA and Frankia, during the growth process of Australian pine (Casuarina equisetifolia). The authors concluded that the growth phase of the pine is positively affected by incorporating microbial inoculants on the substrate.

In addition, Berc *et al* (2004) identified that worm from the genus Eisenia foetida (*i.e.*, Red California) used to fertilize crops such as tobacco, cacao, coffee, and rice have a high potential for agricultural development in tropical areas. Specifically, the authors reported 30% yield improvements by using the mentioned species on potato, banana, tomato, garlic, coffee, and cocoa crops. Similarly, the amount of chemical fertilizers used was reduced by 40%.

Alfonso *et al* (2005) used mineral fertilizers to improve yield performance on tomato crops. The authors reported water savings close to 25%, pesticide use savings of 40%, and 25 to 30% reduction of mineral fertilizers use. Furthermore, increases of 10% yield per hectare were observed for crops treated with Arbuscular Mycorrhizal (AM). The resistance of plants to the action of pathogenic microorganisms was higher with the use of AM.

Treatment of cows' feces through anaerobic and aerobic digestion has been used as a strategy for biological control of fungal species in crops, as reported by Kupper *et al* (2006). The authors used a bio-fertilizer obtained from biological digestion of feces to control the growth of Guignardia citricarpa sp., which, for years, incorporated black spots on the peel of citrus fruits like orange, tangerine, and lemon. It was concluded that compost manure coming from cattle is useful for biological control and obtaining of better fruits, suitable for sale and consumption.

Padilla *et al* (2006) reported increases close to 36% in the production of melons from the incorporation of arbuscular mycorrhizae (AM) in crops. Besides that, 100% saving in the use of phosphate fertilizers, 20% saving on potassium fertilization and nitrogen, 25% saving of water, and 100% reduction in fungicide use were obtained during the research study. In this sense, the authors concluded that polyethylene padding techniques, along with microbial inoculants (*i.e.*, probiotics) in the cultivation of melons, can increase proportions of mycorrhizae in the soil and favorably influence the elimination of pathogenic fungi (*e.g.*, Alternaria , Fusarium and Rhizoctonia).

Classen *et al* (2007) quantified the effect of vermicomposting from pig excreta on growth process of turnips. The researchers assessed the influence of rainfall regimes and the amount of bio-fertilizers applied on the yield per acre. As a result of the experimental design, increases in the size of turnip leaves and higher growth rates were reported. Fresh fruits size obtained by using the technique was superior to that of the control treatment. The authors concluded that productivity increased by a factor between 2 and 5, suggesting benefits for the application of vermicomposting on tropical areas.

Responses on the photosynthetic process of coffee plants by addition of organic fertilizers have been evaluated by Gordillo *et al* (2008). The authors used photo acoustic techniques to compare the photosynthetic activity of plants treated with chemical fertilizers and others with bio-fertilizers (*i.e.*, from microbial type). The procedure consisted in applying white light, from an artificial xenon lamp, on two 7-months-age coffee plant treatments to evaluate the production of oxygen (O_2) and energy storage.

For assessment and monitoring purposes, photo acoustic and pressure detectors were used. The first treatment received chemical fertilizers, while the second received organic fertilizers. According to the authors, organic fertilizers encourage and accelerate photosynthetic activity, allowing it to be faster and minimizing stress in plants. The authors concluded that biological products for fertilization of coffee are a sustainable and a clean alternative that can be replicated in different geographical areas.

Gharib *et al* (2008) evaluated the potential application of compost and bio-fertilizers in oregano crops (Majorana hortensis L). Researchers created a microbial inoculum with a mixture of Azospirillum brasiliense, Azotobacter chroococcum, Bacillus polymyxa, and B. circulans. Main results indicated that combined use of bio-fertilizer on crops provides better yield performance, higher by a factor of two, and better physical characteristics of individual plants than those from nitrogen fixing bacteria or compost extracts. The authors concluded that inoculation of oregano (marjoram) with 15% and 30% compost extracts and bio-fertilizer mixtures has beneficial effects on plant growth, fat content, and dry matter production, as a result of hormonal stimulation and addition of combined forms of nitrogen (N).

Sierra and Moreno (2008) investigated the feasibility of bio-fertilizers prototypes based on native bacteria from rice crops (Oryza sativa). The research used complex mixtures of nitrogen-fixing bacteria with humic substances , and chelates. The formulations rate of application was considered for a 3-month treatment period. The authors reported 10% increases in yield production of the mixtures, passing from 7,625 kg/ha to 8,500 kg/ha. Main outcomes deal with the importance of bio-fertilizers to get higher revenues and increase productivity, in order to obtain, progressively, a sustainable agricultural development.

The effect of applying commercial bio-fertilizers from microbial type, phosphoric, and Cerealien® in tangerine crops was evaluated by Mohamedy and Ahmed (2009). The research consisted on the application of varying doses of bio-fertilizers solely or combined in mandarin crops. The response variables took into account were: type and size of fruit and presence of root diseases (*i.e.*, rotting of the roots). According to the authors, the use of biological fertilizers, combined with humic acids, can reduce the incidence of disease on the roots by 20%, as well as increase productivity by 15% and improve physical characteristics of tangerine fruits.

Bocchi and Malgioglio (2010) used aquatic cyanobacteria Azolla-Anabaena as fertilizer in rice paddies of northern Italy. This species have been frequently used in rice crops in different regions of the world, especially in Asia (*e.g.*, Thailand, China, Indonesia, and India). To monitor the process, dynamics of plant growth, resistance/tolerance to low temperatures, and presence of herbicides in the soil were taken into account. The research allowed obtaining yields close to 40 kg nitrogen/ha during a 3-month period and verifying increases in growth rate of rice. Furthermore, higher resistance of Milan species (named by authors) to the presence of herbicide Propanil was evidenced.

Thenmozhi et al (2010) quantified the increase in biomass production and growth of Amaranthus and hard pea species by using bio-fertilizers (*i.e.*, mix of biogas, vermicomposting, microbial inoculants Azospirillum and Pseudomonas, and a combination of them). Sole or combined additions of organic fertilizers were evaluated and the main outcomes were that the application of organic fertilizers in combination is much more efficient than individually in terms of plant growth. Moreover, the use of biogas and vermicompost permitted to identify better growth of plants (*i.e.*, with larger leaves and healthier roots) and biomass production in a 20-days period.

Conclusions and Recommendations

- Biological fertilization techniques are pertinent strategies for an efficient and rational use of agricultural resources with minimal generation of adverse environmental impacts that may affect water resources, ecosystems or the quality of human life. In addition, biological fertilizers provide a wide range of possibilities for the development of conservative agriculture (CA) in different geographic, economic, and cultural backgrounds.
- Current researches clearly show that biofertilization techniques require less chemical inputs on the soil and facilitate the incorporation of residues that would otherwise go to dumping sites and landfills, which represents relevant reductions on the environmental impacts associated to agriculture activities globally.
- Limitations of biological fertilization require future research focused on identifying the options available to tackle the issues and offer valid frameworks for development of environmentally friendly practices around the world that allows improvements on the efficiency and consequent supply of product for the industry in the global economies.
- Although several options for application of biofertilizers are available, feasibility studies should be carried out by producers and farmers to effectively select the best option that offers better results and allows minimizing environmental impacts.
- Biosolids, animal manures, green manures, composting, microbial inoculants and seaweeds extracts are techniques widely used in today's agriculture, however, their implementation still requires research, investment, and technological development to fully understand their impacts on the soil, flora, fauna and, ultimately, on human health.

MYCORRHIZA: BIO-FERTILIZER FOR THE FUTURE

As different factors like lower-fertility, arsenic contamination, pollution of air and water coupled with draught and attack of pathogens put a dent on harvest Professor Dr Md. Amin Uddni Mridha of Chittagong University comes up with bio-fertilizers and methodology of Mycorrhizal fungi application to overcome the problem and ensure food security with a good yield.

An agricultural scientist of Plant Pathology and Microbiology, especially Mycorrhizal Technology Dr Mridha recently talked to this writer. A pioneer of mycorrhizal research in Bangladesh, he has identified more than 40 different types of endomycorrhiza and 25 different types of ectomycorrhizal fungi used for producing bio-fertilizers that could successfully help crops survive odds and bring forth increased yield.

Dr Mridha also spoke of methodology he recently developed with Arbuscular Mycorrhizal Fungi to reduce the arsenic contamination in crops in collaboration with scientists from USA under a project financed by USDA. Study on biodiversity of wood decay fungi helped him record a number of plant diseases, mainly in medicinal plants, and evolve method for their sustainable management with biopesticides.

Mycorrhizas are highly evolved non-pathogenic symbiotic association between roots of most vascular plants and certain specialized soil fungi (Basidiomycetes, Ascomycetes and Zygomycetes) that colonize the cortical tissues of roots during periods of active plant growth both in natural environment and in cultivation, he said.

Dr Mridha came up with the following findings and suggestions regarding the application of bio-fertilizer producing mycorrhizal fungi.

Plant Growth in Low-fertile or Acid Soil

The beneficial effects of Vesicular-Arbuscular Mycorrhizas (VAM) used in bio-fertilizers to plant growth, especially in soils of low fertility, are well documented, said the professor.

In the tropics, many crops are grown in infertile acid soils, where low levels of available phosphorus frequently limit their establishment. In such soils, an efficient mycorrhizal association can increase phosphorus uptake and crop yield.

In addition to enhanced P (phosphorus) uptake, VAM fungi often enhance acquisition of relatively immobile micronutrient cations, particularly Zn (zink) and Cu (copper). Vesicular-arbuscular mycorrhizas are also important for N (nitrogen) uptake to stimulate the growth and nutrition of plants and are of great ecological importance with regards to N-nutrition of plants, especially non nitrogen-fixing species.

"In mycorrhiza laboratory of Chittagong University we have studied the influence of mycorrhizal inoculation in growth improvement of a large number of agricultural crops, horticultural crops, ornamental plants, forest plants, tea, rubber and other plantation species. In most of the cases we have recorded higher growth," said Dr Mridha.

"During these study we have studied the biodiversity of a large number of Arbuscular Mycorrhizas (AM) fungi belonging to six different identified genera. Out of the 126 species recorded all over the world we have recorded more than 40 different species from our soils in our laboratory for the first time working last 15 years. The fungi recorded in our laboratory have not been reported earlier from Bangladesh," he maintained.

Transfer of Carbon and Nutrients

Vesicular-Arbuscular Mycorrhizas play an important role in the direct transfer of carbon from one plant to another. Nutrient transfer can occur between plants of the same species or between different species and, even in some cases, from trees to herbaceous plants. The plant roots are interconnected by mycorrhizal hyphae through which the nutrients move from one plant to another. Vesicular-Arbuscular Mycorrhizas can mediate interplant transfer of P and N. When there are legumes and non-legumes growing in N deficient soil, this flow of N will likely be from legume to non-legume and may be of considerable significance to the N-economy of the non-legume. The interplant nutrient transfer technology may be useful for Bangladesh agriculture. The implementation of Rhizobium technology available in our country needs mycorrhizal fungi for P nutrition, which is needed as energy of Rhizobium bacteria for better fixation of N from the atmosphere.

Tolerance Against Attack of Root Pathogens

Several studies have demonstrated, the potential of using VAM for establishment and growth of different micropopagated species after transplanting. Many reports have shown that VAM can decrease the severity of diseases caused by root pathogenic fungi, bacteria and nematodes. The VAM appear to decrease plant susceptibility to disease, or increase tolerance against the attack of root pathogens.

"In our laboratory we have recorded plants previously inoculated with fungal symbionts as exhibiting an increased resistance to fungal root diseases like wilts and rots," Dr Mridha said.

Reports have also indicated that VAM increase plant tolerance to salinity, thereby increasing yield in saline soils. The VAM can help plants to become established in nutrient deficient soils or degraded (eroded) habitats and can assist plant establishment in coal waste, eroded desert and disturbed soils, he said.

Land-rehabilitation and Fertility

Land rehabilitation induces the mediation of soil aggregation -- a process known to be carried out by VAM fungi. Vesicular-arbuscular mycorrhizal fungi can improve plant-water relations in many situations. Through maintaining the uptake of slowly diffusing nutrients under water stress situations the VAM fungi can help plants cope with drought stress.

Mycorrhizal fungi play a significant role in the regulation of soil biological activity because of their abundance throughout the uppermost soil layer. It is interacting with other beneficial soil microbes to improve the soil fertility where soil biological activities are reduced because of burning and shifting cultivation. In last fifteen years of mycorrhizal studies it has been established that mycorrhizal inoculation enhance plant growth in degraded soil, saline soils, water stress condition, flooded soils *etc.*

Methodology to Reduce the Arsenic Contamination in Crops

More than 30 million people are already facing the problem of arsenic contamination in our country. The ground water contaminated with Arsenic poses significant risk to animal and human health through soil-crop transfer, through food crops like vegetables and cereals where arsenic entere the food chain. Arsenic poisoning causes skin pigmentation, development of warts, ulcers, cancer *etc.* To reduce the arsenic contamination in crop plants a number of experiments conducted under USDA financed project indicated that using mycorrhizal fungi could reduce the arsenic uptake by the crops.

We have also developed technology by which soil arsenic can be reduced by hyper accumulated plant like Pteris vitata.

Use of VAM Fungi in Agricultural Management

Propagation of inocula for large-scale inoculation is a major problem in popularizing VAM fungi for use in agricultural management because of fungi are obligated symbionts and cannot be multiplied artificially. Although the fungi cannot be cultured in the laboratory, large-scale production techniques under field condition for VAM fungi have been developed in mycorrhiza laboratory in Chittagong University and the inoculum is available for distribution among the farmers.

Dr Mridha said, "The inoculum production technology developed in our laboratory is very useful and unique because we have developed this technology with our own mycotrophic plants available in our country only. The inoculum is ready for distribution among the farmers."

Use of Mycorrhiza Needed to Improve Agricultural Productivity

The reduction of water and air pollution from the leaching loss of chemicals and reduction of soil erosion because of mycorrhizal association can improve the soil health and environmental conditions. It is now recognized that VAM can be harnessed in order to improve productivity in agriculture, fruit culture, and forestry by reducing the input of fertilizers and/or by enhancing plant survival, thus offsetting ecological and environmental concerns, observed Dr Mridha.

For this reason, mycorrhizal research and its practical use as a low input technology for managing soil fertility and plant nutrition in the country is drawing increasing attention, he maintained.

BIO-PESTICIDES AND BIO-FERTILIZERS FOR SUSTAINABLE SERICULTURE

The 20th century has witnessed a slow but steady emergence of bio-pesticides and bio-fertilizers as potential supplementary and environment friendly inputs to their chemical counterparts. With the documentation of nearly 2500 bioactive plant species, over 1000 protozoa pathogenic to insects, 700 species of invertebrates and an array of other micro and macro-bioagents; their role in future crop protection cannot be ignored.

Likewise; bio-inoculants or bio-fertilizers, primarily the nitrogen fixers and the phosphate solubilizers, hold vast potential in meeting plant nutrient requirements while minimizing the use of chemical fertilizers. Furthermore, successful exploitation of genetic engineering technology (metagenomics) to incorporate non-native pesticide producing genes into crop plants has added another dimension to the potential use of these inputs in agriculture. Currently, Asia-Pacific countries are the leading advocates of bio-pesticides and bio-fertilizers for sustainable agriculture.

Agriculture Practices and its Impact on the Earth

In the beginning of 21st century, agriculture accounts for the major share of human use of land. Pasture and crops alone taking up 37 percent of the earth's land area and over two-thirds of human water use is for agriculture (in case of Asia it is four-fifths). Crop and livestock production have a profound effect on the wider environment. They are the main source of water pollution by nitrates, phosphates and pesticides, also the major source of the greenhouse gases methane and nitrous oxide and contribute on a massive scale to other types of air and water pollution.

The overall external costs of all three sectors can be considerable. However, the long-term consequences of these processes are difficult to quantify. If more sustainable production methods are used, the negative impacts of agriculture on the environment can be attenuated. Indeed, in some cases agriculture can play an important role in reversing them, for example by storing carbon in soils (*i.e.*, organic carbon) enhancing the infiltration of water through bio-fertilizers and bio-pesticides.

Role of Fertilizers and Pesticides in Sustainable Environment

Concern has also grown in recent years that the use of fertilizers, particularly inorganic fertilizers, can lead to serious environmental consequences.

Environmental contamination of this type, however, is largely a problem in the developed world and a few regions of the developing world. As fertilizers make up a small share of the total production costs in many developed countries, farmers often apply fertilizers in excess of recommended levels in order to ensure high yields. Over application of inorganic and organic fertilizers is estimated to have boosted nutrient capacity in the soil by about 2000 kg of nitrogen, 700 kg of phosphorus, and 1000 kg of potassium per ha of arable land in Europe and North America during the past 30 years. Such oversupply of nutrients can lead to environmental contamination, which often has negative consequences for humans and animals.

Pollution of groundwater by agricultural chemicals and wastes is a major issue in almost all developed countries, its use efficiency has to be improved or use alternate source of manures. Though the current use in many developing countries is very inefficient, China which is the world's largest consumer of nitrogen fertilizer, up to half the nitrogen applied is lost by volatilization and another 5 to 10 percent by leaching. Nitrogen synthetic fertilizers are considered the most detrimental to the environment, causing leaching and runoff that freshwater habitats and wells. Nitrogen synthetic fertilizers are a major contributor to increased N_2O emissions, which are 300 times more potent than CO_2 as greenhouse gas, which is ominous for global warming as synthetic fertilizer use is forecasted to increase roughly 2.5 times by mid-century.

Insecticides, herbicides and fungicides are also applied heavily in many developed and developing countries, thus polluting the fresh water resources with toxic and other poisons that affect humans and many forms of wildlife. Pesticides also reduce biodiversity by destroying weeds and insects and hence the food species of birds and other animals.

Over the past 3 to 4 decades, usage of insecticides show signs of decline, both in developed countries especially France, Germany and the United Kingdom and in a few developing countries, such as India. In contrast, herbicide use continued to rise in most countries. As concern about pollution and the loss of biodiversity grows, future use of pesticides may grow more slowly than in the past. The future is likely to see increase use of "bio- pesticides and ecological methods of disease and pest control.

A pesticide of biological origin like viruses, bacteria, pheromones, plant or animal compounds is known as bio-pesticide. They are highly specific affecting only the targeted pest or closely related pests and do not harm humans or beneficial organisms while chemical pesticides are broad spectrum known to affect non-target organisms including predators and parasites as well as humans. The striking feature of bio-pesticides is environment friendliness and easy biodegradability, thereby resulting in lower pesticide residues and largely avoiding pollution problems associated with chemical pesticides. Further, use of bio-pesticides as a component of Integrated Pest Management (IPM) program can greatly decrease the use of conventional (chemical) pesticides, while achieving almost the same level of crop yield. However, effective use of bio-pesticides demands understanding of a great deal about managing pests especially by the end users.

In terms of production and commercialization also bio-pesticides have an edge over chemical pesticides like low research expenditure, faster rate of product development as well as flexible registration process. The global weighed average consumption level of bio-pesticides is approximately 1 kg/ha. With the global organic farming area comprising about 24 million hectares, global bio-pesticide consumption is thus estimated at about 24 million kg. The bio-pesticide market is growing very rapidly, in 2005 it accounted for about 2.5% of the total pesticide market, which was merely 0.2% during 2000. This share is expected to a level of about 4.2% by 2010 while the market value is estimated to reach more than US$ 1 billion. However, the overall growth rate of bio-pesticides is estimated to be about 10% per annum for the next 5 years.

Research Experiences Towards Sustainable Sericulture Development

The Sericulture-based Integrated Farming System Management (IFSM), Integrated Nutrient Management (INM) and Integrated Pest Management (IPM) programmes are farmer-centered and farmer-participatory aimed at promotion of eco-friendly approaches for sustaining productivity and also optimization of resources utilization. These two projects were implemented in Andhra Pradesh, Tamilnadu and Karnataka states during 2002 to 2007. Under these projects, a number of technologies were evolved towards sustainable sericulture which can be translated to other crops. The technologies developed were of integrated nature and cost effective serving the purpose of eco-friendliness, sustainable development and reduction in cost of production in silk.

More poor people depend upon sericulture for their livelihood. Poverty alleviation necessarily demands increased sericulture productivity, which is a largely a function of 'technology'. Absence of technological intervention would lead to increased depletion of natural resource base. Poverty subverts efforts to introduce sustained sericulture in ways detrimental to the environment. This would lead to a decline in sericulture production potential, and sustaining capacity of sericulture would also decrease. With the ever-increasing population poverty gets perpetuated widely and deeply, furthering environmental degradation.

'Silk and Milk'

Karnataka, amply termed as silk bowl of India, from three southern states around 1.4 lakh ha. of mulberry, wherein approximately 42 lakh MT of annual production is estimated. The wastage (bed refusal) of silkworm rearing activity is used as cattle feed; silkworm litter or faucal matter as organic manure; pupae for extraction of oil / feed for poultry *etc.* One hectare of mulberry garden can produce about 25- 28 tones of mulberry leaf year from which 1200- 1500 kg of cocoons are produced. During this activity, about 20% (750-10000 kg/ha/crop) of leaf and/or twigs which is considered as bed refusal is fit to use as fodder for cattle. Bed refusal (uneaten leaves and tender stump) produced during rearing is effectively used by the farmers to rear cattle. It constitutes about 50% of the fodder requirement of cattle.

Further, the bed refusal and tender stumps produced out of one hectare of mulberry is almost sufficient for 7 -8 cows. Apart from this, around 5% of the mulberry, which will either be soiled or unfit due to over maturity, can be used as fodder to the cattle. Rearing of eight cattle along with silkworm rearing will in turn meets the FYM requirement of one hectare of mulberry plantation. Valuable FYM will also be produced at the farm itself. Mulberry leaf stalk/stump forms the low cost/no cost alternative fodder for the cattle. By and large it is important that dairy animals have good appetite and mulberry leaf is palatable and the animal eats it very well.

Soil Health

Proper soil management without impairing soil health is the prerequisite for achieving higher productivity in sericulture. The situation has created a renewed interest in the biological transformation of sericultural farm wastes including all organic wastes into valuable manures which can be profitably used in the mulberry fields for the cost-effective production of mulberry leaves. This nutrient-rich compost is a better supplement for farmyard manure, the cost of which is escalating day by day due to shortage of cattle dung.

Integrated Plant Nutrient Supply

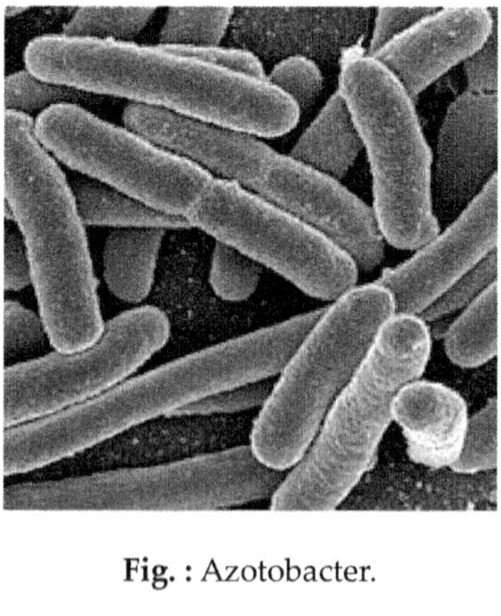

Fig. : Azotobacter.

Heavy use of nitrogen started up showing in increased nitrate concentration in underground waters in states like Maharashtra and Punjab. Due to these alarming signals, Integrated Plant Nutrient Supply (IPNS) has become a sustainability concept. IPNS involves meeting a part of the nutrient need of crops by organic manures, crop residue, green manures, dual-purpose legumes and bio-fertilizers. The need for IPNS was suggested by long-term fertilizer experiments on wheat. Application of N alone reduced the yield of wheat to zero after 13 years. Application of P along with N extended it to 22 years but in NPK plot some yield was obtained even after 22 years.

However, plots receiving a part or all nutrients through FYM not only maintained but also recorded a gradual increase in crop yield over years. Combined application of FYM and inorganic fertilizer was the best.

Bio-fertilizers: In recent years, use of bio-fertilizers holds much promise to improve the yield of crops. Bio-fertilizers sustain soil fertility resulting in increased crop yield without causing any environmental, water, or soil hazards. Out of many microorganisms, which are identified as biofertilizers, *Azotobacter chroococcum, Azospirillum brasilense, A. lipoferum,* Phosphorus Solubilizing Bacteria (PSB) and Vesicular Arbuscular Mycorrhizae (VAM) have proved to be beneficial in mulberry nutrition.

VAM

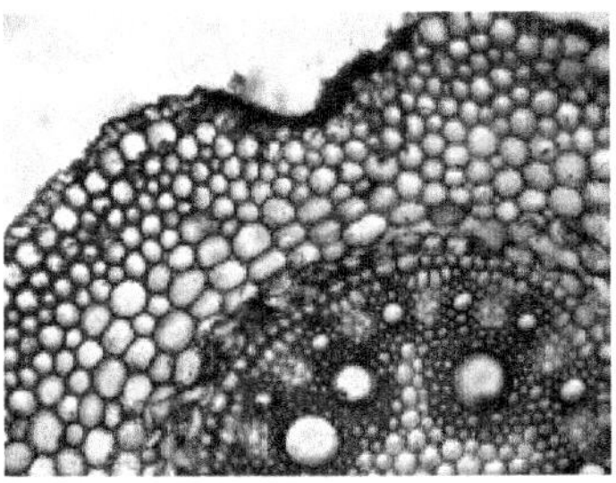

Fig. : VAM

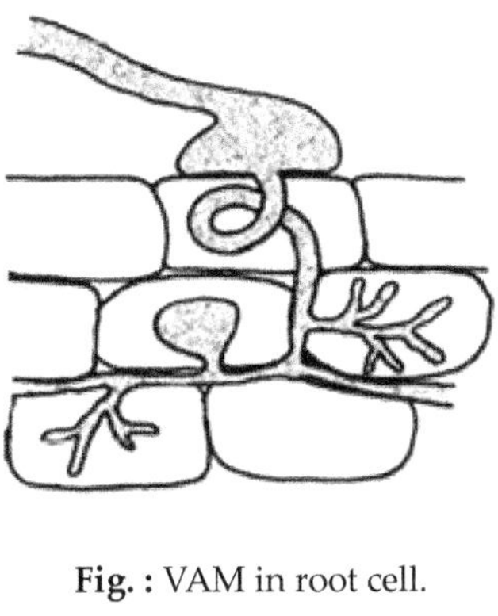

Fig. : VAM in root cell.

VAM fungi in association with higher plants play an important role in phosphorus nutrition and increase plant growth and yield. Application of phosphorus @ 30 kg/ha/year as Mussorie Rock Phosphate (MRP) with *Glomus mossae* inoculation significantly increased the cocoon yield. Moreover, cocoon quality due to application of phosphorus @ 30 kg/ha/year in the form of MRP with *G. mossae* and G *fasciculatum*inoculation was statistically at par with plant receiving full dose of phosphorus without VAM inoculation. Thus curtailing 75% of phosphorus application in mulberry cultivation and reduce of cost of cultivation. Combined inoculation of VAM with diazotropic *Azospirillum brasile*Add caption*nse* yields more mulberry leaf biomass with high nutrient level than inoculation of individual organisms.

Phosphobacterium

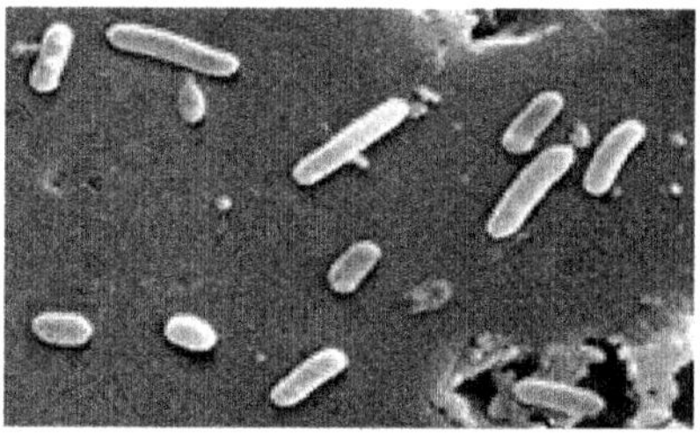

Fig. : Phospho bacterium.

Similarly to VAM and PSM the *Phosphobacterium* when applied to the soil, it solubilizes the insoluble phosphorus in the soil to make it available to the plants for absorption. This helps in efficient use of Phosphorus and save application of Phosphorus. Many of the cultivated soils contain high amount of total phosphorus but availability is limited due to insoluble forms. Only water-soluble phosphorus is useful for the crop plants. Many bacteria are capable of solubilising insoluble soil phosphorus, of which the role played by Phosphorus Solubilizing Bacteria - *Bacilus megatherium* var *phosphaticum* is significant.

Organic Manures

With a view to minimize the use of chemical fertilizers, farmers are able to produce different organic manures and apply as part of IPN system. Recycling of Sericultural Farm Wastes as Compost: It was established that various composts prepared out of sericulture farm wastes on mulberry revealed that application of compost prepared out of silkworm rearing wastes @ 20 MT per hectare per year or application of recommended dose of farm yard manure (20 MT per hectare per year) and chemical fertilizers also produced same quantity of leaf as against the conventional farm yard manure and chemical fertilizers. Two years crop data showed significant increase in leaf yield (61.5 tonne/ha) with application of compost @ 4.5 tonne/ha/year in conjunction with full dose of chemical fertilizers. Compost application @ 4.5 MT/ha/year (Prepared from silkworm rearing waste mixed with biogas spent slurry) + 300: 120: 120 NPK kg/ha/year produced a leaf yield was 57.8 tonne /ha/year. The study indicates that the compost and vermi-compost produced out of sericulture wastes are qualitatively superior over FYM.

Fortification of Vermi-compost

Fortifications carried out by blending the sericultural wastes with microbial inoculum - Azotobacter, phosphate solubilizing microorganisms (PSM) and add rock phosphate and single super phosphate. The process of vermi-composting of sericultural wastes (200 kg/trench) was carried out by introducing a mixed culture of juvenile earthworms *Eudrilus eugeniae, Eisenia fetida* and *Perionyx excavatus* @ 500 grams per trench. The chemical analysis of vermi-compost had organic carbon content ranged from 9.15 to 10.65%. NPK contents (2.45%; 1.75% and 1.17% respectively) was maximum in Semi decomposed sericultural farm wastes fortified with both S.S.P. and R.P. @ 12.5 kg each/MT of wastes and by inoculating with *Azotobacter* and P.S.M. Each cycle of vermi-composting was completed in 60 days compared to all other treatments.

Green Manure Compost

A large number of leguminous crops are used for green manuring in South India. Daincha *(Sesbania aculeata),* sunnhemp (*Crotalaria juncea*), wild indigo *(Tephrosia purpurea),* indigo *(Indigofera tinctoria)* and *Pillipesara (Phaseolus trilobus)* are the most popular crops with farmers. A number of other leguminous crops like cowpea, lablab, red gram; *Crotalaria striata, etc.,* are also to able to fix atmospheric nitrogen and do not necessarily depend on the nitrogen in the soil. The phosphorus present in green manure is in organic combination and becomes available for use by mulberry in the succeeding harvests.Green manure decomposes easily without leaving much of the residue in the soil. The amount of humus added to the stock already in the soil is negligible. In this respect it is different from farm yard manure. If, however, the green manure crop is allowed to mature and is ploughed in, it decomposes slowly and some humus is added to the soil.

Pesticides of Plant Origin

One of the viable alternatives to chemical pesticides is the development and use of better botanical pesticides, which are environmentally safe, effective against pests, easy, and economic. As plants and insects have co-evolved over millions of years, they have accumulated specific secondary plant chemicals (SPCs) to counteract the insect damage. These bioactive chemicals include insecticides, anti-feedants, Insect Growth Regulators (IGRs), juvenile hormones, ecdysones, repellents, attractants, arrest ants, *etc.* Hence, plants are thought to be an important alternative source for chemical pesticides. Over the years more than 6000 species of plants have been screened and nearly 2400 plants belonging to 235 families were found to possess significant biological activity against insect pests.

Towards sustainable environment, health and development certain innovative technologies have been developed recently to control root rot diseases caused by*Rhizoctonia bataticola* (anamorph of *Macrophomina phaseolina*) on crops like mulberry, carrot, banana *etc.*, it is well known that the soil borne pathogens cause severe damage to crops especially in tropical belt where the environment is suitable for the pathogen to survive and damage the host plants and complete their life cycle. A novel method was developed using plant bio-formulations, which are anti-fungistatic, and inhibitor of root rot fungal pathogen in soil. Some of the plant species, which are known to inhibit growth of fungus, like Guizotia *abyssinica* (L. f.) Cass. (ramtilla), *Azadirachta indica* A. Juss. (neem), *Brassica juncea*(L.) Czern. (India mustard) are inhibitors of fungal growth in soil up to 93%. A new formulation was developed 'Navinya' an eco-friendly plant based formulation, which was tested for its efficiency on host plants at hot spot areas where root rot is a major cause of damage to mulberry plantation. The formulation was effective in inhibiting the fungal pathogen in its spread and inhibits the growth and makes the plant recover, to support the host plant small quantity (8%) growth promoting substances like succinic acid, boric acid is added along with inorganic chemicals (12%) in the formulation. Among the 15 villages participated in the trials all found satisfied with the control of the disease in mulberry plantations.

Chapter 13

CONTROL OF NEMATODES BY FUNGI

NEMATODE CONTROL

Fungi utilise many different sources of carbon, some living and some dead. The observation that nematodes can be colonised by fungi, led to the possibility that the interaction might be used to control pest nematodes of plants and animals. Closer examination of the interaction has led to the belief that only partial control of nematodes is possible, and that in a practical sense, populations of pests may be suppressed not eliminated. In the case of plant and animal parasitic nematodes, fungi may be used to regulate nematode populations in conjunction with other processes.

Pest Nematodes

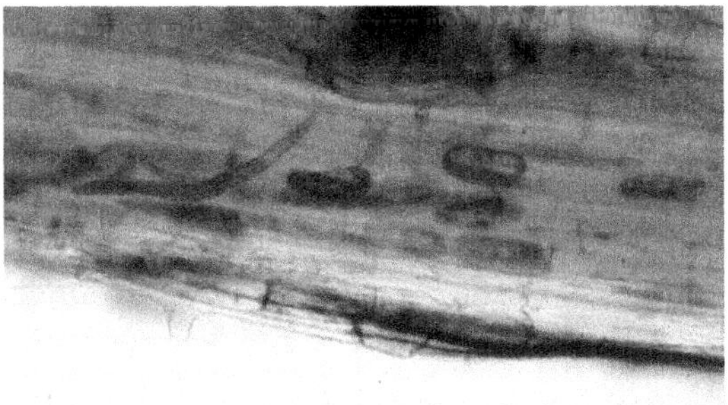

Fig. : *Pratylenchus* in roots.

Nematodes found in plants, animals and soil use various sources of organic energy. Plant parasites feed from plant cells. Sedentary feeders select and then establish a single feeding sites.

Other plant-parasitic nematodes are migratory, moving from site to site on the root, rarely feeding from a single cell for a long time. Migratory plant parasites can be further categorised as internal and external feeders, meaning that they migrate within or outside the root.

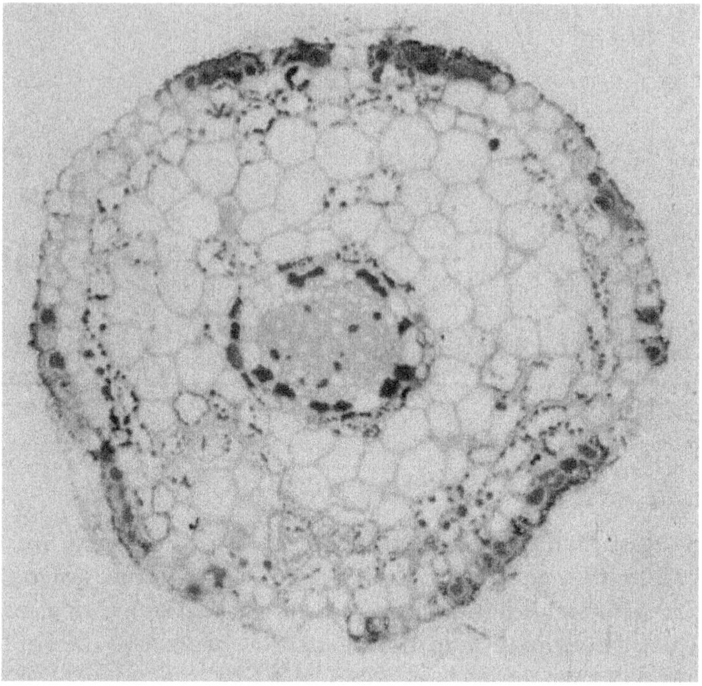

Fig. : Nematode damage in roots.

Animal parasites of agriculturally important ruminants tend to have a life cycle whereby colonisation of the host is followed by a soil phase. Larvae are consumed by grazing animals along with grass. The larvae establish a feeding location within the digestive system of the host. Adults lay eggs within the intestines, and the eggs fall to the ground with the faeces. Eggs subsequently hatch in the faeces and larvae move to blades of grass, where they are consumed.

Fungi that colonise nematodes fall into several different categories, each of which requires a different approach if it is to be used successfully to control pest nematodes. The fungi range from highly specialised nematode parasites, to generalists with no specific host. Some fungi may colonise organic matter, insects and incidentally also colonise eggs, larvae or adult nematodes. Other fungi may trap larvae or adults and colonise their body, using the body as a source of nutritients.

Plant Parasitic Nematodes

Endoparasitic fungi are generally highly specific to single species or genera of nematodes, and are difficult to culture in the absence of their host. *Hirsutella, Meria, Nematophthora* and *Nematoctonus* have all been suggested as ideal biocontrol agents because of the specificity for economically important nematodes. However, the difficulty of culturing them and lack of experimental support for their activity has limited any commercial development.

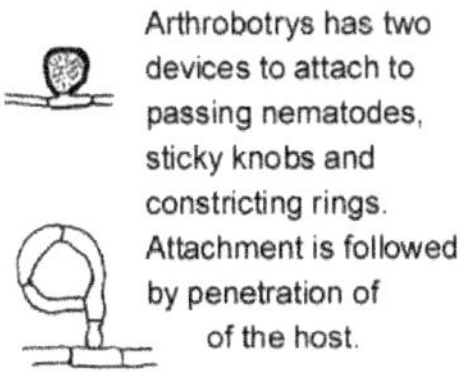

Fig. : Two basic types of trap found in fungi, the sticky nob or the constricting loop.

Nematode trapping fungi utilise either sticky pads or constricting rings to immobilise nematodes. The fungus then digests the internal organs of the nematode and form reproductive cells. The main fungi in this group are *Arthrobotrys* and *Duddingtonia*. These fungi are found in soil where they live on readily-available organic carbon, in the absence of nematodes. They are generalist feeders, being able to trap and digest many nematodes, not just those that infest plants. Thus, the fungi are rare, unable to exist at sufficiently high densities to control nematode populations, and parasitic on beneficial as well as pest species of nematode. Because of these difficulties, exploration of the use of trapping fungi in the field has largely ceased, though see below for a case study of animal pests and *Duddingtonia*.

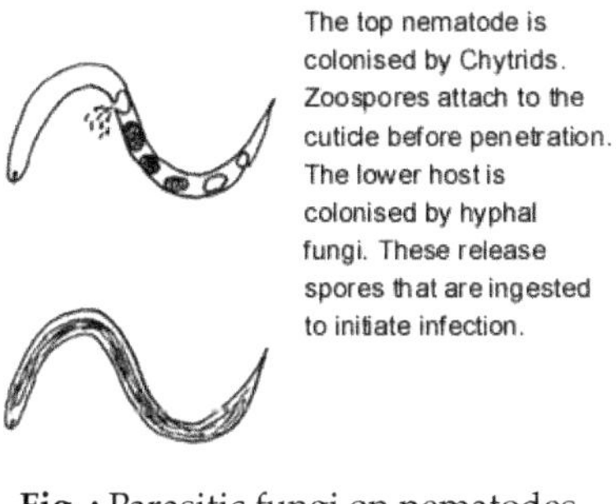

Fig. : Parasitic fungi on nematodes.

Parasitic fungi have been isolated from either eggs, larvae or adults nematodes. Several of these have been found to preferentially parasitise the nematode and thereby reduce the size of the nematode population. Research efforts in some crops are now concentrating on using fungi found locally to control local pests. The general approach is to go to locations where nematodes have reached high densities. Parasitised individuals are extracted from soil, the fungi cultured and then tested as parasites of the pest nematode. The theoretical basis for this approach is that high densities of nematode will enable fungal parasites to increase population size dramatically. Thus by selecting fungi adapted to the host (nematode) and environment, inundative inoculation will control the pest nematode.

Biocontrol in problem fields has been achieved in a surprising number of cases. The fungal species used include *Paecilomyces lilacinus*. The fungus appears to have a wide range of potential hosts, both insect and nematode and plants, and yet isolates have a degree of specificity for each host. The most interesting examples parallel success with control of problematic insects and fungi. Endophytic fungi colonise roots of plants where they access host energy. Hyphae grow from the root where they may contact the parasitic nematodes. The nematode is colonised and damage to the plant limited. Again, sustaining populations of both nematode and parasitic fungus is paramount.

Animal Parasitic Nematodes

Sheep and cattle have traditionally been drenched to control parasitic nematodes. However, nematodes are now developing resistance to the common active ingredients in drenches. Drenches using different active compounds are not being developed because the economical returns are low for the investment. Thus alternative approaches are being sought.

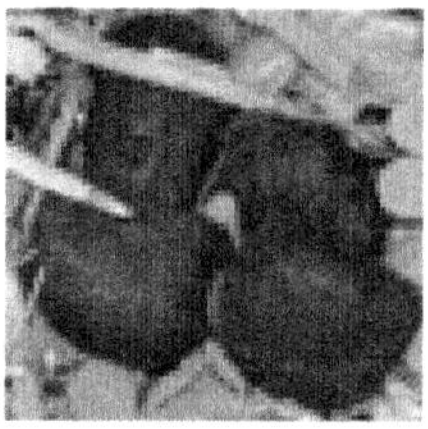

Fig. : Wallaby dung.

In one approach, nematode trapping fungi are placed into nutrient-rich faeces. As nematodes hatch, they become colonised by the fungus. The fungus *Duddingtonia* has been inoculated to the dung via a drench to the animal. Sclerotia of the fungus pass through the intestine, hatch in moist dung, and effectively reduce larval densities IN SOME CASES. While further research is needed before commercial adaptation, this is one example where understanding of the biology of fungi, lateral thinking and some judicious research, has led to the development of a new approach to a serious problem.

Biocontrol of nematodes by fungi appeared to be possible when nematode trapping fungi were first discovered. However, it has taken some considerable time before their potential could be realised. Successful control relies on having sufficiently high densities of fungus where high densities of the pest are maintained. The requirement for high densities of agent to control the pest applies across all biocontrol efforts.

BIOLOGICAL CONTROL OF NEMATODES: PROSPECTS AND OPPORTUNITIES

In the past 20 years three developments have occurred which have had significant effects on the prospects and opportunities for the biological control of plant-parasitic nematodes. First, several nematicides have been withdrawn from the market because of health and environmental problems associated with their production and use. As a result of this, and increasing public concern over the use of pesticides in food production, there has been increased interest in the development of alternative methods of control, including the use of biological agents. Second, it has been demonstrated in several soils that nematophagous fungi and bacteria increase under some perennial crops, and under those grown in monocultures, and so may control some nematode pests, including cyst and root-knot nematodes.

Such nematode-suppressive soils have been reported from around the world and include some of the best documented cases of effective biological control of nematode pests. Finally, a number of commercial products based on nematophagous fungi and bacteria have been developed, but all so far have had only limited success. Their use has been based on empirical research, and it is instructive to consider what might be the key factors for a successful biological control agent for nematodes in order to identify the reasons for the general failure of the products that have been developed.

Biological control is more inconsistent, less effective and slower acting than control normally achieved with chemicals. Although improvements in performance might be expected from more research on individual agents, it seems likely that these limitations are inherent in most biological control agents and that their successful application will depend on integration with other control measures.

The most studied example of natural control of a plant-parasitic nematode concerns the decline of populations of the cereal-cyst nematode, *Heterodera avenae*, under monocultures of susceptible cereals in many soils throughout northern Europe. This is an example of an induced suppression in which the nematode increases to damaging population densities in the second and third cereal crops, but usually declines thereafter to infestations of <5 eggs/g soil, which cause little loss of yield in northern European conditions; it is essential that the nematode is abundant in the early years of the monoculture to support the build-up of the microbial parasites.

The decline in nematode populations is mainly caused by two parasitic fungi, *Nematophthora gynophila* and *Verticillium chlamydosporium*, which attack the developing female on the root surface; in suppressive soils 95 to 97 percent of the females and eggs are destroyed. Thus the natural control of cereal-cyst nematode in a range of soils is predictable and effective, but slow acting. Research on the manipulation of natural control has been limited and attention has concentrated on the introduction of specific agents to provide more rapid control that might be commercially exploitable.

It has proved difficult to develop a biological control agent that is effective worldwide for any soil-borne disease. Despite much research effort only two agents have had widespread success: *Phlebia gigantea* for the control of *Heterobasidion annosum*, which spreads from tree stumps to the roots of adjacent trees, *and*

Agrobacterium radiobacter that is applied as a root dip to transplants for control of crown gall caused by *A. tumefasciens*. For both, the agents are applied in high concentrations as inundative treatments to a readily accessible site of action (the cut surface of a tree stump or the bare roots of a transplant) and protection from the diseases for a relatively short period provides long-term control.

In most situations where nematode control is required, there is a need to provide long-term protection to a relatively inaccessible and growing root system without the use of inundative treatments that are likely to be impractical and uneconomic. Hence the development of biological control agents for plant-parasitic nematodes is likely to be difficult and to require a detailed understanding of the biology and ecology of the agent and the nematode target.

The farming system in which biological control is applied has a marked effect on the way the agent is used. In general, growers in developed agriculture require an agent that can be applied to crops grown in monocultures over large areas using standard application machinery, so a formulated product with a good shelf-life that can be applied at low dosages is required; seed treatments are preferred for most arable crops. Little research has been done on the mass production and formulation of biological control agents for nematodes. Some organisms, such as rhizosphere bacteria, can be applied as seed treatments , but such applications tend to provide short-term control and are only useful in reducing the invasion of roots by nematodes that have a single generation in the growing season.

In subsistence farming systems, crops tend to be grown in mixed stands in relatively small areas (often less than one hectare) and labour inputs are often large. As a consequence, relatively large application rates (up to one tonne per hectare) of an unformulated agent could be mixed into the soil by hand, as long as the organism could be produced cheaply and locally. Thus, the initial exploitation of biological control agents for nematodes may be in developing countries. However, if agents are only effective against specific nematode pests, and their efficacy is dependent on pest densities, then their effective use will require expert advice that may not be available in many developing countries.

Exploitation of Natural Enemies

The biology, ecology and potential of biological control agents for nematodes have been extensively reviewed in recent years. Nematologists have identified natural enemies with a range of modes of action similar to those currently studied by plant pathologists for the control of soil-borne diseases. Comments on the advantages and limitations of the major groups of organisms with potential as biological control agents are summarized. It must be stressed that several organisms that are effective natural enemies of nematodes in the field, may have limited potential as biological control agents for application by growers. For example *Nematophthora gynophila* is a causal agent of cereal-cyst nematode decline in many soils, but its limited host range, complex requirements for *in vitro* culture, and need for soil moisture levels to be at field capacity to ensure infection, mean that its potential use is too limited to warrant further development as a biological control agent.

The stage in the nematode life cycle attacked by the control agent has a profound effect on the damage to the crop and the level of population control. Therefore, trapping fungi and rhizosphere bacteria that attack the second-stage juveniles of cyst and root-knot nematodes may significantly improve crop growth but is unlikely to prevent nematode populations increasing, especially those species that have more than one generation in a growing season. In contrast, those parasites that attack developing females and eggs act like a partially resistant cultivar in that initial nematode invasion and plant damage is not prevented but multiplication of the nematode is significantly reduced. Root colonizing fungi such as the mycorrhizae and endophytic species such as *Fusarium* spp. may reduce both nematode invasion and development.

Sikora (1992) reviewed a range of control measures, including crop rotation, partial soil sterilization, soil amendments and nematicides, that could be combined to increase the activity of naturally occurring biological control agents. Such measures could also be used to improve the performance of agents added to soil. Partial soil sterilization by methods such as solarization reduces nematode infestations and also reduces the competition from the residual soil microflora, enabling the biological control agent to establish more readily. Soil amendments may also reduce nematode infestations and, by providing an energy source, help to increase numbers of facultative parasites; pre-colonized substrates are most effective in establishing nematophagous fungi in soil.

Soils that are naturally suppressive to some nematodes may be used to shorten rotations of susceptible crops and improve the performance of nematicides and resistant cultivars. The use of crop cultivars which are tolerant of nematode attack is likely to be very important for the successful deployment of biological control agents, that provide less effective nematode control than most nematicides. Agents such as rhizosphere bacteria and rhizosphere competent fungi depend on root exudates for their proliferation on roots; exudation from roots differs markedly between plant species and cultivar and affects the efficacy of these agents.

Table: The advantages and limitations of potential biological control agents with different modes of action against plant-parasitic nematodes

Type of agent	Mode of action	Comments
Facultative parasites Trapping fungi	Traps produced on modified mycelium give rise to infective trophic hyphae	*Advantages:* easily produced *in vitro;* some species rhizosphere competent; wide host range. Limitations:difficult to ensure trapping activity coincides with activity of infective stages of cyst and root-knot nematodes; often do not produce resting structures and so may be difficult to formulate.

Paecilomyces lilacinus	*Hyphal penetration*	*Advantages:*easily produced *in vitro;* rhizosphere competent; attacks the eggs of several nematode species; treatment of planting material (*e.g.* seed tubers) can be effective. *Limitations:* requires high soil temperatures; has given variable control in range of conditions; large numbers of propagules (10^6/g soil) required for nematode control; some isolates are pathogenic to humans.
Verticitlium chlamydosporium	*Hyphal penetration*	*Advantages:* easily produced *in vitro;* some isolates rhizosphere competent, and virulent (10^3 propagules/g soil required for nematode control); resistant resting spores produced; survives throughout growing season in soil. *Limitations:* seed treatments ineffective; efficacy dependent on nematode species, density and plant host.
Obligate parasitesPasteuria spp.	Adhesive spores	*Advantages:* most isolates highly virulent; infective spores resistant to drying; good shelf-life; reduce infectivity of nematodes as well as fecundity. *Limitations:* very difficultto culture in *vitro;* isolates are highly specific; no proliferation in soil in absence of nematodes.
Hirsutella spp.	Adhesive spores	*Advantages:* relatively easy to culture in *vitro;* attack infective nematodes in soil. *Limitations:* poor saprophytic competitor; limited spread in soil.
Rhizosphere bacteria	Toxins or modification of root exudates	*Advantages:* easy to culture *in vitro;* can be applied as seed treatments; reduce plant damage. *Limitations:* effective for a relatively short period; activity affected by crop cultivar and nematode species; little effect on nematode multiplication.
Endophytic fungi(non-pathogenic root-infecting fungi and mycorrhizae)	Competition in roots and modification of root exudates	*Advantages:*include agents with potential to control migratory endoparasitic nematodes in roots; may improve plant growth even in absence of nematodes; reduce damage caused by wide range of nematodes and limit their multiplication; can be mass produced and formulated; could be applied to seeds or transplant material; may reduce fungal root rots. *Limitations:*non-mycorrhizal fungi may be difficult to register as closely related to plant pathogens; efficacy affected by plant cultivar and other crops in rotation such as crucifers can reduce survival.

Environmental Concerns

Soil conditions, nematode species, rate of development and density and host plant all have a considerable effect on the establishment and activity of biological control agents for nematodes. It is essential to understand these interactions for effective control.

In temperate zones soil temperatures range between 10° and 15°C for much of the growing season whereas in warmer climates temperatures of 20° to 25 °C are normal at 10 cm depth. Too often, research workers in temperate climates have tested the effects of biological control agents in pot experiments in greenhouses at temperatures considerably above those which occur in the field; data from such tests are of limited value. Temperature has a direct effect on the growth and sporulation of the agent and on the rate of development of the nematode target. Moisture in soil is rarely limiting for the growth of most fungi but affects spore dispersal, especially of zoospores.

Bacteria are more sensitive to soil moisture levels than fungi, but are unlikely to be affected by levels that enable nematodes to remain active. Soil texture and structure influence nematode activity and the growth and spread of micro-organisms. The great bulk of soil (2 500 tonnes per hectare) to plough depth (20 cm) makes the thorough incorporation of biological control agents difficult and broadcast treatments are unlikely to be economic. The use of agents in arable crops is likely to depend on those organisms that proliferate in the rhizosphere from in-row applications or from seed treatments. The residual soil microflora competes with the introduced agent for scarce energy sources and can significantly affect the performance of the agent even when it is added to soil in a pre-colonized substrate.

Most plant-parasitic nematodes remain mobile throughout their life cycle and microbial parasites must produce traps or adhesive spores to infect them. Nematodes with sedentary stages such as cyst and root-knot nematodes may be parasitized from vegetative hyphae in the rhizosphere without the production of specialized infective structures apart from an appressorium.

Fungi such as *Verticillium chlamydosporium* infect female cyst nematodes that are exposed on the root surface for several weeks during their maturation; depending on the time of infection, the fungus has a range of effects including a reduction in the number of mature females and their fecundity, and parasitism of the eggs. Usually only the eggs and not the females of root-knot nematodes are exposed in the rhizosphere, and these complete their embryonic development and hatch within about ten days at 25 °C. Hence, egg-parasitic fungi have less time to kill root-knot nematode eggs compared with those of cyst nematodes. Non-parasitic micro-organisms that degrade soil amendments and release nematicidal compounds, such as the bacterium which degrades chitin to produce ammonia , are likely to kill most nematodes in soil. Similarly, many endophytic fungi may compete with a wide range of nematodes in roots although those which are sedentary are likely to be more susceptible.

The host plant influences the rate of nematode development and fecundity and the establishment of some natural enemies in the rhizosphere. Despite these effects the role of the host plant has rarely been considered in research on the biological control of nematodes.

Development of a Biological Control Agent for Nematodes - An Ecological Approach

Too often, biological control agents have failed because they have been used before a basic knowledge of their ecology and biology has been established. The importance of such knowledge can be seen from work at Rothamsted aimed at the development of *V. chlamydosporium* as a biological control agent for root-knot nematodes. Stirling (1991) describes the design and methods used in the conduct of biological control experiments, which are more complex than those required to assess the efficacy of a nematicide. He identified five key aspects in setting up an experiment to evaluate a biological control agent and these are presented below in a slightly modified form.

- The test organism and any organic amendment should be applied at practical application rates; 0.1 percent w/w soil is equivalent to 2.5 tonnes/ha and should represent a maximum dose. Tests should always be performed in a non-sterilized soil with a natural residual soil microflora.
- Appropriate treatments as well as an untreated control should be included if the organism is added with a substrate. These treatments should include the substrate alone, the organism alone, and the autoclaved colonized substrate. Too often, untreated controls are compared only with large applications of the organism and substrate and this does not allow separation of the effects of the agent from the effects of the substrate. In several tests reported in the literature, application of the substrate alone has decreased nematode populations to the same extent as the substrate colonized by the agent, and there is no clear evidence of biological control.
- Population densities of the agent under test should be monitored to ensure that it has survived in soil throughout the period that activity against the nematode target is required. Such monitoring may require the development of selective media, which can be a difficult and time-consuming task.
- Nematode mortality caused by the organism under test should be measured to assess whether differences between nematode population densities in treated and untreated soil relate to the levels of kill caused by the agent. Infection levels are relatively straightforward to estimate for most parasites, but repeated sampling is required to determine total kills. The effects of agents which produce toxins or have indirect effects on nematodes through competition, the modification of root exudates or the colonization of feeding cells, can only be measured by assessing their impact on nematode development.

- The impact of the soil environment, host plant and nematode should be tested as these are likely to affect the efficacy of the biological control agent, and could account for the lack of activity of potential agents in specific test conditions.

In a review of the literature less than 15 percent of experiments purporting to demonstrate biological control caused by *Paecilomyces lilacinus* satisfied the above criteria. Although this is an unsatisfactory situation that must be remedied, the difficulties in conducting carefully controlled and monitored experiments should not be underestimated.

Isolates of *V. chlamydosporium* differ markedly in their growth and sporulation *in vitro*, and in their virulence, saprophytic competitiveness and rhizosphere competence. Such differences between isolates of the same species of micro-organism are common and there is a need for simple laboratory-based screening methods to select the most promising isolates for further testing. *Verticillium chlamydosporium* isolates were collected from infested nematode females and eggs in suppressive soils around the world. Hence, they were isolated from the niche in which they would be required to be active as a biological control agent and from soils most likely to yield active isolates. Once isolated in pure culture using standard techniques, the *in vitro* growth requirements were determined.

Virulent isolates were selected by counting the number of nematode eggs parasitized after exposure to the fungus on agar in a standard test. Tests for proliferation in the rhizosphere and growth in soil were also used to select promising isolates. Although simple laboratory-based screens help eliminate many isolates that show insufficient activity to justify further testing, selected isolates will not necessarily be active in the field. As it may take 10 to 16 weeks to investigate adequately the performance of different isolates against cyst and root-knot nematodes in pot tests, relatively few can be screened.

The method of mass culturing of *V. chlamydosporium* for experiments can have a marked effect on the subsequent survival and proliferation of the fungus in soil. Inoculum produced in shaken liquid cultures consists mostly of hyphae and conidia, which require an energy source to ensure proliferation in soil, whereas on solid media large numbers of chlamydospores are produced and these can be added to soil in aqueous suspension and rapidly establish the fungus.

The development of a semi-selective medium has enabled detailed studies to be made on changes in relative abundance of the fungus in soil and on roots. Some isolates of *V. chlamydosporium* may be extremely abundant in soil but unless they are capable of colonizing the rhizosphere they do not parasitize the eggs of root-knot nematodes. Growth in the rhizosphere differs markedly between plant species, *e.g.* tomato, cabbage and maize roots support much growth, whereas sorghum, pepper and cotton are poor hosts.

Colonization by the fungus is confined to the rhizosphere and rhizoplane and there is no spread into root tissue; no lesions have been observed on roots grown in soil treated with *V. chlamydosporium* and there have been no detrimental effects on the growth of a range of crop species. A tenfold reduction (from 10^4 to 10^3 chlamydospores/g soil) in the amount of fungus applied to soil had no effect on the extent of colonization in the rhizosphere ; the ability to proliferate on the root surface where the fungus is required to control nematodes is an important characteristic which may allow significant reductions in the amount of inoculum applied to soil.

The efficacy of *V. chlamydosporium* as a biological control agent for root-knot nematodes is affected by three key factors: the amount of fungus in the rhizosphere ; the rate of development of eggs in the egg masses; and the size of the galls in which the female nematodes develop. In large galls female root-knot nematodes may produce egg masses which remain within the gall and are not exposed to parasitism by *V. chlamydosporium,* which is confined to the rhizosphere. Hence, *V. chlamydosporium* is less effective in controlling root-knot nematodes in heavily infested soils and on highly susceptible crops because large galls are formed on the roots and many eggs escape parasitism. *Verticillium chlamydosporium is* unlikely to be useful in these situations where a grower would normally apply a nematicide.

Also, at temperatures above 25°C eggs may complete their embryonic development and hatch before the fungus has completely colonized the egg mass; at 30°C about 30 percent of eggs of three root-knot species hatched and the second-stage juveniles escaped from the egg mass before the eggs were killed. These studies in pot tests, if supported by field experiments, help to define the conditions in which *V. chlamydosporium* might be used successfully for control of root-knot nematodes. It is only from such detailed studies that the limitations and requirements of the fungus can be assessed.

Table: The extent of colonization of the rhizosphere by *Verticillium chlamydosporium* on several plants grown on soil treated with 5 000 chlamydospores/g soil and the control of *Meloidogyne incognita*

Host plant	Rhizosphere colonization (cfu/cm^2 root)[1]	Control (%)
Tomato	229	58
Maize	216	71
Cabbage	199	68
Potato	88	79
Tobacco	43	40
Pepper	29	36
Sorghum	25	0
Wheat	15	5

[1] cfu are the colony-forming units which develop on the semi-selective medium.

Verticillium chlamydosporium is not a replacement for nematicides but, despite its limitations, it may be a useful management tool. In most tests biological control agents have been applied to protect susceptible crops. However, application of the fungus to a relatively poor host for the nematode on which small galls are produced so that most egg masses are exposed in the rhizosphere might provide more effective control; soil population densities of the nematode would be reduced to non-damaging levels before a susceptible crop was planted.

Bridge (1987) recommends the following rotation for the management of root-knot nematodes: susceptible host - poor host - poor host - resistant or non-host -susceptible crop. Application of the fungus before the first or second poor host may permit the shortening of rotations, or the replacement of the resistant or non-host crop, without increasing nematode damage to the susceptible crops. Currently, at Rothamsted Experimental Station, United Kingdom, poor hosts for root-knot nematodes are being screened for their ability to support *V. chlamydosporium* in their rhizospheres. It is hoped that by combining the host plant and the biological control agent, more effective and consistent control can be achieved.

The Next 20 Years

The past 20 years have seen a significant increase in the number of scientists involved in research on the biological control of nematodes. Surveys and empirical tests are being replaced by quantitative experimentation and basic research on the modes of action, host specificity and epidemiology of selected organisms. Such basic information is essential for a realistic appraisal of the impact of molecular biology on the improvement of microbial agents and monitoring the spread and survival of released organisms, and for the development of rational strategies for control.

Current experience suggests that biological control agents will not replace the use of nematicides but, integrated with other control measures including chemicals, they could play an important role in the development of integrated control strategies in both developed and developing agriculture. The urgent need to reduce the dependence on nematicides should provide the necessary impetus for the considerable amount of research and development still required to ensure the successful use of such agents.

NEMATODE TRAPPING FUNGI

Nematode trapping fungi, or "nematophageous fungi," are carnivorous fungi that have developed methods and structures that enable them to successfully trap and consume nematodes. Nematode trapping fungi are responsible for keeping the nematode population in check and are an important part of the subsoil ecosystem. These fungi prey on nematodes and are in turn consumed by organisms on the next trophic level. Nematophageous fungi use several methods

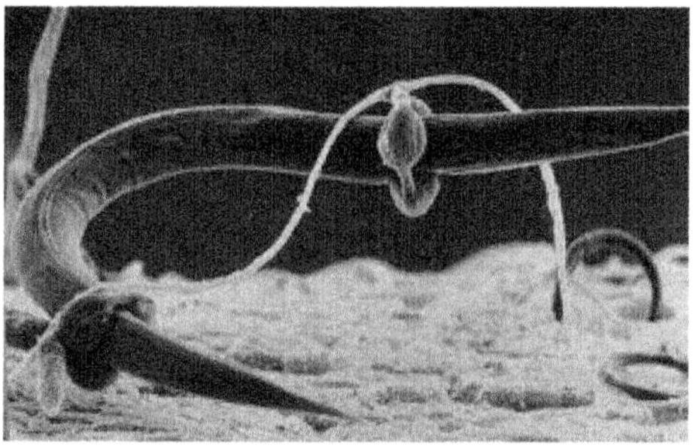

Fig. : Electron micrograph of a nematode being trapped by fungal rings.

to hunt their prey. These methods include living within the nematode and slowly consuming them as well as spreading diseases through nematode populations. The fungi also live in the soil and set traps for the nematodes to squirm into.

Biological Interaction

Nematophagous fungi and nematodes share a special predator-prey relationship. The nematodes, like many other soil inhabitants, secrete chemicals as they travel through the soil. Fungi that prey upon these nematodes have found a way to detect and respond to the existence of these chemicals. The fungi have also developed many sophisticated ways to trap the nematodes by attacking from both outside and within the nematode.

Detecting the Nematodes

In order to effectively hunt its prey, many species of fungi have developed the ability to detect some of the chemicals that nematodes use to communicate and grow. These chemicals are detected when the nematodes wander too close to the fungus. And if the fungus is in need of nutrients that are not provided in its diet, it will create traps in the area in which it detected the chemicals, thus effectively hunting its prey.

Secretion by Nematodes

Nematodes produce a highly conserved family of small molecules called ascarosides. These small molecules are unique to the nematodes and the fungi are able to detect them. Different species of nematode produce different varieties of ascarosides.

Detection by Fungi

Traps to ensnare the nematode are created when these ascarosides are detected. Rather than being proactive and using valuable energy constructing trap structures that may never be used, the fungus in question acts in a reactive manner. Traps are constructed in the area in which the ascarosides were detected.

This act however, is not the norm. This chemical eavesdropping phenomenon is done by a reserved few species of fungi such as *Arthrobotrys oligospora*. And also, only when the fungus is nutrient deprived. Such structures are energy intensive, so the fungus must be hungry and know where its prey is located before it will set its traps.

Trapping the Nematodes

Fungi use many methods in order to ensnare their prey. These fungi can strike from the outside as well as from the inside.

Hyphal Rings

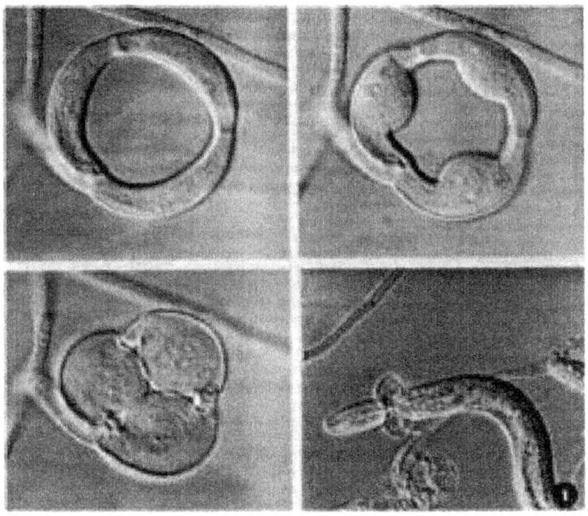

Fig. : Electron micrograph of the constriction of a hyphal loop and a nematode being trapped by such a loop.

One method used by fungi to trap nematodes is the fungal ring. The fungus produces hyphae that end in an open constricting loop. When a nematode swims through this loop, the loop suddenly fills with water. This sudden change in the physiology of the loop causes the diameter of the inside of the loop to narrow and in turn constricts around the nematode. Within 24 hours, hyphae form from within the loop and penetrate the nematode as it begins to digest it.

Adhesive Hyphae

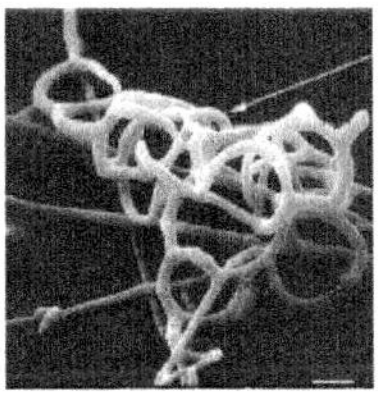

Fig. : Electron micrograph of a network of adhesive hyphae.

Another method commonly used by nematophagous fungi is the use of adhesive hyphae. The fungi who utilize this method create adhesive knobs or nets out of adhesive hyphae. The fungus creates such structures in order to trap the nematode within the structure, and by utilizing the adhesive hyphae the nematode becomes easily ensnared thus providing the fungus with a nitrogen rich meal.

Nematotoxic Compounds

Nematode trapping fungi can use either or both of the methods listed above in order to trap their prey. But some fungi such as the Pleurotaceae family of mushrooms also take it a step further. This family of fungi utilize adhesive knobs to catch nematodes, but once the nematode is caught the fungus secretes a nematacide that kills the nematode and prevents its escape. This extra step is not used by all nematode trapping fungi, but it is utilized by some. It is an extra step of insurance that guarantees that those nematodes that are caught are unable to escape.

Attacking from Within

While many kinds of fungi use traps to catch nematodes, there are also others that live within the nematode and cause damage from there.

Endoparasitic Fungi

Fungi that live within the nematode can infect the nematode at any stage depending on the species of fungus. The infection can be done perorally as in through the mouth or percutanously which means through the skin. From within the nematode, the fungus can consume nutrients meant for the nematode's growth and spread to other nematodes when able. These endoparasitic fungi can spread throughout a community in a measurably short time and weaken a whole population of nematodes throughout the process.

Niche

Due to the overwhelming abundance of nematodes throughout the world it is easier to say where nematodes cannot be found than to list the many places that they can be found. The interaction between fungus and nematode is limited to environments where they can both thrive.

Nematode Habitats

Nematodes survive and thrive in almost every habitat on earth. They are present on both land and in aquatic environments. The only stipulation that can limit the spread of nematodes is availability of moisture and prey.

Aquatic Habitats

Nematodes can be found in great abundance in both fresh and sea water. Nematodes can infect aquatic plants and animals. There is no discernible biological difference between nematodes that live in an aqueous environment and soil nematodes. Soil nematodes also swim but in the water filled pores between soil particles.

Terrestrial Habitats

Nematodes can be found in nearly soil on earth. As long as a minimal amount of soil moisture and some sort of organism is present in the soil, it is certain that nematodes can be found there.

Fungal Habitats

Fungi can be found in almost every soil and water source in the world. Although not as enormously widespread and survivable as nematodes, fungi are nearly ever present. Fungi are important to the breakdown of organic compounds and require organic compounds, moisture, and proper pH to survive.

Aquatic Habitats

Fungi can be easily found in both fresh and saltwater sources. Fungi can live and grow in rivers, ponds, and oceans. Purely terrestrial fungi also often rely on such water sources to aid in spore dispersal as well as dispersal of its mycelial elements.

Terrestrial Habitats

Soil is the most common environment for fungi. Fungi are present in various soil types all over the world. They are vital to innumerable food webs and work to break down organic matter.

Microbial Processes

This microbial interaction is important due to its effect on subsoil trophic levels. By consuming nematodes, the fungi creates biomass that can be consumed by other organisms. And in consuming the nematodes it also helps to control the present population keeping the nematodes in check.

Benefits

Having nematophagous fungi in the soil confers many benefits both environmentally and economically. Nematodes can affect both plants and animals. They cause several diseases through the use of their feeding stylet as well as some speciesof the worm living within the host.

Benefits to Environment

Nematophagous fungi are predators and can be limited by the amount of prey that is available. This means that it's population is constantly in check by its prey. It also keeps the nematode population under control. This allows a wide variety of plants to grow, even those that are especially susceptible to nematodes. The mass that is gained by the fungus by consuming the nematodes is also beneficial. This mass provides a food source to other larger organisms that are higher up the food chain. Having this microbial interaction promotes the cycling of nutrients and diversifies the food web. These nematodes are also innumerable worldwide which results in a stable supply for soil ecosystems. And constant replenishment through the means of soil movement.

Economic Benefits

Nematodes are present in nearly every square foot of soil worldwide. Plant pathogenic nematodes can also be detrimental to crops. The nematodes are ever present in the fields stealing nutrients from the crops. Nematophagous fungi however, can intervene and help to destroy many of these nematodes before they can cause even more damage. If nematophagous fungi can be weaponized even further against this threat, then we will be able to see an even more noticeable difference. Plant parasitic nematodes feed through a needle-like mouthpiece called a stylet. This mouthpiece is injected into the host and the nematode ingests the nutrients of its host. This feeding action not only deprives the host of nutrients but also creates wounds that can be entryways for invading microbes to colonize the plant and spread disease.

The feeding action of nematodes can also spread disease by itself. By feeding from an infected host then moving to an uninfected host, the nematode can directly inject the disease into the new host. The damage done by nematode feeding is oftentimes unnoticed. Seeing that the plant hosts will usually still grow, even though this growth can be stunted or slow. And a drop in yield can often be seen versus uninfested crops. The number of nemataphageous fungi in the soil does not by any means limit the population of these nematodes. But they are a step in the right direction while trying to create an IPM plan to counter the nematodes.

Key Microorganisms

There are many different varieties of nematophagous fungi. And many have yet to be studied extensively. They are present nearly everywhere in the world and have evolved to hunt nematodes in their own way. These nematophagous fungi have evolved alongside the nematodes that they consume and have over time developed methods that are most useful to catch the nematode species that they hunt the most.

Varieties of Nematode Trapping Fungi

Fungal Ring Users

Notable fungi that use a fungal ring include *Drechslerella anchonia, Drechslerella dactyloides*, and *Drechslerella brochopaga.*

Adhesive Hyphae Users

Fungi that utilize adhesive hyphae include *Dactylellina candida, Drechmeria coniospora,* and *Arthrobotrys oligospora.*

Nematotoxin Users

Notable nematotoxin users include *Coprinus comatus, Stropharia rugosoannulata, Arthrobotrys dactyloides,* and the Pleurotaceae family.

Endoparasitic Fungi

Fungi that survive as spores in the soil until the time is right include *Drechmeria coniospora, Pythium caudatum,* and *Harposporium leptospira.*

Current Research

Due to the over abundance of nematodes and the huge economic loss that is caused by them each year, many experiments are under way in order to find a way to alleviate this damage and promote plant health.

Enter summaries of recent research here--at least three required

"The nematophagous fungus ***Verticilium chlamydosporium*** as a potential biological control agent for ***Meloidogyne arenaria***"

The purpose of this experiment was to determine the effectiveness of the fungus *Verticillium chlamydosporium* against the nematode against *eloidogyne arenaria* under greenhouse conditions using tomato plants as a host. It was decided to use three isolates of the fungus. This fungus was rinsed in a shaken liquid culture of Czapek Dox broth for a week at 18 degrees Celsius. The fungal liquid was then added to a peat/sandy soil and allowed to grow in the greenhouse at 23 degrees Celsius for two weeks. The first experiment involved adding second stage juveniles to the soil of the tomato plants. After 50 days, nematode populations were estimated. For the second experiment, sand and sand/bran mixtures were set up and aliquots of 0.5 and 5 mL of the nematodes were added to the soil after they had been mixed with fungal particles.

This experiment was to determine how well fungal particles of a specific fungal isolate affected the nematodes. The last experiment involved infecting the soil of tomato plants with nematodes but also adding a substantial amount of a fungal isolate. This experiment was to determine if the implementation of either added material affected the plant's growth. Throughout these experiments it was plain to see through the data that was collected that the isolate that was used in the experiments did provide an 80 percent reduction of nematode populations. This isolate also did not cause harm to the plants that it had been planted near.

"Study of the nematicidal properties of the culture filtrate of the nematophagous fungus ***Paecilomyces lilacinus***"

This experiment was conducted to determine which kind of culture of *Paecilomyces lilacinu* (a nematophagous fungus) was most effective against which varieties of nematodes. Seventeen varieties of nematodes were used throughout this experiment and all of these nematodes were exposed to the filtered filtrate of the fungus. The test was to conclude whether or not the filtrate paralyzed the nematode. This was conducted by exposing all of the nematodes(50 at a time in 20 microliters of water) to the filtrate. After 24 hrs, the nematodes were probed with a sharp needle in order to test for paralysis.

Estimates were taken after this 24 hr period in order to determine the number of paralyzed individuals. Several types of medium were also tested to see which method produced the most toxin. The study found that toxin production was not influenced by light or pH, and that the toxin only killed efficiently against the Heteroderidae family.

"Attraction of nematodes to living mycelium of nematophagous fungi"

This experiment investigated how well the bacteria-feeding nematode *P. redivivus* was attracted to living mycelia of different nematophagous fungi. Methods were developed that allowed the detection of both attraction and repulsion as well as the magnitude of each interaction. The fungi that were involved in this experiment were allowed to grow to nearly the edge of their plates and were grown on various kinds of agar. The nematophagous fungi were also tested with traps present as well as with out traps. The nematode used in the experiment was the bacteria-feeding nematode *P. redivivus* and it was reared and provided in a way that made it constant throughout the experiment.

The attraction assay consisted of adding a drop of the nematode suspension which contained about 100 specimens onto a small round section of the plate after the plate had been dried for 2-3 days. After 24 hours, the remaining nematodes were counted and the attraction/repulsion towards the hyphae was measured. The attraction intensity assay was conducted in the same way as the other experiment, but the nematodes were counted hourly in order to determine how quickly the nematodes traveled towards or away form the hyphae. The experiment ended with a bank of valuable data. It was found through all of the fungi present whether or not traps were formed spontaneously from the introduction of nematodes, attraction vs repulsion, and also attraction vs repulsion after killing the fungi using U.V. radiation.

THE LIVING SOIL: NEMATODES

Nematodes are non-segmented worms typically 1/500 of an inch (50 μm) in diameter and 1/20 of an inch (1 mm) in length. Those few species responsible for plant diseases have received a lot of attention, but far less is known about the majority of the nematode community that plays beneficial roles in soil.

An incredible variety of nematodes function at several trophic levels of the soil food web. Some feed on the plants and algae (first trophic level); others are grazers that feed on bacteria and fungi (second trophic level); and some feed on other nematodes (higher trophic levels).

Free-living nematodes can be divided into four broad groups based on their diet. *Bacterial-feeders* consume bacteria. *Fungal-feeders* feed by puncturing the cell wall of fungi and sucking out the internal contents. *Predatory nematodes* eat all types of nematodes and protozoa. They eat smaller organisms whole, or attach themselves to the cuticle of larger nematodes, scraping away until the prey's internal body parts can be extracted. *Omnivores* eat a variety of organisms or may have a different diet at each life stage. *Root-feeders* are plant parasites, and thus are not free-living in the soil.

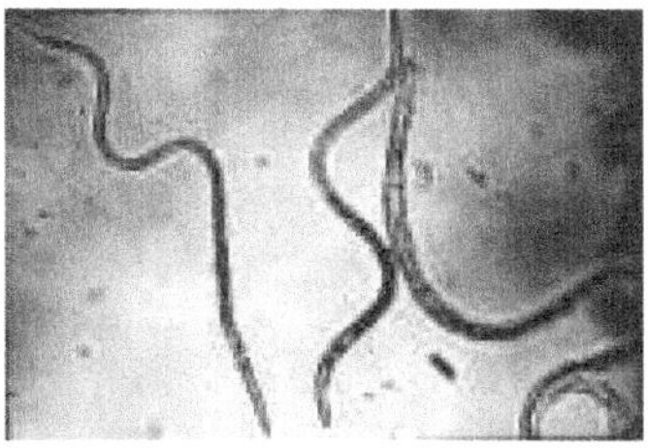

Fig. : Most nematodes in the soil are not plant parasites. Beneficial nematodes help control disease and cycle nutrients.
Credit: *Elaine R. Ingham, Oregon State University, Corvallis*

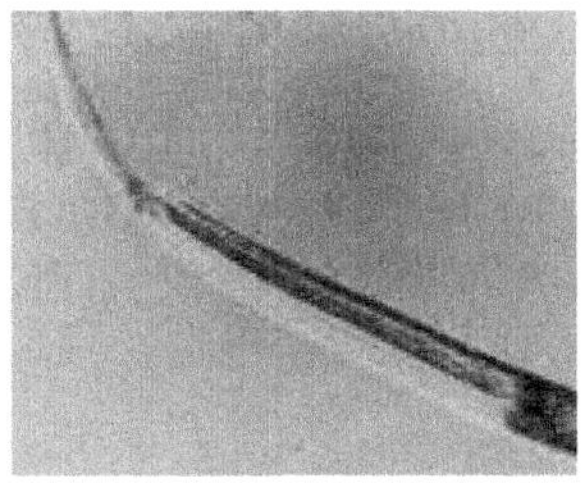

Fig. : A predatory nematode consumes a smaller nematode.
Credit: *Kathy Merrifield, Oregon State University, Corvallis.*

What do Nematodes Do?

Nutrient Cycling

Like protozoa, nematodes are important in mineralizing, or releasing, nutrients in plant-available forms. When nematodes eat bacteria or fungi, ammonium (NH4+) is released because bacteria and fungi contain much more nitrogen than the nematodes require.

Grazing

At low nematode densities, feeding by nematodes stimulates the growth rate of prey populations. That is, bacterial-feeders stimulate bacterial growth, plant-feeders stimulate plant growth, and so on. At higher densities, nematodes will reduce the population of their prey. This may decrease plant productivity, may negatively impact mycorrhizal fungi, and can reduce decomposition and immobilization rates by bacteria and fungi. Predatory nematodes may regulate populations of bacterial-and fungal-feeding nematodes, thus preventing over-grazing by those groups. Nematode grazing may control the balance between bacteria and fungi, and the species composition of the microbial community.

Dispersal of Microbes

Nematodes help distribute bacteria and fungi through the soil and along roots by carrying live and dormant microbes on their surfaces and in their digestive systems.

Food Source

Nematodes are food for higher level predators, including predatory nematodes, soil microarthropods, and soil insects. They are also parasitized by bacteria and fungi.

Disease Suppression and Development

Some nematodes cause disease. Others consume disease-causing organisms, such as root-feeding nematodes, or prevent their access to roots. These may be potential biocontrol agents.

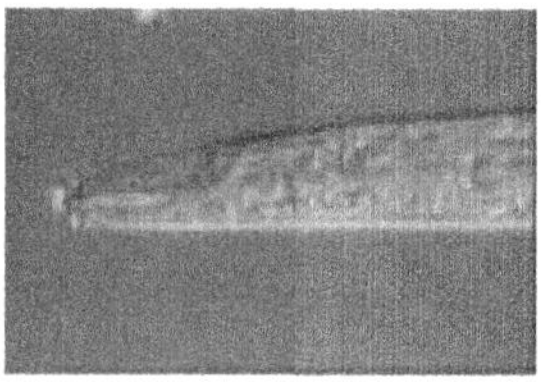

Fig. : Fungal-feeding nematodes have small, narrow stylets, or spears, in their stoma (mouth) which they use to puncture the cell walls of fungal hyphae and withdraw the cell fluid. This interaction releases plant-available nitrogen from fungal biomass.

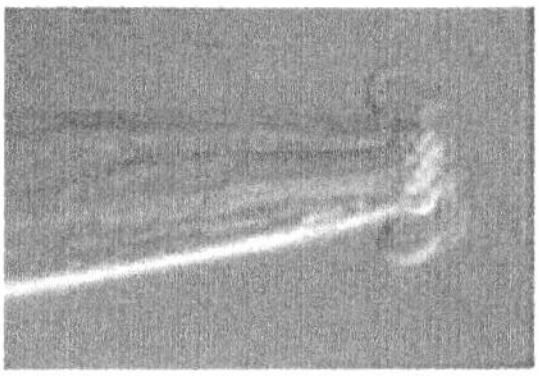

Fig. : This bacterial-feeding nematode, *Elaphonema,* has ornate lip structures that distinguish it from other nematodes. Bacterial-feeders release plant-available nitrogen when they consume bacteria.

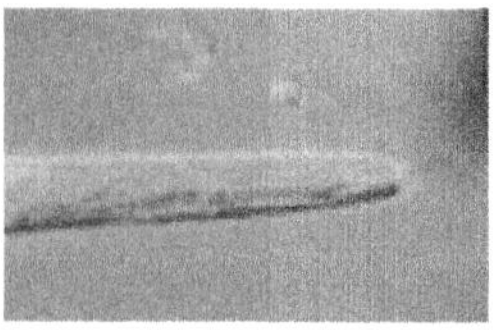

Fig. : The *Pratylenchus,* or lesion nematode, has a shorter, thicker stylet in its mouth than the root feeder in Figure.

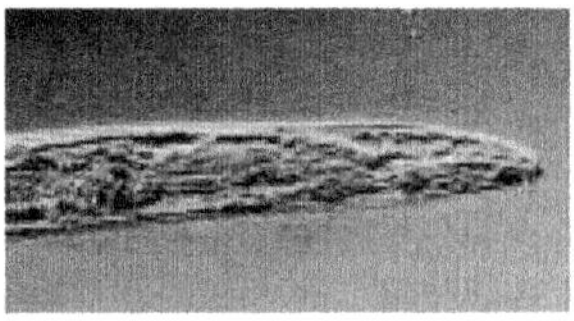

Fig. : Root-feeding nematodes use their stylets to puncture the thick cell wall of plant root cells and siphon off the internal contents of the plant cell. This usually causes economically significant damage to crops. The curved stylet seen inside this nematode is characteristic of the genus *Trichodorus.*

Where are Nematodes?

Nematodes are concentrated near their prey groups. Bacterial-feeders abound near roots where bacteria congregate; fungal-feeders are near fungal biomass; root-feeders are concentrated around roots of stressed or susceptible plants. Predatory nematodes are more likely to be abundant in soils with high numbers of nematodes.

Because of their size, nematodes tend to be more common in coarser-textured soils. Nematodes move in water films in large (>1/500 inch or 50 μm) pore spaces.

Agricultural soils generally support less than 100 nematodes in each teaspoon (dry gram) of soil. Grasslands may contain 50 to 500 nematodes, and forest soils generally hold several hundred per teaspoon. The proportion of bacterial-feeding and fungal-feeding nematodes is related to the amount of bacteria and fungi in the soil. Commonly, less disturbed soils contain more predatory nematodes, suggesting that predatory nematodes are highly sensitive to a wide range of disturbances.

Nematodes and Soil Quality

Nematodes may be useful indicators of soil quality because of their tremendous diversity and their participation in many functions at different levels of the soil food web. Several researchers have proposed approaches to assessing the status of soil quality by counting the number of nematodes in different families or trophic groups. In addition to their diversity, nematodes may be useful indicators because their populations are relatively stable in response to changes in moisture and temperature (in contrast to bacteria), yet nematode populations respond to land management changes in predictable ways. Because they are quite small and live in water films, changes in nematode populations reflect changes in soil microenvironments.

BIOLOGICAL CONTROL OF NEMATODES

Historical Development

The history of biological control of nematodes can be divided into two phases.

Prior to 1975, most research was descriptive. Antagonists of nematodes, especially nematode-trapping fungi, were considered interesting wonders of nature with little relevance to agriculture. *Nematicides* were inexpensive and effective, and little research effort was devoted to studying the effect of antagonists on nematode populations in soil.

The second phase began in 1977 with the loss of *DBCP* (which was followed by the restriction of several other important nematicides). The research in the second phase, which continues into the present, is typified by attempts to replace nematicides with antagonists. Thus far, few of these efforts have resulted ineffective biological control and the research has done little to increase our understanding of how biological control may or may not be achieved.

Our greatest need is for sound, in-depth biological information on how organisms, populations, and communities operate in the soil. The next phase will be typified by basic investigations of organismal, population, and community ecology.

Mechanisms

Because nematodes often occur in high numbers in soil, it is not surprising that a wide variety of soil organisms exploit nematodes as food, *i.e.*, as sources of carbon, nitrogen, and energy. Those organisms that seek out and consume nematodes are called predators.

Predators of plant-parasitic nematodes include mites, collembola, flatworms, protozoa, and other predacious nematodes.

Many exploiters of nematodes have prolonged and specialized interactions with nematodes; these organisms are called parasites. Parasites of plant-parasitic nematodes include fungi, bacteria, and mycoplasma-like organisms.

Other organisms may have a detrimental effect on nematodes without utilizing them as a substrate, by competing for food, space, and necessary resources. Competitors of plant-parasitic nematodes include other nematodes, bacteria, and fungi. However, interaction between plant-parasitic nematodes and other competitors may cause increased damage to a particular food source.

Some organisms may antagonize nematodes by producing nematicidal or nemastatic compounds such as ammonia, certain fatty acids, and avermectins. This mode of action is referred to as antibiosis, and involves bacteria and fungi.

Of these four mechanisms of antagonism (predation, parasitism, competition, and antibiosis), parasitism has received the most research effort. Predators are difficult to handle and manipulate, while competition and antibiosis are less direct and more difficult to quantify than parasitism and predation. With the exception of one bacterium, *Pasteuria penetrans*, most effort has focused on fungal parasites of nematode eggs and females.

Introduction of Antagonists

Introduction of organisms antagonistic to pests into an environment (one form of classical biological control) has been successful in entomology, but not in plant pathology or nematology. This approach involves the one-time or repetitive release of an antagonist in an area where it is not naturally present.

Introduction has been successful against exotic insect pests and weeds that have escaped their natural enemies when introduced to a new area. Most nematode pests are not exotic (or not recently exotic), and it is considered difficult to introduce an organism into soil where all ecological niches are expected to be filled by the wide variety of organisms already present (the soil foodweb). However, further research effort is warranted in this area.

Inundation by Antagonists

A number of researchers have applied large quantities of nematode antagonists to soil with the hope of inducing biological control. Although occasional success has been reported, this approach has generally failed. The reason for failure is unknown because the researchers typically did not determine if the applied organisms became established.

An exception has been the application of rhizosphere bacteria through drip irrigation. Drip irrigation would seem to be a viable method of applying and maintaining large numbers of nematode antagonists (presumably antibiotic producers) in the rhizosphere, although research has shown that the distance microorganisms are delivered from the emitter depends on the size of the organisms; the delivery distance is usually less than 12 inches.

Conservation and Augmentation of Antagonists

The greatest success in utilizing biological control of nematodes has involved the conservation and enhancement of antagonists naturally present in soil.

In England, the cereal cyst nematode, *Heterodera avenae*, is controlled successfully by growing monocultures of small grains which support high biomass of certain nematophagous fungi.

Natural *suppression* of plant-parasitic nematodes has also been documented in peach orchards in the United States and vineyards in Australia. Methods for enhancing natural biological control include chitin amendments (to stimulate chitinolytic organisms that can degrade the eggshells of nematodes), collagen amendments (to stimulate collagenolytic organisms that can degrade the cuticle of vermiform nematodes), and other organic amendments (to increase the C:N ratio and, thus, stimulate the activity of nematode-trapping fungi).

Biopesticides

Biopesticides are materials that are the natural products of organisms that are shown to be toxic to nematodes and other pests. The frequent implication is that such materials are environmentally safe because they are produced naturally rather than synthetically. This assumption is unwarranted, however, as some naturally occurring toxins are known to be extremely carcinogenic.

There are some intriguing possibilities for nematode management in this area. *Becker et al.* (1988) have shown that plant roots can be protected from fungal and, possibly, nematode attack by the presence of certain bacteria in the rhizosphere. The most probable mechanism is antibiosis, since the bacteria produce compounds that are lethal to the plant pests. Considerable work has been conducted on attempting to isolate the effective rhizobacteria, culture them, and inoculate seedling plants, or to introduce the bacteria at appropriate intervals in irrigation water.

An approach of considerable commercial interest involves growing effective bacteria in fermenters, using patented processes and media, and marketing the antibiotic product as a biopesticide. Abbott Laboratories is currently working with a toxin produced by a soil fungus that may have potential for biological control (*DiTera*) of plant-parasitic nematodes.

Researchers at the Mycogen Corporation isolated and patented a strain of *Bacillus thuringiensis* (Bt) that produces a toxin effective against certain plant-parasitic nematodes. Bt crystals, are protein crystals formed during sporulation in some *Bt* strains. *Bt* produces proteins that aggregate to form a crystal. In insects, the crystal proteins bind specifically to certain receptors in the intestine. Humans and other vertebrates do not have these receptors in their bodies and so are unaffected by the toxin.

The Bt gene responsible for toxin production effective against certain insect pests has already been transgenically inserted into the genome of some plants. Introduction of the nematode-effective Bt gene into a plant genome would ostensibly render that cultivar resistant to nematode parasitism through antibiosis, while retaining its desirable agronomic or horticultural characteristics. This approach is especially promising in woody perennial plant species that can be grown on rootstocks. Development of Bt-transgenic rootstocks would potentially confer nematode resistance with no effect on the characteristics of the scion.

Bt toxins relatively easy to make and are safe to humans and vertebrate animals. Wei et al (2003) found that the Bt crystal protein Cry5B destroyed the intestine of *C. elegans*. They also found that Bt crystal proteins, Cry6A and Cry14A, reduced *C. elegans* fecundity. They determined that six bacterivore nematode species, *C. elegans, Pristionchus pacificus, Panagrellus redivivus, Acrobeloides* sp.,*Distolabrellus beechi and* the free-living stage *Nippostrongylus brasiliensis* are susceptible in varying degrees to certain Bt crystal proteins, either by killing them, damaging their intestines or reducing fecundity.

The toxic crystal molecules coded by the Bt gene are large; they range in molecular mass from 40 to >70 kDa. Crystals would need to be ingested by the plant-feeding nematode and there has been some speculation that stylet-aperture exclusion would be a problem. However, *Meloidogyne incognita, Globodera rostochiensis* and *Rotylenchulus reniformis* are able to ingest green flourescent protein molecules of >28 kDa size from plant cells. The size of the Bt crystal is 40-70 kDa and there is currently no evidence to suggest that crystals of that size would be excluded by the stylet aperture. The inability of *Heterodera schachtii* to digest proteins >40 kDa has been attributed to the a zone of exclusion established in the cytoplasm when these nematodes feed in plant cells (Bockenhoff and Grundler, 1994; Urwin, et al, 1997).

Scientists are aware, however, that resistance conferred by Bt toxins may be short-lived. If Bt toxin is not specific to nematodes, it will be important to ensure that it remains in the rootstock; if it is also expressed in the scion, it may select for resistance in lepidopterous pests and jeopardize the effectiveness of a currently useful management approach. There are several examples of insect pests that have developed resistance to this single-gene controlled product. In general, however, nematodes are less migratory than most insect species, and if a line of resistant individuals developed, it would require some time for the whole root system of a tree to be colonized, or for spread through an entire orchard to occur. Certainly, the strategy should allow for establishment of vigorous trees, under minimal nematode pressure, even if the nematode population subsequently developed in the mature orchard.

In the mid-1990s the Mycogen Corporation transferred rights to the nematode-effective Bt to Monsanto. Currently, Monsanto is not pursuing development.

Research Needs - Biological Control

Where biological control of nematode pests appears to be occurring, it is essential that the mechanism of antagonism be established for the system in question. This is relatively easy for parasites and predators, but quite difficult for antibiosis and competition. In addition, we must be able to quantify the antagonist and antagonist activity. Without this information, we cannot understand and remedy the inconsistency of results that is characteristic of biological control research.

Quantitative assays are lacking for most antagonists of nematodes. Biological control occurs when a population or community of antagonists suppresses a population of nematodes. Quantitative assays and models must be developed in order to understand the interaction of pest populations and biological control agents.

The soil environment is physically, chemically, and biologically very complex. It varies in its characteristics and complexity both spatially and temporally. Biologically, it is most complex, and the pace of interactions greatest, around plant roots. Antagonists of nematodes occur in many groups of soil organisms, and depending on the composition of these organisms in the soil or rhizosphere, the dynamics of the nematode population may be substantially buffered. The basic context of interactions among organisms is that each group is a food source for others; each group of organisms in the soil constitutes a source of required carbon and energy for those organisms feeding on it, the soil foodweb.

As their proximal sources, these food chains are dependant on plants and their fixing of carbon dioxide and water into carbohydrate through the process of photosynthesis. Primary consumers, including plant-parasitic nematodes, are among the initial links in the chains. Other initial links are direct release of plant materials and products into the soil that fuel pathways of decomposition; the fuel includes plant litter, rhizodeposition and root sloughings, and root exudates.

Several types of fungi are important antagonists of nematodes. Among them are the trapping fungi that capture nematodes in various forms of traps. In many of these fungi, trap formation is stimulated when nematodes are present, and research indicates that a variety of proteins and amino acids originating from the nematodes may be involved.

The fungi, *Arthrobotrys dactyloides* and *A. brochopaga,* form ring traps that constrict around the body of a nematode passing through them. Other fungi produce sticky nets (*A. oligospora*), sticky knobs (*Dactylellina haptotyla* and *Nematoctonus* spp.), or sticky spores (*Drechmeria coniospora* and *Hirsutella rhossiliensis*). In each case, the fungal trap initially binds to the nematode, a penetration peg grows into the nematode body, and fungal hyphae ramify through the nematode. After the resources of the nematode have been exploited, hyphae emerge from the nematode carcass into the soil, and new traps are formed.

The spores of some fungi are ingested by nematodes, but this would not occur with plant-feeding nematodes. Other fungi have motile zoospores that swim through the soil solution and encyst on natural body openings of the nematode; zoosporangia then fill the body cavity, and new zoospores are released into the soil solution. Questions arise as to whether such water-loving fungi might be delivered to nematode-infested fields through irrigation.

A final, important group of fungal parasites of nematodes are the *egg parasites,* including *Pochonia chlamydosporia, Paecilomyces lilacinus,* and *Dactylella oviparasitica.* These fungi may be of particular significance with those nematodes that aggregate their eggs and fungal food source in masses.

The soil also contains bacteria that are direct parasites of nematodes, including *Pasteuria penetrans,* which has adhesive spores. Considerable research has been conducted with this organism, but consistent application tactics have not yet emerged.

The *predacious nematodes* are formidable natural enemies of plant-parasitic and other nematodes. *Seinura, Labronema,* and *Eudorylaimus* pierce their prey with a hollow stylet and remove the body contents, while *Mononchus* and *Odontopharynx* incise the nematode body with a large tooth, then ingest the body. Predacious nematodes tend to be susceptible to *soil disturbance,* and probably act as important population buffers in relatively undisturbed ecosystems, including perennial crops.

Other soil-dwelling organisms that obtain their source of carbon and energy from nematodes include certain amoebae (phylum Sarcomastigophora), *tardigrades* (phylum Tardigrada), *flatworms* (phylum Platyhelminthes), *mites* (phylum Arthropoda, class Arachnida), and *colembolla* (phylum Arthropoda, class Insecta). For example, the mesostigmatid mite *Gamasellodes vermivorax* has nematodes as a preferred food source and will also feed on soft-bodied mites and colembollans. In culture it devours nematodes voraciously and is considered an important nematode predator in semi-arid grassland soils of the Great Plains.

Clearly, there are myriad interactions and potential interactions among soil organisms that require detailed study. These complex interactions, if better identified and understood, might be manipulated to reduce the population of plant-parasitic nematodes.

Advances in the techniques of biotechnology introduce additional possibilities of biological engineering of nematode antagonists. The strategy would involve selection or induction of variants of potential biological control agents with characteristics that enhance their effectiveness; it would also require development of cultural practices that promote the growth of beneficial biota in the soil. Increased effectiveness might involve the introduction of an organism that is a more successful competitor, a more aggressive predator, a more virulent parasite, or have enhanced survival characteristics. The approach may involve mutagenesis and screening, or the direct insertion of genes for the desired characteristic into the genome of the organism. There is an important arena of basic and applied research opportunities in the genetic tailoring organisms for effectiveness in specific environmental and cultural situations.

Rhizosphere bacteria that are natural colonists and inhabitants of the root surface and adjacent soil, appear to have many important, although little understood, properties. These bacteria may enhance plant growth or might suppress nematode populations by exerting an antibiotic effect. Additional research is needed to determine which organisms are most effective, and also to suggest methods for enhancing their activity in the soil. Such enhancement might include introducing the organisms into the soil at strategic intervals or introducing a food substrate to promote their activity.

The tactics of Introduction, Inundation, and Augmentation as biological controls of soil-inhabiting nematodes raise a series of implementation challenges.

How will organisms most effectively be introduced into soil?

- Possibilities include:
- encapsulating them with seed,
- dipping seedling roots into a suspension of the organism at transplanting,
- direct incorporation into the soil, and
- application through irrigation systems.

In other cases, the organisms will already be present in the soil and will be augmented by alteration of the environment or introduction of a food substrate.

One research opportunity, which has not received adequate field testing, might involve the introduction of a preferred crop or a soil amendment with the antagonistic organism. *Pochonia chlamydosporia,* for example, is damaging to cyst nematodes and also pathogenic to some daisy-type plants. In all situations, the optimal approach will need to be case-specific and will depend on the characteristics of the organism, the biological and physical nature of the soil, and the nature of the crop. Certainly, the organisms must retain their effectiveness through the introduction process. Other important research and implementation questions include quality control in the production of the organisms, and their shipping and shelf-life characteristics.

Given an adequate foundation of basic information on how the system in question functions, researchers may proceed to the third phase of biological control work, in which the understanding achieved through basic research is integrated and tested in field implementation strategies. This may involve formulation and application of specific antagonists-a procedure that has failed thus far due to lack of basic biological information. The basic information may also lead us to specific cultural practices which enhance the activity of native and introduced agents. Such cultural practices could include *crop rotations*, organic amendments, and pH adjustment.

BIOLOGICAL CONTROL OF HELMINTH PARASITES BY PREDATORY FUNGI

During the last 10-15 years there has been an increasing emphasis on the need of development of new alternative to or supplements for chemical control of parasitic nematodes in grazing livestock. The background for this interest is multi-factorial but the major reason is the serious development of anthelmintic resistance in parasitic populations. Anthelmintic resistance involving particularly the gastrointestinal nematodes of small ruminants is escalating globally and is the single most important concern of parasitologists around the world since it threatens the survivability of small ruminant farming as well as the helminthologists. The problem is most severe in the countries of southern hemisphere like South Africa , Australia , New Zealand and many other Latin American countries.

India is slowly and steadily emerging as the resistance epicenter of South Asia. The global tempo of development and extent of anthelmintic resistance in helminths of small ruminants in particular, indicates that the numerous anthelmintics and strategies developed and implemented over the period of last 40-50 years have been incorrectly applied. Other reasons include handling of parasite problems in the organic livestock production, regulation of conventional drug use by legislation, political and consumer pressure for reduced chemical residues in products. Sutherst (1986), among others, has talked about the importance of implementing integrated parasite control based upon host resistance, immunization and non-living vaccines to reduce the use of chemotherapy. To become sustainable, parasite control schemes need to be based on the principles of integrated pest management. Towards this objective, significant advances have recently been made in the development of worm vaccines, in the breeding of animals for parasite resistance and biological control exploiting predacious fungi.

Parasite Control Measures

Gronvold *et al.* (1996) have suggested two major groups of control measures that are in use and would be tried in future.

Chemical Control	Non-Chemical Control
Chemotherapy	Biological Control
Spraying of Poison	Worm Vaccines
Use of Repellants	Selection for Host Resistance
Use of Pheromones	Grazing Management
	Nutritional Management
	Inter-Specific Competition
	Sterile Male Technique

Present Status of Different Control Strategies

Chemical Control

Three crucial reasons for opting alternative parasite control strategies including biological control are drug resistance, residues on food and environmental degradation. Frequent and haphazard use, over-reliance on chemicals are the causes of the drug resistance. As resistance to newer anthelmintics develops, there is a need for control measures alternative to chemotherapy. Chemical residues in foods are now a major concern and a strong driving force for reduced chemical inputs in agriculture.

Consumers increasingly demand that food supply should be free from contaminants of all kinds. Unlike organochlorine ectoparasiticides, residues are not a major problem for anthelmintics. The benzimidazoles and their prodrugs are subjected to close scrutiny because some are known teratogens. The environmental impact of anthelmintics has been generally regarded as un-important. Levamisole and the benzimidazoles pose little cause of concern, but there is greater worry regarding the avermectins. There is evidence of adverse effects on a variety of dung colonizing insects.

Worm Vaccines

Within the last decade, much effort has been directed at the development of a vaccine especially against*Haemonchus contortus*, either based upon naturally exposed or hidden antigens. Despite promising results and mass appraisal over the years, a commercial product is still to be released. Dictol, based on infective *Dictyocaulus viviparous* larvae attenuated by irradiation, is the only marketed vaccine against GI nematodes but it has only a very limited distribution.

Grazing Management

Grazing management strategies have been demonstrated to be useful to alleviate the impact of GI nematodes in livestock. Unfortunately, these strategies have not been adopted to their full extent, perhaps due to the ease for the farmer to use drugs and secondly, the increased demand for land, which makes this proposition less likely in many intensive livestock systems. Where it is used it is often in combination with chemotherapy. In organic livestock production these strategies are widely used, but are primarily based upon the availability of herbage rather than an active measure to control problems with GI nematodes.

Breeding for Resistant Host

Breeding for resistance in host animals to GI nematodes has been attempted with some success, but although breeding programmes are promoted and adopted based upon these principles (*e.g.* Nemesis in Australia, Worm FEC in New Zealand), they are far from widely implemented.

Nutritional Management

This is now a well-established fact that supplementation of the diet with additional protein does not appear to affect initial establishment of nematode infection but the patho-physiological consequences are generally more severe on lower planes of protein nutrition. The main effect of protein supplementation is to increase the rate of acquisition of immunity and increase resistance to re-infection and this has been associated with an enhanced cellular immune response in the gastro-intestinal mucosa.

Studies on the influence of nutrition on the expression of genotype have shown that the benefits of a superior genotype are not lost on a low protein diet whereas a high protein diet can partially ameliorate the disadvantages of an inferior genotype. Although many aspects of the interaction between nutrition and helminth parasites have been established many features remain to be examined. It has now largely been acknowledged that parasite control cannot be achieved by a single method. The available classical and alternate technologies should properly be integrated similar to those practiced in the integrated pest control in agriculture. Biological control seems to be one of such alternate control strategy.

Concept of Biological Control

This area of research has attracted increasing attention, especially within the last 10-12 years. Biological control is defined as the action of natural enemies which maintain a host population at levels lower than would occur in the absence of enemies. This not only includes classical un-exploited organisms but also those that are genetically modified to enhance these properties. Biological control is divided into two major categories, *viz.*, natural and applied. Natural biological control is affected by native or co-evolved natural enemies in the environment without human intervention.

Within the environment, the pre-parasitic stages of nematodes are subjected to a variety of both abiotic and biotic factors that can profoundly influence their development and survival. The most important abiotic factors are temperature; humidity and oxygen-extremes in these can be lethal on the free-living stages. In regard to biotic factors, there exists a vast assemblage of living organisms that can affect the success of worm eggs developing larvae.

From these may emerge candidate(s) for biological control of worm parasites. By definition, biological does not assume to be a substitute for chemotherapy, where the expectation, if not the reality, is that parasites may be eradicated by frequent use of drugs with efficacies approaching 100%. Biological control agents really eliminate the target organism, but reduce the number of expectable levels and maintain a balance between the pathogen and the antagonist. In contrast to chemical control of nematode parasites, which is directed entirely at the parasitic stage within the host, biological control will almost certainly be focused on the free-living stages of parasites on pasture.

The intention of using biological control methods is to lower the density of pest population below the clinical level and perhaps below the economic threshold above which production losses are obvious owing to a high parasitic population density.

Agents for Biological Control

All gastro-intestinal nematode parasites of livestock have a life-cycle which involves not only the parasitic stage within the host, but also a free-living or pre-parasitic stage on pasture. The pre-parasitic stages on the pasture are potentially vulnerable to attack by biological control agents. A number of organisms have been identified to exploit the free-living stages of parasites as food source and are likely to be commercially exploited in the near future. These organisms include micro-arthropods, protozoa, predacious nematodes, virus, bacteria and fungi. Although the all are of intrinsic interest, it is from the last group of organism that breakthroughs in biological control of nematode parasites of livestock are likely to emerge.

Fungi as Biological Control Tool

Fungi that exhibit anti-nematode properties have been known for a long time. They consist of a great variety of species characterized by their ability to capture and exploit nematodes either as the main source of nutrients or supplementary to a saprophytic existence. They are divided into three major groups based on their morphology and types of nematode-destroying apparatus.

Predacious Fungi

They produce specialized nematode-trapping structures (adhesive knobs, networks, rings *etc.*) on the mycelium. The idea of possibly using predacious micro-fungi to control animal nematodes arose in the 1930's. It was not until the mid 1980s before thorough and systematic investigations were undertaken and since then, two lines of work can be clearly distinguished.

Trials performed with mainly with *Arthobotrys* spp. (*A. oligospora*) and *Monacrosporium* spp. as biological control agents. A group of Danish researchers, testing the effect of fungus *A. oligospora* primarily against parasitic nematodes in cattle but also in other livestock species. Testing different doses of spores mixed into faeces, 250 and 2500 conidia per gm of faeces was found to significantly lower the number (70 & 99% reduction, respectively) of developing *C. oncophora* larvae in faecal cultures. The trapping activity of the fungus was influenced by the motility of the infective larvae & there is no specificity for the parasitic species.

Unfortunately various trials performed to test *A. oligospora*mycelium and conidia failed due to the destruction of these structures in the GI tract of the host animals. A high dose (between 470 & 680 gm of fungal material on millet) of one of the three different fungal species (*A. musiformis, A. tortur, Dactylaria candida*) was fed to housed lambs, harboring a mono infection of either *H. contortus* or *O. circumcincta*. This subsequently led to survival of *A. tortur* through the GI tract at a level high enough to significantly reduce the number of *H. contortus* in faecal cultures.

The other line of research is with *Duddingtonia flagrans*. This predacious fungus produces three dimensional, sticky networks on its growing hyphae. It also produces an abundance of intercalary thick walled resting spores, chlamydospores. This fungus is relatively slow growing and as with other predacious fungi growth is strongly influenced by temperature. Many other species of predacious fungi are fast growing but the spores of these fungi are much more sensitive to the stress of the GI tract than that of the chlamydospores of *D. flagrans*. In plot trials *D. flagrans* have shown good reduction of free living larval stages of parasitic nematodes of cattle , sheep and horses. These field trials shown that daily feeding of fungal spores to grazing animals for 3-4 months prevents build-up of dangerous levels of infective larvae on the pasture. In an Australian study Knox and Faedo (2001) found that sheep feed supplement containing *D. flagrans* chlamydospores had lower egg counts and improved liveweight gains compared to untreated animals.

Endo-parasitic Fungi

These invade nematodes either by penetration of cuticle from sticky spores adhering to the cuticle or following ingestion of spores which lodged in the gut. This type of fungi is obligate parasite of nematodes, with very limited capacity to develop outside the prey and density dependent. This has led researchers to believe that there might be better or stronger BC candidate against pest nematodes. *Drechmeria coniospora* is a fungus producing sticky drops on very small conidia, which adhere to the cuticle of the nematode, penetrate the cuticle and destroy the victim.

By applying a very high dose (108 conidia per gm of faeces) to faecal cultures, Santos & Charles (1995) found that only infective third stage parasite larvae stripped of the protective extra cuticle, became infected by the fungus. Another endoparasitic fungus, *Harposporium anguillulae* produce very small, half moon shaped conidia which lodge in the digestive tract of the feeding nematode and after germination totally digest the victim before finally breaking through the cuticle to produce new conidia on the short conidiophores. In a laboratory study it was found that at a dose of 3 lakh conidia/ gm faeces, the number of *H. contortus* larvae recovered was significantly reduced. The requirement of spore dispersion or infection to be more or less directly from one infected individual to the next, severely limits or almost excludes the use of this group of fungi as practical BC agents.

Egg-parasitic Fungi

These have the ability to attack the egg stage and may have a role in the control of animal parasites which have a long development and/or survival time in the egg stage in the environment outside host, *e.g.,Ascaris, Fasciola* spp., amphistomes *etc.*

Eggs of *Ascaris lumbricoides* collected from naturally infected pigs were used to test the effect of mainly the fungus *Verticillium chlamydosporium* but also other *Verticillium* spp. the fungus was shown to be able to degrade the egg shell enzymatically and infect the eggs. Short exposure to high temperature or UV- irradiation rendered the eggs more susceptible to fungal attack. In the USA, Chien and colleagues have shown that *V. chlamydosporium* attacked and destroyed eggs of *Ascaridia galli* and *Parascaris equorum* but only rarely invaded *Trichuris suis*. In Denmark works were done on the predacious fungi *Arthobotrys* spp. and egg parasitic fungi *Paecilomyces lilacinus* for activity against eggs of *T. canis*. *P. lilacinus* showed some activity (16% eggs infected after 7 days) but the predacious species did not attack the eggs.

Present Status

Present Global Status

As a result of renewed interest and intensified research in biological control during the last 15 years, a convincing amount of evidence on the potential of this principle has been gathered. Most of this work has been carried out in Europe (Denmark, Sweden, UK and France) and Australia and recently initiated in USA, Latin America, Africa, South-east Asia and Far East. Among the nematode trapping fungi,*Duddingtonia flagrans* has displayed superior abilities with respect to survival through gastro-intestinal transit as well as subsequent destruction of parasitic larvae in faecal pats. Danish scientists first demonstrated through laboratory and field trials, that biological control against pre-parasitic stages of nematodes could be achieved by feeding chlamydospores of this fungus to cattle , horses , pigs and sheep.

Successful pilot scale trials with *D. flagrans* chlamydospores through feed supplement , feed blocks and slow release devices provide sufficient euphoria for commercial exploitation of fungal delivery devices in future integrated parasite control programmes. Besides being safe for animals and man, it is imperative that new technologies dealing with BC need to be of no negative impact to the grazing environment. Short time impact studies have shown no negative effect of the fungus on earthworms and on soil nematodes. The future appears to be very promising in early outcome of a bio-control product to control nematode parasite of livestock.

Present Indian Status

India initiated the work on biological control of animal nematode parasites using mycological means in 1998 and two species of nematode-trapping fungi, viz., *Arthrobotrys oligospora* and *D. flagrans* and two species of egg parasitic fungi, viz., *Paecilomyces lilacinus* and *Verticillium chlamydosporium* were isolated from organic environment of Gujarat and Chhattisgarh. They were subjected to stringent screening for their suitability as biocontrol agents against nematode parasites of ruminants using growth assay, predatory activity, germination potential and ability to survive ruminant gut passage. The study indicated that the isolates of *D. flagrans* and *V. chlamydosporium* fulfilled all the possible criteria. A strategy is formulated for application of nematode-trapping fungi to control gastrointestinal nematodosis of ruminants.

NEMATODE CONTROL

Nematodes are unsegmented, mostly microscope roundworms that are ubiquitous in soils. More than 80,000 different nematodes have been described. Most species of nematodes rely on bacteria, fungi, or other microscopic organisms for food, and, therefore, are a key feature of healthy soil ecosystems. Additionally, some nematodes are beneficial to plant health by virtue of their ability to feed on plant pathogens or insect pests. Nonetheless, some nematodes parasitize plants, causing more than $50 billion in crop losses annually.

Plant parasitic nematodes have a stylet, or mouth-spear, similar to a hypodermic needle. The stylet is used to puncture plant cells and inject digestive enzymes and other fluids. The nematode then draws plant fluids through the stylets. The most problematic nematodes are the root knot nematode, which feeds on more than 2000 species of plants including most major crops, and the cyst nematode which is an important pest in soybeans and potatoes. More than 50% of all nematode control efforts are aimed at these two types of nematodes.

Historically, the most effective products for nematode control have been fumigants, particularly methyl bromide. With regulatory pressure to reduce or eliminate these fumigants, biopesticides have begun to emerge as alternative treatments to limit nematode damage.

Nematologists have identified several bacterial and fungal products for control of soil nematodes, as well as plant extracts that display nematoxicity or control the pest indirectly by boosting the natural defenses of crop plants. As fumigant nematicides continue to be phased out, biopesticides will play an ever-increasing role in control of these important pests.

NEMATOPHAGOUS FUNGI - AN ILLUSTRATED OVERVIEW

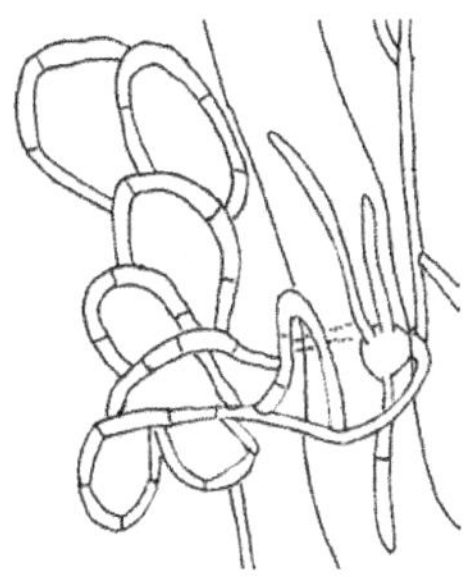

Fig. : Adhesive network of Arthrobotrys oligospora with captured Nematode.

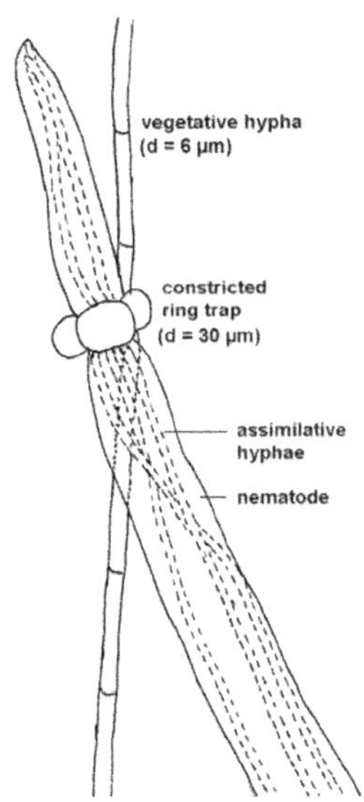

Fig. : Constricting ring of Arthrobotrys dactyloides with captured nematode.

So far we know more than 160 fungus species that live on *Nematodes* partially or entirely. Nematophagous fungi are found in almost every natural soil and additionally in a number of other substrates. They play an important role as antagonists of nematodes. Therefore there is great interest in research of possibilities to use them in biological plant protection.

Nematode-trapping Fungi (Predators, Predaceous Fungi)

Predators we call fungi that develop their *mycelium* predominantly outside their host, which means that they are also able to live as *saprophytes* and trap their prey on vegetative *hyphae* or in special traps. This means that only mobile nematode stages are caught.

The predators are devided into several groups concerning the trapping mechanisms. Except for some species that are able to develop adhesive substance on any part of their mycelium, predacious fungi develop special traps: adhesive three-dimensional networks, adhesive columnar branches, adhesive knobs, constricting rings and non-constricting rings.

At the place of adhesion, a mortiferous excrescence perforates the *cuticle* of the nematode (in the case of most predators - but not of Nematoctonus - forming a globular structure called infection bulb), then giving rise to assimilative hypha growing through the body of the prey, which is digested by enzymes secreted by the fungus and after that resorbed.

Endoparasitic Nematophagous Fungi

Except for the development and spreading of the spores, the entire life cycle of the endoparasites takes place inside the host. Depending on the species, any nematode stages can be attacked. The way of infection can be either *peroral* (*e. g.* Harposporium anguillulae LOHDE) or - in most species - *percutanous*. Only the second group is of significance for biological control of phytopathogenous nematodes, because eelworms with a mouth prick cannot swallow whole spores, instead, some of them are able to suck out the conidia of the fungus.

EARTH-KIND NEMATODE CONTROL DURING SUMMER MONTHS

If you have nematodes in your garden sooner or later they will decrease or end production. Nematodes will severely damage all garden crops except corn, onions, garlic and nematode- resistant tomatoes. Even these nematode-tolerant crops will be adversely affected if a severe nematode population is present.

How does one know if nematodes are a problem in their garden? Above ground symptoms of nematode infestations are similar to many other root diseases or environmental factors limiting water and nutrient uptake. These symptoms consist of wilting during periods of moisture stress, stunted plants, chlorotic or pale green leaves and reduced yields. Most characteristic symptoms are those occurring on underground plant parts. Infected roots swell at the point of infection and form knots or galls. Several infections may occur along the same area resulting in large fleshy galls. The appearance of the galls will depend in part upon the plant being affected and the nematode species involved. Fast growing annuals will have a large fleshy gall and woody perennials, small hard galls. Infected roots are retarded in growth and lack fine feeder roots. Rotting of roots develop late in the season. When tubers, corms or other edible root portions are infected, small swellings or pimpling is evident on the surface.

Vapam was the most effective nematicide available. Unfortunately, the chemical control for nematodes named Vapam is no longer available. Organically, only the planting of cereal rye (Elbon) in the fall to grow during the winter will decrease nematode populations. Excessive drying called solarization (soil pasteurization) of the soil during July will also help. It consists of covering well-tilled, highly moist soil with clear plastic, sealing the edges and leaving it for several weeks. Do not use black plastic because soil heating is not as great as heating under clear plastic. Best results have been reported with fungi control, principally those causing root rots. Certain weeds and nematodes are controlled but not consistently. Solarization also stimulates plant growth possible by creating a good environment for beneficial microorganisms; however, this phenomenon is not fully understood. Solarization works best during the hottest months; July and August are ideal for home garden sites. Continue the treatment at least one month.

Another possible solution may be the solid planting of marigolds for 3 months in areas heavily contaminated with nematodes. The marigold, when grown on soil infested with nematodes, suppresses the population of these nematodes and reduces the numbers found in the roots of susceptible host plants. Three compounds of an a-terthienyl type, toxic to nematodes, have been identified in root exudates from these plants. Terthienyls are released from growing roots, even without their decay, but benefits require three to four months to become clear. There is some evidence that a-terthienyl is inhibitory to some plant-pathogenic fungi too. Marigolds also function as a trap crop since larvae which penetrate the roots do not develop beyond the second larval stage and do not lay eggs.

Marigolds have never been used for a biological nematode control because Vapam was less trouble-some and more economic. Since Vapam is no longer available and since cereal rye does not survive in hot weather, use of the marigold for nematode control must now be examined. Of course, you will want to follow proper planting procedures. Marigolds perform best if planted in a well drained location which receives at least 8-10 hours of sun daily. Incorporate generous amounts of organic material (pine bark mulch, peat moss, or compost), or fill with a commercially available landscape mix. Then add slow-release lawn fertilizer (no herbicide, please!) at 2 pounds per 100 square feet. Soak the bed immediately after transplanting and, depending on the weather, irrigate when the soil becomes dry and the plants begin to wilt slightly. (CAUTION: DO NOT keep these plants wet after they become established! Keep the soil feeling slightly moist to the touch but not wet. Continuously soggy, wet soil in the planting bed will stunt marigolds growth and eventually kill them.)

The new MARI-MUM marigolds are especially popular for planting. Buy transplants which are completely green or those that are just forming buds. Transplants covered with open blossoms will not produce the normally expected growth after planting. Bloom removal and/or fertilization WILL NOT overcome the detrimental affects of transplanting a MARI-MUM in full bloom. MARI-MUMS are among the most durable of transplanted annuals and are also the easiest to transplant. Unlike impatiens, begonias, petunias and bluebonnets, MARI-MUMS cannot be planted too deep. Other annuals rot if planted too deep; MARI-MUMS root. In fact, many times roots can be seen on the stems of transplants when purchased at nurseries. Transplanting depth does not affect ultimate plant height. Because of this never-too-deep transplanting phenomenon, large, leggy transplants can be salvaged by planting deeper to provide support during establishment.

The name marigold conjures up thoughts of pests. Some people think marigolds repel certain pests while other folks fear the devastating spider mite which the beautiful flowering plants seem to attract. Marigolds DO NOT repel rabbits, other rodents or deer. Marigolds also don't repel insects; in fact, they are attacked by and shelter the formidable spider mite.

So why should anyone who fears a spider mite invasion plant MARI-MUMS? MARI-MUMs are recommended for transplanting in August because spider mite populations decline in cooling fall temperatures. The optimum temperature for spider mite reproduction is 97 degrees F. during which the mite population can double every 5 days. As fall temperatures cool to 77 degrees F., mite populations require as many as 20 days to double. Because mite populations are virtually eliminated when planting beds are renovated for MARI- MUM establishment in August and since mite populations do not have an adequate time of favorable temperatures to regenerate their masses, spider mite damage to fall MARI-MUM plantings is minimal, even without a single pesticide application!

MARI- MUMS ARE NOT spider mite resistant; the lack of mites on fall plantings is merely a function of temperature and population control. Interested gardeners who have a nematode infested area should transplant marigolds 12 inches apart in August and allow them to grow there until fall planting of cole crops begins in October. Tops of the marigolds should then be removed and the root system tilled into the soil. By using marigold in the summer, you can rid the planting area of those devastating nematodes without endangering yourself or the environment, a truly EARTH-KIND™ practice.

BIOLOGICAL CONTROL FOR FUNGUS GNATS

Fungus gnats are small flies that infest soil, potting mix, or plants grown in pots or containers. Your customers may complain about fungus gnats resting on or swarming around plants or soil or flying around windows in their houses.

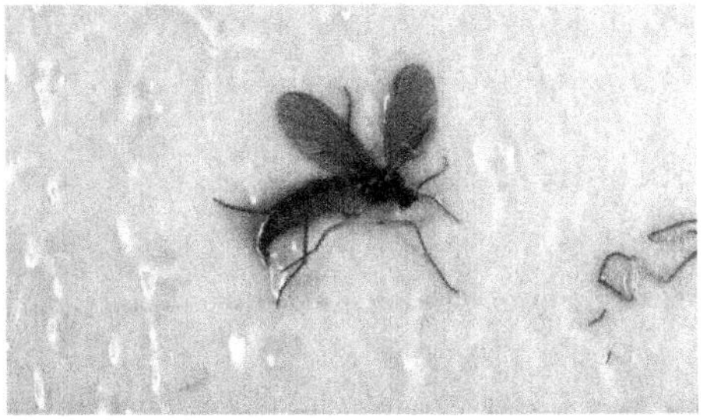

Fig. : Fungus gnat adult trapped on yellow sticky trap. This fly is less than ¼ inch long.

Adult fungus gnats don't harm plants or people, but they can be annoying. On the other hand, fungus gnat larvae , which live in soil and feed on roots, can be damaging when their numbers are very high, sometimes causing plants to wilt. Their feeding may cause seedling plants to die or injure roots, allowing entry of plant pathogens. Larvae are whitish to clear with black heads and, with a length of less than ¼ inch, very difficult to see. When abundant, they may leave shiny trails on the soil surface.

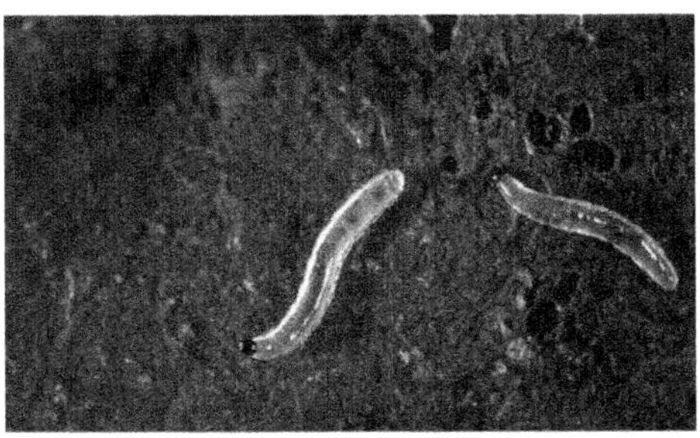

Fig. : Fungus gnat larvae.

Cultural Practices and Trapping

For many customers, changing watering practices may solve fungus gnat problems. For instance, allowing soil surfaces to dry between watering or using pasteurized potting mix and keeping infested plants away from clean ones. In some cases it may be possible to manage the situation by placing small pieces of yellow sticky traps in pots to control adults.

Biological Controls

Where cultural practices haven't effectively controlled the problem, customers may ask about treatment options. Although insecticides such as pyrethrins are available for this use, the safest and often most effective products for home use are the commercially available biological insecticide *Bacillus thuringiensis* subspecies *israelensis* (Bti) and the insect-attacking *Steinernema* nematodes.

Bacillus thuringiensis ssp. *israelensis* is the most common of these products on store shelves. Mosquito Bits and Gnatrol are common trade names. Bti is a bacterial by-product that kills flies and is related to *B. thuringiensis* ssp. *kurstaki* products available for caterpillar control, but Bti is toxic only to fly larvae such as mosquitoes, crane flies and fungus gnats. Repeat applications are often needed for long-term control.

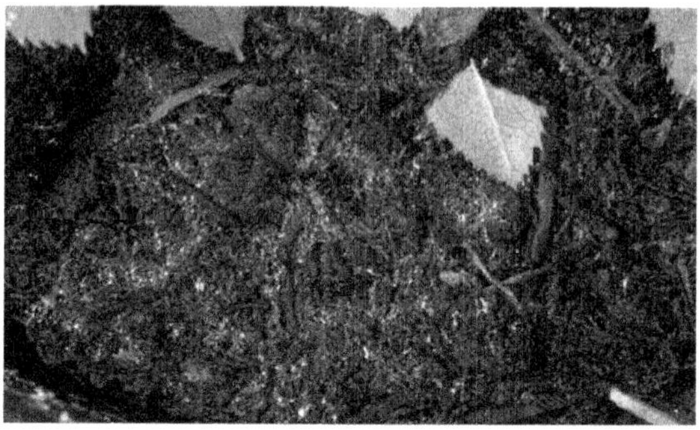

Fig. : Shiny trails left by fungus gnat larvae on the soil surface of a plant pot.

Nematode products must be refrigerated and have a short shelf life so many stores don't carry them, but some stores provide customers with a mail-order voucher to obtain fresh product. There are several species of insect attacking nematodes and it is important to use the correct species. *Steinernema feltia* is the most effective commercially available species for managing fungus gnats. Nematodes are mixed with water and applied as a soil drench. Nematodes reproduce and search for hosts so may provide longer-term control after several applications when soil is moist and temperatures are warm.

SOIL NEMATODES IN ORGANIC FARMING SYSTEMS

Nematodes are microscopic, wormlike organisms that live in water films and water-filled pore spaces in the soil. Typically, they are most abundant in the upper soil layers where organic matter, plant roots, and other resources are most abundant. Nematode abundance in soils—managed and unmanaged—ranges from 1–10 million individuals/m^2.

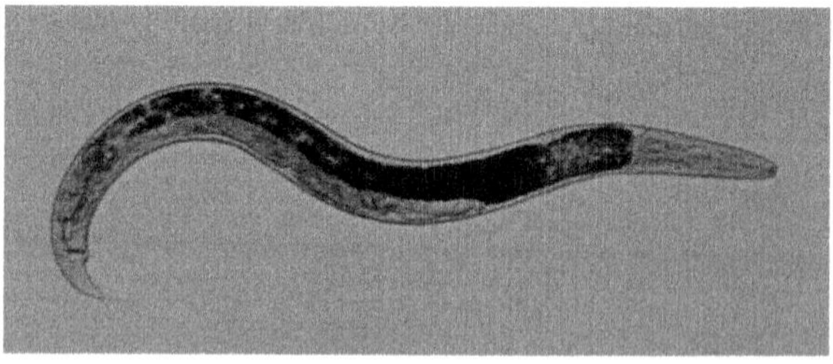

Fig. : A typical free-living, bacterial-feeding nematode, less than 1 mm (0.04 in) in length.

Most research on soil nematodes has focused on the plant-parasitic nematodes that attack the roots of cultivated crops. Less attention has been given to nematodes that are not plant-feeders and play beneficial roles in the soil environment. This chapter describes the important roles played by nematodes in soil ecosystems, as well as their potential to be used as indicators of soil condition in organic farming systems.

Nematode Feeding Habits

Nematodes can be classified into functional groups based on their feeding habits, which can often be deduced from the structure of their mouthparts. In agricultural soils, the most common groups of nematodes are the bacterial-feeders, fungal-feeders, plant parasites, predators, and omnivores. Predatory nematodes feed on protozoa and other soil nematodes. Omnivores feed on different foods depending on environmental conditions and food availability; for example, omnivorous nematodes can be predators, but in the absence of their primary food source, they can feed on fungi or bacteria.

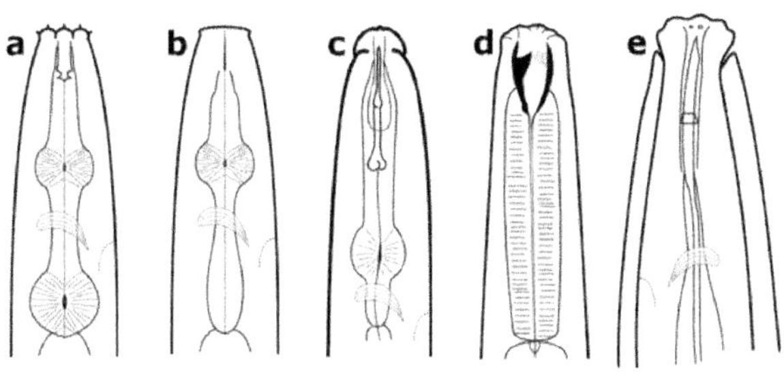

Fig. : Nematodes can be classified into different feeding groups based on the structure of their mouthparts. (a) bacterial feeder, (b) fungal feeder, (c) plant feeder, (d) predator, (e) omnivore.

Importance of Nematodes in Agricultural Systems

Nematodes contribute to a variety of functions within the soil system. In agricultural systems, nematodes can enhance nutrient mineralization and act as biological control agents.

Nematodes and Soil Fertility

Soil nematodes, especially bacterial- and fungal-feeding nematodes, can contribute to maintaining adequate levels of plant-available N in farming systems relying on organic sources of fertility. The process of converting nutrients from organic to inorganic form is termed mineralization; mineralization is a critical soil process because plants take up nutrients from the soil primarily in inorganic forms. Nematodes contribute directly to nutrient mineralization through their feeding interactions.

For example, bacterial-feeding nematodes consume N in the form of proteins and other N-containing compounds in bacterial tissues and release excess N in the form of ammonium, which is readily available for plant use. Indirectly, nematodes enhance decomposition and nutrient cycling by grazing and rejuvenating old, inactive bacterial and fungal colonies, and by spreading bacteria and fungi to newly available organic residues. In the absence of grazers, such as nematodes and protozoa, nutrients can remain immobilized and unavailable for plant uptake in bacterial and fungal biomass.

Bacterial-feeding nematodes are the most abundant nematode group in agricultural soils. Their abundance closely follows that of bacterial populations, which tend to increase when soil disturbances, such as tillage, increase the availability of readily-decomposable organic matter. Nitrogen mineralization in the soil occurs at a higher rate when bacterial-feeding nematodes are present than when they are absent. The contribution of bacterial-feeding nematodes to soil N supply depends, in part, on the quality and quantity of soil organic matter fueling the system.

Net N mineralization from decomposing organic residues takes place when the carbon:nitrogen (C:N) ratio of organic residue is below 20 (that is, 20 parts C to 1 part N). When the C:N ratio is greater than 30, the rate of mineralization decreases because microbes compete for N to meet their nutritional requirements. In this situation, N is immobilized in the microbial biomass. Incorporation of manure, compost, and cover crops with intermediate C:N ratios (ranging from 10 to 18) may stimulate bacterial growth and the abundance of bacterial-feeding nematodes, and increase soil N availability to plants.

Fungal-feeding nematodes are relatively more abundant in less-disturbed (*e.g.* notill systems) and perennial systems, where conditions for fungal growth are promoted, than in disturbed systems. Like bacterial feeding nematodes, fungal-feeding nematodes contribute to the process of nutrient mineralization by releasing N and other plant nutrients from consumed fungal tissue. However, in agricultural systems, bacterial-feeding nematodes typically release more inorganic N than fungal-feeding nematodes.

Nematodes as Natural Enemies and Biological Control Agents

Predatory nematodes are of interest because of their role in regulating the populations of other organisms. They generally feed on smaller organisms like protozoa and other nematodes. Thus they can help moderate population growth of bacterial- and fungal-feeding nematodes and protozoa, and help regulate populations of plant-parasitic nematodes.

Insect-parasitic nematodes are species of bacterial-feeding nematodes that live in close association with specific species of bacteria; together, they can infect and kill a range of insect hosts. The infective juvenile stage of insect-parasitic nematodes seeks out insect hosts to continue its development into adults. Once a host is found, the nematodes penetrate the insect body and release their bacterial associates into the insect's body cavity. These bacteria multiply and overwhelm the immune response of the host insect, ultimately killing the host. The nematodes feed on these bacteria, mature, and reproduce until all the resources within the insect host are consumed; then, infective juvenile nematodes escape the insect host's body and disperse in the soil to seek new hosts. Insect-parasitic nematodes are available commercially for use in inundative releases to manage the populations of a variety of insect pests.

Plant-Parasitic Nematodes

Most plant-parasitic nematodes feed on the roots of plants. Some species attach to the outside surface of plant roots , piercing the root tissue to suck up the cellular content; other species pierce and penetrate the roots of plants, living and reproducing entirely within the root itself. A relatively small number of important plant-parasitic nematode species are known to cause substantial economic damage in cropping systems around the world. The determination of tolerance limits or economic thresholds for plant-parasitic nematodes varies with many factors like species, plant tolerance, and soil type. Because plant parasitic nematodes show varying degrees of host specificity, carefully designed crop rotations are usually a powerful tool for reducing nematode-associated yield losses.

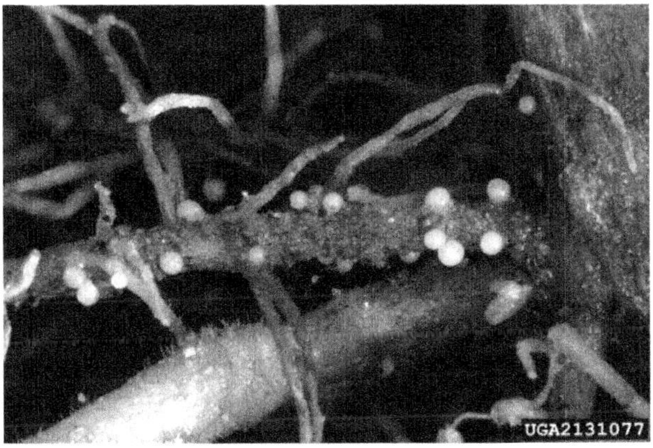

Fig. : White potato cyst nematode, Globodera pallida (Stone) Behrens, on plant roots. Cyst nematode females attach to root systems with their mouthparts to feed, and then their bodies swell into egg-filled cysts that can be visible to the naked eye.

Soil Nematode Communities

The proportions of the different feeding groups in the soil nematode community vary between systems and seasons, and they are influenced by a variety of factors, including crop and soil management practices and the presence and abundance of natural enemies. Management practices like tillage, crop rotation, and the use of organic amendments influence the physical and biological characteristics of the soil that influence the abundance of nematodes. Fungal-feeding, predatory, and omnivorous nematodes are very sensitive to soil disturbances , and agricultural systems with fewer physical and chemical disturbances, such as pastures, hay fields, and orchards, tend to support larger populations of these nematodes than more frequently disturbed systems like vegetable- and row-crop fields.

On the other hand, tillage and incorporation of organic residues increase the proportion of some bacterial-feeding nematodes, often offsetting declines in the numbers of other feeding groups and increasing the total abundance of nematodes. The wide variety of natural enemies that feed on or infect nematodes—predatory nematodes, predatory microarthropods, and nematode-trapping fungi, for example—may have a considerable impact on nematodes in agricultural systems.

Implications for Farming System Management

Agricultural management may increase the abundance of soil nematodes, primarily through the increase in the abundance of bacterial-feeding nematodes associated with tillage and the incorporation of organic residues. Soil conditions in agricultural production systems can be improved by enhancing nutrient availability and providing habitat for beneficial soil organisms. Maintenance of large populations of bacterial-feeding nematodes with practices that promote N mineralization throughout the growing season may enhance crop productivity, but a surplus of mineral N is not desirable from the environmental point of view because of an increased risk of nitrate leaching. In an ideal production system, N supply would be synchronized with plant demand. On the other hand, cultural practices like tillage or cultivation may reduce the complexity of the soil food web. Thus, a decrease in the frequency and intensity of tillage may promote the conservation of predatory nematodes and contribute to improved farming system performance.

FIGHTING FUNGUS GNATS WITH PARASITIC NEMATODES

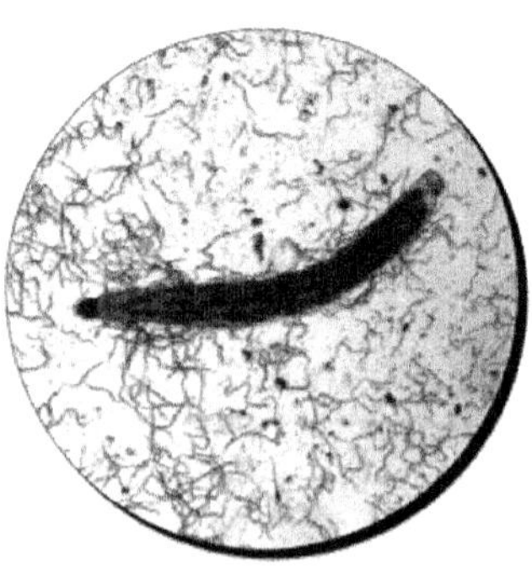

Conventional insecticides have been used to control pests since the early 1900s. In the 1960s, additional options were developed with the establishment of integrated pest management (IPM). Biological control agents (BCAs) are one of the oldest and most effective means for achieving insect control. Although

BCAs didn't catch on right away, leaps and bounds have been made in their acceptances into pest management programs in the past two decades and are now commonly used tools in the pest management toolbox.

Fungus Gnat Larvae Surrounded by Parasitic Nematode S. Feltiae.

Fungus Gnat Biology

Fungus gnats are a common and widespread greenhouse pest with cultural, physical, biological and chemical control options that readily lend themselves to IPM programs. Adult fungus gnats are small (1/8 in. long), black-gray flies. Easily monitored with yellow sticky cards, fungus gnat adults have long legs, many-segmented antennae and a single pair of grayish transparent wings with a Y-shaped vein near the wing tip. Adults are weak fliers and often are found near light sources or along the soil surface. Females can lay up to 200 eggs in areas where adults are active.

The eggs hatch within three to six days, depending on the temperature, and will then go through four larval instars.

The fungus gnat larvae are the most destructive life stage. Larvae are about ¼ in. long and are translucent, legless and worm-like, with a distinctive shiny black head. Larvae are usually concentrated in the top 3 in. of the soil, but also can be found throughout the soil profile. Fungus gnat larvae prefer to eat fungus, which is needed in their diet for optimal development; however, if fungus is not readily available, larvae will feed on plant material.

Damage and Detection

Larval feeding can cause direct root damage, as well as provide an entry into the roots for fungal pathogens like Thielaviopsis, Pythium, Botrytis, Verticillium and Fusarium. Fungus gnat larvae tunnel into thick, succulent stems and roots, causing the plant to wilt and, in severe infestations, die. Seedlings, young plants and unrooted cuttings lacking fully developed root systems are particularly vulnerable to larvae. Larval populations can be monitored by using potato slices, cut 1 in. thick, that are then placed on the surface of growing media. Potato slices should be examined weekly for incidence of fungus gnat larvae.

Fungus gnats thrive in moist, saturated conditions, so cultural practices should be utilized to limit freestanding water and over-saturated growing media. Irrigate only as needed and allow the surface of growing media to dry slightly between waterings. Always repair irrigation system leaks and avoid over-fertilization. Because fungus gnats are weak fliers, physical barriers like mesh screens can be used in vents and doorways to minimize pest movement.

How Nematodes Can Help

If populations of fungus gnats are observed on yellow sticky cards or potato slices that exceed grower-defined action thresholds, biological or chemical control options are warranted. Several BCAs, such as insect parasitic nematodes, predatory mites and beetles, parasitic wasps and beneficial fungi and bacteria can be used to reduce pest populations.

Steinernema feltiae, an insect parasitic nematode, has been used extensively for fungus gnat control for many years. *S. feltiae* is a microscopic, non-segmented roundworm that seeks out and enters fungus gnat larvae through natural body openings like the mouth or anus. *S. feltiae* uses carbon dioxide, movement and vibrations to locate larvae. Once inside, the nematodes release symbiotic bacteria that quickly kill targeted insects. Reproduction inside the insect releases new generations of infective juvenile nematodes that disperse in search of further prey.

S. feltiae can be used as a stand-alone control option, in combination with other BCAs such as *Bacillus thuringiensis* subspecies*israelensis, Hypoaspis aculeifer, H. miles* and *Atheta coriaria*, or in rotation with conventional insecticides. Rotation with conventional insecticides affords the opportunity to greatly reduce the number of chemical applications needed for acceptable levels of pest control, reducing the likelihood that pests will develop pesticide resistance.

If using *S. feltiae* as a component of a resistance management program, treatments of *S. feltiae* should begin early in the growth cycle when pest pressure is low. If new plant material is introduced into a greenhouse area, or if yellow sticky card monitoring indicates that fungus gnat populations exceed action thresholds, conventional insecticides can be used to quickly reduce adult populations. By combining chemicals and biologicals, more stages of the fungus gnats' life cycle will be targeted, providing more complete pest control.

One of the most effective and easiest ways to apply nematodes is through irrigation injection. Solutions emitting from dosing units can be applied through boom sprayers, overhead and drip irrigation, hand watering and other commonly used watering methods. Nematodes also can be applied through spray tanks, backpack sprayers or by dip applications. When nematodes are in solution, it's important to ensure stock solutions are constantly agitated to avoid settling in tanks and to prevent nematode solutions from getting above 86°F (30°C). Air circulation, mechanical circulation and hand-mixing can be used to prevent settling.

Applications of nematodes should be made to moist soil and applied in the early morning or late evening to avoid high temperatures, UV exposure and nematode desiccation. Remove sprayer filters of 50 mesh or finer and prevent pump pressures from exceeding 300 psi. If nematodes are applied over foliage, irrigate immediately with plain water to wash nematodes off the foliage and into the soil. Soil temperatures should be between 50°F and 86°F (10°C and 30°C) in order for nematodes to actively seek insect hosts.

NEMATODES: ALTERNATIVE CONTROLS

Nematodes are tiny, worm-like, multicellular animals adapted to living in water. The number of nematode species is estimated at half a million, many of which are "free-living" types found in the oceans, in freshwater habitats, and in soils. Plant-parasitic species form a smaller group. Nematodes are common in soils all over the world As a commentator in the early twentieth century wrote:

If all the matter in the universe except the nematodes were swept away, our world would still be dimly recognizable, and if, as disembodied spirits, we could investigate it, we should find its mountains, hills, vales, rivers, lakes and oceans represented by a thin film of nematodes.

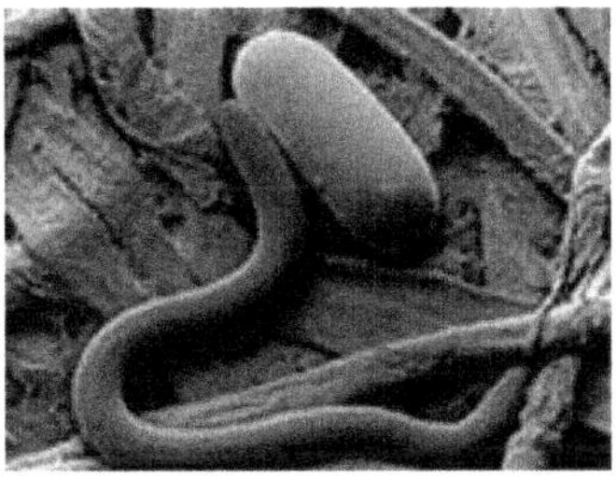

Fig. : Pratylenchus sp. larva and egg.

An important part of the soil fauna, nematodes live in the maze of interconnected channels—called pores—that are formed by soil processes. They move in the films of water that cling to soil particles. Many genera and species have particular soil and climatic requirements. For example, certain species do best in sandy soils, while others favor clay soils. Nematode populations are generally denser and more prevalent in the world's warmer regions, where longer growing seasons extend feeding periods and increase reproductive rates. In the southern United States, as many as ten generations are produced in one season.

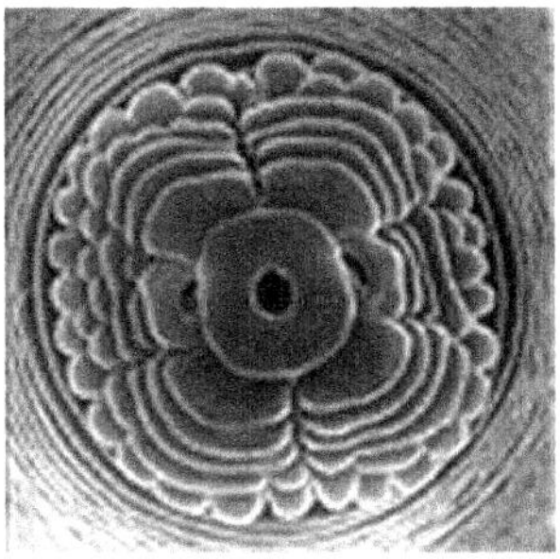

Fig. : Face view of lance nematode, Hoplolaimus sp.

Light, sandy soils generally harbor larger populations of plant-parasitic nematodes than clay soils. This is attributable to the more efficient aeration of sandy soil, the presence of fewer organisms that compete with and prey on nematodes, and the ease with which nematodes can move through the root zone. Also, plants growing in readily drained soils are more likely to suffer from intermittent drought, and are thus more vulnerable to damage by parasitic nematodes. Desert valleys and tropical sandy soils are particularly challenged by nematode overpopulation.

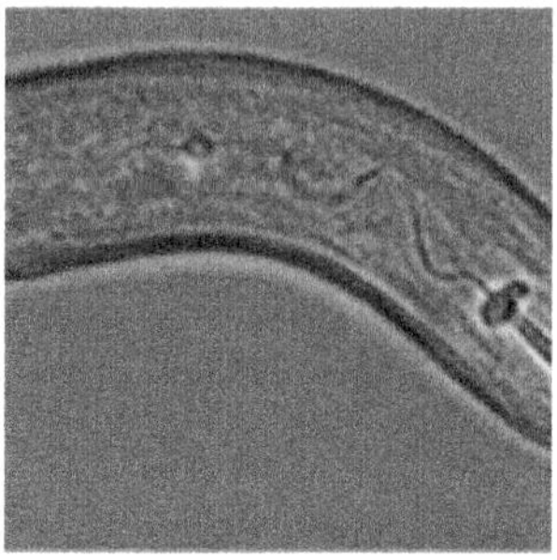

Fig. : Sugarbeet cyst nematode juvenile.

Plant-parasitic nematodes—the majority of which are root feeders, completing their lifecycles in the root zone—are found in association with most plants. Some are endoparasitic, living and feeding within the tissue of roots, tubers, buds, seeds, *etc.* Others are ectoparasitic, feeding externally through plant walls. A single endoparasitic nematode can kill a plant or reduce its productivity, while several hundred ectoparasitic nematodes might feed on a plant without seriously affecting production. A few species are highly host-specific, such as *Heterodera glycines* on soybeans and *Globodera rostochiensis* on potatoes. But in general, nematodes have a wide host range.

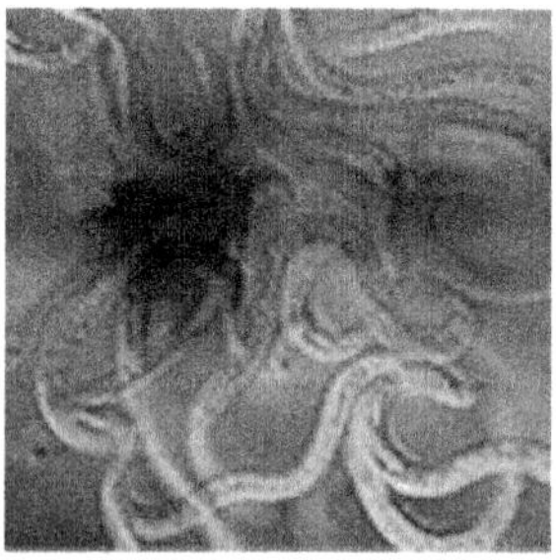

Fig. : Lesion nematodes penetrating a root.

Endoparasitic root feeders include such economically important pests as the root-knot nematodes (*Meloidogyne species*), the cyst nematodes (*Heterodera species*), and the root-lesion nematodes (*Pratylenchus species*). Important ectoparasitic root feeders include: root (*Paratrichodorus* and *Trichodorus*), dagger (*Xiphinema*), needle (*Longidorus, Paralongidorus*), ring (*Criconemella, Macroposthhonia*), stunt (*Tylenchorhynchus* and *Merlinius*), pin (*Paratylenchus*), and spiral (*Helicotylenchus, Rotylenchus,* and *Scutellonema*) nematodes. Direct feeding by nematodes can drastically decrease a plant's uptake of nutrients and water.

Nematodes have the greatest impact on crop productivity when they attack the roots of seedlings immediately after seed germination. Nematode feeding also creates open wounds that provide entry to a wide variety of plant-pathogenic fungi and bacteria. These microbial infections are often more economically damaging than the direct effects of nematode feeding.

Major Plant-parasitic Nematode Damage to Plants:

- Root-knot nematodes (*Meloidogyne species*) form galls on injured plant tissue. The galls block water and nutrient flow to the plant, stunting growth, impairing fruit production, and causing foliage to yellow and wilt. Roots become rough and pimpled and susceptible to cracking.
- Cyst nematodes (*Heterodera species*) give plants an unthrifty or malnourished appearance, and cause them to produce smaller-than-normal tops. Foliage is liable to wilt and curl, while roots become thick and tough and take on a red or brown coloring.
- Sting nematodes (*Belonolaimus species*) are found mainly in the South, especially in sandy soils with meager organic-matter content. Areas of stunted plants are an early indicator. As these areas grow larger and finally meet, the plants that were first affected will start to die at the margins of older leaves.
- Root-lesion or meadow nematodes (*Pratylenchus species*) cause internal browning in potato tubers and in the roots of corn, lettuce, peas, carrots, tomatoes, and brassicas.

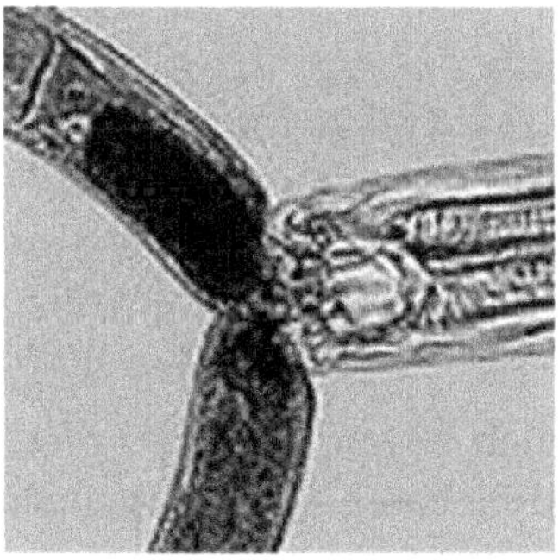

Fig. : Mononchoid nematode feeding on another nematode.

Nematode control is essentially *prevention*, because once a plant is parasitized it is impossible to kill the nematode without also destroying the host. The most sustainable approach to nematode control will integrate several tools and strategies, including cover crops, crop rotation, soil solarization, least-toxic pesticides, and plant varieties resistant to nematode damage. These methods work best in the context of a healthy soil environment with sufficient organic matter to support diverse populations of microorganisms. A balanced soil ecosystem will support a wide variety of biological control organisms that will help keep nematode pest populations in check.

Symptoms and Sampling

Usually, sampling is done because the grower observes a section of field with unhealthy plants, or notices an unexplained yield reduction. Because nematodes damage roots, any condition that stresses the plant—such as drought (or even hot spells), flooding, nutrient deficiencies, or soil compaction—will tend to amplify the damage symptoms noted above. Failure to respond normally to fertilizers and slower-than-normal recovery from wilting are signs of nematode infestation. In the undisturbed soil of groves, turf, and pastures, visible symptoms of nematode injury normally appear as round, oval, or irregular areas that gradually increase in size year by year. In cultivated land, nematode-injured spots are often elongated in the direction of cultivation, because nematodes are moved by machinery.

It is important to note that species of nematode are present in all soils; their mere presence does not necessarily mean that they are damaging plants. Harmless or even beneficial species are found in proximity to plants, right along with the parasitic species. Beneficial nematodes feed on such pests as Japanese beetle grubs and plant-parasitic nematodes, and release nutrients into the soil by eating bacteria and fungi. An experienced nematologist can identify species, and determine which, if any, are responsible for the observed damage.

Nematode sampling techniques vary depending on the crop, the root depth, the type of nematode causing damage, and the time of the season. The procedure presented here is a generic sampling technique for annual crops. Soil samples taken in the late summer are best when testing for the presence of nematodes. Root-zone soil samples are best taken immediately after harvest, or just prior to harvest if the crop showed signs of damage. First, fields should be divided into 20-acre blocks that have similar damage, soil texture, or cropping history. From each block take several sub-samples, mixing them well to create a single one-quart sample for each block. Soil samples should be kept cool, but not frozen.

Samples for established perennial crops are best taken from the feeder root zone, which is usually located around the canopy drip line. Your county or state Cooperative Extension Service can provide names of commercial labs that have nematode-identification services.

Preventing Further Spread of Nematodes

Preventing nematodes from entering uninfested areas is important; under their own steam they can spread across a field at a rate of three feet per year. The following measures will help prevent human-assisted spread of nematodes to uninfested fields:

- Use certified planting material
- Use soilless growing media in greenhouses

- Clean soil from equipment before moving between fields (washing equipment—including tires—with water is most effective)
- Keep excess irrigation water in a holding pond so that any nematodes present can settle out, pump water from near the surface of the pond; plan irrigation to minimize excess water
- Prevent or reduce animal movement from infested to uninfested fields
- Compost manure to kill any nematodes that might be present, before applying it to fields
- Eliminate important weed hosts such as crabgrass, ragweed, and cocklebur

Managing Soil Biology

The basis of sustainable nematode control is the maintenance of a healthy soil food-web. This begins with routine application of organic matter. There is substantial evidence that the addition of organic matter in the form of compost or manure will decrease nematode pest populations and associated damage to crops. This could be a result of improved soil structure and fertility, alteration of the level of plant resistance, release of nemato-toxins, or increased populations of fungal and bacterial parasites and other nematode-antagonistic agents. Reduced nematode damage from increased organic matter in soil is likely a combination of these interaction. Higher organic matter content increases soil's water-holding capacity, and supports thriving communities of the decomposers and predators that make up the soil's "digestive system."

Nematodes are important participants in this underground energy-transfer system. They consume living plant material, fungi, bacteria, mites, insects, and each other, and are themselves consumed in turn. Some fungi, for example, capture nematodes with traps, sticky knobs, and other specialized structures. Nematodes and protozoa regulate mineralization processes.

Evidence suggests that between 30 and 50 percent of the nitrogen present in crop plants was made available by the activity of bacteria-consuming nematodes. Research in Denmark has indicated that nematodes convert about as much energy as earthworms in certain forest soils. Don't forget, the vast majority of nematodes found in the soil are not plant parasites.

The food-web's stability is challenged by the yearly turning of the soil, which reduces the numbers of organisms that displace or prey on plant-parasitic nematodes, while bringing more nematodes to the surface from deeper soil. If the same host crop is planted year after year, plant-parasitic nematodes may increase to damaging levels. Root-feeding nematodes are very opportunistic, and are among the first organisms to invade after a disturbance.

Keeping these facts in mind, it is important to actively manage soil biology using minimum-tillage practices, compost, animal manures, green manures, cover crops, and crop rotations. These practices help promote the growth of beneficial organisms while suppressing plant parasites. Certain organisms that are associated with well-managed crop soils—*e.g.*, Rhizobacteria and mycorrhizae—may induce systemic host resistance to nematodes and to some foliar diseases. For further information see the

Soil Amendments for Nematode Control

Some sources of organic matter known to be nematode-suppressive include oilcakes, sawdust, sugarcane bagasse, bone meal, horn meal, compost, and certain green manures.

Most nematode species can be significantly reduced by tilling in chitinous materials such as crushed shells of crustaceans (shrimp, crab, *etc.*). This is effective because several species of fungi that "feed" on chitin also attack chitin-containing nematode eggs and nematodes. Increasing the amount of chitin in the soil will also increase the population of these fungi. A shrimp-shell-based fertilizer called Eco Poly 21™ Micro shrimp fertilizer is available from Peaceful Valley Farm Supply. At 2002 catalog prices, it would cost between $87 and $216 to treat an acre with this product (the suggested application rate is 20 to 50 lbs. per acre). Clandosan™, a nematicide made of crab shells and agricultural-grade urea, can be used as a pre-plant treatment (it should not be used on plants because the amount of urea in it can "burn" or kill them).

Crop Rotations and Cover Crops

Crop rotation to a non-host crop is often adequate by itself to prevent nematode populations from reaching economically damaging levels. However, it is necessary to positively identify the species of nematode in order to know what plants are its host(s) and non-hosts. A general rule of thumb is to rotate to crops that are not related to each other. For example, pumpkin and cucumbers are closely related and rotating between them would probably not be effective to keep nematode populations down. A pumpkin/bell pepper rotation might be more effective. Even better is a rotation from a broadleaf to a grass. Asparagus, corn, onions, garlic, small grains, Cahaba white vetch, and Nova vetch are good rotation crops for reducing root-knot nematode populations. Crotalaria, velvet bean, and grasses like rye are usually resistant to root-knot nematodes. Rotations like these will not only help prevent nematode populations from reaching economic levels, they will also help control plant diseases and insect pests.

Allelochemicals are plant-produced compounds (other than food compounds) that affect the behavior of other organisms in the plant's environment. For example, sudangrass (and sorghum) contain a chemical, dhurrin, that degrades into hydrogen cyanide, which is a powerful nematicide. Some cover crops have exhibited nematode-suppressive characteristics equivalent to aldicarb, a synthetic chemical pesticide.

Farmers in Alabama have added sesame into rotation with cotton, peanuts, and soybeans. Nematode levels are reduced and yields significantly increased among those crops in fields previously planted in sesame. Sesame yields averaged 1500 lbs per acre, well above the world average of 500 to 600 lbs per acre. Research shows that sesame may be an effective rotation crop to control peanut root knot nematode (*Meloidogyne arenaria*) and southern root knot nematode (*Meloidogyne incognita*). Sesame rotation is not effective, however, for the Javanese root knot nematode (*Meloidogyne javanica*). Commercial nematode control products derived from sesame include Dragonfire™ (oil), Ontrol™ (seed meal)—both manufactured by Poulenger USA—and Nemastop™ (ground up sesame plant) from Natural Organic Products.

In South Texas, soybean varieties were shown as possible alternatives to grain sorghum in cotton cropping sequences. Eighteen soybean varieties of maturity group 5, 6, 7, and 8 were tested in *Rotylenchulus reniformis*-infested soil, either nonfumigated or fumigated with 1,3- dichloropropene. Reproductive rates of R. *reniformis* were compared in the first year. Both experiments were planted with cotton in the second year to measure the rotational effects of soybean on cotton yield compared with grain sorghum and fallow. The high-yielding soybean cultivars with potential to suppress reniform nematode were "HY574," "Padre," "DP7375RR," and "NK83-30."

A 2000-2002 Maryland study evaluated crop rotations and other cultural practices to manage southern root-knot nematodes and lesion nematodes. Researchers grew nematode-susceptible potatoes and cucumbers, and compared the effect of several summer rotations on nematode problems. A summer rotation of sorghum sudangrass (*Sorghum bicolor x Sorghum arundinaceum var. sudanense*) reduced the root knot nematode population as effectively as the control treatment (soyabean cultivar with no known root-knot resistance and one nematicide application). Poultry litter/tillage (Year 1) and fallow (Year 2) were equally effective in managing the lesion nematode population. To maintain the effect, the rotations had to be included annually. Either summer or early-autumn sampling dates were more effective than midspring to identify threshold levels of the pests.

Nematodes and pH

Cyst nematodes do not hatch well in very acid soils (pH 4) or alkaline soils (pH 8). They do best in soil with a near-neutral pH of 6. This can be used to some advantage. For example, potatoes may be safest from nematode damage in an acid soil, while cabbage and beets can be planted in alkaline soil. But most plants do best at the pH that favors nematodes.

Researchers have observed that brassicas (*e.g.*, rapeseed, mustard) have a nematode-supressive effect that benefits the following crop in a rotation. This "mustard effect" is attributed to glucosinolate compounds contained in brassica residues. Toxicity is attributed to enzymatically induced breakdown products

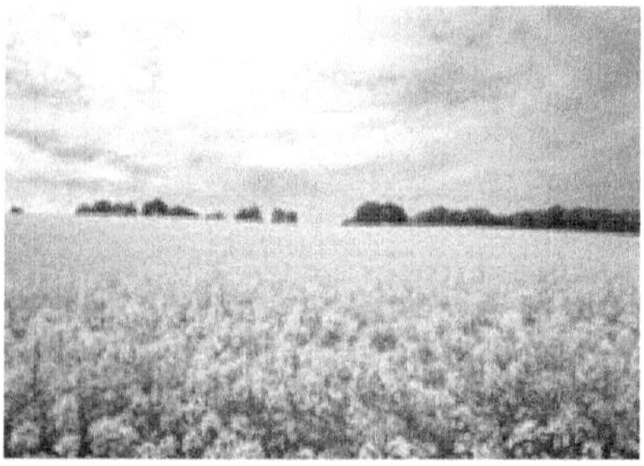

Fig. : Mustard.

of glucosinolates, a large class of compounds known as isothiocyanates and nitriles that suppress nematodes by interfering with their reproductive cycle. These glucosinolate breakdown products are similar to the chemical fumigant VAPAM® , which degrades in soil to methyl isothiocyanate. Glucosinolate compounds are also responsible for the pungent flavors and odors of mustards and horseradish. Jack Brown, PhD, a plant breeder specializing in brassicas at the University of Idaho, has released two biofumigant varieties, "Humus" rapeseed and "IdaGold" mustard, each containing elevated levels of glucosinolates. Cover crop seed for mustards, rapeseed, and oilseed radish are available from a variety of sources. Several Extension Service bulletins describe the use of brassica cover crops in greater detail.

Allelopathic Cover Crops

Some plants produce allelochemicals that function as nematode-antagonistic compounds, such as polythienyls, glucosinolates, cyanogenic glycosides, alkaloids, lipids, terpenoids, steroids, triterpenoids, and phenolics, among compounds from these plants—*e.g.*, castor bean, chrysanthemum, partridge pea, velvetbean, sesame, jackbean, crotalaria, sorghum-sudan, indigo, tephrosia—are exuded during the growing season or released during green manure decomposition. Sunn hemp, a tropical legume, and sorghum-sudan, a prolific grass plant grown for its biomass, are popular nematode-suppressive cover crops that produce the allelochemicals known as monocrotaline and dhurrin, respectively.

Here are some examples of how brassica crops are being used to manage nematodes:

- Oil radish as a green manure has dramatically reduced stubby root nematode (*Trichodorus*) and root lesion nematode (*Pratylenchus*) in Idaho potato fields.
- Oil radish is used as a "trap crop" for the sugarbeet cyst nematode, its roots exude chemicals that stimulate hatching of nematode eggs. The larvae that emerge are unable to develop into reproductive females—reducing the population densities for the following crop.

- Rape or mustard plantings in rotation with strawberries have checked the increase of some nematodes.
- Rapeseed and sudangrass green manures grown prior to potatoes at Prosser, Washington, provided between 72 and 86 percent control of the root-knot nematode in that crop. In the same study, on-farm research in western Idaho showed that rapeseed green manures decreased soil populations of rootlesion nematodes to a greater extent than did sudangrass green manures. Fall sudangrass should be plowed down after it is stressed (*i.e.*, the first frost, stopping irrigation). Winter rapeseed and canola should be incorporated in very early spring.

Rotation

The best rotation to control the Columbia root-knot nematode in potatoes involves planting a summer non-host crop, followed by a winter cover crop (rapeseed) incorporated as a green manure. Non-host crops include supersweet corn , pepper, lima bean, turnip, cowpea, muskmelon, watermelon, squash, rapeseed, canola, mustard, and sudangrass.

For root lesion nematode control on potatoes, researchers found that forage pearl millet and marigold (Crakerjack) as rotation crops with potatoes resulted in fewer root lesion nematodes and increased potato yields than rotation with rye.

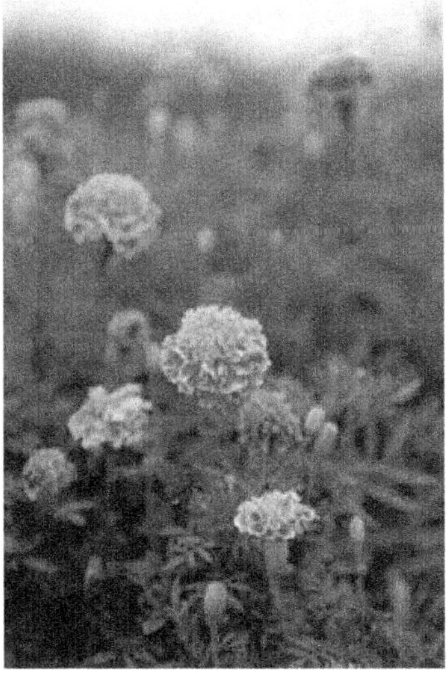

Fig. : Marigolds.

Marigold (*Tagetes species*) is one of the most highly studied crops for its ability to suppress nematodes with antagonistic phytochemical exudates, namely the polythienyls. Research also demonstrates that rhizobacteria living in association with marigold roots are suppressive to root lesion and other nematodes. These multiple effect nematode-control properties can benefit other crops when marigolds are grown in rotation. African marigold (*Tagetes erecta*) and French marigold (*Tagetes patula*) are popular ornamentals in the horticultural trade with several nematode-suppressive varieties each. Muster John Henry or little marigold (*Tagetes minuta*) is sold as the "Nematicidal" marigold, but it controls a relatively limited range of nematode species and readers should note that it is classified as a noxious weed in California.

Tomatoes planted two weeks after African marigolds (*Tagetes erecta*) were disked into the soil showed a 99 percent reduction in root-lesion nematode damage compared to a tomato-tomato or fallow-tomato rotation. The French marigold cultivar "Single Gold" provided 99 percent control of nematodes in Dutch tests. Burpee Seed Co. has carried a French marigold variety known as "Nema-gone." The most effective marigold cultivars are those that germinate quickly, grow vigorously, and have deep root penetration. Cover crops exhibit tremendous variability in their susceptibility to or suppression of the four major types of plant-parasitic nematodes. For example, cover crops that suppress root-knot nematodes may be susceptible to sting nematodes. It is important to identify the nematode species in the field—and know what their plant hosts and antagonists are—before planning a cover-cropping strategy.

Allies From the Prairie

In Ontario, certain prairie species have been found to provide excellent nematode control when used as a cover crop, including blackeyed susan, gaillardia, and switchgrass, according to Marvin Pritts, PhD, of Cornell University. Another North American native known as "Indian Blanket" or "Blanket Flower" (*Gaillardia pulchella*) was effective in controlling southern root knot nematode (*Meloidogyne incognita*) on sweet potato. Tissue extracts of Indian Blanket were lethal to various plant-parasitic nematodes but were innocuous to free-living nematodes. Root exudates of Indian Blanket were lethal to mobile juvenals of *M. incognita* and were inhibitory to the hatch of eggs at concentrations of 250 parts per million or higher. Indian Blanket could be used to manage southern root knot nematode as a rotation crop, a co-planted crop, or a soil amendment to control root-knot nematode.

Fields left fallow but kept weed-free for one to two years usually have an 80 to 90 percent per-year reduction in root-knot populations. This host-free period can be achieved in one season, rather than two years, by disking every ten days all summer. While such disking offers the added advantage to reduce perennial weeds, it is expensive in terms of fuel costs, possible erosion, and loss of organic matter through oxidation.

Botanical Nematicides

Certain plants are able to kill or repel pests, disrupt their lifecycle, or discourage them from feeding. Some of these—marigolds, sesame, castorbean, and various brassicas—have been discussed previously as nematode-suppressive cover crops. In this section we will look at plants whose extracts or essential oils can be applied as nematicides.

Biocontrols

Several microbial pathogens have been developed into commercial formulations against nematodes. These include the bacteria *Pasteuria penetrans* (formerly known as *Bacillus penetrans*), *Bacillus thuringiensis* (available in insecticidal formulations) and*Burkholderia cepacia*. Nematicidal fungi include *Trichoderma harzianum*, *Hirsutella rhossiliensis*, *Hirsutella minnesotensis*, *Verticillium chlamydosporum*, *Arthrobotrys dactyloides*, and *Paceilomyces lilacinus*. Another fungus, *Myrothecium verrucaria*, found to be highly effective in the control of nematodes , is available in a commercial formulation, DiTera™, from Abbott Laboratories. Circle One, Inc. offers a combination of several mycorrhizal fungal spores in a nematode-control product called Prosper-Nema™. Stein Microbial products offers the bacterium *Burkholderia cepacia* in a product called Deny™ and Blue Circle™. Rincon-Vitova offers a product called Activate™ whose active ingredient is the bacterium *Bacillus chitinosporus*.

The insect-attacking nematode *Steinernema riobravis* can provide root-knot nematode control comparable to that achieved with chemical nematicides. Although the exact mechanisms of control are not known, researchers hypothesize that an allelochemical is involved (perhaps manufactured by symbiotic bacteria that live within*S. riobravis*) that repels plantparasitic nematodes. Recent research measured the effect of beneficial nematodes on root-knot nematodes (*Meloidogyne species*) infecting tomatoes and peanuts. In the laboratory, peanut seedlings treated with the beneficial nematodes *Steinernema feltiae* and *Steinernema riobrave* showed resistance to pest nematodes.

In the greenhouse, scientists tested application levels and timing on peanut and tomato plants. On peanuts, pre- and post-infestation applications of *S. feltiae* suppressed *M.hapla* penetration but not egg production. Only pre-infestation applications of *S. riobrave* suppressed *M. hapla*. The tomatoes were infested with *Meloidogyne incognita* eggs and treated with *Steinernema glaseri* or *Heterorhabditis megidis* applied at the same times as the tomato treatments. The low rate of *S. glaseri* suppressed *M. incognita* penetration into tomato roots and the high rate of *S. glaseri* reduced egg production. Those interested in using this biocontrol will need to experiment with application rates and techniques to develop methods best suited to their operations.

A soil-dwelling predatory mite, *Hypoaspis miles*, preys primarily on fungus-gnat larvae but will also attack spring tails, thrips, and nematodes. These mites are available commercially for the control of fungus gnats in greenhouse production of tomatoes, peppers, cucumbers, flowers, and foliage plants. The mites are applied to the planting media.

It is clear that a wide range of organisms feed on, kill, or repel nematodes. These organisms are most effective, and are found most commonly, in healthy, well-managed soils.

Plant Resistance

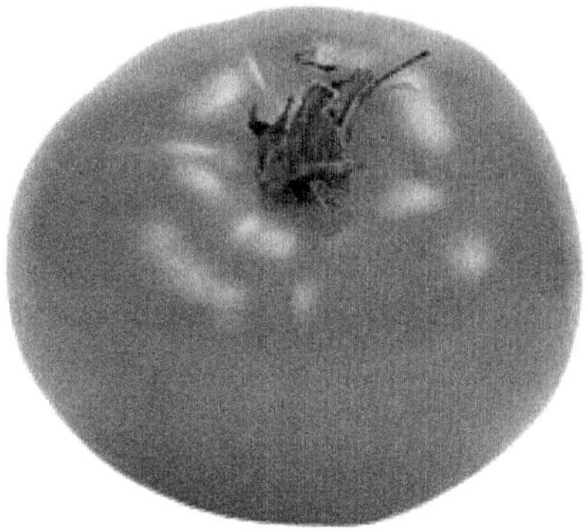

Generally speaking, a resistant cultivar is more effective against sedentary endoparasitic species such as root-knot and cyst nematodes than against "grazing" ectoparasitic species. Root-knot and cyst nematodes spend most of their lifecycle within the root, relying on specialized cells for feeding. Upon entering the roots of resistant cultivars, these nematodes become trapped as the feeding cells necessary for their survival fail to develop.

Many crop cultivars—tomatoes and soybeans in particular—have been specifically bred for nematode resistance. The "N" designation on tomato seed packages (usually as part of "VFN") refers to nematode resistance. A few cultivars of potatoes are resistant to the golden nematode, which is a pest only in a small area of the northeastern U.S. Although most cultivars of potatoes are susceptible to infection by nematodes, some varieties tolerate infection better than others. For example, population densities of root-lesion nematodes (*Pratylenchus penetrans*) that would affect yield in "Superior" are tolerated with little effect by "Russet Burbank."

Richard L. Fery, PhD, a geneticist at the USDA's Agricultural Re search Service in Charleston, South Carolina, developed two nematode-resistant varieties of bell pepper, "Charleston Belle" and "Carolina Wonder," available from commercial seed companies. Charleston Belle and its susceptible parent, "Keystone Resistant Giant," were compared as spring crops to manage the southern root-knot nematode (*Meloidogyne incognita*) in autumn-cropped cucumber and squash. Cucumber grown in plots following Charleston Belle had lower root gall severity indices than in crops following Keystone Resistant Giant. Cucumber yields were 87 percent heavier and numbers of fruit 85 percent higher in plots previously planted to Charleston Belle than to Keystone Resistant Giant. Squash grown in plots following Charleston Belle had lower root gall severity indices than those following Keystone Resistant Giant. Squash yields were 55 percent heavier and numbers of fruit 50 percent higher in plots previously planted to Charleston Belle than to Keystone Resistant Giant.

These results demonstrate that root-knot nematode-resistant bell pepper cultivars such as Charleston Belle are useful tools to manage *M. incognita* in double-cropping systems with cucurbit crops. Nematode-tolerant or resistant cultivars of snap beans ("Harvester" and "Alabama #1"), lima beans ("Nemagreen"), and sweet potatoes ("Carolina Bunch," "Excel," "Jewel," "Regal," "Nugget," and "Carver") also exist and may be used in a similar strategy to reduce nematode levels for crops that follow.

The choice of nematode-resistant rootstock for perennial fruit production is important to ensure protection of trees and vines against these unseen pests. Consult with a local farm advisor to confirm that the rootstock you choose is appropriate for the area.

Breeding for nematode resistance in most crops is complicated by the ability of the nematode species (primarily cyst nematodes and root-knot nematodes) to develop races or biotypes that overcome the genetic resistance factors in the crop. In order to maintain resistant crop cultivars on farms, researchers suggest that susceptible and resistant cultivars be planted in rotation. When a nematode-resistant cultivar is planted, nematode populations generally decrease, but over the course of the growing season the few nematodes in a particular population capable of overcoming this resistance begin to increase. If in the following season the farmer plants a susceptible cultivar, the overall nematode numbers will still be low enough to avoid significant yield reduction, but more importantly, the selective pressure favoring the increase of the "counter-resistant" bio types is removed. As long as the farmer continues to alternate susceptible and resistant cultivars (and, better yet, incorporate non-host crops into the rotation), the nematodes can be kept at non-damaging levels.

Red Plastic Mulch

Springtime field tests at the Agricultural Research Service in Florence, South Carolina, indicate that red plastic mulch suppresses root-knot nematode damage in tomatoes. According to Michael Kasperbauer, one of the researchers, "The red mulch reflects wavelengths of light that cause the plant to keep more growth above ground, which results in greater yield. Meanwhile, the plant is putting less energy into its root system—the very food the nematodes feed on. So reflection from the red mulch, in effect, tugs food away from the nematodes that are trying to draw nutrients from the roots." The research team planted tomatoes in sterilized soil, mulched them with red or black plastic, and inoculated the roots with nematodes. Plants inoculated with 200,000 nematode eggs and mulched with black plastic produced 8 pounds of tomatoes, while those mulched with red plastic produced 17 pounds. The red mulch is available commercially from Ken-Bar, Inc., of Reading, Massachusetts.

Solarization

Soil solarization, a method of pasteurization, can effectively suppress most species of nematode. However, it is consistently effective only where summers are predictably sunny and warm. The basic technique entails laying clear plastic over tilled, moistened soil for approximately six to eight weeks. Solar heat is trapped by the plastic, raising the soil temperature. The incorporation of poultry litter prior to solarization, or use of a second layer of clear plastic, can reduce effective solarization time to 30 days.

Brassica residues are also known to increase the solarization effect, in a process known as biofumigation. The plastic holds in the gaseous breakdown products of the brassica crop (or food processing wastes), thereby increasing the fumigation-like effect. Large-scale field experiments using cabbage residues with solarization obtained results comparable to solarization combined with methyl bromide.

Solarization is well documented as an appropriate technology for control of soilborne pathogens and nematodes, but the economics of purchasing and applying plastic restrict its use to high-value crops. Further information on solarization is available from ATTRA on request.

Soil Steaming

Steaming the soil suppresses nematodes in a manner similar to solarization. There are prototype steam machines capable of performing field applications, but steaming is probably economical only for greenhouse operations or small plantings of high-value crops.

Flooding

In certain parts of the country (*e.g.*, Tule Lake in California) where water is usually available and water pumping equipment and dikes already exist, and for certain large-scale monocultures (*e.g.*, potatoes), flooding is sometimes used as a management tool to control nematodes. But for most farms, it is probably not an option. Flooding the soil for seven to nine months kills nematodes by reducing the amount of oxygen available for respiration and increasing concentrations of naturally occurring substances—such as organic acids, methane, and hydrogen sulfide—that are toxic to nematodes. However, it may take two years to kill all the nematode egg masses. Flooding works best if both soil and air temperatures remain warm. An alternative to continuous flooding is several cycles of flooding (minimum two weeks) alternating with drying and disking. But note that insufficient or poorly managed flooding can make matters worse, as water is also an excellent means of nematode dispersal.

EFFECTIVE CONTROL OF FUNGUS GNATS WITH ENTOMOPATHOGENIC NEMATODES

Fungus gnats scientifically known as *Bradysia* spp are one of the most common pests of greenhouse ornamentals including African violets, carnations, chrysanthemums, cyclamen, lilies, geraniums, impatiens and poinsettias. Fungus gnat maggots primarily feed on fungi and organic matter but they can also feed on the plant roots and stems by chewing/stripping and tunneling, respectively.

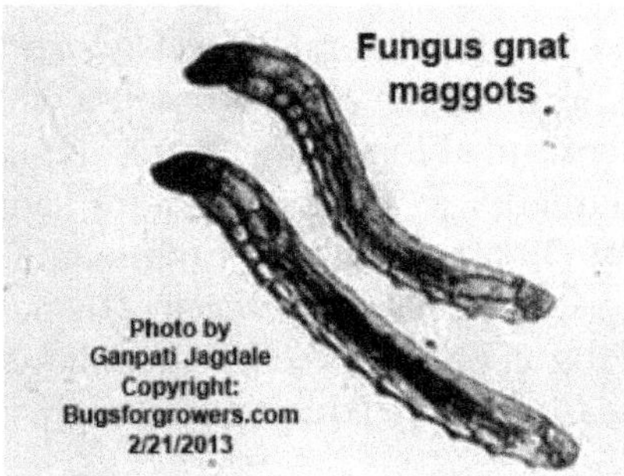

Fig. : Beneficial Entomopathogenic nematodes can infect and kill fungus gnat maggots.

Severely injured plants look unhealthy, discolored and eventually they dry. Fungus gnat maggots can also transmit fungal diseases such as *Fusarium, Phoma, Pythium* and *Verticillium* during feeding. Fungus gnat adults are nuisance to people and known to disseminate fungal spores from plant to plant when they migrate through the greenhouse.

Overlapping generations of fungus gnats in the greenhouse have made most control strategies difficult. Most growers generally rely on insecticides to manage fungus gnats in greenhouses but their use is restricted due to the detrimental effects on the environment and human health. Therefore, the environmentally safer and effective alternatives are required for the management of fungus gnats in greenhouses.

Three beneficial entomopathogenic nematodes including *Steinernema feltiae* and *Heterorhabditis bacteriophora* and *Heterorhabditis indica* have provided a potential alternative to chemical pesticides. The activity of these beneficial nematodes (also called as entomopathogenic nematodes) is strongly influenced by temperature because *S. feltiae* nematode is a cold adapted species whereas *H. bacteriophora* and *H. indica* are warm adapted species.

Beneficial Nematode, Steinernema Felitiae

Beneficial nematode, *S. feltiae* is a cold adapted nematode which is proved to be very effective in controlling fungus gnats in the European greenhouses where temperatures are maintained between 20 and 24°C. In these cool greenhouses, cold adapted *S. feltiae* nematodes when applied at the rate of 1 billion nematodes/ acre (2.5 X 10^5 infective juveniles/m^2) can provide over 66% control of fungus gnats. However, if *S. feltiae* is applied at the same rate but at 25°C temperature it can only reduce 23% population of fungus gnats. This suggests that the *S. feltiae* works effectively against fungus gnats in the green houses where temperature is below 25°C or 77°F.

Beneficial Nematode, Heterorhabditis Bacteriophora

The warm adapted beneficial nematode *H. bacteriophora* can reduce about 52% population of fungus gnats infesting poinsettia when applied at the rate of 1 billion nematodes/ acre and at the temperature ranging from 25°C to 29°C. In contrast, when cold adapted *S. feltiae* applied at the same rate and temperature range, it can cause only 38% mortality of fungus gnats. This proves that the warm adapted entomopathogenic *H. bacteriophora* nematode will perform better than *S. felitae* against fungus gnats in the US greenhouses where temperatures are often very warm (≥30°C or 86°F) during summer.

Beneficial Nematode Heterorhabditis Indica

Another warm adapted beneficial nematode *H. indica* is considered as good as *H. bacteriophora* and better than *S. felitae* in reducing populations of fungus gnats when applied in the greenhouse at the rate of 1 billion nematodes/acre and at temperatures above 25°C or 77°F. This nematode can reduce about 54% populations of fungus gnats infesting poinsettia in the greenhouse.

INDEX

R

S

T

W